Advances in Remote Sensing of Post-fire Environmental Damage and Recovery Dynamics

Advances in Remote Sensing of Post-fire Environmental Damage and Recovery Dynamics

Editors

Alfonso Fernández-Manso
Carmen Quintano

MDPI • Basel • Beijing • Wuhan • Barcelona • Belgrade • Manchester • Tokyo • Cluj • Tianjin

Editors
Alfonso Fernández-Manso
University of León
Spain

Carmen Quintano
University of Valladolid
Spain

Editorial Office
MDPI
St. Alban-Anlage 66
4052 Basel, Switzerland

This is a reprint of articles from the Special Issue published online in the open access journal *Remote Sensing* (ISSN 2072-4292) (available at: https://www.mdpi.com/journal/remotesensing/special_issues/Post-fire_Environmental_Damage_Recovery_Dynamics).

For citation purposes, cite each article independently as indicated on the article page online and as indicated below:

LastName, A.A.; LastName, B.B.; LastName, C.C. Article Title. *Journal Name* **Year**, *Volume Number*, Page Range.

ISBN 978-3-0365-5667-3 (Hbk)
ISBN 978-3-0365-5668-0 (PDF)

Contents

About the Editors

Alfonso Fernández-Manso

Dr. Alfonso Fernández-Manso is a full professor in Applied Remote Sensing at the University of León and a member of the Consolidated Research Unit 210 (UIC) of Applied Ecology and Remote Sensing (GEAT). His research interests are focused on the use of new geoinformation technologies in forest fires (remote sensing and GIS), territorial innovation (sensu lato), and the study of biocultural systems. In relation to geoinformation technologies, his research focuses on estimating the damage caused by forest fires from satellite data and has allowed for the establishment of bases for forest management at the local, regional, and national levels and for the application of advanced models based on spectral mixtures for the study of forest resources. His contributions to the generation of knowledge can be summarized as follows: (1) he has derived new remote sensing methods from surface energy balance to diagnose the severity of large forest fires, and (2) he has applied classifiers with a higher level of complexity to the study of forest fires, thus obtaining a higher reliability (MESMA, RLM, RF, or RN). The scientific–technical transfer of his knowledge has been applied to contracts with different companies and Regional Administrations, as well as transfer projects, such as the FIREMAP program. He has conducted research stays, promoting the exchange of scientific knowledge, in various institutions, among which include the University of California, the International Institute for Geoinformation Science and Earth Observation (Enschede, The Netherlands), and the National Institute for Space Research.

Carmen Quintano

Dr. Carmen Quintano is a senior lecturer at the University of Valladolid (Spain) and a member of the Institute of Sustainable Forest Management of the University of Valladolid and of the Consolidated Research Unit 210 (UIC) of Applied Ecology and Remote Sensing (GEAT) of the University of León. Her research interests are focused on the use of new geoinformation technologies (remote sensing and GIS) applied to natural resources, specifically the study of the impact of forest fires. Her research focuses on the estimation of damage caused by forest fires from satellite data and has allowed for the establishment of bases for forest management at the local, regional, and national levels and for the application of advanced models based on spectral mixtures for the study of forest resources. Her contributions have generated knowledge in the following: (1) the implementation of innovative techniques for the measurement of burn severity from the use of biophysical variables derived from satellite data; (2) the application of new remote sensing techniques, specifically SMA and MESMA, in the study of forest fires; (3) the design of novel spectral indices from satellite images to quantify burn severity with high reliability; (4) the use of classifiers with high levels of complexity in the study of forest fires (multinomial logistic regression, random forest, neural networks, or MAXENT); and (5) participation in the design and programming of the RS-EducA software, which allows satellite images to be processed in a very intuitive way.

remote sensing

MDPI

Article

The Potential of Multispectral Imagery and 3D Point Clouds from Unoccupied Aerial Systems (UAS) for Monitoring Forest Structure and the Impacts of Wildfire in Mediterranean-Climate Forests

Sean Reilly [1,*], Matthew L. Clark [2], Lisa Patrick Bentley [3], Corbin Matley [2,3], Elise Piazza [4] and Imma Oliveras Menor [1]

1 Environmental Change Institute, School of Geography and the Environment, University of Oxford, Oxford OX1 3QY, UK; imma.oliveras@ouce.ox.ac.uk
2 Center for Interdisciplinary Geospatial Analysis, Department of Geography, Environment, and Planning, Sonoma State University, Rohnert Park, CA 94928, USA; matthew.clark@sonoma.edu (M.L.C.); matley@sonoma.edu (C.M.)
3 Department of Biology, Sonoma State University, Rohnert Park, CA 94928, USA; lisa.bentley@sonoma.edu
4 US Army Corps of Engineers, Regulatory Division, South Branch, San Francisco, CA 94102, USA; elise.h.piazza@usace.army.mil
* Correspondence: sean.reilly@ouce.ox.ac.uk

Citation: Reilly, S.; Clark, M.L.; Bentley, L.P.; Matley, C.; Piazza, E.; Oliveras Menor, I. The Potential of Multispectral Imagery and 3D Point Clouds from Unoccupied Aerial Systems (UAS) for Monitoring Forest Structure and the Impacts of Wildfire in Mediterranean-Climate Forests. *Remote Sens.* **2021**, *13*, 3810. https://doi.org/10.3390/rs13193810

Academic Editors: Alfonso Fernández-Manso and Carmen Quintano

Received: 29 July 2021
Accepted: 17 September 2021
Published: 23 September 2021

Publisher's Note: MDPI stays neutral with regard to jurisdictional claims in published maps and institutional affiliations.

Abstract: Wildfire shapes vegetation assemblages in Mediterranean ecosystems, such as those in the state of California, United States. Successful restorative management of forests in-line with ecologically beneficial fire regimes relies on a thorough understanding of wildfire impacts on forest structure and fuel loads. As these data are often difficult to comprehensively measure on the ground, remote sensing approaches can be used to estimate forest structure and fuel load parameters over large spatial extents. Here, we analyze the capabilities of one such methodology, unoccupied aerial system structure from motion (UAS-SfM) from digital aerial photogrammetry, for mapping forest structure and wildfire impacts in the Mediterranean forests of northern California. To determine the ability of UAS-SfM to map the structure of mixed oak and conifer woodlands and to detect persistent changes caused by fire, we compared UAS-SfM derived metrics of terrain height and canopy structure to pre-fire airborne laser scanning (ALS) measurements. We found that UAS-SfM was able to accurately capture the forest's upper-canopy structure, but was unable to resolve mid- and below-canopy structure. The addition of a normalized difference vegetation index (NDVI) ground point filter to the DTM generation process improved DTM root-mean-square error (RMSE) by ~1 m with an overall DTM RMSE of 2.12 m. Upper-canopy metrics (max height, 95th percentile height, and 75th percentile height) were highly correlated between ALS and UAS-SfM ($r > +0.9$), while lower-canopy metrics and metrics of density and vertical variation had little to no similarity. Two years after the 2017 Sonoma County Tubbs fire, we found significant decreases in UAS-SfM metrics of bulk canopy height and NDVI with increasing burn severity, indicating the lasting impact of the fire on vegetation health and structure. These results point to the utility of UAS-SfM as a monitoring tool in Mediterranean forests, especially for post-fire canopy changes and subsequent recovery.

Keywords: UAS; structure-from-motion; Mediterranean; California; fire; forest structure; fire management; airborne laser scanner; ALS; lidar

1. Introduction

While widespread, mixed-severity wildfire regimes play a crucial role in shaping the ecosystems of Mediterranean climatic regions [1–4], current climate change [5] and historic wildfire suppression have increased the frequency of high-intensity and stand-replacing fires [6], as well as 90th percentile fire size [7], in California (United States).

Consequently, there is a pressing need to quantify forest structure and fuel loading at multiple spatial scales to understand and forecast relationships between wildfire hazards and forest structure attributes [8–10]. While ground-based surveys may provide accurate data, their limited spatial coverage hampers their usefulness at broader scales [11,12]. Remote sensing methods for mapping forest structure and fuels provide a useful alternative since they have high temporal resolution and large spatial coverage. However, due to its passive, top-down collection method, imagery from multispectral satellite sensors generally fail to detect forest properties below the canopy, nor is it optimal for determining vegetation height [13], particularly in forests with high canopy cover. Therefore, mapping forest structure from multispectral satellite data is often done indirectly. For example, in the case of estimating the spatial distribution of fire fuels, fuel density models are applied to satellite-derived maps of vegetation characteristics [14]. However, this indirect method cannot detect variations in vertical forest structure or fuel continuity masked below the canopy within an area possessing similar overstory vegetation characteristics.

Airborne laser scanners (ALS), or lidar sensors, provide an alternative remote sensing method not hindered by these shortcomings as they actively send energy pulses that penetrate below the canopy [13]. Canopy structure metrics derived from ALS point clouds have a strong relationship with field measurements [15] and can be used to determine vertical fuel structures [16]. Notably, ALS can detect the presence of ladder fuels [17]. Ladder fuels are live and dead vegetation that exists between ground vegetation and the base of the canopy above, allowing fire to climb from the forest floor and ignite the canopy [18]. However, management of yearly or seasonal events such as wildfires require frequent repeated monitoring [8]. While the advent of the NASA Surface Topography and Vegetation program and future lidar satellite systems may address this shortcoming e.g., [19], ALS is currently excessively cost prohibitive and effort intensive to be conducted with sufficient regularity for this purpose [20].

As opposed to the expense and effort required for an ALS survey, forest structure metrics can also be obtained relatively inexpensively from digital aerial photogrammetry (DAP) using unoccupied aerial systems (UAS) carrying commercially available cameras e.g., [21]. Three-dimensional point clouds can be generated from overlapping aerial images using structure from motion (SfM) processing techniques, providing an alternative method for mapping forest structure at a potentially lower cost than ALS depending on spatial extent e.g., [11,22–24], as well as UAS lidar at stand scales [25]. The low cost and relative ease of the UAS DAP SfM (hereafter UAS-SfM) method make it particularly useful for repeated monitoring during rapidly changing events at plot to stand scales [26], such as interannual wildfires. Of additional benefit over ALS, when used with a multispectral or hyperspectral sensor and downwelling irradiance sensor, UAS-SfM can provide calibrated spectral measurements as well as structural data. However, UAS-SfM technology is not without its own limitations. Due to the passive nature of an imaging sensor and the need for a point to appear in multiple overlapping photographs for its spatial location to be determined, DAP has minimal ability to detect structure below canopy cover as compared to ALS [27,28]. Therefore, DAP may be limited in the assessment of terrain height, which is needed for accurate point cloud height normalization, and detection of below-canopy forest structure.

To the best of our knowledge, the capabilities of multispectral UAS-SfM for mapping forest structure and below-canopy fuels in the mixed-oak and conifer woodlands of California have not been explored. Focused on a study site in a Mediterranean-climate mixed forest of Northern California, USA, this study seeks to address this research gap with the following objectives:

1. Evaluate the accuracy, as compared with ALS, of multispectral UAS-SfM in estimating ground elevation, a fundamental component of estimating accurate forest heights.
2. Determine the ability, as compared with ALS, of UAS-SfM to measure different metrics of canopy structure.

3. Demonstrate the utility of multispectral UAS-SfM in assessing the impact of wildfire on changes in photosynthetic productivity (greenness) and canopy height relative to ALS baseline conditions.

Findings from these objectives seek to answer the broad question: Can accurate measurements of forest attributes be obtained from UAS-SfM so it can be used as a stand-scale (e.g., 1 to 500 ha) remote-sensing monitoring tool to assess fire fuels and post-fire changes?

2. Materials and Methods

2.1. Study Site

This study took place at Pepperwood Preserve (38.57° N, 122.68° W), a 1261 ha nature preserve in Sonoma County, California, USA (Figure 1A). Located in the northern Coast Range, the property has predominantly rolling hills (61–475 m elevation) vegetated with a mosaic of Douglas-fir forest, oak woodland, chaparral, and grassland (Figure 1B) [29,30]. The area is characterized by a Mediterranean climate with dry, hot summers (April to October) reaching over 37 °C and mild, wet winters (November to March) infrequently dropping below 0 °C. The preserve experiences high interannual precipitation variability but receives, on average, around 86 cm of precipitation per year.

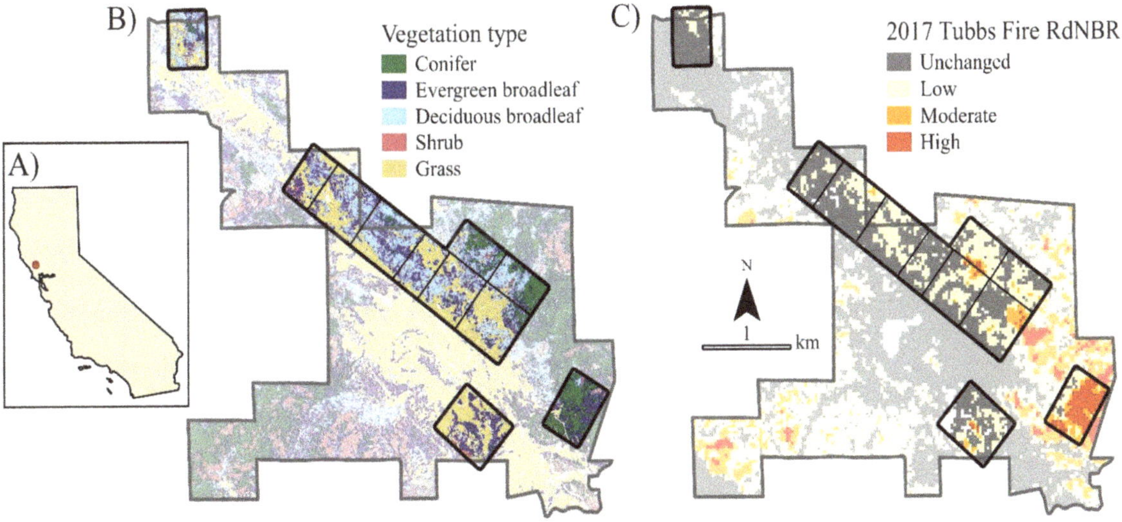

Figure 1. (**A**) Study site location at Pepperwood Preserve (red point) in California, USA. Maps of (**B**) vegetation assemblages and (**C**) the 2017 Tubbs Fire burn severity (MTBS) across the study site. Unoccupied aerial systems structure from motion (UAS-SfM) data collection extent is represented by black boxes.

As is common throughout California, this climatic regime generates conditions for wildfires by late summer, due to sustained drying of fuels and vegetation, that are further exacerbated by strong autumnal winds [31]. In 2017, the Tubbs Fire swept through Pepperwood (Figure 1C), burning over 95% of the preserve [32]. This fire was the fourth deadliest and second most destructive in California history prior to 2020 [33,34]. The fire ignited on 9 October 2017 4:45 UTC as part of a complex of fires in the region north of San Francisco, CA. Strong easterly Diablo winds rapidly spread the fire during the first 12 h after ignition [31]. During its 23-day span, the Tubbs fire burned 14,895 ha across all vegetation types in the region.

2.2. Data Sources

2.2.1. Unoccupied Aerial System (UAS) Structure from Motion (SfM) Multispectral Data Collection and Processing

This study employed a SenseFly eBee X fixed-wing UAS platform with a MicaSense RedEdge-MX sensor onboard. This multispectral sensor collects five spectral bands: blue (475 nm center, 20 nm bandwidth), green (560 nm center, 20 nm bandwidth), red (668 nm center, 10 nm bandwidth), red edge (717 nm center, 10 nm bandwidth), and near-IR (840 nm center, 40 nm bandwidth). Additionally, an onboard MicaSense Downwelling Light Sensor (DLS) 2 collected ambient sunlight and sun angle. Due to variations in solar light intensity, sensor radiance values must be standardized to reflectance and calibrated based on a sample of known reflectance values in order for multispectral measurements to be comparable across flight missions performed on different days or under varying light conditions, as well as to be compared against reflectance data from other sensors, such as multispectral satellite images. Radiance data were converted to reflectance using downwelling radiance measured from the DLS2 and calibrated with images of a MicaSense calibration panel (RP04-1926247-OB) captured on the ground immediately before or after each flight.

We flew the UAS for 13 days between 27 September 2019 and 8 October 2019 under leaf-on conditions (i.e., leaves present on deciduous trees) and clear-sky conditions, almost exactly two years after the Tubbs fire. All flights were conducted within two hours of solar noon to minimize shadowing. We divided the study coverage area (Figure 1B,C, black boxes) into 11 flight mission zones based on battery flight-time limitations. These 11 boxes had an average size of 33 ha and total area of 332 ha. SenseFly's eMotion 3 software (version 3.7) and a ground station controlled the UAS. The UAS flew autonomously using pre-planned missions to ensure consistent coverage and flight characteristics. Flight plans utilized eMotion's built-in flight control function and a raster digital surface model (DSM) from the 2013 ALS mission (Section 2.2.2). This function adjusts flight altitude in order to maintain a 122-m (400-ft) flying height above the DSM. Flights were conducted with a 75% latitudinal and longitudinal image overlap. We flew each mission zone twice, employing a perpendicular "lawn mower" pattern to form a cross hatch grid. To improve georeferencing, at least three ground control points (GCPs, i.e., visible targets) were distributed throughout the mission zone prior to each flight. We obtained GPS coordinates of each GCP from >100 readings using a Trimble Juno SB and differential correction with Pathfinder Office. GCPs had a mean horizontal accuracy of 2.35 m and vertical accuracy of 2.65 m.

We processed the resulting images using Pix4Dmapper (Pix4D, Lausanne, Switzerland, version 4.4.12), which performed the reflectance calibration and generated reflectance orthomosaics for each spectral band with an 8 cm pixel size and a three-dimensional point cloud from the green spectral band. The point cloud had an average point density of 180 points/m^2. The resulting point clouds had a mean RMS error from the GCPs of 1.5 m, with the exception of one highly erroneous GCP. This point was not found to affect the overall georeferencing of the resulting point cloud. We calculated pixel-level Normalized Difference Vegetation Index (NDVI) values using the red and near-IR orthomosaic images and merged the spectral data, including NDVI, with the point cloud in R (version 3.6.3) using the lidR and raster packages [35–37].

2.2.2. Airborne Laser Scanner (ALS) Data

We obtained ALS data from the 2013 Sonoma County Vegetation Mapping and Lidar Program website (www.sonomavegmap.org, accessed on 30 September 2019). Watershed Sciences, Inc. (Portland, OR, USA) collected these data between 28 September 2013 and 26 November 2013 using aircraft-mounted ALS50 and ALS70 sensors at a pulse density of >8 pulses/m^2 [38]. The point cloud had an average point density of 15 points/m^2 and an average point density in vegetation of 9.2 points/m^2. The survey employed 9685 GCPs with a maximum x, y, z RMSE of 0.2 cm. Watershed Sciences conducted ground point classification using TerraSolid software and refined the point classification using a subsequent statistical surface algorithm.

2.2.3. Vegetation Distribution Data

We obtained vegetation distribution data for Pepperwood Preserve from Ackerly et al. [29]. They generated their vegetation distribution map from imagery obtained by the National Aeronautics and Space Administration's Jet Propulsion Lab using the Airborne Visible Infrared Spectrometer-Next Generation (AVIRIS-NG) on 5 June 2014. Ackerly et al. classified this hyperspectral imagery with a support vector machine algorithm, first with broad land-cover classes, then at the tree species level in forest pixels. For the purposes of this study, we grouped tree species pixels into three generalized forest types: conifer (*Pseudotsuga menziesii* and *Sequoia sempervirens*), evergreen broadleaf (*Arbutus menziesii, Notholithocarpus densiflorus, Quercus agrifolia,* and *Umbellularia californica*), and deciduous broadleaf (*Acer macrophyllum, Aesculus california, Quercus douglasii, Quercus garryana, Quercus kelloggii,* and *Quercus lobata*).

2.2.4. 2017 Tubbs Fire Burn Severity Data

We obtained burn severity data for the 2017 Tubbs Fire from the Monitoring Trends in Burn Severity (MTBS) 2017 burn severity mosaic [39,40] (Figure 1C). This dataset utilizes the relative delta normalized burn ratio (RdNBR) [41] which analysts classify by fire into burn severity categories [39,40].

2.3. Data Analysis

2.3.1. Digital Terrain Model (DTM) Generation Capability Analysis

DTMs are commonly generated from point cloud data by filtering for ground points and then interpolating a raster surface from these points [42]. A number of ground-filtering methods and algorithms exist and perform comparably well [43], including the cloth simulation filter (CSF) from Zhang et al. [44]. The CSF classifies ground points based on the behavior of a simulated cloth draped over an inverted version of the point cloud. Four parameters determine the performance of the cloth: rigidness, the ability of the cloth to bend; grid resolution, the spacing of cloth particles; distance threshold, the maximum distance from the cloth for a point to be classified as ground; and, time step, the movement between iterations. Rigidness only has three possible states, corresponding to the number of times a simulated cloth particle can be moved per step, while the other parameters can take any positive value. Compared to other ground-finding algorithms, the accessible parameters of the CSF algorithm render it well suited to applications in challenging forest environments where algorithm optimization must be conducted to reduce DTM generation error.

In our study site, the ground in September to October is generally covered with dirt or senesced grass, both of which have low NDVI values relative to understory vegetation. We employed the CSF in conjunction with a novel post-processing NDVI filter that takes advantage of the NDVI contrast between ground and understory components. After CSF classification of ground points, we applied a filter to reclassify all ground points with an NDVI above a threshold value. This filter removes misclassified photosynthetic understory vegetation points from the preexisting set of ground points. We generated DTMs from the identified and filtered ground points using a triangulated irregular network (TIN) to grid interpolation at one-meter resolution. Ground finding and DTM generation were conducted using the lidR package in R [38].

To optimize DTM generation for the varied forests of Pepperwood, we iterated each CSF parameter and the post-processing NDVI filter threshold over an extreme range of values while holding all other parameters constant (Table 1). Due to the relatively high accuracy of ALS DTMs, we considered any deviation of the UAS-SfM DTM from the ALS DTM to be erroneous. We produced error rasters comparing the resulting sequence of DTMs to the ALS-derived benchmark. We registered the UAS-SfM point cloud to the ALS data using the iterative closest point (ICP) algorithm in the software Lidar360 (GreenValley International, Berkeley, CA, USA, version 4.0; RMSE range: 1.57–2.34 m), and applied the resulting transformation matrices to the UAS-SfM point cloud in R. This ICP step was conducted prior to DTM error analysis using the complete UAS-SfM and ALS point clouds in order to align the ground and ensure any detected deviance in the UAS-SfM DTM produced was due to ground finding error and the DTM generation process, rather than due to a systematic offset between the point clouds. We limited this analysis to flight mission interiors at least 50 m from the mission edge to reduce the effect of edge artifacts. We calculated root mean squared error (RMSE) for the entirety of the study site as well as by vegetation type.

Table 1. Cloth simulation filter (CSF) parameter and Normalized Difference Vegetation Index (NDVI) threshold testing ranges. Range and step define the limits and spacing of the values over which each parameter was iterated.

Parameter	Round One			Round Two			Final
	Constant	Range	Step	Constant	Range	Step	Constant
Rigidness	1	1–3	1	3			3
Grid resolution	0.5	0.1–2.5	0.1	1	0.5–1.1	0.01	0.45
Distance threshold	0.5	0.1–2.5	0.1	0.1	0–0.2	0.01	0.01
Time step	0.65	0.1–2.5	0.1	0.6	0.5–1	0.01	0.58
NDVI threshold	off	0.1–1	0.1	0.5	0.4–0.6	0.01	0.55

We conducted a second round of CSF parameter testing across the range of minimum RMSE values from the first round using a smaller step between iterations to further refine parameter values. As in round one, each parameter was run through its range while the other parameters were held constant. As mentioned previously, rigidness only has three distinct levels. Therefore, we did not repeat the iteration process for rigidness in round two. Unlike the first round, we conducted this second round in a stepwise manner, whereby the constant value for a parameter was updated to its newly identified optimum before proceeding with the next parameter assessment. A parameter's relative effect on RMSE in round one determined the order of optimization for this second round. We optimized grid resolution first, holding the other parameters at the constant corresponding to the minimum RMSE obtained in round one. Time step was optimized second, using the updated round two optimized grid resolution value, along with the other constants from round one. Distance threshold was parameterized third and NDVI threshold last, again updating parameters to their new values before proceeding. Final optimized CSF parameter values corresponded to the minimum RMSE obtained in round two. We produced a final UAS-SfM DTM from this set of optimized parameters using TIN-to-grid interpolation. We then calculated absolute error (relative to the ALS DTM) for each raster cell and separated this error by vegetation type.

2.3.2. Comparison of Forest Structure from UAS-SfM and ALS Point Clouds

We first compared the actual vertical height point cloud structures of UAS-SfM and ALS height-normalized point cloud data for each forest type. We then compared vertical height structure metrics between these data types. The general workflow for extracting vegetation structure metrics from point cloud data involves: (1) height normalizing the data against a digital terrain model (DTM) to isolate canopy height from terrain features, and (2) extracting metrics from these normalized data (Figure 2). We conducted this height normalization process on the original, unregistered UAS-SfM point cloud in two ways. First, to determine the capability of UAS-SfM as a standalone method for measuring forest structure, we height normalized the UAS-SfM point cloud using the UAS-SfM DTM produced through the previously described CSF optimization (Section 2.3.1). Second, to determine the enhanced capability of UAS-SfM in areas with preexisting high-resolution DTMs, we height normalized the UAS-SfM point cloud using the ALS DTM. To correct for any systematic vertical and horizontal location bias separating the UAS-SfM point cloud and the ALS DTM, we registered the UAS-SfM ground points to the ALS DTM prior to height normalization. We generated transformation matrices for the alignment of the UAS-SfM ground points to the ALS DTM using the ICP algorithm in Lidar360, and applied the matrices to the entire UAS-SfM point cloud in R. This method allowed for registration based solely on the ground points to avoid overfitting of canopy metric comparisons as would occur if the entire point clouds had been used for registration purposes.

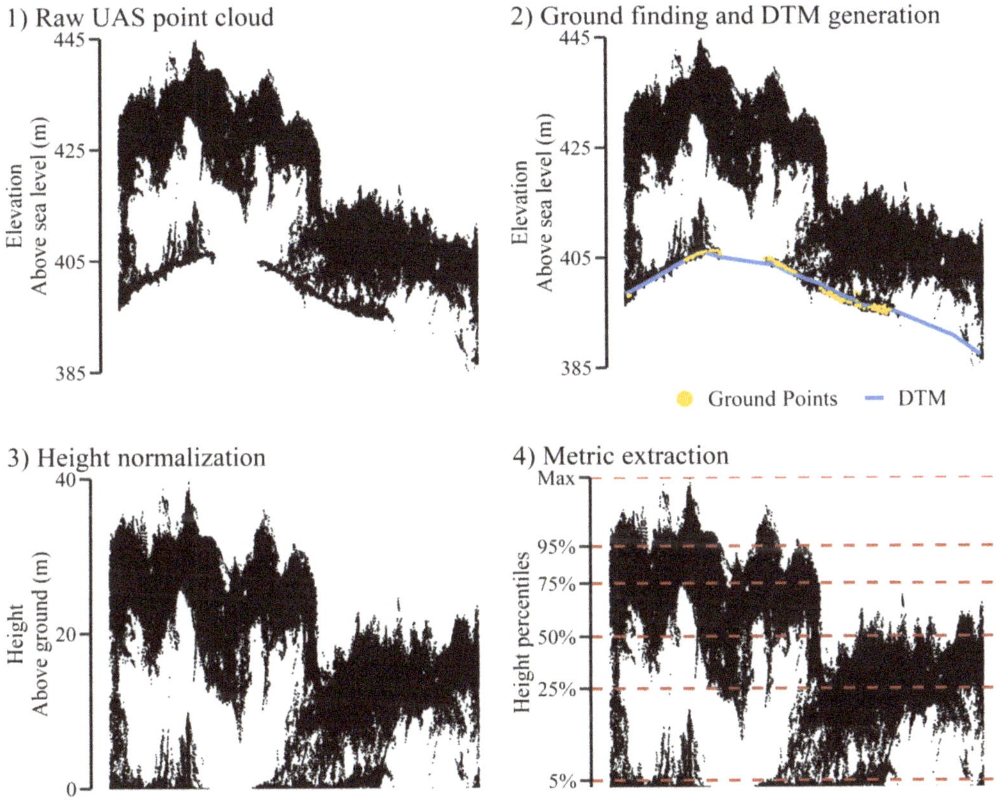

Figure 2. Point cloud processing workflow.

Using these height normalized data, we compared area-based height metrics between both UAS-SfM datasets and ALS. A subset of metrics from Filippelli et al. [8] were computed at a 20 × 20-m grid scale. While the high resolution of UAS-SfM data permits analysis at finer spatial scales, this larger grid scale was selected to align with a preexisting plot network at Pepperwood. Additionally, the MTBS burn severity dataset is assessed at 30 m resolution, limiting finer scale analysis of differences between burn severities. Density metrics represented the percentage of 3D points within a grid cell that fell within a particular vertical height band. Other metrics from each cell included height percentiles (P_5, P_{25}, P_{50}, P_{75}, and P_{95}), mean, maximum, standard deviation, skewness, and kurtosis of point cloud heights. To reduce confounding factors of mixed species and structural change due to fire, respectively, we filtered grid cells for only those cells covered by at least 75% of one forest type and with a burn severity of unchanged or low. From this filtered dataset, we computed the slope linear regression coefficient and Pearson's correlation coefficient for the relationship between UAS-SfM and ALS values for each of these metrics. This was done using 10,000 bootstrap replicates of 100 grid cells per forest type to correct for spatial autocorrelation and maintain balanced sample sizes. In addition to these standard metrics of point cloud height, we compared the ladder fuel metric from Green et al. [45] between the two datasets in the same way as the other metrics. This metric is calculated as the number of points between 1 and 4 m divided by the total number of points below 4 m and was shown to significantly contribute to canopy damage in local non-wind-driven fires, including the Tubbs Fire.

2.3.3. Utility of UAS-SfM for Detecting Post-Fire Forest Change

We took a random sample of 20 grid cells per forest type (conifer, broadleaf evergreen, broadleaf deciduous) from each of the four burn severity classes (none, low, medium, and high). High burn severity data in deciduous broadleaf forest was excluded from all Tubbs Fire analysis due to insufficient sample size (n = 8 grid cells). These data were found to violate normality, so non-parametric statistical tests were implemented in subsequent analyses.

We determined the influence of ladder fuels on Tubbs Fire burn severity by forest type. Ladder fuels from the ALS data were used in this analysis since these data represented pre-fire conditions (four-year difference). We conducted a Kruskal–Wallis one-way analysis of variance to test differences in mean rank ladder fuel density among burn severity classes for the three forest types. Significant results were followed by Dunn's post-hoc test to examine the nature of between-group differences. We adjusted *p*-values via the Bonferroni method to account for multiple comparisons.

We determined the impact of fire severity in the Tubbs Fire on forest canopy structure using the 95th percentile height (P_{95}) metric and 75th percentile height (P_{75}) metrics. Since these metrics are highly positively correlated between UAS-SfM and ALS data with regression coefficients close to one (Table 2), the two can be directly compared for assessing change in canopy structure [8]. In this case, ALS represents pre-fire canopy conditions and UAS-SfM represents post-fire conditions. We also tested differences in mean rank of P_{95} and P_{75} canopy metrics, respectively, among burn severity classes by forest type using a Kruskal–Wallis one-way analysis of variance test. Significant results were followed by Dunn's post-hoc tests to examine the nature of between-group differences. We adjusted *p*-values via the Bonferroni method to account for multiple comparisons. For this analysis, we used the UAS-SfM dataset height normalized using the ALS DTM.

Lastly, we used mean NDVI at the P_{75} canopy height to determine the impact of fire severity on canopy health. In the absence of pre-fire NDVI values, we used the unburned forest stands as the control reference for the purpose of detecting the fire's impact. As before, we tested the differences in mean rank P_{75} canopy metrics among burn severity classes by forest type using a Kruskal–Wallis test with a Dunn's post-hoc test for between-group differences and Bonferroni adjusted *p*-values.

3. Results

3.1. DTM and CHM Generation Capability Comparisons

Compared to the baseline ALS DTM, two rounds of CSF parameter optimization achieved a UAS-SfM-derived DTM with a minimum RMSE of 2.12 m across the study site (Figure 3I). This was a reduction of 1.68 m of error relative to the DTM generated from CSF default parameters (Figure 3A). Responses to parameter changes were largely consistent among vegetation types. Therefore, parameter optimization could be conducted across the full study site without sacrificing accuracy in any given vegetation type. Across the wide range of the first round of parameter testing, the rigidness and distance threshold parameters had minimal effect (Figure 3C,D), while the grid resolution and time step parameters greatly affected both DTM accuracy and processing time (Figure 3A,B). In particular, grid resolution processing times rose sharply as the resolution approached zero.

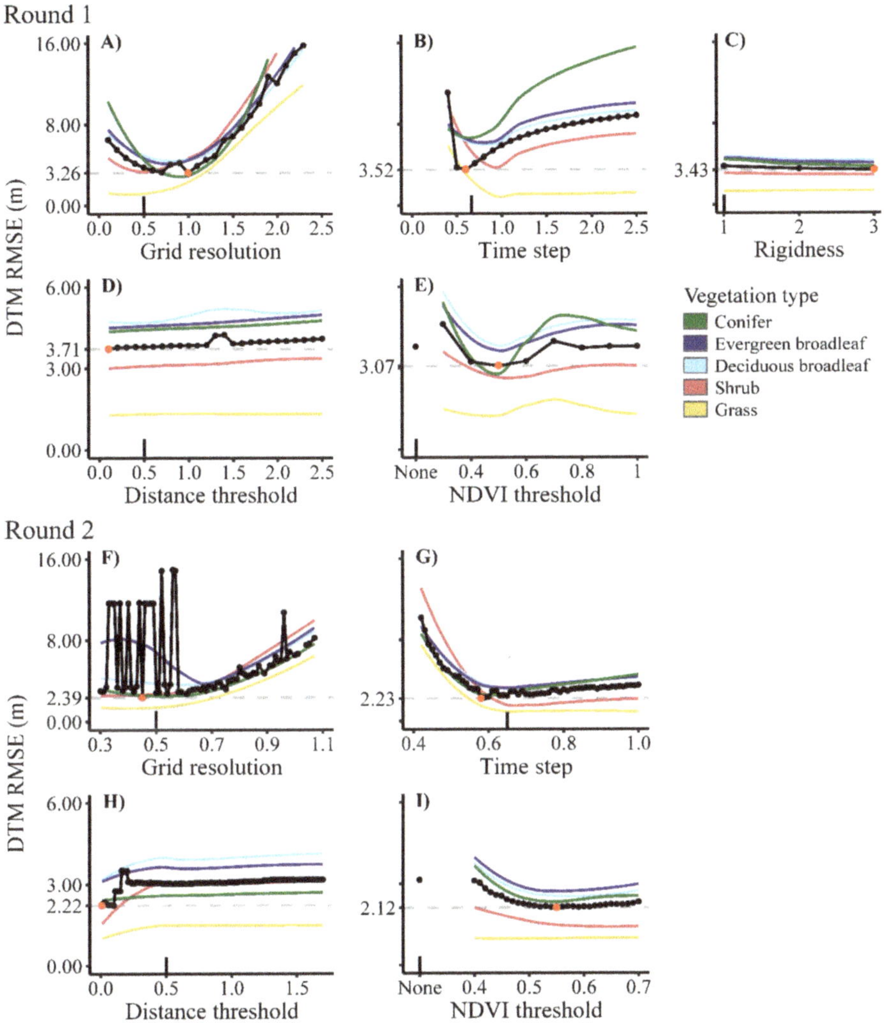

Figure 3. CSF parameter effect on DTM accuracy for Round 1 (top) and Round 2 (bottom) of optimization. The DTM root-mean-square error (RMSE, m) for the entire study site is shown by black points and black line. Colored lines depict RMSE by vegetation type smoothed using local polynomial regression. Red dot and horizontal grey dashed line denote minimum achieved RMSE. Vertical black line intersecting the x-axis indicates default parameter value.

The post-processing NDVI threshold filter reduced UAS-SfM DTM error in both rounds of processing (Figure 3E,I). Compared to the DTM produced from the CSF default parameters alone, the NDVI filter reduced site-wide RMSE by 0.73 m, down to 3.07 m (Figure 3E). In comparison, two rounds of parameter optimization without the NDVI filter reduced site-wide RMSE by only 0.66 m. The performance of the NDVI filter was even more pronounced in the second round. With optimal values for other CSF parameters in place, the addition of the NDVI filter reduced site-wide RMSE by 1.02 m (Figure 3I). Therefore, of the final achieved optimization improvement of 1.68 m, ~61% was due to the NDVI threshold filter. Additional processing time for this filter was negligible. When the error analysis was limited from the entire DTM to just the identified ground points, regression results between these ground points and the corresponding elevation of the

ALS DTM show a strong positive relationship (r ≈ +1, slope ≈ +1). This confirms the CSF process accurately identifies ground points.

However, the distribution of absolute error from all 1 m cells of the UAS-SfM DTM, when compared to an ALS DTM, showed a substantial number of extreme outliers across all vegetation types, and, in some cases, higher than the canopy height (Figure 4). While the ground finding process improved overall DTM accuracy, the measurement limitations of UAS-SfM meant that it was unable to detect any ground points in some regions with substantial changes in terrain height, subsequently resulting in DTM errors observed in those areas. Based on their respective interquartile ranges, UAS-SfM DTM generation appears to demonstrate greater consistency in accuracy for grasslands than forested areas. However, due to the number of outliers across vegetation types, statistical analyses could not be conducted to differentiate DTM accuracy by vegetation type. Errors over 15 m were observed even in a few grasslands, indicating that large errors can appear even in areas without canopy cover. Errors over 4 m represent locations where the DTM inaccuracy would be so great as to render detection of ladder fuels largely impossible. Eight percent of the combined forest UAS-SfM DTM cells had errors greater than 4 m, while only 1% of grass and shrub cells had errors above this threshold, respectively. When separated by forest type, the forest cell error percentages ranged from 7–10%.

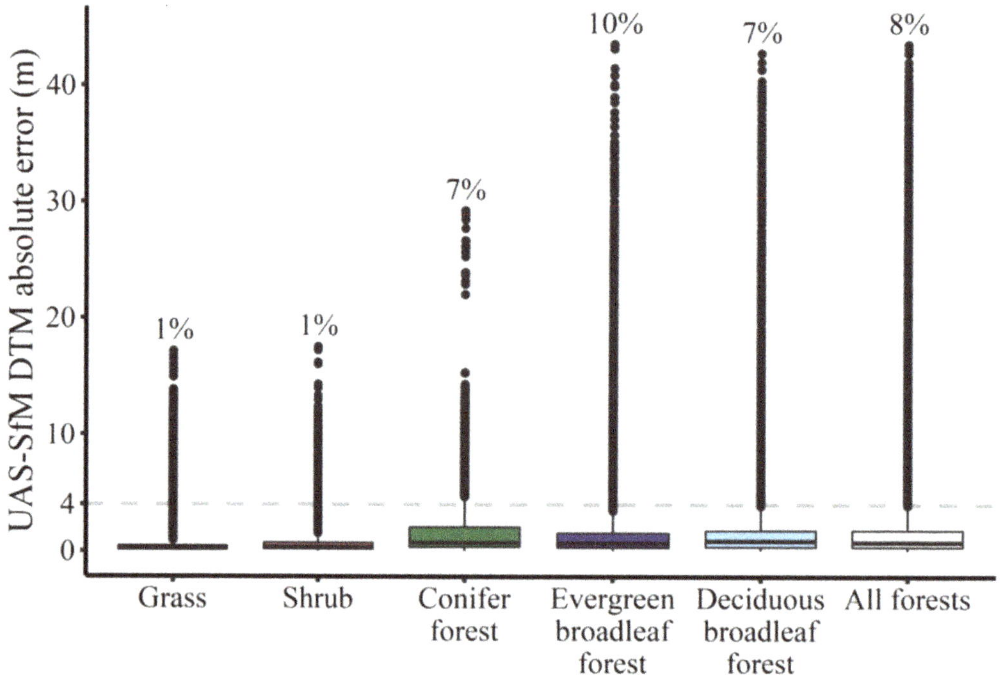

Figure 4. Boxplot of UAS-SfM DTM 1-m cell error by vegetation type for DTM obtained from optimized CSF parameters as compared to the ALS DTM. Numbers above each boxplot represent percentage of DTM cells with greater than 4 m of error.

3.2. *Comparison of Forest Structure from UAS-SfM and ALS Point Clouds*

The vertical structure of UAS-SfM and ALS point clouds reflect the ability of each sensor to detect structure below the canopy. Across all forest types for the complete site-wide dataset, ALS point clouds had a negatively tapered vertical density from the middle to lower part of forest canopies and a high density of ground points (Figure 5). The UAS-SfM point cloud upper-canopy density structure closely resembled that of ALS data, reaching a maximum canopy point density at a similar height. However, the lower canopy structure for UAS data tapered rapidly, indicating that the UAS point cloud was generally unable to detect the ground under heavy tree canopy. However, an exception to this pattern was observed in the evergreen broadleaf forest, which exhibited an increased point density at ground level. The unburned stand of evergreen broadleaf forest in this analysis contained trees spatially separated from one another, and it is from these gaps between the trees that the ground points were measured. Therefore, the increase in ground points does not actually represent an increased ability of UAS-SfM to detect below canopy structure in this forest type.

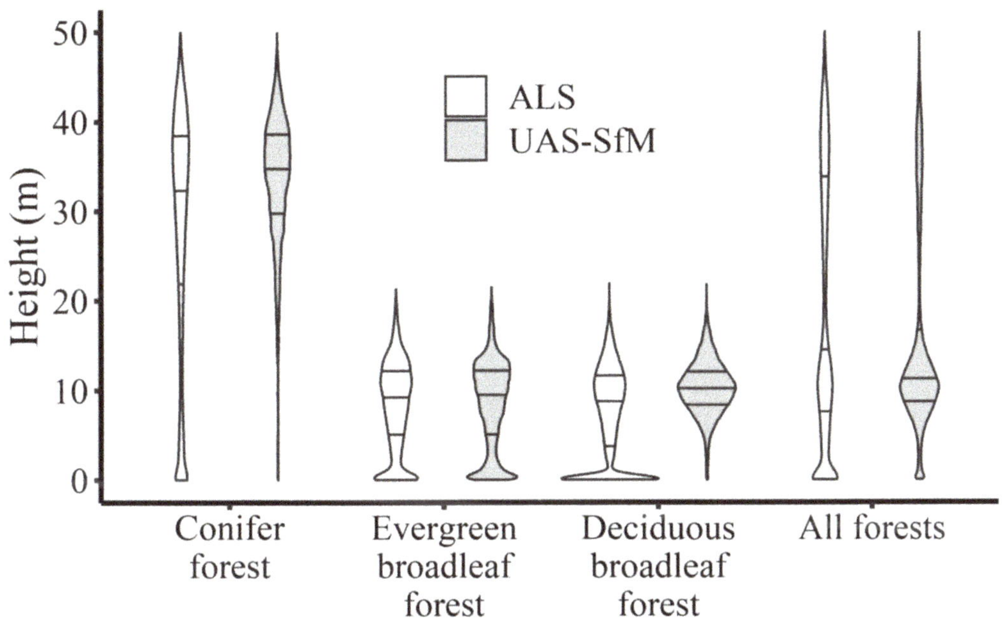

Figure 5. Above-ground point cloud height distributions (meters) from complete study-site data by forest type as measured by ALS (white) and UAS-SfM height normalized using the ALS DTM (shaded). Plots scaled for equal area. Horizontal lines represent 25%, 50% and 75% percentiles, respectively.

The area-based metric comparisons between UAS-SfM and ALS data reflect the preceding observations on sub-canopy detection across vegetation types (Tables 2 and 3). For both the UAS-SfM height metrics extracted using the UAS-SfM derived DTM and the ALS DTM for height normalization, only metrics associated with the upper half of the canopy (e.g., P_{50}, P_{75}, P_{95}, mean height, and max height) exhibited Pearson's $r > +0.80$. No relationships of this type exist between UAS-SfM and ALS for any of the metrics of height variation, nor those of height density for either of the height normalization approaches, with the exception of upper-canopy density (10–15 m) from the ALS DTM normalized dataset. As expected from the errors identified during the UAS-SfM DTM generation analysis (Section 3.1), the metrics extracted from the UAS-SfM data height normalized using the ALS DTM systematically performed better than the UAS-SfM data height normalized using the UAS-SfM DTM. However, these differences were small, especially for percentile metrics associated with the upper half of the canopy (Figure 6).

Table 2. Regression slope coefficients and Pearson's r correlation coefficient for point cloud grid metrics comparison between airborne laser scanner (ALS) data and unoccupied aerial system structure from motion (UAS-SfM) height normalized using the UAS-SfM digital terrain model (DTM). Metrics P_5, P_{25}, etc. represent height percentiles. Values given are mean (standard error) of 10,000 bootstrap replicas of 100 unburned plots per forest type. Coefficient and r values greater than 0.8 are shown in bold.

	Conifer		Evergreen Broadleaf		Deciduous Broadleaf		All Forest Types	
Height Metric	**Coefficient**	**r**	**Coefficient**	**r**	**Coefficient**	**r**	**Coefficient**	**r**
Ladder fuels	+0.09 (0.07)	+0.13 (0.10)	+0.05 (0.09)	+0.08 (0.13)	+0.16 (0.04)	+0.39 (0.07)	+0.16 (0.04)	+0.23 (0.06)
Density								
2–5 m	+0.12 (0.18)	+0.07 (0.08)	+0.09 (0.05)	+0.24 (0.11)	+0.07 (0.06)	+0.18 (0.12)	+0.08 (0.04)	+0.16 (0.08)
5–10 m	+0.36 (0.13)	+0.45 (0.13)	+0.21 (0.05)	+0.39 (0.09)	+0.25 (0.04)	+0.61 (0.09)	+0.33 (0.03)	+0.63 (0.04)
10–15 m	+0.38 (0.08)	+0.64 (0.10)	+0.22 (0.05)	+0.47 (0.10)	+0.21 (0.04)	+0.53 (0.09)	+0.32 (0.02)	+0.67 (0.04)
Distribution								
Max	+0.52 (0.12)	+0.67 (0.07)	+0.64 (0.08)	+0.76 (0.07)	+0.62 (0.11)	+0.69 (0.11)	**+0.81 (0.03)**	**+0.89 (0.02)**
Mean	+0.52 (0.06)	**+0.80 (0.05)**	+0.38 (0.08)	+0.60 (0.10)	+0.35 (0.09)	+0.56 (0.11)	+0.66 (0.02)	**+0.88 (0.02)**
P_5	+0.01 (0.01)	+0.17 (0.12)	−0.00 (0.02)	+0.00 (0.13)	+0.00 (0.00)	+0.07 (0.10)	+0.01 (0.01)	+0.19 (0.08)
P_{25}	+0.44 (0.06)	+0.66 (0.07)	+0.12 (0.11)	+0.17 (0.15)	+0.03 (0.04)	+0.09 (0.12)	+0.53 (0.04)	+0.74 (0.04)
P_{50}	+0.62 (0.08)	**+0.80 (0.05)**	+0.39 (0.08)	+0.58 (0.10)	+0.41 (0.12)	+0.41 (0.12)	+0.73 (0.03)	**+0.86 (0.02)**
P_{75}	+0.61 (0.09)	+0.77 (0.05)	+0.44 (0.07)	+0.68 (0.09)	+0.47 (0.11)	+0.59 (0.12)	+0.75 (0.03)	**+0.89 (0.02)**
P_{95}	+0.56 (0.10)	+0.73 (0.06)	+0.55 (0.07)	+0.75 (0.07)	+0.55 (0.11)	+0.67 (0.11)	+0.78 (0.03)	**+0.89 (0.02)**
Variation								
SD	+0.24 (0.08)	+0.29 (0.09)	+0.47 (0.05)	+0.72 (0.07)	+0.28 (0.12)	+0.33 (0.15)	+0.71 (0.04)	+0.72 (0.03)
Skewness	+0.29 (0.04)	+0.63 (0.08)	+0.16 (0.09)	+0.28 (0.11)	+0.05 (0.04)	+0.15 (0.11)	+0.15 (0.04)	+0.33 (0.07)
Kurtosis	+0.05 (0.02)	+0.42 (0.12)	−0.02 (0.04)	−0.05 (0.08)	−0.03 (0.01)	−0.33 (0.08)	+0.01 (0.01)	+0.04 (0.08)

Table 3. Regression slope coefficients and Pearson's r correlation coefficient for point cloud grid metrics comparison between ALS data and UAS-SfM height normalized using an ALS DTM. Metrics P_5, P_{25}, etc. represent height percentiles. Values given are mean (standard error) of 10,000 bootstrap replicas of 100 unburned plots per forest type. Coefficient and r values greater than 0.8 are shown in bold.

	Conifer		Evergreen Broadleaf		Deciduous Broadleaf		All Forest Types	
Height Metric	**Coefficient**	**r**	**Coefficient**	**r**	**Coefficient**	**r**	**Coefficient**	**r**
Ladder fuels	+0.16 (0.06)	+0.26 (0.10)	+0.02 (0.10)	+0.04 (0.14)	+0.08 (0.04)	+0.19 (0.10)	+0.08 (0.04)	+0.13 (0.07)
Density								
2–5 m	+0.40 (0.13)	+0.35 (0.10)	+0.16 (0.08)	+0.33 (0.14)	+0.18 (0.07)	+0.43 (0.10)	+0.20 (0.05)	+0.40 (0.08)
5–10 m	+0.54 (0.08)	+0.74 (0.07)	+0.35 (0.05)	+0.61 (0.07)	+0.23 (0.04)	+0.61 (0.07)	+0.36 (0.03)	+0.70 (0.03)
10–15 m	+0.62 (0.06)	**+0.83 (0.04)**	+0.35 (0.04)	+0.72 (0.07)	+0.26 (0.03)	+0.68 (0.07)	+0.39 (0.02)	**+0.80 (0.03)**
Distribution								
Max	+0.72 (0.18)	**+0.81 (0.09)**	+0.71 (0.10)	+0.79 (0.08)	+0.73 (0.11)	**+0.85 (0.12)**	**+0.92 (0.04)**	**+0.93 (0.02)**
Mean	+0.65 (0.09)	**+0.84 (0.07)**	+0.41 (0.10)	+0.60 (0.15)	+0.49 (0.06)	**+0.81 (0.06)**	+0.75 (0.03)	**+0.91 (0.03)**
P_5	+0.01 (0.01)	+0.23 (0.05)	−0.01 (0.03)	−0.03 (0.15)	−0.00 (0.01)	−0.03 (0.11)	+0.01 (0.01)	+0.17 (0.10)
P_{25}	+0.57 (0.10)	+0.67 (0.08)	+0.13 (0.12)	+0.17 (0.15)	+0.02 (0.02)	+0.11 (0.10)	+0.60 (0.05)	+0.75 (0.04)
P_{50}	+0.76 (0.12)	**+0.82 (0.07)**	+0.41 (0.10)	+0.57 (0.13)	+0.66 (0.09)	+0.70 (0.07)	**+0.83 (0.04)**	**+0.89 (0.03)**
P_{75}	+0.78 (0.14)	**+0.83 (0.07)**	+0.49 (0.10)	+0.69 (0.12)	+0.64 (0.08)	**+0.82 (0.08)**	**+0.86 (0.03)**	**+0.92 (0.02)**
P_{95}	+0.75(0.16)	**+0.84(0.08)**	+0.61(0.10)	+0.77(0.09)	+0.69(0.10)	**+0.85(0.10)**	**+0.89(0.03)**	**+0.93(0.02)**
Variation								
SD	+0.22(0.10)	+0.21(0.10)	+0.50(0.06)	+0.69(0.07)	+0.26(0.10)	+0.35(0.14)	**+0.80(0.05)**	+0.67(0.03)
Skewness	+0.30(0.07)	+0.48(0.09)	+0.25(0.05)	+0.45(0.07)	+0.12(0.04)	+0.35(0.11)	+0.18(0.03)	+0.34(0.05)
Kurtosis	+0.06(0.02)	+0.28(0.10)	−0.01(0.03)	−0.04(0.08)	−0.02(0.01)	−0.25(0.15)	−0.01(0.01)	−0.03(0.06)

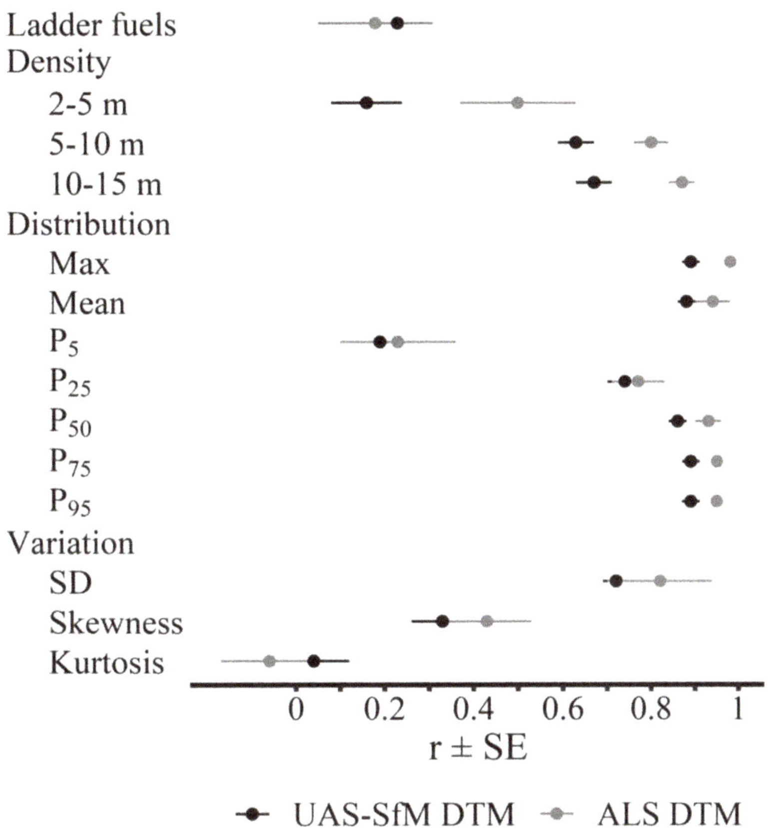

Figure 6. Comparison of Pearson correlations for point cloud height metrics extracted from UAS-SfM height normalized using either UAS-SfM DTM or ALS DTM across all forest types. Coefficients and their standard errors computed using 10,000 bootstrap replicates between UAS-SfM and ALS.

3.3. Utility of UAS-SfM for Detecting Post-Fire Forest Change

Ladder fuels: Across all forest types, the mean rank ladder fuel density was significantly higher in the high burn severity class, as compared to all other areas (Table 4, Figure 7). In evergreen broadleaf forests, the mean rank ladder fuel density was significantly lower in the medium burn severity class, as compared to the unburned and high burn severities classes (Table 4, Figure 7). The ladder fuel densities did not differ among burn severities for the other forest types.

Table 4. Kruskal–Wallis one-way analysis of variance results for 2017 Tubbs fire burn severity impact metrics. Significance at the 95% level denoted in bold.

Metric and Sample	χ^2	p	df	N
Pre-fire ladder fuel				
All forests	18.99	**<0.001**	3	220
Conifer	2.59	0.46	3	80
Evergreen broadleaf	13.68	**0.003**	3	80
Deciduous broadleaf	0.41	0.82	2	60
ΔP_{95}				
All forests	7.16	0.06	3	220
Conifer	0.79	0.85	3	80
Evergreen broadleaf	10.82	**0.01**	3	80
Deciduous broadleaf	0.73	0.69	2	60
ΔP_{75}				
All forests	35.01	**<0.001**	3	220
Conifer	10.98	**0.01**	3	80
Evergreen broadleaf	30.41	**<0.001**	3	80
Deciduous broadleaf	1.17	0.56	2	60
P_{75} NDVI				
All forests	89.81	**<0.001**	3	220
Conifer	54.49	**<0.001**	3	80
Evergreen broadleaf	35.31	**<0.001**	3	80
Deciduous broadleaf	9.21	0.01	2	60

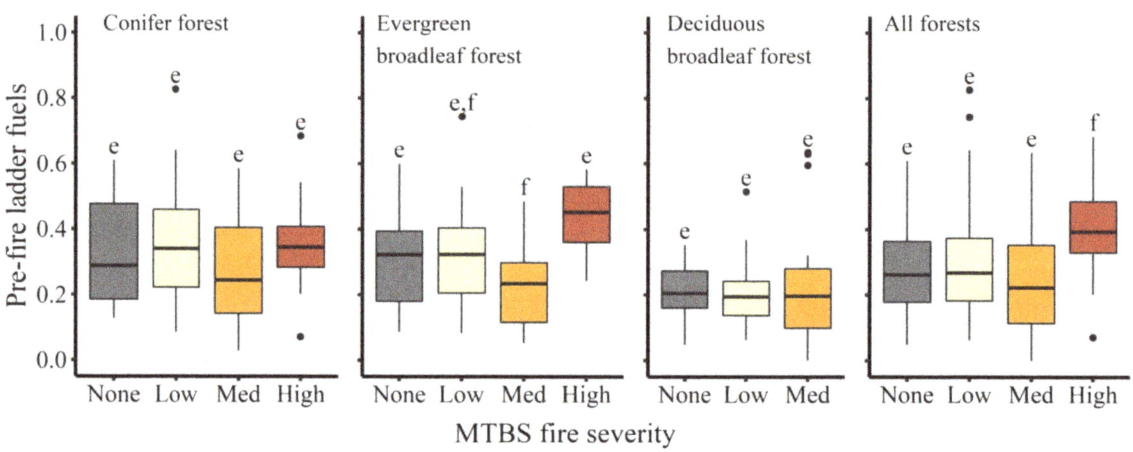

Figure 7. Relationship among pre-fire ladder fuels derived from 2013 ALS data and 2017 Tubbs fire burn severity by forest types. Lower case letters represent statistically similar groups at the 95% confidence level using a Dunn's post-hoc test.

ΔP_{95}: In evergreen broadleaf forests, the mean rank in pre-fire (ALS) to post-fire (UAS-SfM) P_{95} canopy height change (ΔP_{95}) was significantly more negative (i.e., greater decrease in heights) in the high burn severity class as compared to the low and medium burn severity classes (Table 4, Figure 8B). The mean rank ΔP_{95} was not significantly different across burn severities in the other forest types or for all forests combined (Table 4, Figure 8B–D).

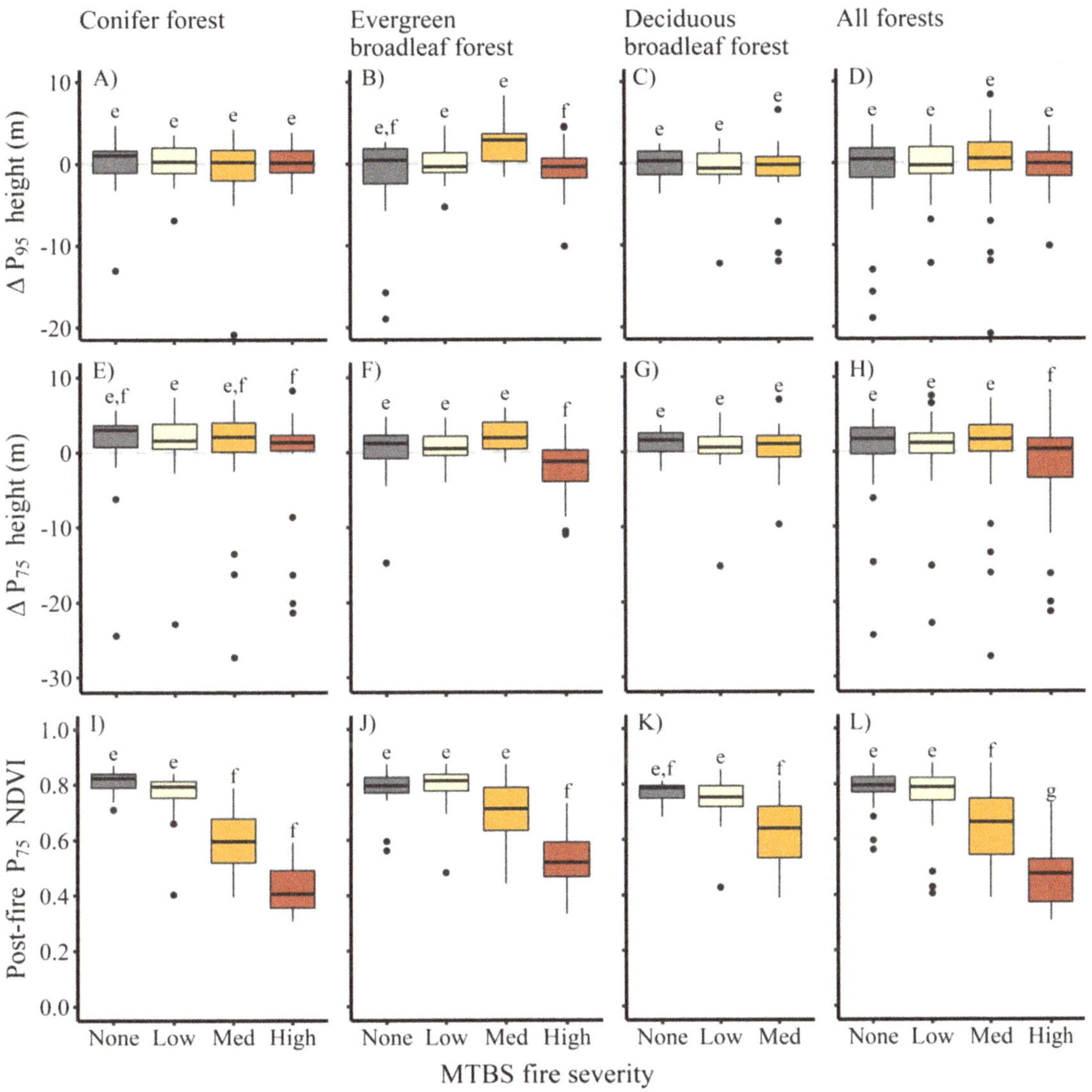

Figure 8. The impact of the 2017 Tubbs Fire on 95th percentile (P_{95}) and 75th percentile heights (P_{75}) and P_{75} mean NDVI by MTBS fire severity in forest types. Lower case letters represent statistically similar groups at the 95% confidence level. Change in canopy height metrics calculated as pre-fire ALS percentile height minus post-fire UAS-SfM percentile height.

ΔP_{75}: Across all forest types and in evergreen broadleaf forests, the mean rank ΔP_{75} was significantly more negative in the high burn severity class as compared to all others (Table 4, Figure 8F,H). In conifer forests, the mean rank ΔP_{75} was significantly more negative in the high burn severity class as compared to the low burn severity (Table 4, Figure 8E). The mean rank ΔP_{75} did not differ in deciduous broadleaf forests.

P_{75} NDVI: Across all forest types and in evergreen broadleaf forests, the mean rank in NDVI at post-fire (UAS) P_{75} canopy height (P_{75} NDVI) was significantly lower in the high burn severity class as compared to all others (Table 4, Figure 8L,J). In addition, across all forest types, the mean rank P_{75} was significantly lower in the medium burn severity class as compared to the low and unburned classes (Table 4, Figure 8L). In conifer forests, the mean

rank in P_{75} NDVI was significantly lower in the medium and high burn severity classes as compared to all others (Table 4, Figure 8I). In deciduous broadleaf forests, the mean rank in P_{75} NDVI was significantly lower in the medium burn severity class as compared to the low burn severity class (Table 4, Figure 8K).

In summary, only evergreen broadleaf forests exhibited a change in 95th percentile canopy height at higher burn severities. Bulk canopy height, represented by the 75th height percentile, decreased at high burn severity when considering all forests together, and within evergreen broadleaf and conifer forests. We also found that NDVI values in the bulk canopy decreased (i.e., lower greenness) with increasing burn severity, regardless of forest type.

4. Discussion

4.1. DTM and CHM Generation Capability Comparisons

This study sought to assess the capacity of UAS-SfM as a stand-alone forest monitoring tool to be employed in Mediterranean forests without pre-existing ALS data. This requires that a sufficiently accurate DTM can be estimated from UAS-SfM 3D point clouds, considering both detection of ground points and the classification of these points as ground relative to understory vegetation. This terrain finding process enables the isolation of canopy height from topography. While it is possible to achieve sub-centimeter DTM accuracy from UAS-SfM for bare ground conditions [46], increasing vegetation cover heavily impacts DTM accuracy [47–49]. Through two rounds of ground finding parameter optimization, we achieved a site wide DTM RMSE of 2.12 m. However, 8% of DTM coverage in forest areas contained errors over 4 m. This height threshold represents the upper limit of the ladder fuel metric from Green et al. [45], so errors of this magnitude connote an inability to detect surface or ladder fuels.

While the proliferation of error of this magnitude is serious, deciding whether this represents an unacceptable level of error largely depends on the context in which these data will be used. To determine canopy height and other metrics in areas of dense forest, having a baseline DTM from a more reliable source, such as airborne or UAS lidar, will always be best practice [50,51], assuming terrain does not change considerably over a study time period. However, in situations where these data do not exist, 8% error coverage may have to represent an acceptable margin. Of perhaps more importance among this erroneous area are the individual DTM sites with extreme errors, linked with gaps in the distribution of detected ground points. Future research should analyze the role of topography and canopy type in producing these areas of extreme error magnitude in order to determine conditions under which UAS-SfM cannot function without a preexisting DTM. Another future avenue of research to avoid these errors entirely is to determine if DTM-independent methods for extraction of forest metrics, which have been validated in forestry growing stock volume measurements [52], can be applied to the mapping of fuel loads.

This study employed the Cloth Simulation Filter (CSF) from Zhang et al. [44] in conjunction with a subsequent NDVI filter to classify ground points from which a DTM could be generated through interpolation. In Mediterranean shrublands, Carvajal-Ramirez et al. [53] found the CSF performance was weaker relative to an NDVI classifier, and the authors called for the need to assess the impact of CSF parameter setting on ground-finding accuracy. In this study, CSF parameter optimization alone only marginally improved RMSE over the default parameters. However, the addition of our NDVI threshold filter resulted in a large (>60%) improvement in RMSE. Drawing on the concept of the NDVI classifier from Carvajal-Ramirez et al. [53], the combination of CSF with an NDVI filter leverages the strengths of both methods. The CSF identifies an initial set of ground points from which the NDVI filter removes misclassified vegetation. When multispectral data exist, this filter provides a computationally simple way to improve ground finding based on point cloud structure alone. While methods for ground finding are continually being developed or improved e.g., [54], this filter provides an additional tool in mitigating the subcanopy detection shortcomings of UAS-SfM. However, one important caveat to our results is that our study area benefits from relatively low senesced grass (depending on cattle rotations and deer grazing) and higher live shrubs in late summer, thereby providing a strong NDVI contrast that aids the ground filter. Further tests of the CSF and NDVI filter combination in other ecosystems are certainly warranted.

4.2. Comparison of Forest Structure from UAS-SfM and ALS Point Clouds

Linear regression comparisons between UAS-SfM and ALS confirm past findings from other forest types and applications that UAS-SfM can only reliably resolve upper-canopy structure, but does so with a high degree of accuracy e.g., [49,55]. The analysis presented here closely matches Filippelli et al. [8], who found similar relationships in a montane coniferous and lodgepole pine forest between ALS and SfM DAP collected using an airplane flown at high altitude and with lower image overlap than used in this study. This suggests that SfM DAP has the same strength and limitations irrespective of image acquisition method and forest type. In particular, SfM DAP provides a powerful tool for measuring upper-canopy forest structure. The strength of this tool is dependent on the accuracy of the DTM underlying the analysis. While the use of the ALS DTM for height normalization systematically improved the accuracy of extracted canopy metrics as compared to the UAS-SfM, both methods performed well in measuring all metrics associated with the upper half of the canopy.

For SfM DAP to determine a point's 3D location, it must appear in multiple overlapping images. Therefore, the technique often fails to detect gaps in canopy cover, let alone resolve what lies below the canopy [28]. In our study, this shortcoming was most acutely demonstrated in our attempt to measure ladder fuels. No relationship existed between the ALS and UAS-SfM derived metrics. SfM DAP overestimates metrics of density, such as the ladder fuels metric, and canopy cover due to this lack of subcanopy detection [50]. The analysis of the point cloud's respective vertical structure lends additional weight to the conclusion that UAS-SfM cannot measure ladder fuels using the same metric as is employed for ALS. However, this study was limited in this analysis due to the six years between the collection of ALS and UAS-SfM data. While we restricted this analysis to areas of the study site unaffected by the Tubbs Fire, forest structure may have changed during the interim period (e.g., tree growth, disease, or mortality), which may account for some of the measured UAS-SfM error relative to the ALS baseline. Despite the subcanopy detection limitations of SfM DAP, Filippelli et al. [8] found DAP models of canopy structure performed comparably to those derived from ALS. Future studies with coincident ALS and UAS-SfM data should investigate this relationship further to determine if other methods can measure ladder fuel density from UAS-SfM despite its limitations under high canopy cover.

4.3. Utility of UAS-SfM for Detecting Post-Fire Forest Change

Two years after the Tubbs Fire swept through Pepperwood, UAS-SfM found minimal changes in canopy height relative to pre-fire metrics. In the aggregate analysis of all forest types, measures of maximum height (95th percentile) did not change, while bulk canopy height (75th percentile) decreased significantly in regions with the highest burn severity. We observed significant decreases in NDVI with increasing burn severity across all forests regardless of structural change. While we did not have calibrated pre-fire measures of NDVI on the same spatial scale to which we could compare, the unburned areas were used as a control reference in this comparison. This demonstrates the long-lasting impact a severe wildfire has on Mediterranean forest health.

The role of NDVI in understanding forest health indicates that multispectral sensors have an important role in forest monitoring for fire hazard and post-fire recovery. Change in NDVI can detect fine-scale forest physiological stress [56]. Change in NDVI also provides the best estimator of post-fire severity in Mediterranean vegetation [45,57]. The long-term trend of NDVI over a stand or ecosystem can be used to measure the degree to which different vegetation types manage to recover from a severe fire event [58]. In our study ecosystem, UAS-SfM-based measures of NDVI or other greenness indices over multiple times post-fire could provide valuable crown-scale information on tree response to fire [59,60], such as initiation of epicormic sprouting and subsequent canopy growth. As demonstrated here, this information can provide insight on impacts beyond what structural information alone could provide. In addition, since multispectral UAS-SfM provides coupled spectral and structural information, it offers a means to target analysis at particular levels of the canopy in a way that is not possible with aerial or satellite imagery alone. For example, here we were able to use this technology to disentangle bulk canopy changes in NDVI from other post-fire ecosystem changes, such as ground vegetation recovery. While our results targeted change detection at the 20 m grid level, future studies should leverage the high resolution of this technology to analyze change at finer scales or even at the individual tree level.

The analysis of post-fire canopy changes would not have been possible without the existence of the pre-fire ALS data. While the unburned areas could be used as a control in the absence of such data, as done here with the NDVI analysis, change analysis of this type greatly benefits from ongoing monitoring. It is in this capacity that UAS-SfM provides a particularly strong benefit due to the ease of data acquisition at plot to stand scales.

Although UAS-SfM could not estimate ladder fuels, our analysis with ALS-based pre-fire ladder fuels and burn severity revealed a pattern that merits further investigation. In the aggregate analysis of all forest types, the high burn severity class exhibited greater pre-fire ladder fuel density than the low burn severity class. We found a significant relationship between increasing pre-fire ladder fuel densities and medium to high burn severity in evergreen broadleaf forest, but not in the conifer forest. Neither forest type had a significant difference between low and high burn severities. Conifers in particular showed no relationship between burn severity and ladder fuel density. Two explanations are possible for this. First, the majority of conifer forest that burned at high severity during the Tubbs Fire did so during the initial "Diablo" wind-driven portion of the event. Green et al. [45] identified these winds to be an even stronger driver of burn severity than ladder fuels. Therefore, the mixed signal across forest types and in conifer forests may be attributable to wind strength overwhelming any difference that ladder fuel densities may have otherwise caused in burn severity [61]. An alternative explanation, however, is that the ladder fuel metric employed here performs well in shorter evergreen broadleaf forests, especially in determining if higher intensities fires are able to cross the high severity threshold, but does not apply in tall conifer forests where canopy base height may be well above 4 m. Further work is needed to investigate the utility of different measures of vertical fuel continuity, such as ladder fuel metrics, in different forest types. Furthermore, the estimation of such sub-canopy metrics at a stand scale using UAS may best be achieved with an attached lidar sensor instead of the SfM approach used here [62]. Although UAS-SfM is currently more economical, prices for UAS lidar are expected to decrease over time [25].

5. Conclusions

For fire hazard mitigation land management treatments to be effective, land managers must implement them at broad spatial scales, from single forest stands to larger regions [63]. This requires an improved understanding of forest structure and fuel loading to inform management practices.

Due to the comparatively low investment in time or resources involved in conducting repeated data collection missions, UAS-SfM provides a unique opportunity to monitor ongoing changes and recovery of forest structure and physiological health following major disturbance events at plot to stand scales [26,64]. Our study in a Mediterranean forest site in Northern California found UAS-SfM able to accurately detect ground points. However, ground points in some areas were too sparsely distributed to reliably estimate ground elevation, a limitation that affects subsequent canopy height retrieval and limits its use in these areas as a stand-alone tool in the absence of more accurate terrain data.

This shortcoming arose from UAS-SfM's limited subcanopy detection ability, which also hindered its ability to detect ladder fuels and other below-canopy metrics. Across the study site, UAS-SfM excelled at mapping the upper canopy of forests and, when combined with a multispectral sensor, can provide information on vegetation productivity and physiological stress. This accuracy can be improved further when the UAS-SfM is paired with an accurate baseline measurement of ground elevation (e.g., from an airborne lidar sensor). UAS-SfM, therefore, provides a valuable tool for monitoring post-fire impacts and recovery through the use of upper-canopy metrics of structure and health, especially at sites with pre-fire data as presented here. Much work remains to be done in the evaluation of UAS-SfM as a fire fuel monitoring tool in Mediterranean ecosystems. Future work should focus on leveraging the strengths of this method to improve management outcomes.

Author Contributions: Conceptualization, S.R., M.L.C., L.P.B. and I.O.M.; methodology, S.R., M.L.C., L.P.B. and I.O.M.; investigation, S.R., M.L.C., L.P.B., C.M. and E.P.; software, S.R. and M.L.C.; formal analysis, S.R. and M.L.C.; resources, M.L.C. and L.P.B.; writing—original draft preparation, S.R.; writing—review and editing, I.O.M., M.L.C., L.P.B., C.M. and E.P.; visualization, S.R.; supervision, I.O.M., M.L.C. and L.P.B.; funding acquisition, L.P.B. and M.L.C. All authors have read and agreed to the published version of the manuscript.

Funding: Funding for this research was supported by CAL FIRE Forest Health and Forest Legacy (8GG18806) and California State University, Agricultural Research Institute (20-01-106) awards. SR was funded by the Rhodes Trust and through the University of Oxford Environmental Change Institute Small Grant Scheme.

Institutional Review Board Statement: Not applicable.

Informed Consent Statement: Not applicable.

Data Availability Statement: ALS data used in this study are available through the Sonoma Veg Map online portal (sonomavegmap.org). MTBS burn severity data used in this study are available through the MTBS online portal (mtbs.gov). UAS data are available upon request.

Acknowledgments: We thank the staff of Pepperwood Preserve for allowing us access to their beautiful site. This work is a product of the 3D Forests research group.

Conflicts of Interest: The authors declare no conflict of interest.

References

1. Rundel, P.W.; Arroyo, M.T.K.; Cowling, R.M.; Keeley, J.E.; Lamont, B.B.; Vargas, P. Mediterranean Biomes: Evolution of Their Vegetation, Floras, and Climate. *Annu. Rev. Ecol. Evol. Syst.* **2016**, *47*, 383–407. [CrossRef]
2. Beaty, R.M.; Taylor, A.H. Fire History and the Structure and Dynamics of a Mixed Conifer Forest Landscape in the Northern Sierra Nevada, Lake Tahoe Basin, California, USA. *For. Ecol. Manag.* **2008**, *255*, 707–719. [CrossRef]
3. Cowling, R.; Rundel, P.; Lamont, B.; Arroyo, M.; Arianoutsou, M. Plant Diversity in Mediterranean-Climate Regions. *Trends Ecol. Evol.* **1996**, *11*, 362–366. [CrossRef]
4. Odion, D.C.; Moritz, M.A.; DellaSala, D.A. Alternative Community States Maintained by Fire in the Klamath Mountains, USA: Fire and Alternative Community States. *J. Ecol.* **2010**, *98*, 96–105. [CrossRef]
5. Westerling, A. *Wildfire Simulations for California's Fourth Climate Change Assessment: Projecting Changes in Extreme Wildfire Events with a Warming Climate*; California's Fourth Climate Change Assessment; California Energy Commission: Sacramento, CA, USA, 2018; p. 57.
6. Miller, J.D.; Safford, H.D.; Crimmins, M.; Thode, A.E. Quantitative Evidence for Increasing Forest Fire Severity in the Sierra Nevada and Southern Cascade Mountains, California and Nevada, USA. *Ecosystems* **2009**, *12*, 16–32. [CrossRef]
7. Dennison, P.E.; Brewer, S.C.; Arnold, J.D.; Moritz, M.A. Large Wildfire Trends in the Western United States, 1984-2011. *Geophys. Res. Lett.* **2014**, *41*, 2928–2933. [CrossRef]
8. Filippelli, S.K.; Lefsky, M.A.; Rocca, M.E. Comparison and Integration of Lidar and Photogrammetric Point Clouds for Mapping Pre-Fire Forest Structure. *Remote Sens. Environ.* **2019**, *224*, 154–166. [CrossRef]
9. Flannigan, M.D.; Krawchuk, M.A.; de Groot, W.J.; Wotton, B.M.; Gowman, L.M. Implications of Changing Climate for Global Wildland Fire. *Int. J. Wildland Fire* **2009**, *18*, 483–507. [CrossRef]
10. Westerling, A.L.; Bryant, B.P. Climate Change and Wildfire in California. *Clim. Chang.* **2008**, *87*, 231–249. [CrossRef]
11. Alonzo, M.; Andersen, H.-E.; Morton, D.; Cook, B. Quantifying Boreal Forest Structure and Composition Using UAV Structure from Motion. *Forests* **2018**, *9*, 119. [CrossRef]

12. Berra, E.F.; Gaulton, R.; Barr, S. Assessing Spring Phenology of a Temperate Woodland: A Multiscale Comparison of Ground Unmanned Aerial Vehicle and Landsat Satellite Observations. *Remote Sens. Environ.* **2019**, *223*, 229–242. [CrossRef]
13. Arroyo, L.A.; Pascual, C.; Manzanera, J.A. Fire Models and Methods to Map Fuel Types: The Role of Remote Sensing. *For. Ecol Manag.* **2008**, *256*, 1239–1252. [CrossRef]
14. Stefanidou, A.; Gitas, I.Z.; Katagis, T. A National Fuel Type Mapping Method Improvement Using Sentinel-2 Satellite Data *Geocarto Int.* **2020**, 1–21. [CrossRef]
15. Andersen, H.-E.; McGaughey, R.J.; Reutebuch, S.E. Estimating Forest Canopy Fuel Parameters Using LIDAR Data. *Remote Sens Environ.* **2005**, *94*, 441–449. [CrossRef]
16. Marino, E.; Montes, F.; Tomé, J.L.; Navarro, J.A.; Hernando, C. Vertical Forest Structure Analysis for Wildfire Prevention Comparing Airborne Laser Scanning Data and Stereoscopic Hemispherical Images. *Int. J. Appl. Earth Obs. Geoinf.* **2018**, *73* 438–449. [CrossRef]
17. Kramer, H.A.; Collins, B.M.; Kelly, M.; Stephens, S.L. Quantifying Ladder Fuels: A New Approach Using LiDAR. *Forests* **2014**, *5* 1432–1453. [CrossRef]
18. Menning, K.M.; Stephens, S.L. Fire Climbing in the Forest: A Semiqualitative, Semiquantitative Approach to Assessing Ladder Fuel Hazards. *West. J. Appl. For.* **2007**, *22*, 88–93. [CrossRef]
19. Donnellan, A.; Harding, D.; Lundgren, P.; Wessels, K.; Gardner, A.; Simard, M.; Parrish, C.; Jones, C.; Lou, Y.; Stoker, J.; et al *Observing Earth's Changing Surface Topography and Vegetation Structure: A Framework for the Decade*; NASA Surface Topography and Vegetation Incubation Study; NASA: Washington, DC, USA, 2021; p. 210.
20. Chuvieco, E.; Aguado, I.; Salas, J.; García, M.; Yebra, M.; Oliva, P. Satellite Remote Sensing Contributions to Wildland Fire Science and Management. *Curr. For. Rep.* **2020**, *6*, 81–96. [CrossRef]
21. Dandois, J.P.; Olano, M.; Ellis, E.C. Optimal Altitude, Overlap, and Weather Conditions for Computer Vision UAV Estimates of Forest Structure. *Remote Sens.* **2015**, *7*, 13895–13920. [CrossRef]
22. Alonzo, M.; Dial, R.J.; Schulz, B.K.; Andersen, H.-E.; Lewis-Clark, E.; Cook, B.D.; Morton, D.C. Mapping Tall Shrub Biomass in Alaska at Landscape Scale Using Structure-from-Motion Photogrammetry and Lidar. *Remote Sens. Environ.* **2020**, *245*, 111841 [CrossRef]
23. Dandois, J.P.; Ellis, E.C. High Spatial Resolution Three-Dimensional Mapping of Vegetation Spectral Dynamics Using Computer Vision. *Remote Sens. Environ.* **2013**, *136*, 259–276. [CrossRef]
24. Tompalski, P.; White, J.C.; Coops, N.C.; Wulder, M.A. Quantifying the Contribution of Spectral Metrics Derived from Digital Aerial Photogrammetry to Area-Based Models of Forest Inventory Attributes. *Remote Sens. Environ.* **2019**, *234*, 111434. [CrossRef]
25. Hu, T.; Sun, X.; Su, Y.; Guan, H.; Sun, Q.; Kelly, M.; Guo, Q. Development and Performance Evaluation of a Very Low-Cost UAV-Lidar System for Forestry Applications. *Remote Sens.* **2020**, *13*, 77. [CrossRef]
26. Campbell, M.J.; Dennison, P.E.; Tune, J.W.; Kannenberg, S.A.; Kerr, K.L.; Codding, B.F.; Anderegg, W.R.L. A Multi-Sensor Multi-Scale Approach to Mapping Tree Mortality in Woodland Ecosystems. *Remote Sens. Environ.* **2020**, *245*, 111853. [CrossRef]
27. Cunliffe, A.M.; Brazier, R.E.; Anderson, K. Ultra-Fine Grain Landscape-Scale Quantification of Dryland Vegetation Structure with Drone-Acquired Structure-from-Motion Photogrammetry. *Remote Sens. Environ.* **2016**, *183*, 129–143. [CrossRef]
28. White, J.C.; Tompalski, P.; Coops, N.C.; Wulder, M.A. Comparison of Airborne Laser Scanning and Digital Stereo Imagery for Characterizing Forest Canopy Gaps in Coastal Temperate Rainforests. *Remote Sens. Environ.* **2018**, *208*, 1–14. [CrossRef]
29. Ackerly, D.D.; Kling, M.M.; Clark, M.L.; Papper, P.; Oldfather, M.F.; Flint, A.L.; Flint, L.E. Topoclimates, Refugia, and Biotic Responses to Climate Change. *Front. Ecol. Environ.* **2020**, *18*, 288–297. [CrossRef]
30. Evett, R.R.; Dawson, A.; Bartolome, J.W. Estimating Vegetation Reference Conditions by Combining Historical Source Analysis and Soil Phytolith Analysis at Pepperwood Preserve, Northern California Coast Ranges, U.S.A: Estimating Vegetation Reference Conditions. *Restor. Ecol.* **2013**, *21*, 464–473. [CrossRef]
31. Nauslar, N.; Abatzoglou, J.; Marsh, P. The 2017 North Bay and Southern California Fires: A Case Study. *Fire* **2018**, *1*, 18. [CrossRef]
32. Pepperwood Preserve. What Does Resilience Look like after Two Fires in Two Years? In *Pepperwood Field Notes*; Pepperwood Preserve: Santa Rosa, CA, USA, 2019.
33. CalFire. *Top 20 Deadliest California Wildfires*; CalFire: Sacramento, CA, USA, 2019.
34. CalFire. *Top 20 Most Destructive California Wildfires*; CalFire: Sacramento, CA, USA, 2019.
35. Roussel, J.-R.; Auty, D.; De Boissieu, F.; Sánchez Meador, A.; Jean-François, B. *LidR: Airborne LiDAR Data Manipulation and Visualization for Forestry Applications*; R Foundation for Statistical Computing: Vienna, Austria, 2020.
36. R Core Team. *R: A Language and Environment for Statistical Computing*; R Foundation for Statistical Computing: Vienna, Austria 2019.
37. Hijmans, R.J.; van Etten, J.; Sumner, M.; Cheng, J.; Bevan, A.; Bivand, R.; Busetto, L.; Canty, M.; Forrest, D.; Ghosh, A.; et al. *Raster: Geographic Data Analysis and Modeling*, R Foundation for Statistical Computing: Vienna, Austria, 2019.
38. Watershed Sciences. *Sonoma County Vegetation Mapping and Lidar Program: Technical Data Report*; Watershed Sciences: Portland OR, USA, 2016.
39. Eidenshink, J.; Schwind, B.; Brewer, K.; Zhu, Z.-L.; Quayle, B.; Howard, S. A Project for Monitoring Trends in Burn Severity. *Fire Ecol.* **2007**, *3*, 3–21. [CrossRef]
40. Picotte, J.J.; Bhattarai, K.; Howard, D.; Lecker, J.; Epting, J.; Quayle, B.; Benson, N.; Nelson, K. Changes to the Monitoring Trends in Burn Severity Program Mapping Production Procedures and Data Products. *Fire Ecol.* **2020**, *16*, 16. [CrossRef]

41. Miller, J.D.; Thode, A.E. Quantifying Burn Severity in a Heterogeneous Landscape with a Relative Version of the Delta Normalized Burn Ratio (DNBR). *Remote Sens. Environ.* **2007**, *109*, 66–80. [CrossRef]
42. Bigdeli, B.; Amini Amirkolaee, H.; Pahlavani, P. DTM Extraction under Forest Canopy Using LiDAR Data and a Modified Invasive Weed Optimization Algorithm. *Remote. Sens. Environ.* **2018**, *216*, 289–300. [CrossRef]
43. Klápště, P.; Fogl, M.; Barták, V.; Gdulová, K.; Urban, R.; Moudrý, V. Sensitivity Analysis of Parameters and Contrasting Performance of Ground Filtering Algorithms with UAV Photogrammetry-Based and LiDAR Point Clouds. *Int. J. Digit. Earth* **2020**, *13*, 1–23. [CrossRef]
44. Zhang, W.; Qi, J.; Wan, P.; Wang, H.; Xie, D.; Wang, X.; Yan, G. An Easy-to-Use Airborne LiDAR Data Filtering Method Based on Cloth Simulation. *Remote Sens.* **2016**, *8*, 501. [CrossRef]
45. Green, K.; Tukman, M.; Loudon, D.; Schichtel, A.; Gaffney, K.; Clark, M. Sonoma County Complex Fires of 2017: Remote Sensing Data and Modeling to Support Ecosystem and Community Resiliency. *Calif. Fish Wildl. J.* **2020**, *Fire Special Issue*, 14–45. Available online: https://nrm.dfg.ca.gov/FileHandler.ashx?DocumentID=184827&inline (accessed on 29 July 2021).
46. Adjidjonu, D.; Burgett, J. Assessing the Accuracy of Unmanned Aerial Vehicles Photogrammetric Survey. *Int. J. Constr. Educ. Res.* **2021**, *17*, 85–96. [CrossRef]
47. Goodbody, T.R.H.; Coops, N.C.; Hermosilla, T.; Tompalski, P.; Pelletier, G. Vegetation Phenology Driving Error Variation in Digital Aerial Photogrammetrically Derived Terrain Models. *Remote Sens.* **2018**, *10*, 1554. [CrossRef]
48. Graham, A.; Coops, N.C.; Wilcox, M.; Plowright, A. Evaluation of Ground Surface Models Derived from Unmanned Aerial Systems with Digital Aerial Photogrammetry in a Disturbed Conifer Forest. *Remote Sens.* **2019**, *11*, 84. [CrossRef]
49. Wallace, L.; Bellman, C.; Hally, B.; Hernandez, J.; Jones, S.; Hillman, S. Assessing the Ability of Image Based Point Clouds Captured from a UAV to Measure the Terrain in the Presence of Canopy Cover. *Forests* **2019**, *10*, 284. [CrossRef]
50. Jayathunga, S.; Owari, T.; Tsuyuki, S. Evaluating the Performance of Photogrammetric Products Using Fixed-Wing UAV Imagery over a Mixed Conifer–Broadleaf Forest: Comparison with Airborne Laser Scanning. *Remote Sens.* **2018**, *10*, 187. [CrossRef]
51. Prata, G.A.; Broadbent, E.N.; de Almeida, D.R.A.; St. Peter, J.; Drake, J.; Medley, P.; Corte, A.P.D.; Vogel, J.; Sharma, A.; Silva, C.A.; et al. Single-Pass UAV-Borne GatorEye LiDAR Sampling as a Rapid Assessment Method for Surveying Forest Structure. *Remote Sens.* **2020**, *12*, 4111. [CrossRef]
52. Giannetti, F.; Chirici, G.; Gobakken, T.; Næsset, E.; Travaglini, D.; Puliti, S. A New Approach with DTM-Independent Metrics for Forest Growing Stock Prediction Using UAV Photogrammetric Data. *Remote Sens. Environ.* **2018**, *213*, 195–205. [CrossRef]
53. Carvajal-Ramírez, F.; Serrano, J.M.P.R.; Agüera-Vega, F.; Martínez-Carricondo, P. A Comparative Analysis of Phytovolume Estimation Methods Based on UAV-Photogrammetry and Multispectral Imagery in a Mediterranean Forest. *Remote Sens.* **2019**, *11*, 2579. [CrossRef]
54. Cai, S.; Zhang, W.; Liang, X.; Wan, P.; Qi, J.; Yu, S.; Yan, G.; Shao, J. Filtering Airborne LiDAR Data through Complementary Cloth Simulation and Progressive TIN Densification Filters. *Remote Sens.* **2019**, *11*, 1037. [CrossRef]
55. Vastaranta, M.; Wulder, M.A.; White, J.C.; Pekkarinen, A.; Tuominen, S.; Ginzler, C.; Kankare, V.; Holopainen, M.; Hyyppä, H. Airborne Laser Scanning and Digital Stereo Imagery Measures of Forest Structure: Comparative Results and Implications to Forest Mapping and Inventory Update. *Can. J. Remote Sens.* **2013**, *39*, 382–395. [CrossRef]
56. Dash, J.P.; Pearse, G.D.; Watt, M.S. UAV Multispectral Imagery Can Complement Satellite Data for Monitoring Forest Health. *Remote Sens.* **2018**, *10*, 1216. [CrossRef]
57. Carvajal-Ramírez, F.; Marques da Silva, J.R.; Agüera-Vega, F.; Martínez-Carricondo, P.; Serrano, J.; Moral, F.J. Evaluation of Fire Severity Indices Based on Pre- and Post-Fire Multispectral Imagery Sensed from UAV. *Remote Sens.* **2019**, *11*, 993. [CrossRef]
58. Díaz-Delgado, R.; Lloret, F.; Pons, X. Influence of Fire Severity on Plant Regeneration by Means of Remote Sensing Imagery. *Int. J. Remote Sens.* **2003**, *24*, 1751–1763. [CrossRef]
59. Larrinaga, A.R.; Brotons, L. Greenness Indices from a Low-Cost UAV Imagery as Tools for Monitoring Post-Fire Forest Recovery. *Drones* **2019**, *3*, 6. [CrossRef]
60. Pádua, L.; Guimarães, N.; Adão, T.; Sousa, A.; Peres, E.; Sousa, J.J. Effectiveness of Sentinel-2 in Multi-Temporal Post-Fire Monitoring When Compared with UAV Imagery. *IJGI* **2020**, *9*, 225. [CrossRef]
61. Keeley, J.E.; Syphard, A.D. Twenty-First Century California, USA, Wildfires: Fuel-Dominated vs. Wind-Dominated Fires. *Fire Ecol.* **2019**, *15*, 1–15. [CrossRef]
62. Hillman, S.; Wallace, L.; Lucieer, A.; Reinke, K.; Turner, D.; Jones, S. A Comparison of Terrestrial and UAS Sensors for Measuring Fuel Hazard in a Dry Sclerophyll Forest. *Int. J. Appl. Earth Obs. Geoinf.* **2021**, *95*, 102261. [CrossRef]
63. Collins, B.M.; Kramer, H.A.; Menning, K.; Dillingham, C.; Saah, D.; Stine, P.A.; Stephens, S.L. Modeling Hazardous Fire Potential within a Completed Fuel Treatment Network in the Northern Sierra Nevada. *For. Ecol. Manag.* **2013**, *310*, 156–166. [CrossRef]
64. Goodbody, T.R.H.; Coops, N.C.; Tompalski, P.; Crawford, P. Assessing the Status of Forest Regeneration Using Digital Aerial Photogrammetry and Unmanned Aerial Systems. *Int. J. Remote Sens.* **2018**, *39*, 5246–5264. [CrossRef]

remote sensing

MDPI

Article

A Framework for Multi-Dimensional Assessment of Wildfire Disturbance Severity from Remotely Sensed Ecosystem Functioning Attributes

Bruno Marcos [1,2,*], João Gonçalves [1,3], Domingo Alcaraz-Segura [4,5,6], Mário Cunha [2,7] and João P. Honrado [1,2]

1 Research Centre in Biodiversity and Genetic Resources, Research Network in Biodiversity and Evolutionary Biology (CIBIO-InBIO), Campus Agrário de Vairão, Universidade do Porto, Rua Padre Armando Quintas, 4485-661 Vairão, Portugal; joao.goncalves@cibio.up.pt (J.G.); jhonrado@fc.up.pt (J.P.H.)
2 Faculty of Sciences, University of Porto, Rua Campo Alegre s/n, 4169-007 Porto, Portugal; mccunha@fc.up.pt
3 proMetheus—Research Unit in Materials, Energy and Environment for Sustainability, Instituto Politécnico de Viana do Castelo (IPVC), Avenida do Atlântico, no. 644, 4900-348 Viana do Castelo, Portugal
4 Department of Botany, Faculty of Sciences, University of Granada, Av. Fuentenueva, 18071 Granada, Spain; dalcaraz@ugr.es
5 iEcolab. Interuniversity Institute for Earth System Research (IISTA), University of Granada, Av. del Mediterráneo, 18006 Granada, Spain
6 Andalusian Center for the Assessment and Monitoring of Global Change (CAESCG), Universidad de Almería, Crta. San Urbano, 04120 Almería, Spain
7 Institute for Systems and Computer Engineering, Technology and Science (INESC TEC), Campus da Faculdade de Engenharia da Universidade do Porto, Rua Dr. Roberto Frias, 4200-465 Porto, Portugal
* Correspondence: bruno.marcos@cibio.up.pt

Abstract: Wildfire disturbances can cause modifications in different dimensions of ecosystem functioning, i.e., the flows of matter and energy. There is an increasing need for methods to assess such changes, as functional approaches offer advantages over those focused solely on structural or compositional attributes. In this regard, remote sensing can support indicators for estimating a wide variety of effects of fire on ecosystem functioning, beyond burn severity assessment. These indicators can be described using intra-annual metrics of quantity, seasonality, and timing, called Ecosystem Functioning Attributes (EFAs). Here, we propose a satellite-based framework to evaluate the impacts, at short to medium term (i.e., from the year of fire to the second year after), of wildfires on four dimensions of ecosystem functioning: (i) primary productivity, (ii) vegetation water content, (iii) albedo, and (iv) sensible heat. We illustrated our approach by comparing inter-annual anomalies in satellite-based EFAs in the northwest of the Iberian Peninsula, from 2000 to 2018. Random Forest models were used to assess the ability of EFAs to discriminate burned vs. unburned areas and to rank the predictive importance of EFAs. Together with effect sizes, this ranking was used to select a parsimonious set of indicators for analyzing the main effects of wildfire disturbances on ecosystem functioning, for both the whole study area (i.e., regional scale), as well as for four selected burned patches with different environmental conditions (i.e., local scale). With both high accuracies (area under the receiver operating characteristic curve (AUC) > 0.98) and effect sizes (Cohen's $|d| > 0.8$), we found important effects on all four dimensions, especially on primary productivity and sensible heat, with the best performance for quantity metrics. Different spatiotemporal patterns of wildfire severity across the selected burned patches for different dimensions further highlighted the importance of considering the multi-dimensional effects of wildfire disturbances on key aspects of ecosystem functioning at different timeframes, which allowed us to diagnose both abrupt and lagged effects. Finally, we discuss the applicability as well as the potential advantages of the proposed approach for more comprehensive assessments of fire severity.

Keywords: ecological disturbance; ecosystem functioning; EFAs; fire severity; satellite image time-series; wildfires

Citation: Marcos, B.; Gonçalves, J.; Alcaraz-Segura, D.; Cunha, M.; Honrado, J.P. A Framework for Multi-Dimensional Assessment of Wildfire Disturbance Severity from Remotely Sensed Ecosystem Functioning Attributes. *Remote Sens.* **2021**, *13*, 780. https://doi.org/10.3390/rs13040780

Received: 21 January 2021
Accepted: 18 February 2021
Published: 20 February 2021

1. Introduction

Wildfire disturbances are considered an integral part of the natural dynamics of ecosystems in several biomes [1,2]. Notwithstanding, wildfire events can modify the composition, structure, and functioning of ecosystems and, consequently, the provision of ecosystem services to humankind [3]. Among those alterations, changes in ecosystem functioning are of particular interest, as fire can cause rapid modifications in the matter and energy budgets of ecosystems [4]. This is because attributes of ecosystem functioning have a shorter response time to disturbances than structural or compositional attributes of landscapes and are more directly connected to ecosystem services [5].

Wildfires can have profound impacts on many key aspects of the flows of matter and energy. These include dimensions of ecosystem functioning related to the biogeochemical cycles of carbon (e.g., primary productivity, biomass), and water (e.g., vegetation water content, soil moisture), as well as energy balances (e.g., albedo, latent heat, sensible heat). Namely, fires play an important role in the terrestrial biosphere carbon cycle [6], as they can cause substantial losses in carbon storage [7], including both aboveground biomass [8] as well as soil organic matter [9]. Moreover, fire-mediated changes in nutrient concentrations can ultimately limit productivity [10] as well as induce phenological changes [11]. Wildfires can also affect both water supply [12] and quality [13,14], as well as induce vegetation mortality due to prolonged loss of foliar moisture [15]. Additionally, wildfires can have negative impacts on evapotranspiration [16,17], which, in turn, also influences radiation exchanges, as evapotranspiration is related to latent heat flux [18]. Changes in wildfire frequency and/or severity may also have strong impacts on Earth's surface radiative budget [19,20]. Burn severity influences the magnitude of changes in land surface albedo [21,22], which can undergo a darkening and/or a brightening that can persist in time [23,24]. Such alterations cause an increase in land surface temperature immediately following fires [25], which influences both burned area and the duration of fires [26], since vegetation is a regulator of land surface energy fluxes [27]. Although the effects of wildfires span across multiple dimensions of ecosystem functioning, most studies evaluating those effects seldom address those multiple dimensions at the same time, thus failing to fully depict spatiotemporal changes caused by these disturbances and limiting the assessment of environmental impacts.

As wildfire events have been increasing in recent decades, both in terms of frequency and intensity [28,29], exacerbated by global climate change and by shifts in land use and forest management [30], there is an increasing need for methods to assess and monitor the ecological consequences of such disturbances [31]. Over the last decades, remote sensing (RS) has revolutionized the way environmental changes are monitored, with new satellite sensors (e.g., Sentinel-2 MSI, Landsat-8 OLI/TIRS) providing valuable data at increasingly higher temporal, spatial, and spectral resolutions. Other satellites provide long-term archives of images that are useful to support temporal studies (e.g., Terra/Aqua MODIS, Landsat 5/7 missions). This growing variety and quality of remotely sensed data enables a wide range of fire-related applications, from the detection of fire occurrences [32] to an evaluation of fire severity [33] and monitoring of post-fire recovery [31,34,35] and resilience [36]. The ability to assess and map the heterogeneous (both spatially and temporally) effects of wildfires on ecosystems with remotely sensed data is a major asset not only for risk assessment and governance but also for post-fire management and restoration [30,37–39]. In this regard, spectral indices and metrics derived from satellite multi-spectral images have been used to assess different aspects of fire severity, usually through the comparison of pre-fire versus post-fire values. For instance, spectral indices sensitive to photosynthetically active radiation and/or water content, such as the Normalized Difference Vegetation Index (NDVI) and the Normalized Burn Ratio (NBR) have been used, in combination with data collected in-field (e.g., through the Composite Burn Index; CBI), to assess "burn severity" (e.g., [39–41]), which is a post-fire measure of organic matter loss and soil alteration [30,37,42]. Furthermore, both satellite-derived land surface temperature (LST; [43]) and albedo [22] have been used for the same end. Nevertheless, to fully characterize and understand the effects of wildfire disturbances on multiple dimensions of ecosystems, a better translation of spectral indices into informative ecosystem variables is needed. More diverse sets of indicators are thus required to extend "burn severity" assessments to a wider variety of fire effects, i.e., towards more comprehensive approaches in "fire severity" assessment.

In recent decades, Ecosystem Functioning Attributes (EFAs) derived from satellite image time-series (SITS) have been increasingly used in a wide range of ecological applications due to their strong relation to biophysical properties and processes of ecosystems [5]. These include wildfire-related applications, such as improving the detection of wildfire disturbances in space and time [44], but also a wide range of other ecological applications, such as describing major ecological patterns [45–47], predicting and projecting species distributions [48–50], and supporting the definition of conservation priorities [51]. More specifically, EFAs extracted from SITS can provide information on inter-annual quantity (e.g., mean, minimum, maximum), seasonality (i.e., seasonal range or variability), and timing (e.g., dates of specific moments) components over multiple dimensions of ecosystem functioning. These include frequently used satellite-derived descriptors of the intra-annual dynamics of carbon gains, such as primary productivity, vegetation seasonality, and phenology (e.g., [46,52,53]). However, analogous measures can also be extracted from RS-derived proxy descriptors related to other dimensions of ecosystem functioning, such as water content, albedo, and sensible heat (e.g., [50,54]). By combining these four complementary dimensions of ecosystem functioning across each component (i.e., quantity, seasonality, and timing), remotely sensed EFAs can enable in-depth and integrative characterizations of fire severity [3,44].

The main objective of this study is to propose, describe, and showcase a framework to support enhanced assessments of fire severity, encompassing multiple dimensions of ecosystem functioning. Our approach is based on the effects of wildfire disturbances, at short (i.e., the year of the fire) to medium term (i.e., up to the second year after the fire), on descriptors of the intra-annual dynamics (i.e., EFAs) of four essential dimensions of the flows of matter and energy in ecosystems: (i) primary productivity; (ii) vegetation water content; (iii) albedo; and (iv) sensible heat. To illustrate our approach, we assessed the potential of inter-annual anomalies (i.e., deviations from the normal inter-annual variability) of a large set of EFAs extracted from MODIS data, for the period between 2000 and 2018 in the NW Iberian Peninsula, as estimators of the effects of wildfire disturbances. To this end, we compared, ranked, and analyzed the main patterns of those inter-annual anomalies at the regional scale. We then showcased the translation of inter-annual anomalies of the selected EFAs into indicators of wildfire disturbance severity for four selected burned patches within the study area and compared and analyzed their main spatial and temporal patterns at the local scale. Finally, we discussed the potential and added value of the proposed approach to improve RS-based assessment, mapping, and monitoring of wildfire disturbance severity on multiple dimensions of ecosystem functioning.

2. Materials and Methods

2.1. Generic Framework

2.1.1. General Workflow

The proposed approach to evaluate the effects of wildfire disturbances on ecosystem functioning consists of using indicators derived from Ecosystem Functioning Attributes (EFAs) for each of the four key dimensions of ecosystem functioning (i.e., primary productivity, vegetation water content, albedo, and sensible heat). Figure 1 presents the general workflow of the proposed approach, which is composed of five steps, described in the following sub-sections.

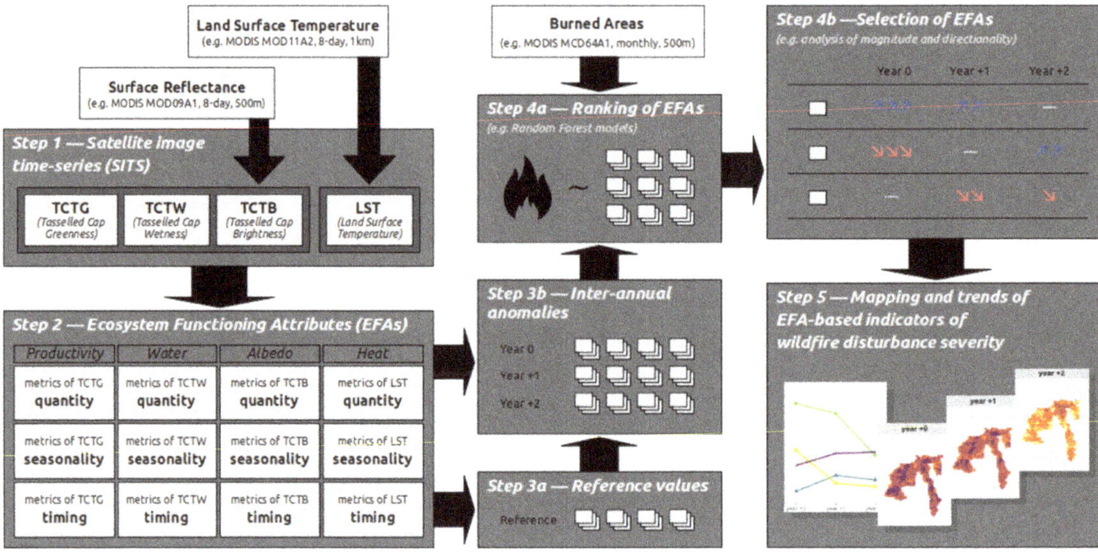

Figure 1. General workflow of the proposed approach to assess wildfire disturbance severity using indicators based on Ecosystem Functioning Attributes (EFAs) derived from satellite image time-series (SITS). TCTB: Tasseled Cap Transformation Brightness feature; TCTG: Tasseled Cap Transformation Greenness feature; TCTW: Tasseled Cap Transformation Wetness feature; and LST: land surface temperature.

2.1.2. Step 1—Satellite Time-Series

First, satellite image time-series (SITS) for each of the four dimensions of ecosystem functioning have to be collected from available satellite products and preprocessed (Figure 1—"Step 1"). From the large number of spectral band combinations and vegetation indices available that have been successfully used for fire applications (e.g., [43,44,55–57]), we propose, for that end, the use of land surface temperature (LST) for the "heat" dimension, as well as the Tasseled Cap Transformation (TCT) features of "Brightness" (TCTB), "Greenness" (TCTG), and "Wetness" (TCTW) for the "albedo", "primary productivity", and "vegetation water content" dimensions, respectively. LST is a measure of the thermal energy emitted by different surfaces, calibrated from thermal emissivity measured by the satellite instrument [58]. On the other hand, TCT features are well-known, sensor-specific, linear combinations of bands in the visible, near-infrared, and short-wave infrared regions of the electromagnetic spectrum [59]. These three TCT features of "Brightness", "Greenness", and "Wetness" result from a rotation of principal component axes, derived from a global sample, to be aligned with the biophysical parameters of albedo, amount of photosynthetically active vegetation, and soil moisture, respectively [55].

2.1.3. Step 2—Extraction of EFAs

Next, metrics describing aspects of the intra-annual dynamics of ecosystem functioning (i.e., EFAs) have to be extracted from the SITS for each target year (Figure 1—"Step 2"). These types of metrics are commonly used to describe intra-annual aspects of ecosystem functioning related to quantity, seasonality, and timing (e.g., [5,46,48]). Examples of "quantity" metrics are the annual mean or median, the maximum or minimum values, or an integrated cumulative sum during a particular part of the year (e.g., the growing season). For "seasonality", metrics such as the intra-annual standard deviation or range can be used. Examples of "timing" metrics are the dates of the annual maximum, minimum, or other important moments (e.g., the start/end of the growing season).

2.1.4. Step 3—Computation of EFA Anomalies

To obtain inter-annual EFA anomalies, it is necessary to first compute reference values for each pixel (Figure 1—"Step 3a"). To this end, EFAs are extracted for a multi-annual reference profile, which is obtained by averaging all years, after excluding values for years with any record of wildfire occurrences (this procedure is similar to the one used in [34,60], except for the exclusion of values in years with fire occurrences). Then, the anomalies are calculated by subtracting the reference values to those of each year, for each EFA (Figure 1—"Step 3b"). More precisely, in the case of linear (i.e., non-circular) metrics, the anomalies are calculated using the following expression:

$$A_m = X_m - R_m, \tag{1}$$

where A_m represents the annual anomalies of EFA m; X_m is the annual values of EFA m, and R_m represents the values of the reference year for EFA m. In the case of circular metrics—e.g., a day of the year (DOY)—anomalies are calculated through the smallest differences between angles, in the following manner:

$$A_m = \left(\frac{365}{2\pi} \times \theta\right) + 35, \tag{2}$$

where A_m represents the annual anomalies of EFA m, and theta is defined as follows:

$$\theta = \begin{cases} arctan(\frac{y}{x}), & if x > 0 \\ arctan(\frac{y}{x}) + \pi, & if x < 0 \text{ and } y \geqslant 0 \\ arctan(\frac{y}{x}) - \pi, & if x < 0 \text{ and } y < 0 \\ \frac{\pi}{2}, & if x = 0 \text{ and } y > 0 \\ \frac{-\pi}{2}, & if x = 0 \text{ and } y < 0 \\ undefined, & if x = 0 \text{ and } y = 0 \end{cases} \tag{3}$$

with:

$$x = cos(X_m - R_m), \tag{4}$$

and:

$$y = sin(X_m - R_m), \tag{5}$$

where X_m and R_m represent the same as in Equation (1).

2.1.5. Step 4—EFA Ranking and Selection Procedures

Since multiple aspects of the intra-annual dynamics of ecosystem functioning can be obtained through the computation of EFAs, a selection process is necessary to reduce the initial set of metrics to an essential (i.e., relevant and less correlated) parsimonious set. In this selection process, classification models are used to rank all the EFAs considered in terms of their potential to measure the remotely sensed effects (as proxies of severity) of wildfire disturbances on the different EFAs (Figure 1—"Step 4a"). To this end, burned-area maps considered the response variable, while the anomalies of all EFAs are considered the predictive variables. The resulting ranking of EFA anomalies is used to compare all EFAs across (i) base SITS, with each corresponding to a different dimension of ecosystem functioning (i.e., "primary productivity", "vegetation water content", "albedo", and "sensible heat"); and (ii) the specific metric (e.g., mean vs. median) and (iii) component of the inter-annual dynamics of ecosystem functioning (i.e., "quantity", "seasonality", and "timing") used, each corresponding to an EFA. Furthermore, potential temporally lagged effects can also be evaluated by including values of EFA anomalies of different years concerning the year of the wildfire disturbance event (e.g., the anomalies for the year of the fire event, plus the first year after, the second year after, etc.).

The resulting ranking can be coupled with additional selection criteria, such as pairwise correlations, to minimize potential redundancy between EFAs so that a compact set of twelve EFAs (i.e., one for each dimension and each component of the inter-annual dynamics of ecosystem functioning) are selected (Figure 1—"Step 4b"). Then, both the magnitude and the direction of the pairwise relationship between the response variable (e.g., burned areas) and each of the selected EFAs are assessed (e.g., using Cohen's d effect size statistics [61]). This allows for the inter-comparison of selected EFAs across all the dimensions and components of ecosystem functioning, as well as other factors considered (e.g., the year of the early post-fire period). Based on these procedures, a final selection of EFAs is carried out to support the analysis of the main patterns of wildfire disturbance severity on the dimensions of ecosystem functioning considered.

2.1.6. Step 5—Translation into Indicators of Wildfire Disturbance Severity

To support a multi-dimensional assessment of the effects of wildfire disturbances on ecosystem functioning, the final set of selected EFAs are then converted into indicators of wildfire disturbance severity (Figure 1—"Step 5"). For obtaining these indicators, the absolute values of the corresponding EFA anomalies are used, as these constitute measures of the magnitude of deviations from reference, thus allowing easier comparison across different dimensions. The resulting compact set of indicators is then used to support more detailed analyses.

2.2. Test Case

2.2.1. Study Area

To illustrate our proposed approach, we chose the northwest of the Iberian Peninsula (NW-IP) as the overall study area (Figure 2), since this is a hotspot in terms of wildfire occurrences within the European context, being a highly fire-prone landscape [3]. This area extends to the north and the west to the Atlantic Ocean, while its southern boundary was delimited by the edge of MODIS tile h17v04. Its eastern boundary, on the other hand was defined to include the northern half of mainland Portugal, as well as the Spanish autonomous community of Galicia.

The NW-IP features strong environmental gradients and includes a major biogeographic transition, from an Atlantic climate with temperate deciduous broadleaf and mixed forests in the north and northwest, to a Mediterranean climate with evergreen sclerophyllous vegetation towards the southeast. Land cover and uses are well diversified and highly heterogeneous (Table 1), including cropland areas, urban areas, plantation forests (mostly maritime pine, eucalyptus, and mixed stands), and shrublands in different successional stages, along with historical use of fire for agrosilvopastoral purposes. Wildfire occurrences are more frequent during the months between July and October—but especially in August—with burned areas over 10,000 ha being increasingly common. Wildfire events have been increasing in recent decades because of changes in land use, which mainly resulted from extensive abandonment of farming and husbandry, increasing the fuel load and its continuity in the landscape [30,62]. The frequency of extreme fire events, also known as mega-fires (i.e., complex catastrophic events in terms of size, intensity, resistance to control, severity, etc. [30]), has also been increasing, despite the huge investments in fire suppression [28], making this one of the regions with both the highest annual values of burnt area in Europe and the highest density of ignitions among southern European countries [63].

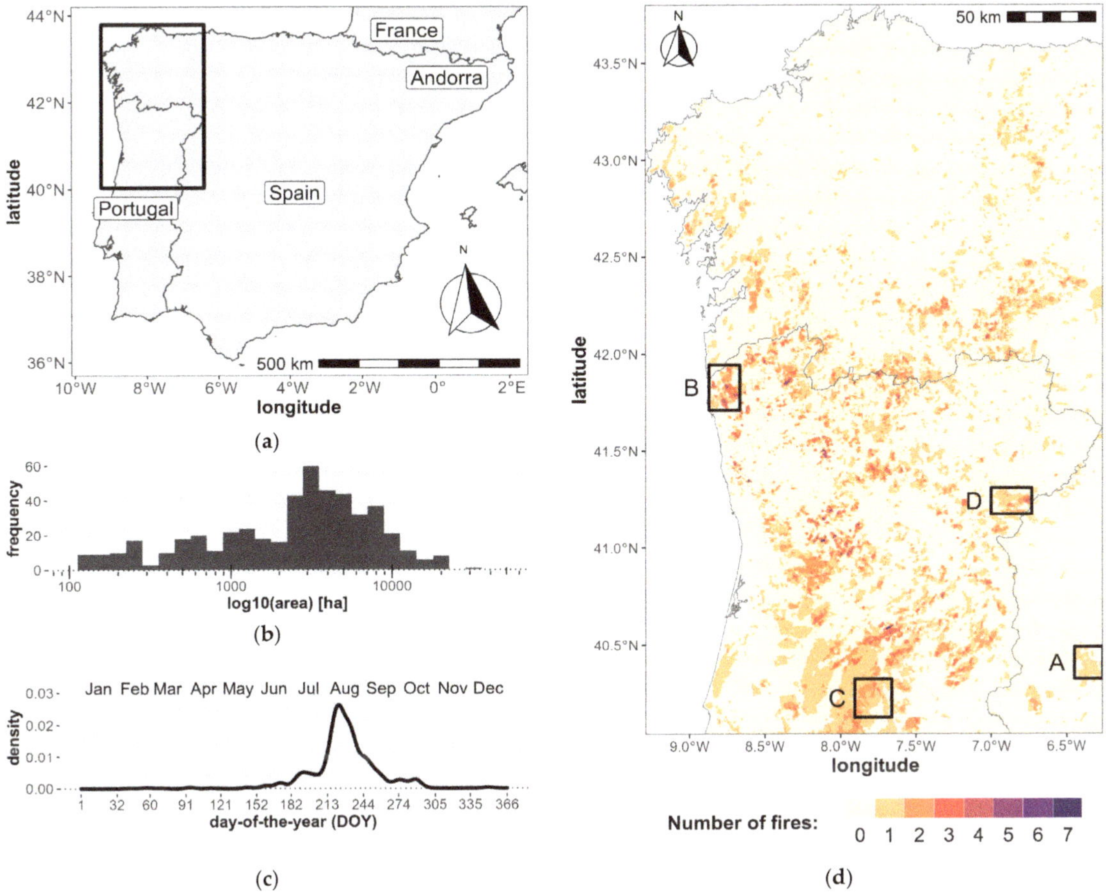

Figure 2. The study area for the illustrative test case, with the (**a**) location of the overall study area for regional-scale analyses within the Iberian Peninsula; (**b**) size distribution of all fires; (**c**) distribution of the area burned, by day of the year (DOY); and (**d**) location of the four individual burned patches selected for local-scale analyses (labels A–D), overlapped with the number of fires that occurred. For these figures, all wildfires with area > 100 ha which occurred between 2000 and 2018 (according to the MODIS MCD64A1 Burned Areas product) were considered.

To showcase our proposed framework, the full study area in NW-IP, including both burned and unburned areas of NW-IP between 2000 and 2018, was used for regional-scale analyses. Then, for local-scale analyses, four individual burned patches (see boxes labelled A–D in Figure 2) were selected from different geographical locations within the study area, featuring diverse baseline environmental characteristics (e.g., altitude, distance to coast, climate, proportions of major vegetation types), and with the focal wildfire event having occurred at different years within the considered period (A: 2003; B and C: 2005; D: 2013). Table 1 presents a summary of the characteristics of both the full study area, used for regional-scale analyses (i.e., NW-IP), as well as the four individual burned patches used for local-scale analyses.

2.2.2. Satellite Data Preprocessing

In our test case, we used SITS extracted from MODIS products. Most tasks were undertaken within the R statistical programming environment [67], using mainly the "raster" package [68], with additional R packages complemented by other software for more particular tasks, as specified whenever relevant.

Table 1. Summary of baseline environmental characteristics of the overall study area for regional-scale analyses (northwest Iberian Peninsula, NW-IP), as well as the four selected burned patches selected for local-scale analyses (A–D). Elevation was extracted from MERIT DEM [64]; climate variables were extracted from the CHELSA Bioclim dataset [65]; the percentages of land-cover classes were extracted from CORINE Land Cover (CLC) 2000 [66].

	NW-IP	A	B	C	D
Year of fire	—	2003	2005	2005	2013
Average burned area (ha)	3617	14,625	17,600	19,325	14,850
Distance to coast (km)	—	212	9	91	153
Average elevation (m a.s.l.)	580	801	261	712	505
Average temperature (°C)	12.7	13.0	14.1	12.5	14.1
Minimum temperature (°C)	3.1	1.2	7.5	1.0	2.8
Maximum temperature (°C)	2.5	28.9	21.8	26.7	29.1
Average total precipitation ($mm \cdot yr^{-1}$)	1139	1075	1747	1229	620
% of Urban	1.8	0.0	1.1	0.1	0.2
% of Agricultural	29.0	9.3	8.0	5.8	26.0
% of Broad-leaf forests	7.1	0.8	1.6	0.2	4.9
% of Coniferous forests	5.9	34.5	17.7	32.8	10.8
% of Mixed forests	8.7	4.5	9.1	3.4	1.5
% of Natural grasslands	5.6	0.2	4.0	0.0	17.4
% of Shrublands	20.2	50.1	32.7	57.1	38.2
% of Bare rocks or sparsely vegetated	2.0	0.6	25.7	0.6	1.0

We started by downloading and preprocessing all available images encompassing NW-IP (i.e., MODIS tile h17v04), between 2000 and 2018, from the following MODIS products, using the "MODIStsp" R package [69]:

- MODIS/Terra+Aqua Burned Area Monthly L3 Global 500m (MCD64A1), Collection 6 [70];
- MODIS/Terra Land Surface Temperature/Emissivity 8-Day L3 Global 1km (MOD11A2), Collection 6 [71];
- MODIS/Terra Surface Reflectance 8-Day L3 Global 500m (MOD09A1), Collection 6 [72].

Preprocessing consisted of converting to the GeoTIFF file format, re-projecting to a common reference system (WGS84/UTM29N), and resampling to 500-m pixel size using the nearest neighbor interpolation algorithm. Additionally, a filter based on the Hampel outlier identifier [73,74] with a window size of 5 was applied to the resulting SITS to reduce residual noise that hinders time-series data.

Next, LST and TCTB, TCTG, and TCTW were extracted from the MOD11A2 and MOD01A1 products, respectively, using "GDAL" [75] and the "rasterio" Python package [76]. For LST, only the day temperatures were used, which were then rescaled to degrees Celsius. The three TCT features of "Brightness", "Greenness", and "Wetness" were obtained by combining the 7 bands available in the MOD09A1 product, using the coefficients presented in Table 2 [59].

Burned areas extracted from the MCD64A1 product were filtered by size, eliminating all burned patches smaller than 100 ha, following a set of criteria detailed in previous work (see [44]). To further reduce the remaining noise in the data, a Whittaker–Henderson smoother [77,78] was applied to the four extracted SITS, with a lambda parameter equal to 2 (a rather low value, to conserve the characteristics of the time-series while still reducing extreme noise).

Table 2. MODIS-specific coefficients to calculate the Tasseled Cap Transformation (TCT) features "Brightness", "Greenness", and "Wetness". These features result from a rigid rotation of principal component axes, which are aligned with the biophysical parameters of albedo, the amount of photosynthetically active vegetation, and soil moisture, respectively [59].

Band			Coefficients		
No.	**Name**	**Range (nm)**	**Brightness**	**Greenness**	**Wetness**
1	Red	620–670	0.4395	–0.4064	0.1147
2	NIR 1	841–876	0.5945	0.5129	0.2489
3	Blue	459–479	0.2460	–0.2744	0.2408
4	Green	545–565	0.3918	–0.2893	0.3132
5	NIR 2	1230–1250	0.3506	0.4882	–0.3122
6	SWIR 1	1628–1652	0.2136	–0.0036	–0.6416
7	SWIR 2	2105–2155	0.2678	–0.4169	–0.5087

NIR: near-infrared; SWIR: short-wavelength infrared.

2.2.3. EFA Anomalies Computation

A total of 60 EFAs were extracted, using 15 different intra-annual metrics (Table 3), from each of the four SITS—LST, TCTB, TCTG, and TCTW—for each year between 2000 and 2018 for NW-IP. Among these, well-known measures were computed for "quantity"—mean, median, maximum, and minimum annual values—and "seasonality"—standard deviation, median absolute deviation, and absolute range. In addition to those, the relative range and a non-parametric relative range were also computed using the following expressions:

$$rrl = log_{10}\left(\left|\frac{rng}{avg}\right|\right) \tag{6}$$

and

$$npr = log_{10}\left(\left|\frac{rng}{mdn}\right|\right), \tag{7}$$

where *rrl* is the relative range metric, while *rng* is the absolute range, *avg* is the mean, *npr* is the non-parametric relative range, and *mdn* is the median.

As for "timing", the times (of the year) of the maximum and minimum values were used. However, as variables of time of the year are circular by definition, we also tested two linearized versions for each, called "winterness" and "springness", using the following expressions:

$$\text{winterness} = cos\left(\frac{2\pi}{365} \times (x - 35)\right) \tag{8}$$

and:

$$\text{springness} = sin\left(\frac{2\pi}{365} \times (x - 35)\right), \tag{9}$$

where x is the day of the year (DOY) of either the maximum or minimum value (tmx or tmn, respectively). Note that these last two expressions (i.e., Equations (8) and (9)) include the conversion from DOY to radians, assuming that a full year has 365 days; as well as a rotational realignment of -35 days so that the maximum and minimum values of winterness and springness (i.e., 1 and -1, respectively) correspond to the mid-point of each season (in the case of NW-IP, as it is located in the northern hemisphere, the mid-point of winter is in early February). These metrics can be interpreted as corresponding to the temporal proximity, within the year, to either the start of winter or spring, respectively.

Finally, anomalies were then obtained for all EFAs for each of the four SITS by calculating either linear differences (Equation (1)), for all metrics except "tmx" and "tmn", or the smallest differences between angles, for those two circular metrics (Equations (2)–(5)).

Table 3. List of metrics extracted from each of the four satellite image time-series used in this study (i.e., land surface temperature and the Tasseled Cap Transformation features of "Brightness", "Greenness", and "Wetness"); as well as the number of predictive variables for models extracted from each one.

Component	Metric	Abbreviation	No. of Variables for Models
Quantity	Mean (or average)	avg	12
	Median	mdn	12
	Maximum	max	12
	Minimum	min	12
Seasonality	Standard deviation	std	12
	Median absolute deviation	mad	12
	Absolute range	rng	12
	Relative range	rrl	12
	Non-parametric relative range	rnp	12
Timing	Time (of the year) of maximum	tmx	12
	"Winterness" of maximum	wmx	12
	"Springness" of maximum	smx	12
	Time (of the year) of minimum	tmn	12
	"Winterness" of minimum	wmn	12
	"Springness" of minimum	smn	12
Total			180

2.2.4. Ranking and Selection of EFAs

A ranking of EFA anomalies was obtained for NW-IP through a classification modeling procedure, in which the response variable was the binary burned class (i.e., burned vs unburned) extracted from the MODIS burned area product. As predictive variables for these models, we used the inter-annual anomalies in each one of the 60 EFAs (i.e., each of the 15 intra-annual metrics described in Table 3, for each of the four SITS), for each one of three years since the fire occurrence: the year of the fire, as well as the first and second years after (i.e., years 0, +1, and +2). This was to allow for the evaluation of potential temporally lagged effects (for which only the years from 2003 to 2016 were used). In total, the number of predictive variables used in these models was 180 (see Supplementary Information S1 for additional information on the predictive variables used in these models). To this end the Random Forest (RF) "ensemble learning" algorithm was used, since this technique scales well on large and/or high-dimensional data [79], thus allowing for the full set of variables to be tested without the need for a selection before modeling.

As the number of pixels identified as "burned" represented only ca. 1.3% of the total overall study area (i.e., NW-IP), data balancing through down-sampling of the majority class (i.e., "unburned") was applied. To that end, 100 random sample sets were defined consisting of all the pixels identified as "burned", as well as an equal number of randomly selected "unburned" pixels for each year, in a total of 122,658 observations (i.e., n = 61,329 per class). Then, an RF classification model was calibrated for each one of the 100 sample sets, using the "caret" R package [80], with the hyperparameters set to the default values (including a fixed value for "mtry" equal to the default squared root of the number of predictive variables), and under a leave-group-out cross-validation (LGOCV) scheme with 50:50 train/test split in each one of 10 repetitions.

Model performance was then evaluated using measures specific to two-class problems—sensitivity (or true positive rate), specificity (or true negative rate), and the area under the receiver operating characteristic curve (AUC)—which were then averaged across the 100 resulting models. Finally, variable importance scores—a common approach to extract interpretable information on the contribution of different variables from RF models [79]—based on the mean decrease in accuracy, were extracted and also aggregated to the median values across the 100 sample sets to support the EFA ranking. Based on the ranking of variable importance scores obtained from the RF models, one EFA was selected for each dimension and each component, giving a total of 12, to support the analysis of the main patterns of wildfire disturbance severity on multiple dimensions of ecosystem functioning in NW-IP.

2.2.5. Analysis of Indicators of Wildfire Disturbance Severity

To support the analysis of both the directionality (i.e., increasing or decreasing) and magnitude of the pairwise relationships between the regional-scale burned areas and each of the selected EFAs, Cohen's d effect size statistics [61] and Spearman pairwise rank correlations were calculated for the 100 sample sets. From this, as well as from the distributions of the selected EFA anomalies, the main overall patterns of the effects of wildfire disturbances on ecosystem functioning were analyzed for NW-IP and compared across the four dimensions and three components of ecosystem functioning, as well as across years of the early post-fire period.

It should be noted that effect sizes were applied here to avoid the use of statistical hypothesis testing based on significance (i.e., *p*-values), which have important known limitations such as not measuring the effect size or the importance of a result [81] or the strong influence of sample size, which is hugely influential in determining significance levels [82] (i.e., when sample sizes are large enough, any effect can produce a small *p*-value if the sample size or measurement precision is high enough [81]; with large sample sizes, often happening at a pixel-level analysis of SITS).

We then computed four indicators of the effects of wildfire disturbances on ecosystem functioning (i.e., one for each dimension) for four selected burned areas (see Figure 2) at a local scale to illustrate the proposed approach for RS-based assessment, mapping, and monitoring of wildfire disturbance severity on multiple dimensions of ecosystem functioning. The selection of EFAs for these final indicators took into consideration the results from both the Random Forest model-based ranking and the pairwise correlations between EFAs. The translation of the selected EFAs into indicators was done by taking the absolute values of the anomalies of the selected EFAs for the four burned areas. Finally, spatial patterns between all indicators, across years, were further analyzed, local-scale, through local Spearman correlations, with a neighborhood window size of 5 pixels.

3. Results

3.1. EFA Ranking

In terms of the performance of the Random Forest (RF) models for our test case, the results obtained for leave-group-out cross-validation can be considered very good across all sample sets, with around 0.9815 ± 0.0004 for AUC, 0.9080 ± 0.0015 for sensitivity, and 0.9647 ± 0.0010 for specificity.

As for variable importance scores, EFAs extracted from TCTG had overall higher values than the other dimensions (Figure 3; see Supplementary Information S2 for the full plot of variable importance scores for all non-aggregated variables). Considering each of the three different EFA components, the ones measuring aspects of "quantity"—particularly the mean and either one of the extreme values (minimum or the maximum, depending on the specific case)—obtained overall higher variable importance scores than "seasonality" or "timing" EFAs, with the notable exception of seasonality EFAs extracted from TCTG at year 0. Conversely, the same exception was observed within all seasonality EFAs, where otherwise, LST was the dimension from which EFA anomalies obtained higher variable importance scores overall, especially for standard deviation. As for timing EFAs, the scores obtained were generally lower than for the two other groups, except for the ones extracted from either TCTG or LST at year 0, particularly the springness of the time of the year of the minimum TCTG value and the winterness of the time of the year of the maximum LST value, closely following the corresponding quantity EFAs.

Across the four different dimensions of ecosystem functioning, those extracted for the year of fire (i.e., year 0) generally obtained better scores than those of the two subsequent years. However, for quantity EFAs extracted from TCTW, the first year after the fire (i.e., year +1) obtained better scores. Furthermore, quantity EFAs extracted from TCTB also achieved higher scores for the second year after the fire (i.e., year +2), although not as high as for year 0. Finally, in terms of individual EFAs, the top-ranked ones were extracted from TCTG, with the minimum value (TCTG-min) and the standard deviation of TCTG (TCTG-std) at year 0 holding the highest scores among all EFAs tested (see Supplementary Information S1 for additional information on the ranking results).

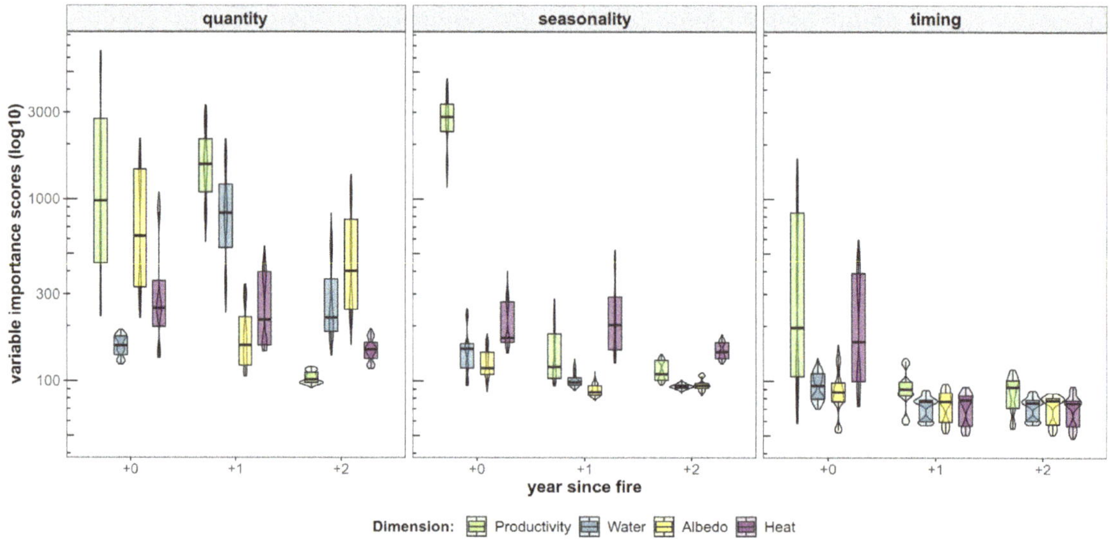

Figure 3. Distributions of variable importance scores obtained from Random Forest models to assess the effects of wildfire disturbances on four dimensions of ecosystem functioning: primary productivity ("productivity"), vegetation water content ("water"), albedo, and sensible heat ("heat"), in the same year as the respective fire ("year 0") and the two years after ("year +1" and "year +2"), for each of the three components considered: quantity (**left**); seasonality (**center**), and timing (**right**) (Note that the scale of the *y*-axis is logarithmic).

3.2. Analysis of Effects

Considering the RF-based importance ranking, 12 EFAs (Table 4) were selected, one for each dimension (i.e., "primary productivity", "vegetation water content", "albedo", and "sensible heat") and each component of intra-annual dynamics (i.e., "quantity", "seasonality", and "timing"). As for "timing" EFAs, it should be noted that the selected metrics do not correspond to the highest-ranked ones for each case. In this instance, parsimony and interpretability criteria were also considered, since linearized versions of timing EFAs (i.e., "winterness" and "springness") often require the use of both at the same time to better understand in which month/season the minima/maxima occur (see Supplementary Information S3 for partial dependence plots of the predictive variables of the models, extracted from the 12 selected EFAs).

Table 4 summarizes the results for the effect sizes (both magnitudes and directions) between the response variable and each of the selected EFA anomalies (see Supplementary Information S4 for additional information on the effect size results). These showed overall strong negative effects (i.e., lower values associated with wildfire disturbances) for TCTG-min and TCTW-avg (especially at years 0 and +1 and years +1 and +2, respectively), while positive (i.e., raised values associated with wildfire disturbances) for LST-max (especially at years 0 and +1). For TCTB-avg, however, the effects changed from strongly negative at year 0 to strongly positive at year +2. As for seasonality EFAs, the effects were, overall, positive, with the strongest magnitude associated with TCTG-std and LST-std. Finally, timing EFAs had overall weaker effects, except for TCTG-tmn (with large negative effects) and LST-tmx (with large positive effects), both only at year 0.

Table 4. List of selected EFAs with the corresponding effects categories, based on Cohen's d effect size statistic. The number of arrows in the "Effect category" columns symbolizes the magnitude of the effect size, with | d | < 0.2 "negligible", | d | < 0.5 "small", | d | < 0.8 "medium", and otherwise "large". The direction and color of the arrows symbolize the sign of the effect size, with ascending arrows depicting positive (non-"negligible") effect sizes while descending arrows depict negative (non-"negligible") effect sizes ("negligible" effects are symbolized by a horizontal dash).

Dimension	Component	Attributes (EFA)	Effect Category		
			Year 0	**Year +1**	**Year +2**
Primary productivity	Quantity	TCTG-min	↘↘↘	↘↘↘	—
	Seasonality	TCTG-std	↗↗↗	↗↗	—
	Timing	TCTG-tmn	↘↘↘	—	—
Vegetation water content	Quantity	TCTW-avg		↘↘↘	↘↘↘
	Seasonality	TCTW-std	—	↗↗	—
	Timing	TCTW-tmn	—	—	—
Albedo	Quantity	TCTB-avg	↘↘↘	↗	↗↗↗
	Seasonality	TCTB-std	—	↗	—
	Timing	TCTB-tmx	—	—	—
Sensible heat	Quantity	LST-max	↗↗↗	↗↗↗	↗
	Seasonality	LST-std	↗↗	↗↗↗	↗
	Timing	LST-tmx	↗↗↗	—	—

3.3. Main Patterns in EFA Anomalies

The distributions of the deviations from the normal variation (i.e., anomalies) for each of the twelve selected EFAs, grouped by year after fire (i.e., years 0, +1, and +2) and by burned vs. unburned pixels, are shown in Figure 4 (see also Supplementary Information S5 for the distributions of the reference values of all pixels in the NW-IP).

Overall, there were notable differences in the distributions obtained for burned vs. unburned pixels, with the most notable contrast having been observed for TCTG-tmn at year 0, with median anomalies corresponding to occurrences of 80–88 days earlier than the reference. Within the same dimension, differences were also notable for TCTG-min (especially at years 0 and +1), and also for TCTG-std (especially at year 0), with values of burned pixels considerably below/above (respectively) those of unburned pixels. Considerable differences between burned vs. unburned were also obtained for TCTW-avg, but only at years +1 and +2, while for TCTB-avg, they were observable at years 0 and +2, with opposite directions (i.e., negative vs. positive). As for heat-related EFAs, differences in the distributions of anomalies between burned and unburned pixels were notable for LST-max and LST-std across all years after fire, but especially at years 0 and +1. Finally, LST-tmx showed a contrast between burned and unburned pixels, with median anomalies corresponding to delays of 8–16 days.

3.4. Multi-Dimensional Assessment of Wildfire Disturbance Severity

For the RS-based assessment of wildfire disturbance severity on the four dimensions of ecosystem functioning considered, at the local level, one indicator for each dimension was used, based on the previously selected anomalies in quantity EFAs (i.e., TCTG-min, TCTW-avg, TCTB-avg, and LST-max). This selection took into consideration the results from both the model-based ranking (i.e., quantity EFAs ranked better overall than those of seasonality or timing) and pairwise correlations (up to ±0.83; see Supplementary Information S6).

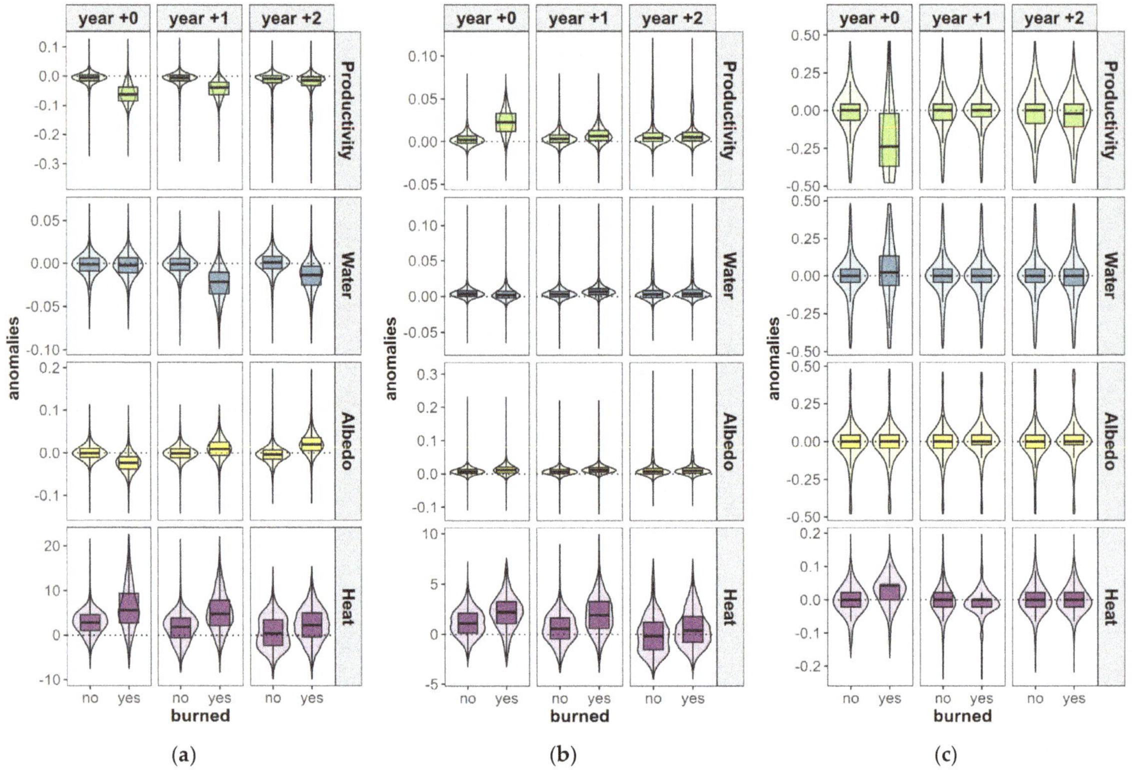

Figure 4. Distributions of the anomalies of the 12 selected EFAs, in the same year as the respective fire ("year 0") and the two years after ("year +1" and "year +2"), for each of the three components considered: (**a**) quantity; (**b**) seasonality, and (**c**) timing. The specific intra-annual metrics selected for each of the four dimensions of ecosystem functioning considered were the following: Primary productivity ("Productivity")—Tasseled Cap Transformation "Greenness" feature minimum value (TCTG-min), standard deviation (TCTG-std), and time of the year of the minimum value (TCTG-tmn); Vegetation water content ("Water")—Tasseled Cap Transformation "Wetness" feature average value (TCTW-avg), standard deviation (TCTW-std), and time of the year of the minimum value (TCTW-tmn); Albedo—Tasseled Cap Transformation "Brightness" feature average value (TCTB-avg), standard deviation (TCTB-std), and time of the year of the maximum value (TCTB-tmx); Sensible heat ("heat")—land surface temperature maximum value (LST-max), standard deviation (LST-std), and time of the year of the maximum value (LST-tmx).

Figure 5 shows the median (±0.5 median absolute deviations) profiles of each of the four selected indicators, across the three years after fire, for each of the four burned areas of interest (see Supplementary Information S7 for additional information on the spatial and temporal patterns of the indicators of wildfire disturbance severity in the four burned areas A–D). This figure illustrates different post-fire trajectories, with varying relations between the four dimensions of ecosystem functioning across individual burned areas Specifically, different areas had different ratios between "Productivity" and "Heat" (e.g., A vs. D). Furthermore, "Water" seemed to be less affected in area D than in the other areas, while "Albedo" in area D exhibited a different post-fire trajectory relative to the other areas, with the value for year +1 being slightly higher than for year 0.

Finally, Figure 6 shows the variation in the spatial and temporal patterns of each of the four selected indicators of primary productivity, vegetation water content, albedo, and sensible heat, for each of the four individual burned areas of interest (A–D). Overall, some wildfire disturbance severity "hotspots" are observable, while not necessarily coinciding across dimensions (see also Supplementary Information S7). Moreover, different burned areas exhibited different patterns across time between dimensions of ecosystem functioning, with temporal trends being divergent in some cases while convergent in others (e.g., "Water" and "Albedo" dimensions in the areas of interest C vs. D), or with different relative preponderances (e.g., "Productivity" vs. "Heat" in the areas of interest A vs. D), further illustrating the patterns observed in Figure 5.

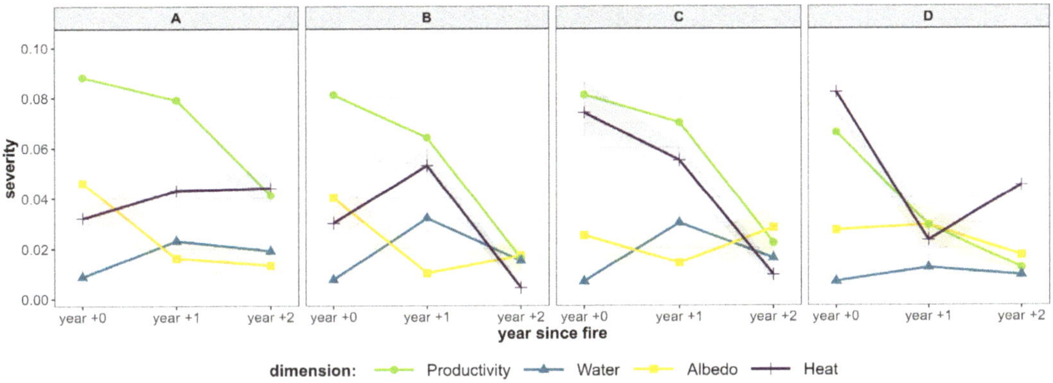

Figure 5. Profiles of wildfire disturbance severity on the four dimensions of ecosystem functioning: Primary productivity ("Productivity"), vegetation water content ("Water"), albedo, and sensible heat ("Heat"), at short to medium term (i.e., between years 0 and +2 after the fire event), for four individual burned areas (letters A–D; see Figure 2 for their location within the study area). Dots connected by thick lines represent median values across all pixels of each individual burned area, while shaded areas of the corresponding color represent ±0.5 × median absolute deviation. Values for the "heat" dimension are scaled by a factor of 0.005 for visual comparability purposes.

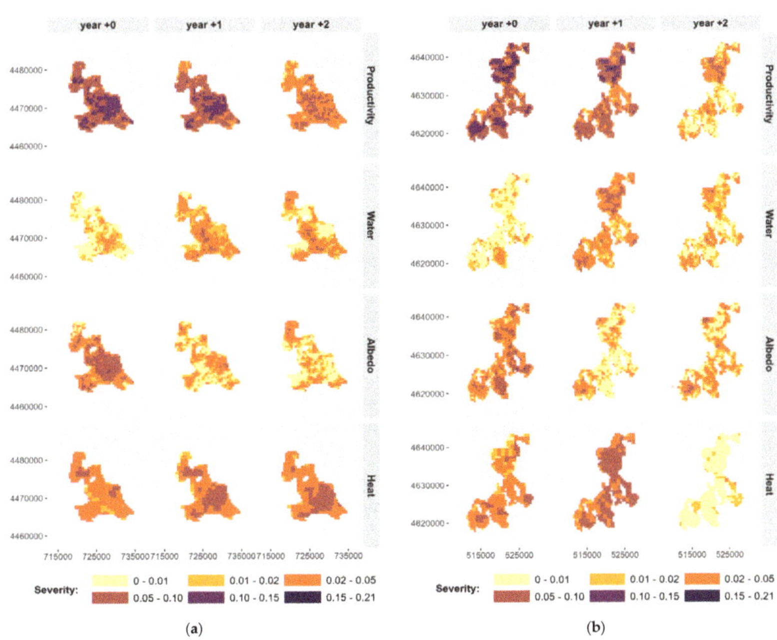

Figure 6. *Cont.*

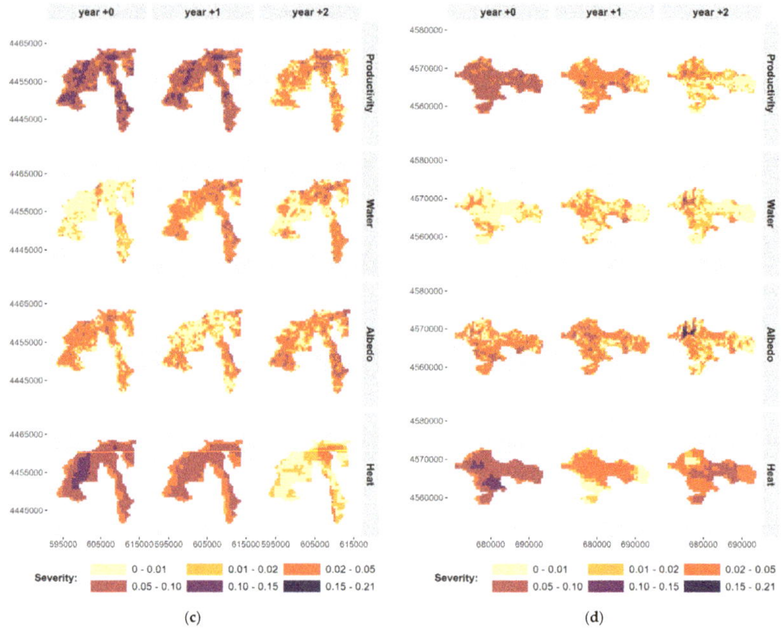

Figure 6. Maps of wildfire severity on primary productivity ("Productivity"), vegetation water content ("Water"), albedo ("Albedo"), and sensible heat ("Heat"), at short to medium term (i.e., between years 0 and +2 after the fire event), for four individual burned areas (letters **a–d**; see Figure 2d, labels A–D), based on indicators derived from inter-annual anomalies in quantity EFAs extracted from satellite image time-series.

4. Discussion

4.1. Fire Severity Patterns in the NW Iberian Peninsula

4.1.1. Effects of Wildfires across Dimensions and Components

In our test case, we found that the associations between burned areas and inter-annual EFA anomalies were particularly strong for primary productivity and sensible heat, although albedo and vegetation water content also revealed important effects caused by wildfire disturbances. This is in line with previously published research, as the sudden removal of green vegetation due to fire usually translates into abrupt breaks that are commonly observable in time-series of spectral vegetation indices (e.g., TCTG). This removal can result in substantial carbon losses, both in terms of aboveground biomass [8] as well as soil organic matter [9,83]. Moreover, fire-mediated changes in nutrient concentrations can ultimately limit productivity over long periods [10]. Because vegetation is a regulator of land surface energy fluxes [27], the removal of vegetation causes an increase in observed LST induced by a reduction in evapotranspiration [25], thus increasing the sensible-to-latent heat ratio [17]. Furthermore, the observed effects in albedo are also consistent with studies reporting either a darkening or a brightening (or both) effect after a fire (e.g., [22,84]) due to the vegetation removal as well as the presence (or not) of black carbon in soot, which absorbs visible solar radiation [84]. Finally, effects on vegetation water content could be related to loss of moisture in canopy foliage due to damage caused by fire, eventually leading to vegetation mortality [15]. Indeed, our results show that using information from multiple dimensions of ecosystem functioning can yield improved results over the use of fewer—or only one—indices for enhanced detection, mapping, and evaluation of ecological effects of disturbances such as wildfires.

Across the three different components of ecosystem functioning considered, anomalies in quantity EFAs (e.g., mean, minimum, maximum) had overall stronger associations with wildfire disturbances than seasonality (e.g., standard deviation) or timing (e.g., time of minimum or maximum) ones. For severity assessment purposes, quantity and seasonality EFAs seemed to be somewhat redundant, as there can be, in some cases, high pairwise correlations between EFA anomalies of those two types. Conversely, for other types of disturbances such as land clearing for agriculture and ranging, the impact on seasonality (i.e., seasonal variability) can be much stronger than on quantities (see [85]). While some studies show that wildfire disturbances can have profound impacts on aspects of timing of key dimensions of ecosystem functioning (e.g., by inducing phenological changes [11]), our approach may not be best-suited to assess such modifications. This is because the intra-annual descriptors used (i.e., EFAs) seemed to more likely capture the location of extreme values (i.e., minimum/maximum) within the year corresponding to direct effects of the wildfire disturbance rather than translating changes (e.g., delays or lags) in the annual cycles of the underlying ecological processes.

As for the specific statistical measures used to extract EFAs, stronger effects can result from the use of parametric measures over that of their non-parametric equivalents. This may be because parametric measures are generally more sensitive (and susceptible) to extreme or abnormal values than non-parametric ones, which may or may not be desirable, depending on the specific application purpose. On the other hand, non-parametric measures may be more adequate for extracting long-term inter-annual trends than parametric ones (e.g., [5]).

The results obtained point to a strong added value of the proposed approach to discriminate the effects of wildfire disturbances on different aspects of ecosystem functioning.

4.1.2. Temporal Effects of Wildfires

Short-term effects of wildfire disturbances on ecosystem functioning seemed to attain better overall performance than medium-term effects, since inter-annual anomalies in intra-annual EFAs for the year of the fire occurrence obtained much better scores, overall, than those for the first and second years following fire. This is partially in line with other studies, as the majority of studies analyzing RS-based effects of fires focus on either short-term (i.e., abrupt) wildfire disturbance severity (e.g., [32,56]) or on longer-term effects more related to post-fire recovery (e.g., [3,34]), while the medium-term effects (i.e., from one or two years to five years after the fire) are often overlooked.

EFAs derived from TCTW (i.e., vegetation water content) seemed to be more affected by wildfire disturbances in the first year after the fire than the year in which the fire occurred, suggesting a temporal lag associated with this dimension relative to both the fire event and other dimensions. This observation (which, to our knowledge, has not been previously reported) could be linked to changes in hydrological dynamics, such as increased soil water repellency, precipitation run-off, and increased quantities of impervious materials such as ashes and soot [84], which can clog soil pores [86]. Alternatively, this effect could be due to post-fire changes in foliar moisture leading to mortality occurring over an extended period after fire [15], since TCTW is quite sensitive to the moisture content of vegetation [59].

As for the observed effects of wildfire disturbances on albedo (i.e., TCTB), this dimension of ecosystem functioning seemed to exhibit a shift in directionality, from negative in year 0 to positive in years +1 and +2, denoting a change in the relationships with the wildfire disturbance at short vs. medium term. This is also in line with the findings reported in other studies [22,87], which observed both a decrease in albedo (i.e., a darkening effect) immediately after the fire, and, usually, for a brief period, as well as a small increase (i.e., brightening) one year after the fire, which can be persistent. On the other hand, this darkening, which can be due to changes in the relative abundance of surfaces with distinct reflective properties [88] (e.g., ash, char, soot, bare soil), can, in some cases, persist for multiple years after a fire [23].

Finally, the effects of wildfire disturbances on both primary productivity and sensible heat (through TCTG and LST, respectively) seemed to be reflected across the three time periods analyzed. However, they were especially noticeable for the year of the fire occurrence. This also goes in the same direction as other studies addressing the impacts of fire on ecosystems, which report usually short-term effects for satellite-derived proxies of primary productivity (e.g., [41,89]) after fire. An increase in observed LST immediately following fires is also frequently reported in the literature [25]. This can sometimes still be observed one year following the fire [19], after which LST anomalies tend to become smaller, and seasonality starts governing the LST time-series [27].

Together, these results highlight the importance of taking into consideration both short and medium-term effects of wildfire disturbances on multiple dimensions of ecosystem functioning.

4.1.3. General Patterns

Overall, EFA anomalies were able to efficiently discriminate burned and unburned areas, with very high values obtained for both effect sizes and performance measures of the Random Forest (RF) models between EFA anomalies (predictive variables) and burned areas (response variable). This allowed supporting the selection of EFAs from which a compact set of indicators were employed to assess the main patterns of wildfire disturbance severity in our test area.

Diverse spatio-temporal patterns of wildfire disturbance severity on the four dimensions of ecosystem functioning—primary productivity, vegetation water content, albedo and sensible heat—could be observed both between as well as within individual burned areas across years (i.e., from the year of the fire to two years after). These results showcase the added value of (i) analyzing the effects of wildfire disturbances for multiple dimensions of ecosystem functioning, as opposed to addressing only overall burn severity in commonly used approaches (e.g., based on the Differenced Normalized Burn Ratio, ΔNBR) and (ii) taking into consideration both the short and medium term, thus enabling to obtain insights on potentially lagged effects of fires on ecosystem functioning, as different dimensions can exhibit different severities, and with different timings.

4.2. General Considerations about the Proposed Framework

4.2.1. Satellite Image Time-Series

While in the presented test case, we used data extracted from MODIS products exclusively, other sources of satellite image time-series (SITS) are available and can equally be applied using the proposed approach. Such data sources include platforms with diverse characteristics in terms of spatial, temporal, and spectral resolutions and length of the historical archive, such as SPOT-VEGETATION, PROBA-V, Landsat 5/7/8, or Sentinel-2/3 (e.g., [90]).

As for the base SITS to be extracted from those sources, we recommend, in the proposed approach, the use of LST and the TCT features of "Brightness", "Greenness", and "Wetness" as RS-based proxies of albedo, primary productivity, and vegetation water content, respectively. This option allows to derive information on the four above-mentioned dimensions of ecosystem functioning, from the smallest number of sources (the three TCT features are derived from the same source), compactly and coherently, and is often available for the major satellite data sources (e.g., [91]). However, in the case of primary productivity and vegetation water content, if data are not available to derive all four RS-based variables, alternative spectral indices could be used instead, such as the Normalized Difference Vegetation Index (NDVI) or the Enhanced Vegetation Index (EVI) for "productivity", or the Normalized Difference Water Index (NDWI) or Land Surface Water Index (LSWI) for "water".

4.2.2. Additional Data Sources

Well-designed field-based measurements could be used, if available, to validate the information provided by SITS, further enhancing the proposed approach, although it is not a mandatory requirement. To that end, data can be collected in the field, following robust sampling designs, such as spectral/radiometric readings, aerial surveys using cameras and sensors mounted on unmanned aerial vehicles (UAVs), or data collected under already existing and commonly used protocols (or adapted versions from these), such as the Composite Burn Index (CBI; e.g., [92]) or the improved geometrically structured Composite Burn Index (GeoCBI; e.g., [93]), to complement the information derived from SITS. Such data can, however, be difficult and/or costly to obtain, even more so if multiple observations are required to produce a coherent time-series. These types of field-level measurements are usually compared to commonly used indicators of burn severity (e.g., [40]) such as the differenced normalized burn ratio (ΔNBR)—or any of its enhanced forms, such as the relative normalized burn ratio (RdNBR) and the relativized burn ratio (RBR).

Unlike the more commonly used RS-derived methods to estimate burn severity, the approach proposed in this study can provide information on the effects of wildfire disturbances on multiple dimensions of ecosystem functioning—namely primary productivity, vegetation water content, albedo, and sensible heat—simultaneously, towards a comprehensive evaluation of fire severity.

4.2.3. Applicability and Future Directions

Multi-dimensional approaches to wildfire disturbances using EFA-like metrics are, to our knowledge, scarce (e.g., [94]). Very few studies have adopted a multi-dimensional approach using EFAs, regardless of the specific application (e.g., [48,50,95]). More studies should be conducted in the future, with diverse contexts in terms of baseline environmental conditions and fire disturbance regimes, to further test and improve the proposed approach. It is important to highlight that although the results obtained for the test case presented in this study may be somewhat specific to the study area and the analyzed period, the general approach could be applied, with relative ease, to other contexts and timeframes to conduct informed selections of the most efficient and informative EFA-based indicators for wildfire disturbance severity assessment.

The proposed approach makes use of annual descriptors of the dynamics of multiple dimensions of ecosystem functioning—so-called Ecosystem Functioning Attributes (EFAs)—extracted from SITS, to derive indicators of wildfire disturbance severity. However, another possible approach could be to derive indicators directly from the time-series, without the need to compute EFAs first. This would open the possibilities for other kinds of indicators, with finer temporal (i.e., <1 year) resolutions, on a (semi-)continuous basis, to be derived from SITS (e.g., [3]). On the other hand, although additional types of indicators could be obtained in this way, others would not—e.g., metrics of timing based on annual cycles, such as the date of the maximum or minimum, would only make sense on an annual basis. Besides, the annual nature of EFAs—and the derived indicators—facilitates both their computation as well as their interpretation, as it offers a better translation of spectral indices into informative ecosystem variables. This can be advantageous for some application purposes, such as for reporting from official authorities. Notwithstanding, approaches adopting either an annual or continuous basis could complement each other to characterize the effects of wildfire disturbances on ecosystem functioning.

Finally, in this study, we focused on short- to medium-term impacts of wildfire disturbances. However, long(er)-term effects of wildfire disturbance severity have been observed (e.g., [96]). Indeed, long-term impacts of wildfire disturbances on ecosystem functioning could also be monitored using the approach proposed in this study, provided that long-enough SITS are available. However, this was not tested in this study, as that would considerably increase both computational requirements as well as methodological complexity.

5. Conclusions

Here, we described a framework to assess the effects of wildfire disturbances on four key dimensions of ecosystem functioning—primary productivity, vegetation water content, albedo, and sensible heat. This approach is based on indicators derived from several descriptors of the intra-annual dynamics of ecosystem functioning—called Ecosystem Functioning Attributes (EFAs)—extracted from remotely sensed satellite image time-series

(SITS). We found that all four dimensions of ecosystem functioning suffered important effects of wildfire disturbances—especially primary productivity and sensible heat. We also found that quantity EFAs held the highest potential for indicating wildfire severity. Finally, we found important differences in short- and medium-term effects between the four different dimensions of ecosystem functioning, suggesting temporally lagged effects in vegetation water content as well as directionality shifts in albedo. Together, these results highlight the added value of the proposed framework to enhance RS-based assessment, mapping, diagnostics, and monitoring of wildfire disturbances. Using this approach, a more comprehensive evaluation of fire severity can be attained, which is not captured by indicators derived from only one spectral index, measuring only overall burn severity. As such, this multi-dimensional framework can contribute to a deeper and more detailed understanding of the impact of fires in ecosystems, with the potential to support assessments of severity, recovery, and resilience and their interactions with biodiversity and ecosystem services.

Supplementary Materials: The following are available online at https://www.mdpi.com/2072-4292/13/4/780/s1: Supplementary Information S1—Summary table of results of the Random Forest models, for all 180 variables obtained from EFA anomalies, tested against burned vs. unburned areas; Supplementary Information S2—Distributions of variable importance scores from Random Forest models, for individual predictive variables; Supplementary Information S3—Partial dependence plots of the 36 predictive variables extracted from the 12 selected EFAs; Supplementary Information S4—Summary table of results, for each of the twelve EFAs selected for further analyses, tested against burned vs. unburned areas, of both the Random Forest models, and the effect sizes analysis; Supplementary Information S5—Distributions of the reference values (i.e., inter-annual median of values from unburned years) of the selected EFAs; Supplementary Information S6—Summary of maximum Spearman rank pairwise correlations between the 12 selected EFA metrics; Supplementary Information S7—Main spatial and temporal patterns obtained for each of the four burned areas of interest.

Author Contributions: Conceptualization, B.M., J.G., M.C., D.A.-S. and J.P.H.; methodology, B.M., J.G.; formal analysis, B.M.; data curation, B.M.; writing, B.M., with the assistance of J.G., M.C., D.A.-S. and J.P.H. All authors have read and agreed to the published version of the manuscript.

Funding: This research was supported by Portuguese national funds through FCT—Foundation for Science and Technology, I.P., under the GreenRehab project (PCIF/RPG/0077/2017). This work was partially supported by the collaboration between D.A.S. and J.H. under the projects RESISTE (P18-RT-1927), funded by Consejería de Economía, Conocimiento, Empresas y Universidad from the Junta de Andalucía, and DETECTOR (A-RNM-256-UGR18), with the contribution of the European Union Funds for Regional Development. B.M. was supported by FCT, the Ministry of Education and Science, and the European Social Fund, within the 2014–2020 EU Strategic Framework, through FCT (PhD scholarship SFRH/BD/99469/2014). J.G. was funded by the Individual Scientific Employment Stimulus Program (2017), through FCT (contract nr. CEECIND/02331/2017).

Acknowledgments: We acknowledge the use of MODIS imagery obtained from NASA's Land Processes Distributed Active Archive Center (LP DAAC), available free of charge. The authors would like to thank the anonymous reviewers for providing comments and suggestions that helped to improve the quality of the original manuscript.

Conflicts of Interest: The authors declare no conflict of interest. The funders had no role in the design of the study; in the collection, analyses, or interpretation of data; in the writing of the manuscript, or in the decision to publish the results.

References

1. Adámek, M.; Hadincová, V.; Wild, J. Long-term effect of wildfires on temperate Pinus sylvestris forests: Vegetation dynamics and ecosystem resilience. *For. Ecol. Manag.* **2016**, *380*, 285–295. [CrossRef]
2. San-Miguel-Ayanz, J.; Moreno, J.M.; Camia, A. Analysis of large fires in European Mediterranean landscapes: Lessons learned and perspectives. *For. Ecol. Manag.* **2013**, *294*, 11–22. [CrossRef]
3. João, T.; João, G.; Bruno, M.; João, H. Indicator-based assessment of post-fire recovery dynamics using satellite NDVI time-series. *Ecol. Indic.* **2018**, *89*, 199–212. [CrossRef]

4. Petropoulos, G.; Carlson, T.; Wooster, M.; Islam, S. A review of Ts/VI remote sensing based methods for the retrieval of land surface energy fluxes and soil surface moisture. *Prog. Phys. Georg. Earth Environ.* **2009**, *33*, 224–250. [CrossRef]
5. Alcaraz-Segura, D.; Cabello, J.; Paruelo, J.M.; Delibes, M. Trends in the surface vegetation dynamics of the national parks of Spain as observed by satellite sensors. *Appl. Veg. Sci.* **2008**, *11*, 431–440. [CrossRef]
6. Wei, X.; Hayes, D.J.; Fraver, S.; Chen, G. Global Pyrogenic Carbon Production During Recent Decades Has Created the Potential for a Large, Long-Term Sink of Atmospheric CO 2. *J. Geophys. Res. Biogeosci.* **2018**, *123*, 3682–3696. [CrossRef]
7. Dunnette, P.V.; Higuera, P.E.; Mclauchlan, K.K.; Derr, K.M.; Briles, C.E.; Keefe, M.H. Biogeochemical impacts of wildfires over four millennia in a Rocky Mountain subalpine watershed. *New Phytol.* **2014**, *203*, 900–912. [CrossRef]
8. Sparks, A.M.; Kolden, C.A.; Smith, A.M.S.; Boschetti, L.; Johnson, D.M.; Cochrane, M.A. Fire intensity impacts on post-fire temperate coniferous forest net primary productivity. *Biogeosciences* **2018**, *15*, 1173–1183. [CrossRef]
9. Pellegrini, A.F.A.; Ahlström, A.; Hobbie, S.E.; Reich, P.B.; Nieradzik, L.P.; Staver, A.C.; Scharenbroch, B.C.; Jumpponen, A.; Anderegg, W.R.L.; Randerson, J.T.; et al. Fire frequency drives decadal changes in soil carbon and nitrogen and ecosystem productivity. *Nat. Cell Biol.* **2018**, *553*, 194–198. [CrossRef]
10. Leys, B.; Higuera, P.; Mclauchlan, K.K.; Dunnette, P.V. Wildfires and geochemical change in a subalpine forest over the past six millennia. *Environ. Res. Lett.* **2016**, *11*, 125003. [CrossRef]
11. Wang, J.; Zhang, X. Impacts of wildfires on interannual trends in land surface phenology: An investigation of the Hayman Fire. *Environ. Res. Lett.* **2017**, *12*, 054008. [CrossRef]
12. Carvalho-Santos, C.; Marcos, B.; Nunes, J.P.; Regos, A.; Palazzi, E.; Terzago, S.; Monteiro, A.T.; Honrado, J.P. Hydrological Impacts of Large Fires and Future Climate: Modeling Approach Supported by Satellite Data. *Remote Sens.* **2019**, *11*, 2832. [CrossRef]
13. Santos, R.; Fernandes, L.S.; Pereira, M.; Cortes, R.; Pacheco, F. Water resources planning for a river basin with recurrent wildfires. *Sci. Total Environ.* **2015**, *526*, 1–13. [CrossRef] [PubMed]
14. Smith, H.G.; Sheridan, G.J.; Lane, P.N.; Nyman, P.; Haydon, S. Wildfire effects on water quality in forest catchments: A review with implications for water supply. *J. Hydrol.* **2011**, *396*, 170–192. [CrossRef]
15. Senf, C.; Seidl, R. Mapping the forest disturbance regimes of Europe. *Nat. Sustain.* **2021**, *4*, 63–70. [CrossRef]
16. Poon, P.K.; Kinoshita, A.M. Spatial and temporal evapotranspiration trends after wildfire in semi-arid landscapes. *J. Hydrol.* **2018**, *559*, 71–83. [CrossRef]
17. Vlassova, L.; Perez-Cabello, F.; Mimbrero, M.R.; Llovería, R.M.; Garcia-Martin, A. Analysis of the Relationship between Land Surface Temperature and Wildfire Severity in a Series of Landsat Images. *Remote Sens.* **2014**, *6*, 6136–6162. [CrossRef]
18. Sun, X.; Zou, C.B.; Wilcox, B.; Stebler, E. Effect of Vegetation on the Energy Balance and Evapotranspiration in Tallgrass Prairie: A Paired Study Using the Eddy-Covariance Method. *Bound. Layer Meteorol.* **2018**, *170*, 127–160. [CrossRef]
19. Liu, Z.; Ballantyne, A.P.; Cooper, L.A. Biophysical feedback of global forest fires on surface temperature. *Nat. Commun.* **2019**, *10*, 1–9. [CrossRef]
20. Rother, D.; De Sales, F. Impact of Wildfire on the Surface Energy Balance in Six California Case Studies. *Bound. Layer Meteorol.* **2021**, *178*, 143–166. [CrossRef]
21. French, N.H.F.; Whitley, M.A.; Jenkins, L.K. Fire disturbance effects on land surface albedo in Alaskan tundra. *J. Geophys. Res. Biogeosci.* **2016**, *121*, 841–854. [CrossRef]
22. Quintano, C.; Fernandez-Manso, A.; Marcos, E.; Calvo, L. Burn Severity and Post-Fire Land Surface Albedo Relationship in Mediterranean Forest Ecosystems. *Remote Sens.* **2019**, *11*, 2309. [CrossRef]
23. Gatebe, C.K.; Ichoku, C.M.; Poudyal, R.; Román, M.; Wilcox, E. Surface albedo darkening from wildfires in northern sub-Saharan Africa. *Environ. Res. Lett.* **2014**, *9*, 065003. [CrossRef]
24. Saha, M.V.; D'Odorico, P.; Scanlon, T.M. Albedo changes after fire as an explanation of fire-induced rainfall suppression. *Geophys. Res. Lett.* **2017**, *44*, 3916–3923. [CrossRef]
25. Liu, H.; Zhan, Q.; Yang, C.; Wang, J. Characterizing the Spatio-Temporal Pattern of Land Surface Temperature through Time Series Clustering: Based on the Latent Pattern and Morphology. *Remote Sens.* **2018**, *10*, 654. [CrossRef]
26. Maffei, C.; Alfieri, S.M.; Menenti, M. Relating Spatiotemporal Patterns of Forest Fires Burned Area and Duration to Diurnal Land Surface Temperature Anomalies. *Remote Sens.* **2018**, *10*, 1777. [CrossRef]
27. Veraverbeke, S.; Verstraeten, W.W.; Lhermitte, S.; Van De Kerchove, R.; Goossens, R. Assessment of post-fire changes in land surface temperature and surface albedo, and their relation with fire—burn severity using multitemporal MODIS imagery. *Int. J. Wildland Fire* **2012**, *21*, 243–256. [CrossRef]
28. Koutsias, N.; Allgöwer, B.; Kalabokidis, K.; Mallinis, G.; Balatsos, P.; Goldammer, J.G. Fire occurrence zoning from local to global scale in the European Mediterranean basin: Implications for multi-scale fire management and policy. *iForest Biogeosci. For.* **2015**, *9*, 195–204. [CrossRef]
29. Bowman, D.M.J.S.; Balch, J.K.; Artaxo, P.; Bond, W.J.; Carlson, J.M.; Cochrane, M.A.; D'Antonio, C.M.; DeFries, R.S.; Doyle, J.C.; Harrison, S.P.; et al. Fire in the Earth System. *Science* **2009**, *324*, 481–484. [CrossRef] [PubMed]
30. Tedim, F.; Remelgado, R.; Borges, C.; Carvalho, S.; Martins, J. Exploring the occurrence of mega-fires in Portugal. *For. Ecol. Manag.* **2013**, *294*, 86–96. [CrossRef]
31. Van Leeuwen, W.J.D.; Casady, G.M.; Neary, D.G.; Bautista, S.; Alloza, J.A.; Carmel, Y.; Wittenberg, L.; Malkinson, D.; Orr, B.J. Monitoring post-wildfire vegetation response with remotely sensed time-series data in Spain, USA and Israel. *Int. J. Wildland Fire* **2010**, *19*, 75–93. [CrossRef]

32. Verbesselt, J.; Hyndman, R.; Zeileis, A.; Culvenor, D. Phenological change detection while accounting for abrupt and gradual trends in satellite image time series. *Remote Sens. Environ.* **2010**, *114*, 2970–2980. [CrossRef]
33. Veraverbeke, S.; Lhermitte, S.; Verstraeten, W.; Goossens, R. A time-integrated MODIS burn severity assessment using the multi-temporal differenced normalized burn ratio (dNBRMT). *Int. J. Appl. Earth Obs. Geoinf.* **2011**, *13*, 52–58. [CrossRef]
34. Bastos, A.; Gouveia, C.M.; Dacamara, C.C.; Trigo, R.M. Modelling post-fire vegetation recovery in Portugal. *Biogeosciences* **2011** *8*, 3593–3607. [CrossRef]
35. Alcaraz-Segura, D.; Chuvieco, E.; Epstein, H.E.; Kasischke, E.S.; Trishchenko, A. Debating the greening vs. browning of the North American boreal forest: Differences between satellite datasets. *Glob. Chang. Biol.* **2010**, *16*, 760–770. [CrossRef]
36. Frazier, A.E.; Renschler, C.S.; Miles, S.B. Evaluating post-disaster ecosystem resilience using MODIS GPP data. *Int. J. Appl. Earth Obs. Geoinf.* **2013**, *21*, 43–52. [CrossRef]
37. Keeley, J.E. Fire intensity, fire severity and burn severity: A brief review and suggested usage. *Int. J. Wildland Fire* **2009**, *18*, 116–126 [CrossRef]
38. Smith, A.M.; Kolden, C.A.; Tinkham, W.T.; Talhelm, A.F.; Marshall, J.D.; Hudak, A.T.; Boschetti, L.; Falkowski, M.J.; Greenberg, J.A. Anderson, J.W.; et al. Remote sensing the vulnerability of vegetation in natural terrestrial ecosystems. *Remote Sens. Environ.* **2014** *154*, 322–337. [CrossRef]
39. Parks, S.A.; Holsinger, L.M.; Koontz, M.J.; Collins, L.; Whitman, E.; Parisien, M.-A.; Loehman, R.A.; Barnes, J.L.; Bourdon, J.-F. Boucher, Y.; et al. Giving Ecological Meaning to Satellite-Derived Fire Severity Metrics across North American Forests. *Remote Sens* **2019**, *11*, 1735. [CrossRef]
40. Cardil, A.; Mola-Yudego, B.; Blázquez-Casado, Á.; González-Olabarria, J.R. Fire and burn severity assessment: Calibration of Relative Differenced Normalized Burn Ratio (RdNBR) with field data. *J. Environ. Manag.* **2019**, *235*, 342–349. [CrossRef]
41. Veraverbeke, S.; Lhermitte, S.; Verstraeten, W.; Goossens, R. Evaluation of pre/post-fire differenced spectral indices for assessing burn severity in a Mediterranean environment with Landsat Thematic Mapper. *Int. J. Remote Sens.* **2011**, *32*, 3521–3537. [CrossRef]
42. Bowman, D.M.; Perry, G.L.; Marston, J. Feedbacks and landscape-level vegetation dynamics. *Trends Ecol. Evol.* **2015**, *30*, 255–260 [CrossRef] [PubMed]
43. Quintano, C.; Fernández-Manso, A.; Calvo, L.; Marcos, E.; Valbuena, L. Land surface temperature as potential indicator of burn severity in forest Mediterranean ecosystems. *Int. J. Appl. Earth Obs. Geoinf.* **2015**, *36*, 1–12. [CrossRef]
44. Marcos, B.; Gonçalves, J.; Alcaraz-Segura, D.; Cunha, M.; Honrado, J.P. Improving the detection of wildfire disturbances in space and time based on indicators extracted from MODIS data: A case study in northern Portugal. *Int. J. Appl. Earth Obs. Geoinf.* **2019** *78*, 77–85. [CrossRef]
45. Coops, N.C.; Wulder, M.A.; Duro, D.C.; Han, T.; Berry, S. The development of a Canadian dynamic habitat index using multi-temporal satellite estimates of canopy light absorbance. *Ecol. Indic.* **2008**, *8*, 754–766. [CrossRef]
46. Alcaraz, D.; Paruelo, J.; Cabello, J. Identification of current ecosystem functional types in the Iberian Peninsula. *Glob. Ecol. Biogeogr* **2006**, *15*, 200–212. [CrossRef]
47. Duro, D.C.; Coops, N.C.; Wulder, M.A.; Han, T. Development of a large area biodiversity monitoring system driven by remote sensing. *Prog. Phys. Geogr. Earth Environ.* **2007**, *31*, 235–260. [CrossRef]
48. Arenas-Castro, S.; Gonçalves, J.; Alves, P.; Alcaraz-Segura, D.; Honrado, J.P. Assessing the multi-scale predictive ability of ecosystem functional attributes for species distribution modelling. *PLoS ONE* **2018**, *13*, e0199292. [CrossRef]
49. Gonçalves, J.; Alves, P.; Pôças, I.; Marcos, B.; Sousa-Silva, R.; Lomba, Â.; Honrado, J.P. Exploring the spatiotemporal dynamics of habitat suitability to improve conservation management of a vulnerable plant species. *Biodivers. Conserv.* **2016**, *25*, 2867–2888 [CrossRef]
50. Regos, A.; Gómez-Rodríguez, P.; Arenas-Castro, S.; Tapia, L.; Vidal, M.; Domínguez, J. Model-Assisted Bird Monitoring Based on Remotely Sensed Ecosystem Functioning and Atlas Data. *Remote Sens.* **2020**, *12*, 2549. [CrossRef]
51. Cazorla, B.P.; Cabello, J.; Peñas, J.; Garcillán, P.P.; Reyes, A.; Alcaraz-Segura, D. Incorporating Ecosystem Functional Diversity into Geographic Conservation Priorities Using Remotely Sensed Ecosystem Functional Types. *Ecosystems* **2020**, 1–17. [CrossRef]
52. Paruelo, J.M.; Piñeiro, G.; Escribano, P.; Oyonarte, C.; Alcaraz, D.; Cabello, J. Temporal and spatial patterns of ecosystem functioning in protected arid areas in southeastern Spain. *Appl. Veg. Sci.* **2005**, *8*, 93–102. [CrossRef]
53. Cazorla, B.; Cabello, J.; Reyes, A.; Guirado, E.; Peñas, J.; Pérez-Luque, A.J.; Alcaraz-Segura, D. A remote sensing-based dataset to characterize the ecosystem functioning and functional diversity of a Biosphere Reserve: Sierra Nevada (SE Spain). *Earth Syst. Sci Data Discuss.* **2020**, *2020*, 1–20. [CrossRef]
54. Arenas-Castro, S.; Regos, A.; Gonçalves, J.F.; Alcaraz-Segura, D.; Honrado, J. Remotely Sensed Variables of Ecosystem Functioning Support Robust Predictions of Abundance Patterns for Rare Species. *Remote Sens.* **2019**, *11*, 2086. [CrossRef]
55. Mildrexler, D.J.; Zhao, M.; Running, S.W. Testing a MODIS Global Disturbance Index across North America. *Remote Sens. Environ.* **2009**, *113*, 2103–2117. [CrossRef]
56. Verbesselt, J.; Hyndman, R.; Newnham, G.; Culvenor, D. Detecting trend and seasonal changes in satellite image time series *Remote Sens. Environ.* **2010**, *114*, 106–115. [CrossRef]
57. Healey, S.; Cohen, W.; Zhiqiang, Y.; Krankina, O. Comparison of Tasseled Cap-based Landsat data structures for use in forest disturbance detection. *Remote Sens. Environ.* **2005**, *97*, 301–310. [CrossRef]
58. Duan, S.-B.; Li, Z.-L.; Li, H.; Göttsche, F.-M.; Wu, H.; Zhao, W.; Leng, P.; Zhang, X.; Coll, C. Validation of Collection 6 MODIS land surface temperature product using in situ measurements. *Remote Sens. Environ.* **2019**, *225*, 16–29. [CrossRef]

59. Lobser, S.E.; Cohen, W.B. MODIS tasselled cap: Land cover characteristics expressed through transformed MODIS data. *Int. J. Remote Sens.* **2007**, *28*, 5079–5101. [CrossRef]
60. Gouveia, C.A.M.; Dacamara, C.C.; Trigo, R.M. Post-fire vegetation recovery in Portugal based on spot/vegetation data. *Nat. Hazards Earth Syst. Sci.* **2010**, *10*, 673–684. [CrossRef]
61. Cohen, J. A power primer. *Psychol. Bull.* **1992**, *112*, 155–159. [CrossRef]
62. Caon, L.; Vallejo, V.R.; Ritsema, C.J.; Geissen, V. Effects of wildfire on soil nutrients in Mediterranean ecosystems. *Earth Sci. Rev.* **2014**, *139*, 47–58. [CrossRef]
63. Catry, F.X.; Rego, F.C.; Bação, F.L.; Moreira, F. Modeling and mapping wildfire ignition risk in Portugal. *Int. J. Wildland Fire* **2009**, *18*, 921–931. [CrossRef]
64. Yamazaki, D.; Ikeshima, D.; Tawatari, R.; Yamaguchi, T.; O'Loughlin, F.; Neal, J.C.; Sampson, C.C.; Kanae, S.; Bates, P.B. A high-accuracy map of global terrain elevations. *Geophys. Res. Lett.* **2017**, *44*, 5844–5853. [CrossRef]
65. Karger, D.N.; Conrad, O.; Böhner, J.; Kawohl, T.; Kreft, H.; Soria-Auza, R.W.; Zimmermann, N.E.; Linder, H.P.; Kessler, M. Climatologies at high resolution for the earth's land surface areas. *Sci. Data* **2017**, *4*, 170122. [CrossRef]
66. European Environment Agency (EEA). CLC 2000—Copernicus Land Monitoring Service. Available online: https://land.copernicus.eu/pan-european/corine-land-cover/clc-2000 (accessed on 11 February 2021).
67. R Development Core Team R. *A Language and Environment for Statistical Computing v 3.5.1*; R Foundation for Statistical Computing: Vienna, Austria, 2018; Volume 2, Available online: https://www.R-project.org (accessed on 2 July 2018).
68. Hijmans, R.J. *Raster: Geographic Data Analysis and Modeling, R Package Version 2.5–8*; R Foundation for Statistical Computing: Vienna, Austria, 2016; Available online: https://cran.microsoft.com/snapshot/2016-08-05/web/packages/raster/index.html (accessed on 2 July 2018).
69. Busetto, L.; Ranghetti, L. MODIStsp: An R package for automatic preprocessing of MODIS Land Products time series. *Comput. Geosci.* **2016**, *97*, 40–48. [CrossRef]
70. Giglio, L.; Boschetti, L.; Roy, D.P.; Humber, M.L.; Justice, C.O. The Collection 6 MODIS burned area mapping algorithm and product. *Remote Sens. Environ.* **2018**, *217*, 72–85. [CrossRef] [PubMed]
71. Wan, Z.; Hook, S.; Hulley, G. MOD11A2 MODIS/Terra Land Surface Temperature/Emissivity 8-Day L3 Global 1km SIN Grid V006. *NASA EOSDIS Land Process. DAAC* **2015**, *10*. Available online: https://lpdaac.usgs.gov/products/mod11a2v006/ (accessed on 19 February 2021).
72. Vermote, E. MOD09A1 MODIS/Terra Surface Reflectance 8-Day L3 Global 500m SIN Grid V006. *NASA EOSDIS Land Process. DAAC* **2015**, *10*. Available online: https://lpdaac.usgs.gov/products/mod09a1v006/ (accessed on 19 February 2021).
73. Hampel, F.R. A General Qualitative Definition of Robustness. *Ann. Math. Stat.* **1971**, *42*, 1887–1896. [CrossRef]
74. Hampel, F.R. The Influence Curve and its Role in Robust Estimation. *J. Am. Stat. Assoc.* **1974**, *69*, 383–393. [CrossRef]
75. GDAL. *Geospatial Data Abstraction Library v 3.0.4*. 2020. Available online: https://gdal.org/ (accessed on 28 January 2020).
76. Gillies, S.; Ward, B.; Petersen, A.S. Rasterio: Geospatial Raster I/O for Python Programmers. 2013. Available online: https://github.com/mapbox/rasterio (accessed on 15 July 2018).
77. Whittaker, E.T. On a New Method of Graduation. *Proc. Edinb. Math. Soc.* **1922**, *41*, 63–75. [CrossRef]
78. Eilers, P.H.C. A Perfect Smoother. *Anal. Chem.* **2003**, *75*, 3631–3636. [CrossRef]
79. Couronné, R.; Probst, P.; Boulesteix, A.-L. Random forest versus logistic regression: A large-scale benchmark experiment. *BMC Bioinform.* **2018**, *19*, 1–14. [CrossRef] [PubMed]
80. Kuhn, M.; Weston, S.; Keefer, C.; Engelhardt, A.; Cooper, T.; Mayer, Z.; Kenkel, B.; Team, R.C.; Benesty, M.; Lescarbeau, R.; et al. *Caret: Classification and Regression Training, R Package v 6.0–70*; R Foundation for Statistical Computing: Vienna, Austria, 2016; Available online: https://CRAN.R-project.org/package=caret (accessed on 2 July 2018).
81. Wasserstein, R.L.; Lazar, N.A. The ASA Statement on *p*-Values: Context, Process, and Purpose. *Am. Stat.* **2016**, *70*, 129–133. [CrossRef]
82. Hubbard, R.; Lindsay, R.M. Why P Values Are Not a Useful Measure of Evidence in Statistical Significance Testing. *Theory Psychol.* **2008**, *18*, 69–88. [CrossRef]
83. Mack, M.C.; Bret-Harte, M.S.; Hollingsworth, T.N.; Jandt, R.R.; Schuur, E.A.G.; Shaver, G.R.; Verbyla, D.L. Carbon loss from an unprecedented Arctic tundra wildfire. *Nat. Cell Biol.* **2011**, *475*, 489–492. [CrossRef] [PubMed]
84. Ramanathan, V.; Carmichael, G. Global and regional climate changes due to black carbon. *Nat. Geosci.* **2008**, *1*, 221–227. [CrossRef]
85. Volante, J.; Alcaraz-Segura, D.; Mosciaro, M.; Viglizzo, E.; Paruelo, J. Ecosystem functional changes associated with land clearing in NW Argentina. *Agric. Ecosyst. Environ.* **2012**, *154*, 12–22. [CrossRef]
86. Bodí, M.B.; Martin, D.A.; Balfour, V.N.; Santín, C.; Doerr, S.H.; Pereira, P.R.G.; Cerdà, A.; Mataix-Solera, J. Wildland fire ash: Production, composition and eco-hydro-geomorphic effects. *Earth Sci. Rev.* **2014**, *130*, 103–127. [CrossRef]
87. Saha, M.V.; D'Odorico, P.; Scanlon, T.M. Kalahari Wildfires Drive Continental Post-Fire Brightening in Sub-Saharan Africa. *Remote Sens.* **2019**, *11*, 1090. [CrossRef]
88. Lentile, L.B.; Holden, Z.A.; Smith, A.M.S.; Falkowski, M.J.; Hudak, A.T.; Morgan, P.; Lewis, S.A.; Gessler, P.E.; Benson, N.C. Remote sensing techniques to assess active fire characteristics and post-fire effects. *Int. J. Wildland Fire* **2006**, *15*, 319–345. [CrossRef]

89. García-Llamas, P.; Suárez-Seoane, S.; Taboada, A.; Fernández-Manso, A.; Quintano, C.; Fernández-García, V.; Fernández-Guisuraga, J.M.; Marcos, E.; Calvo, L. Environmental drivers of fire severity in extreme fire events that affect Mediterranean pine forest ecosystems. *For. Ecol. Manag.* **2019**, *433*, 24–32. [CrossRef]
90. Mallinis, G.; Mitsopoulos, I.; Chrysafi, I. Evaluating and comparing Sentinel 2A and Landsat-8 Operational Land Imager (OLI) spectral indices for estimating fire severity in a Mediterranean pine ecosystem of Greece. *Geosci. Remote Sens.* **2017**, *55*, 1–18. [CrossRef]
91. Shi, T.; Xu, H. Derivation of Tasseled Cap Transformation Coefficients for Sentinel-2 MSI At-Sensor Reflectance Data. *IEEE J. Sel. Top. Appl. Earth Obs. Remote Sens.* **2019**, *12*, 4038–4048. [CrossRef]
92. Marcos, E.; Fernández-García, V.; Fernández-Manso, A.; Quintano, C.; Valbuena, L.; Tárrega, R.; Luis-Calabuig, E.; Calvo, L. Evaluation of Composite Burn Index and Land Surface Temperature for Assessing Soil Burn Severity in Mediterranean Fire-Prone Pine Ecosystems. *Forests* **2018**, *9*, 494. [CrossRef]
93. De Santis, A.; Chuvieco, E. GeoCBI: A modified version of the Composite Burn Index for the initial assessment of the short-term burn severity from remotely sensed data. *Remote Sens. Environ.* **2009**, *113*, 554–562. [CrossRef]
94. Landi, M.A.; Bella, C.M.D.; Bravo, S.J.; Bellis, L.M. Structural resistance and functional resilience of the Chaco forest to wildland fires: An approach with MODIS time series. *Aust. Ecol.* **2020**. [CrossRef]
95. Fernández, N.; Paruelo, J.M.; Delibes, M. Ecosystem functioning of protected and altered Mediterranean environments: A remote sensing classification in Doñana, Spain. *Remote Sens. Environ.* **2010**, *114*, 211–220. [CrossRef]
96. Cocking, M.I.; Varner, J.M.; Knapp, E.E. Long-term effects of fire severity on oak-conifer dynamics in the southern Cascades. *Ecol. Appl.* **2014**, *24*, 94–107. [CrossRef]

remote sensing

MDPI

Article

Wildland Fire Tree Mortality Mapping from Hyperspatial Imagery Using Machine Learning

Dale A. Hamilton *, Kamden L. Brothers, Samuel D. Jones, Jason Colwell and Jacob Winters

Department of Mathematics and Computer Science, Northwest Nazarene University, 623 S University Blvd, Nampa, ID 83686, USA; kbrothers@nnu.edu (K.L.B.); samueljones@nnu.edu (S.D.J.); jcolwell@nnu.edu (J.C.); jwinters@nnu.edu (J.W.)

* Correspondence: dhamilton@nnu.edu; Tel.: +1-(208)-467-8769

Abstract: The use of imagery from small unmanned aircraft systems (sUAS) has enabled the production of more accurate data about the effects of wildland fire, enabling land managers to make more informed decisions. The ability to detect trees in hyperspatial imagery enables the calculation of canopy cover. A comparison of hyperspatial post-fire canopy cover and pre-fire canopy cover from sources such as the LANDFIRE project enables the calculation of tree mortality, which is a major indicator of burn severity. A mask region-based convolutional neural network was trained to classify trees as groups of pixels from a hyperspatial orthomosaic acquired with a small unmanned aircraft system. The tree classification is summarized at 30 m, resulting in a canopy cover raster. A post-fire canopy cover is then compared to LANDFIRE canopy cover preceding the fire, calculating how much the canopy was reduced due to the fire. Canopy reduction allows the mapping of burn severity while also identifying where surface, passive crown, and active crown fire occurred within the burn perimeter. Canopy cover mapped through this effort was lower than the LANDFIRE Canopy Cover product, which literature indicated is typically over reported. Assessment of canopy reduction mapping on a wildland fire reflects observations made both from ground truthing efforts as well as observations made of the associated hyperspatial sUAS orthomosaic.

Keywords: mask region-based convolutional neural network; small unmanned aircraft system; canopy cover; tree mortality

Citation: Hamilton, D.A.; Brothers, K.L.; Jones, S.D.; Colwell, J.; Winters, J. Wildland Fire Tree Mortality Mapping from Hyperspatial Imagery Using Machine Learning. *Remote Sens.* **2021**, *13*, 290. https://doi.org/10.3390/rs13020290

Received: 2 December 2020
Accepted: 8 January 2021
Published: 15 January 2021

1. Introduction

This study investigates the use of machine learning to identify burn severity as a measure of canopy reduction. Tree crowns are classified from hyperspatial (sub-decimeter resolution) imagery using a mask region-based convolutional neural network (MR-CNN). The tree crown classification is then summarized at 30-m resolution, resulting in a canopy cover map. A post-fire canopy cover mapping product was compared to a pre-fire canopy cover mapping layer from the LANDFIRE project. The difference between the pre-fire and post-fire canopy cover enables a calculation of canopy reduction resulting as an effect of the fire, which gives an indication of tree mortality which is a primary indicator of post-fire effects resulting from a wildland fire.

Forest management practices, including fire suppression efforts, have led to forests with higher density than were found under pre-European settlement conditions. As a result, wildlands are experiencing a higher frequency of catastrophic fires, with millions of acres burned across the western US every year, often burning with much higher severity than experienced under historic conditions. These higher intensity fires result in increased incidence of soil erosion, hydrologic degradation, and deteriorating ecosystem resilience. The ability to detect trees in hyperspatial imagery enables the calculation of canopy reduction, which is a measure of wildland fire severity. The availability of canopy reduction data enables better decision making for ecosystem management and fire effect mitigation.

Background

The term "wildland fire severity" can refer to many different effects observed through a fire cycle, from how intensely an active fire is burning to the response of the ecosystem to the fire over the subsequent years. As opposed to the intensity of an active fire, which is often referred to as fire severity, this study investigates direct or immediate effects of a fire as observed in the days and weeks after the fire is contained, which we refer to as burn severity [1,2]. Previous studies in this effort examined the effect of the fire on biomass consumption which indicates how much surface vegetative material has been consumed by the fire [1,3–5]. By contrast, this study investigates the effect that fire has had on canopy vegetation in forested ecosystems. In situations where the amount of current post-fire canopy has not been significantly altered from what can be observed from the pre-fire conditions, it can be assumed that a fire was only consuming surface vegetation (or possibly the lower canopy) but did not consume the upper portion of the tree crowns in the canopy [6]. This low intensity fire low intensity surface fire results in no mortality of the trees in the upper canopy would result in no noticeable change of canopy cover between pre-fire and post-fire conditions. Another situation results from where all the upper tree crowns in the forest canopy have been either been consumed by a fire or the crown has been scorched by a more intense fire on the surface which extends into the tree crowns [6] In these active crown fires, the canopy has been consumed or scorched by the fire, resulting in the death of all the trees provided that the trees are not resprouting deciduous species This high severity active crown fire will result in a complete reduction of canopy cover as compared from pre-fire conditions. In between surface fires and crown fires, lie a situation where fire burning through surface vegetative material having enough intensity to burn or scorch the crowns of individual trees or clusters of trees, resulting in a moderate severity passive crown fire where some of the tree crowns are completely consumed or scorched [7] While the post-fire canopy cover will be reduced from pre-fire canopy conditions as a result of these passive crown fires, the post-fire canopy cover will not be reduced to zero because some of the trees that made up the pre-fire canopy will survive, having been neither burned nor scorched.

Most land managers that use canopy cover mapping layers currently rely on the canopy cover layers that are available through the LANDFIRE project, a collaborative effort between the United States Forest Service and the US Department of the Interior. LANDFIRE produces geospatial products that describe wildland fire fuel conditions across the United States. From the LANDFIRE portal, land managers have access seamless nation-wide geospatial layers with a spatial resolution of 30 m (30 m) [8]. These canopy cover layers were generated from imagery acquired by the Enhanced Thematic Mapper Plus sensor on board the Landsat 7 satellite using a decision tree algorithm that was trained using plot data stored in the LANDFIRE Reference Database which was collected from across the US [9].

The utilization of small unmanned aircraft systems (sUAS) as a remote sensing tool provides land managers an affordable way to acquire imagery with much higher spatial resolution (sub-decimeter) than was previously available. This hyperspatial imagery produces homogeneous pixels which are comprised of a single class, resulting in higher mapping accuracy. By contrast, heterogeneous pixels which are large enough to spatially contain more than one class are common with lower resolution imagery [1,4].

Burned vegetation, unburned canopy vegetation (conifer and deciduous trees), and unburned surface vegetation (brush and grass) have been shown to have distinct spectral responses in both the visible and near infra-red spectra, though the spectral difference between canopy vegetation and surface vegetation is not as large as that between burned and general unburned vegetation [10]. This spectral difference is apparent when observing the two unburned vegetation classes in hyperspatial imagery. Persons who had participated in the acquisition of the sUAS imagery found that they could observe the visual difference between surface and canopy vegetation, with this difference in vegetation type observed in the imagery matching what they had observed in the field while conducting image acquisition flights with the sUAS.

Hamilton [1] found that tree crowns obstruct surface features in imagery acquired with a sUAS, preventing classification of surface features. Obstructed features include surface biomass consumption, a primary indicator of surface fire severity. When a tree crown is surrounded by burned surface fuel in the image, it is reasonable to assume that the ground under the tree has also burned but is obscured by the unburned crown [1,11]. Unburned tree crowns within a burned area cause an under-reporting of fire extent due to the misclassification of unburned tree crowns within the fire perimeter as being unburned areas. The first step to identifying and mapping obscured burned surface fuels is to identify trees within the image. When a tree surrounded by burned surface fuels is identified, it can be inferred that the surface fuels under the tree are in fact burned and should be included within the fire extent. Wildland fire can also affect an ecosystem through tree mortality. Mortality may result from either the ignition and consumption of a tree crown or from scorching of the tree crown caused by heat from surface fire beneath the tree. The extent of tree mortality can be estimated by calculating the change in canopy cover before and after the fire. Post-fire canopy cover can be calculated as the percentage of an image that consists of tree crowns. Pre-fire canopy cover can be retrieved from the Canopy Cover rasters that were published by the LANDFIRE Fire Fuels project [8].

2. Materials and Methods: Mapping Canopy Cover and Tree Mortality

Canopy cover describes the percentage area occupied by the upper tree crowns in forested ecosystems. In this effort, canopy cover is calculated as the percent of upper layer canopy cover within a stand, which is represented as a 30-m pixel. Once the tree crown classification has been completed, canopy cover is determined by summarizing the percentage of hyperspatial (5 cm) pixels that were classified as canopy vegetation within each 30-m Landsat pixel. Finally, the post-fire canopy cover is compared against pre-fire canopy cover from the LANDFIRE project to determine how much upper layer canopy cover reduction occurred due to the fire, giving an indication of tree mortality, which is a primary indicator of post-fire effects in forested ecosystems.

2.1. Classification of Tree Crowns from Hyperspatial Imagery

The initial step in mapping canopy cover and tree mortality enables the mapping of individual tree crowns within a hyperspatial orthomosaic acquired with the use of an sUAS. By classifying the tree crowns within an orthomosaic, we will be able to distinguish between landcover lifeforms that are indicative of canopy vegetation as opposed to surface vegetation.

The mask region-based convolutional neural network (MR-CNN) is an algorithm for instance segmentation. It detects objects in an image and determines which pixels comprise each object. When classifying an image, MR-CNN begins by applying a convolutional neural network (CNN) to extract features from the image. The features extracted from the image are used as input for a region proposal network, which slides over the feature map and detects regions of interest (RoIs) which are likely to contain objects. Each RoI is evaluated further to determine what type of object, if any, is inside (in this case, tree and non-tree); the object's bounding box; and what pixels inside the bounding box comprise the object [12].

Training data for the MR-CNN model consists of images in which the boundaries of all trees are marked with individual polygon. Unmarked areas are assumed to be non-tree. The model is trained using mini-batch gradient descent to minimize a loss function that measures the difference between the model's output in response to a training image and the training data.

A MR-CNN model with a ResNet-101 backbone that had been pre-trained on the Microsoft COCO dataset [13] was retrained to detect trees. Training images were cropped from orthomosaics of study areas flown with sUAS over the Boise National Forest. These forested sites primarily contained Pinus Ponderosa (Ponderosa Pine), Populus Trichocarpa (Black Cottonwood), Pseudotsuga Menzieii (Douglas Fir), and Pinus Contorta (Lodgepole Pine). A polygon was traced around each tree occurring in each training image using the VGG Image Annotator [14] as shown in Figure 1. A total of 1148 trees were labeled over 14 training images. Questions of whether visually ambiguous objects were trees or surface vegetation were answered by referring to a 3D model created from the same imagery as the orthomosaics. To discourage overfitting, training images were rotated and mirrored randomly during the training process. The training data was created using many different forest types which also included images that contained scorched trees. These scorched trees were put in the "nontree" class to help the MR-CNN distinguish between burnt and non-burnt trees.

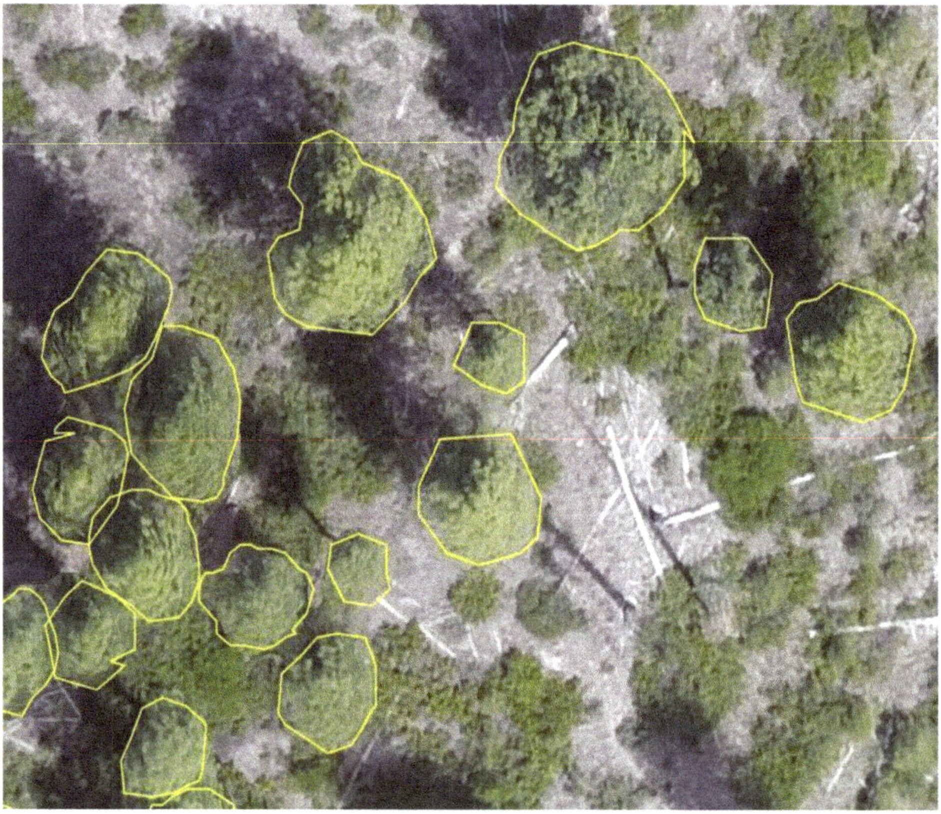

Figure 1. Mask Region-based Convolutional Network training data.

As an instance segmentation algorithm, MR-CNN can answer questions that a pixel-based approach cannot, such as, "How many trees are in the image?" and "What is the distribution of tree sizes?" To facilitate calculation of canopy cover, the MR-CNN's output was used to perform semantic segmentation. All pixels that were determined to be part of a tree were marked as tree crown and all other pixels were marked as surface, an example of which is shown in Figure 2. Only photosynthesizing foliage was annotated in the training data for the MR-CNN, allowing the trained MR-CNN to distinguish between live trees and scorched trees, whose foliage was no longer photosynthesizing, during classification. Scorched trees would be classified as "no tree" by the MR-CNN.

(a) (b)

Figure 2. (**a**) is pre-classified image. (**b**) is an image that has been classified by the MR-CNN.

2.2. Calculating Canopy Cover

Canopy cover describes the vertical projection of the tree canopy onto a horizontal plane in forested ecosystems. In this effort, canopy cover is calculated as the percent canopy cover within a stand. Once the tree crown classification has been completed as described earlier, canopy cover is determined by summarizing the percentage of 5 cm pixels that were classified as canopy vegetation within each 30-m (30 m) Landsat pixel. Because the MR-CNN is trained to only classify live trees, scorched canopy vegetation is not included in the canopy cover calculations.

2.2.1. Calculate Five Centimeter Class Densities for Each 30 Meter Pixel

The density of the tree class is calculated for each 30-m Landsat pixel from the 5 cm tree classification pixels obtained from the MR-CNN. A tool was written which calculates the density of 5 cm pixels classified as tree crown within each 30 m pixel, outputting a 30 m resolution grayscale TIFF image. This Density Tool counts the total number of hyperspatial tree class pixels occurring in the 30 m pixel. The total number of tree pixels that occurred is then divided by the total number of 5 cm pixels within the 30 m pixel and multiplied by 100. The class densities for each 30 m pixel are recorded in the class density output raster. This class density raster contains the percentage of each 30-m pixel that is comprised of 5 cm pixels that are classified as being part of a tree crown, which effectively records Canopy Cover for each 30-m pixel.

2.2.2. Canopy Cover Analysis

Previous efforts including Hamilton [1,4] and Matese [15] have shown that mapping with higher resolution imagery results in increased accuracy than using lower resolution imagery. This section summarizes the analysis performed to assess the effectiveness of the methods specified previously for calculating canopy cover.

Setting and notation: Suppose that there is a 30-m pixel depicting a region of ground. The same 30 m $\times$ 30 m region of ground is viewed as a 600 $\times$ 600 region of 5 cm pixels in the hyperspatial orthomosaic. Then the canopy cover is calculated from the hyperspatial orthomosaic, and this calculation is subject to error. The probability of correctly identifying a "tree" pixel (sensitivity) is observed from training data to be at least 0.70, and the probability of correctly identifying a "non-tree" pixel (specificity) is observed from training data to be at least 0.95. Both values are usually higher (examples are shown in Section 3.2), but we will use the values of 0.70 and 0.95 as conservative estimates.

Question: What level of accuracy can one expect for the value of canopy cover calculated from the 600 × 600 pixel region of the hyperspatial orthomosaic corresponding to a Landsat pixel? Precisely, what is the standard error of this value? Is this significantly smaller than what one might hope to obtain from LANDFIRE data?

Answer: Using The value of canopy cover can be estimated with bias close to 0 and standard error less than 0.001080. In most cases, the standard error is less than 0.000824 These values of standard error are significantly smaller than what one might hope to obtain from LANDFIRE data. They are obtained using mathematics beyond just routine statistical inference. The details are omitted here so as not to disturb the flow of the article. However they are included in the Appendix A for the interested reader.

2.3. Canopy Cover: sUAS Derived vs. LANDFIRE

The LANDFIRE project is a collaborative effort between the United States Forest Service and the US Department of the Interior, producing geospatial products that describe wildland fire fuel conditions supporting wildland fire modeling across the United States LANDFIRE's Existing Vegetation Cover and Forest Canopy Cover products record the vertically projected percentage cover of the upper vegetation canopy layer [8]. For this effort in evaluating tree mortality, the Existing Vegetation Cover raster was updated to only contain canopy cover in forested ecosystems. Additionally, the Existing Vegetation Cover was updated (via the ArcGIS Spatial Analyst Con tool) with the Forest Canopy Cover for forested pixels where the predominant lifeform appeared to be either shrub or herbaceous. This conversion was accomplished by reclassing the vegetation cover for all non-forested lifeforms to zero, then substituting the Forest Canopy Cover for the Existing Vegetation Cover on pixels that had a non-forested life form, but also had a Forest Canopy Cover value. This resulted in a geospatial layer with spatial resolution of 30 m representing vertically projected percentage forest canopy cover.

The comparison of the hyperspatial derived canopy cover (HDCC) as accomplished using the ArcGIS combine raster tool. This resulted in an attribute table that had each combination of canopy cover that occurred within pixels in the two rasters as well as how many pixels the combination occurred in. From the combine table, the difference between the HDCC and LANDFIRE forest Canopy Cover (LFCC) can be calculated. This difference can be used to analyze the statistical difference between the sUAS derived and LANDFIRE products as well as the determination of canopy reduction as a measure of tree mortality in areas where the sUAS canopy cover was calculated from imagery acquired after a wildland fire.

2.3.1. Canopy Cover Comparison Methods

When classifying sUAS imagery through the Mask Regional-Convolutional Neural Network (MR-CNN), a new image is created as an output image in which pixels are based on the original image is part of a tree crown (tree) or does not comprise part of a tree crown (no tree). During this process, however, spatial referencing and null data (where a pixel represents no data) values are lost and must be added back.

Spatial referencing is added back using an ArcPy script which takes the world file spatial referencing from the original images that were aligned with the 30 m imagery inputted into the MR-CNN and adds the spatial referencing to the MR-CNN output.

Band 4 of the classified output from the MR-CNN contains the tree/no tree classification. This band is extracted using the Copy Raster Tool on Band 4 of the image. This newly extracted classification raster is given as input to the density tool, sums the number of 5 cm tree pixels within the 30 m pixel and divides it by the total number of 5 cm pixels within a 30 m pixel. Multiplying this number by one hundred gives the percent of tree pixels or the canopy cover in the 30 m pixel area as expressed in Equation (1). This process is repeated until each pixel in the image has been analyzed giving a 30 m resolution canopy cover (CC) output raster.

$$\% \text{ of CC} = \left(\frac{\text{tree pixels}}{\text{total pixels}}\right) * 100 \quad (1)$$

Before adding the null data back into the orthomosaic, the 5-cm resolution imagery is resampled to 30 m resolution. Resampling before adding the null data speeds up the process of adding the null data back in. Null data is recorded with a pixel value of 0 (not a tree) by the MR-CNN. To restore its value to NULL, ArcPy's conditional toolset is used to compare pixels in the source image (the MR-CNN output) and with the original data to check if the value was NULL. The density tools output is changed to 255 for every pixel where the source image's value is NULL. To represent NULL, 255 is used because it is outside the range of possible values (0% to 100%).

The various methods of processing the data outlined in the previous sections creates a HDCC raster that is aligned with the LFCC raster. This allows the rasters to be compared pixel by pixel, using the ArcGIS Combine Raster tool. This creates a new raster with a raster attribute table column for each of the combined rasters. Next, the LFCC column was subtracted from the HDCC column into a newly added Difference column. With sUAS data of unburned areas, this shows how close HDCC and LFCC data are. When comparing pre-fire LFCC against post-fire HDCC, this difference can be used in the calculation of canopy reduction, an indicator of tree mortality which is a primary effect of wildland fire.

The workflow which enable the calculation of the HDCC and its camparison to the LFCC is summarized in Figure 3.

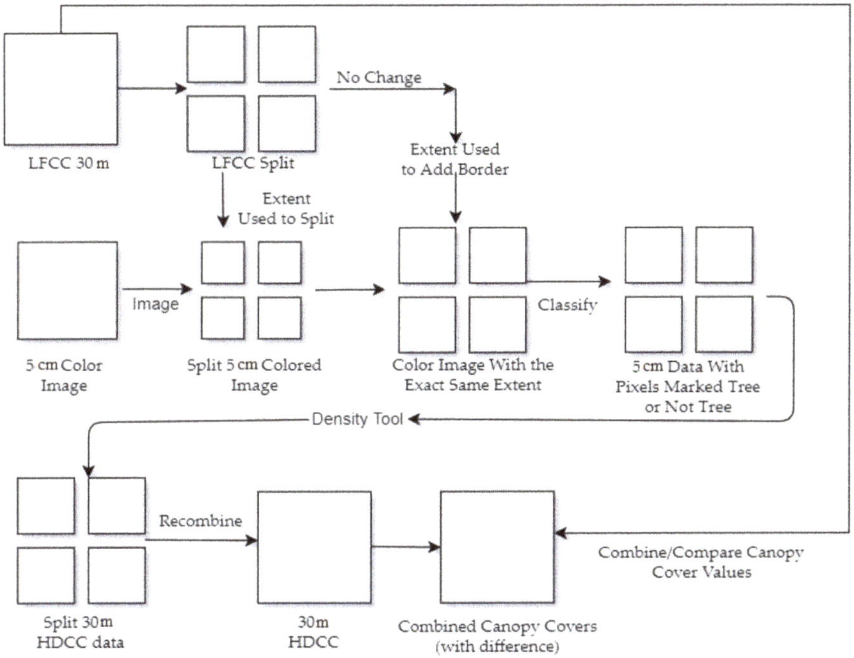

Figure 3. An overview of the processes used to create the HDCC.

2.3.2. Canopy Cover Comparison Analysis

Working assumption: In Section 2.2.2, it was observed that (assuming 5-cm pixel identification better than 0.80 correct for pixels representing ground with canopy cover, better than 0.95 for pixels representing ground without canopy cover, and no more than 0.80 canopy cover) using adjusted hyperspatial derived data, canopy cover can be calculated with a standard error of less than 0.000824. This is a much smaller standard error than we could hope to obtain from 30 m LANDFIRE pixels. Accordingly, it is reasonable, when comparing LANDFIRE data to hyperspatial derive data, to make the working assumption that the adjusted HDCC value is correct.

Objective: The method of maximum-likelihood estimation is used to construct a model of the error of the value of canopy cover derived from LANDFIRE data.

Setting and notation: There are n LANDFIRE 30 m pixels, and for each, the estimated value $B \in [0,1]$ of canopy cover. For each there is a corresponding region of 5 cm pixels in the hyperspatial orthomosaic, and the adjusted value of canopy cover $C \in [0,1]$ calculated from the 600^2 pixels in corresponding region of the hyperspatial orthomosaic. As stated above, the latter value (C) is presumed correct, and the objective is to specify a model of the error in the former value (B). Write

$$X = B - C \tag{2}$$

for the error. The data consists of n pairs $(b_1, c_1), (b_2, c_2), \ldots, (b_n, c_n)$, from which are obtained n error values

$$x_1 = b_1 - c_1, \ldots, x_n = b_n - c_n. \tag{3}$$

Model for error: A normal model for X is used, whose density is

$$fX(x) = \frac{1}{\sigma\sqrt{2\pi}} e^{-\frac{1}{2}\left(\frac{x-\mu}{\sigma}\right)^2} \tag{4}$$

with parameters μ (mean) and σ (standard deviation). (Note that μ and σ are the respective mean and standard deviation of the error $B - C$ only, and not of the canopy cover itself or any estimate thereof). The values of the parameters are unknown, and the next task is to choose the best values for them (so that the model best accounts for the present data).

Maximum-likelihood estimators for parameters: The maximum-likelihood estimations for μ and σ are

$$\hat{\mu} = \overline{X} = \frac{\sum X}{n} \tag{5}$$

and

$$\hat{\sigma} = \hat{S} = \sqrt{\frac{\sum (X - \overline{X})^2}{n}} \tag{6}$$

respectively, the latter being the "population standard deviation" $\hat{S}$ with denominator n instead of the "sample standard deviation" $\overline{S}$ with denominator $n - 1$.

Conclusion: The maximum-likelihood estimation for a normal error X of the LANDFIRE value for canopy cover has mean $\overline{X}$ and standard deviation $\hat{S}$, calculated from the n values $x_1 = b_1 - c_1, \ldots, x_n = b_n - c_n$ of the difference $X = B - C$ between LANDFIRE Canopy Cover and (adjusted) Hyperspatial Derived Canopy Cover.

2.4. *Mapping Tree Mortality via Canopy Reduction*

While it is much more accurate to map canopy cover from hyperspatial sUAS imagery, the mapping extent is limited to how large of a study area can be flown with a sUAS. Current regulatory and technical constraints limit flight operations with a single sUAS in forested environments to under 400 hectares in a single day [16]. For mapping post-fire canopy cover, this constraint can be overcome with more flights on all but the largest fires. Due to the unpredictable nature of fires resulting from unplanned ignitions, it is not feasible to obtain hyperspatial sUAS derived canopy cover data reflecting pre-fire conditions. Fortunately, the LANDFIRE products are generated for the entire United States and are updated every two years.

For mapping canopy reduction, which is a proxy for tree mortality, a comparison was made between the LFCC (pre-fire canopy cover) against the HDCC (post-fire) layer, as described in the previous section.

Once the difference between pre-fire LFCC and post-fire HDCC, canopy cover is calculated, canopy cover reduction (CCR) can be calculated by calculating the difference between LFCC and HDCC, minus the estimated error standard deviation (multiplied by the Standard Normal density) and estimated error mean as shown in Equation (7).

2.4.1. Canopy Reduction Methods

Due to LANDFIRE's canopy cover being less accurate than the HDCC, it cannot simply be stated that reduction is the discrepancy between sUAS Derived data and LANDFIRE data. Instead, we must use the data from unburned areas to see how much this discrepancy varies. Then any numbers outside of the normal variance is a change in canopy cover to a 95% degree of certainty. Once it is determined that there has been a change, there are two possibilities for what that change could be, active crown fire or passive crown fire. If the HDCC is 0% it was an active crown fire in this area otherwise it was a passive crown fire.

The main difficulty with calculating the type of fire based on this method is that there is some uncertainty due to LANDFIRE being 30-m data. For this reason, it is difficult to tell what happened to pixels where there was not enough of a difference to be certain if there was a change. The discrepancy between the data could just be due to LANDFIRE's overreporting or it could be due to a passive crown fire.

2.4.2. Canopy Reduction Analysis

Working assumption: As before, a 30 m-by-30 m square of land is being considered, which corresponds to a single LANDFIRE pixel as well as to a 600×600 pixel region in a hyperspatial orthomosaic. Again, $B \in [0,1]$ is the canopy cover according to LANDFIRE, and $C \in [0,1]$ is the adjusted value of canopy cover obtained from the hyperspatial orthomosaic, which is assumed to be correct. In the previous section, maximum-likelihood estimation was used to construct a model of the error $X = B - C$, which was a normal density with estimated parameter values $\mu = \hat{\mu} = \overline{X}$ and $\sigma = \hat{\sigma} = \hat{S}$.

Objective: A criterion is developed by which it can be concluded, with the desired level of confidence $1 - \alpha$, that canopy cover, from past LANDFIRE image to present hyperspatial orthomosaic, has decreased by at least some particular value r.

The difficulty: The difficulty here is that there is no available pre-fire value of C (canopy cover calculated from a hyperspatial orthomosaic), but only a past value of B (canopy cover calculated from LANDFIRE) to compare with the present value of C. The notation $X = B - C$, is still used, while now it is recognized that B is a past value and C is a present value. This means that X (difference in LFCC and (adjusted) HDCC values) may include change in canopy cover (where a positive value might result from a crown fire), or inaccuracy of LANDFIRE data, or both.

The null hypothesis: The null hypothesis H_0 that we wish to test is that there was a reduction in canopy cover of no more than r. This H_0 will be rejected if, even allowing for a canopy reduction of r, the value of $B - C$ is too high to be explained by chance; that is, if

$$B - C > r + \mu + z_\alpha \sigma, \tag{7}$$

or, equivalently,

$$X > r + \mu + z_\alpha \sigma. \tag{8}$$

Replacing the inequality with an equality and solving for r, the following criterion is obtained.

Criterion: It may be inferred that canopy cover has decreased by at least

$$B - C - \mu - z_\alpha \sigma, \tag{9}$$

where B is past canopy cover according to LANDFIRE, C is the adjusted value of canopy cover from a hyperspatial orthomosaic, μ and σ are the parameter values obtained as described previously, and z_α is the critical value of the Standard Normal density at significance level α. Intuitively, the amount that we can conclude by which canopy cover has been reduced is the difference "past LANDFIRE minus present hyperspatial" $B - C$, minus the average discrepancy μ between the LANDFIRE B and hyperspatial-derived C, minus an additional margin of error $z_\alpha \sigma$ which is larger, the higher the level of confidence we wish to have.

3. Results

3.1. Tree Crown Classification Validation

To assess the performance of the MR-CNN in mapping tree crowns, validation data was prepared using ArcGIS. Sample areas of canopy and surface vegetation were recorded as polygons in shapefiles digitized from existing orthomosaics. Validation data were made for all orthomosaics used in the research.

The MR-CNN was run on the orthomosaics used for validation. Agreement between a classifier's output and the validation data was calculated using ArcMap's Tabulate Area tool as shown in Table 1. A classifier's accuracy on an orthomosaic was computed as the area where the classifier agreed with the class marked in the validation data divided by the total area marked in the validation data.

Table 1. Confusion matrices of MR-CNN tree crown classifications. Classified surface and canopy are shown in the rows while human labeled surface and canopy are in columns.

	South Placerville		**NW Placerville**		**Edna Creek**		**South Exp. Forest**	
	Surface	Canopy	Surface	Canopy	Surface	Canopy	Surface	Canopy
Surface	47.5%	19.9%	46.6%	6.2%	49.0%	0.6%	49.1%	10.6%
Canopy	0.6%	32.0%	2.3%	44.9%	0.0%	50.3%	0.4%	40.0%
	North Exp. Forest		East Placerville		West Placerville			
	Surface	Canopy	Surface	Canopy	Surface	Canopy		
Surface	50.0%	10.5%	49.29%	4.50%	49.98%	7.29%		
Canopy	0.3%	39.2%	0.00%	46.21%	0.00%	42.73%		
	Belshazzar		Newstart		Grimes Creek			
	Surface	Canopy	Surface	Canopy	Surface	Canopy		
Surface	47.1%	3.6%	50.1%	16.9%	51.5%	2.9%		
Canopy	3.0%	46.3%	0.2%	32.8%	0.2%	45.4%		

Orthomosaics where choose with a variety of different conditions to find any anomalies where the classifier underperforms. Trees in the chosen orthomosaics were comprised primarily of Pinus Ponderosa (Ponderosa Pine), Populus Trichocarpa (Black Cottonwood), Pseudotsuga Menzieii (Douglas Fir) and Pinus Contorta (Lodgepole Pine). All orthomosaics were from flights conducted with the Boise National Forest in southwestern Idaho [16,17]. The distinguishing characteristics of the ten chosen orthomosaics were as follows:

- NewStart: Orthomosaic consisted of 474 acres. Surface vegetation comprised of brush and herbaceous vegetation in the riparian zone along Grimes Creek. Low canopy cover.
- GrimesCreek: Orthomosaic consisted of 249 acres. Surface vegetation comprised of brush and herbaceous vegetation both inside and outside the riparian zone along Grimes Creek. Moderate to low canopy cover.
- South Placerville: Orthomosaic consisted of 262 acres. Surface vegetation comprised of brush and herbaceous vegetation both inside and outside the riparian zone. Low canopy cover.
- East Placerville: Orthomosaic consisted of 627 acres. Mix of herbaceous vegetation and canopy cover. Moderate to low canopy cover.
- West Placerville: Orthomosaic consisted of 565 acres. Wide open area covered in trees with surface vegetation along the ground around trees. Moderate to high canopy cover

- Northwest Placerville: Orthomosaic consisted of 1268 acres. Flat area with trees and tree-like brush. Moderate canopy cover.
- Edna Creek: Orthomosaic consisted of 1084 acres. Lots of unforested areas or areas with dead trees. Patches of forest throughout the orthomosaic. Low canopy cover.
- Belshazzar: Orthomosaic consisted of 265 acres. Very dense surface vegetation. Not in a riparian zone. Certain areas experienced larger-than-normal orthomosaic stitching artifacts, which we hypothesized could cause trouble for object detection techniques. Predominantly high canopy cover with some meadows.
- North Experimental Forest: Orthomosaic consisted of 800 acres. Herbaceous vegetation around water sources. Sparse forest. Moderate to low canopy cover.
- South Experimental Forest: Orthomosaic consisted of 642 acres. Not much vegetation other than trees. Ground is white in many areas. Moderate canopy cover.

The MR-CNN model's overall accuracy was 90%. The specificity (correctly identified tree pixels/total number of tree pixels) was 84% and the sensitivity (correctly identified non-tree pixels/total number of non-tree pixels) was 99%. The MR-CNN was found to correctly identify a high number of tree pixels while only missing a low number of tree pixels as shown in Table 2.

Table 2. MR-CNN tree crown classification statistical summary.

	Accuracy	Specificity	Sensitivity
South Placerville	79.5%	98.2%	70.5%
NW Placerville	91.5%	95.1%	88.3%
Edna Creek	99.3%	100.0%	98.8%
South Exp. Forest	89.1%	99.0%	82.2%
North Exp. Forest	89.2%	99.2%	82.6%
East Placerville	95.5%	100.0%	91.6%
West Placerville	92.7%	100.0%	87.3%
Belshazzar	93.4%	93.9%	92.9%
Newstart	82.9%	99.4%	74.8%
Grimes Creek	96.9%	99.6%	94.7%

3.2. Canopy Cover Analysis

Running an orthomosaic through the Snapper Tool, MR-CNN, and Density Tool gave a 30 m orthomosaic with canopy cover that also is coregistered with LANDFIRE data. Then the HDCC numbers were adjusted based on the bias of the MR-CNN. The values used to calculate the bias were the accuracy of correctly identifying tree pixels (sensitivity) and the odds of a false positive (specificity). These numbers ended up being 84% and 99%, respectively. Figure 4 shows the HDCC layer for South Placerville.

Figure 4. Values are in percent canopy cover.

3.3. Canopy Cover Comparison: Hyperspatial Derived vs. LANDFIRE

Ten orthomosaics where there was no fire were compared using the methods stated above. LANDFIRE was over reporting by an average of 2.5% over the HDCC layer. This agrees with what Scott [18] stated, that the "canopy cover values [for LANDFIRE] are too high." The standard deviation for the difference between each pixel was 14. This number is a little high and will hopefully be driven down by future improvements. Figure 5 shows a piece of the results for Edna Creek. White represents a higher canopy cover for the HDCC layer while white represents a higher canopy cover reported by LANDFIRE. The grey composing most of the figure is for when the difference is equal or close to zero.

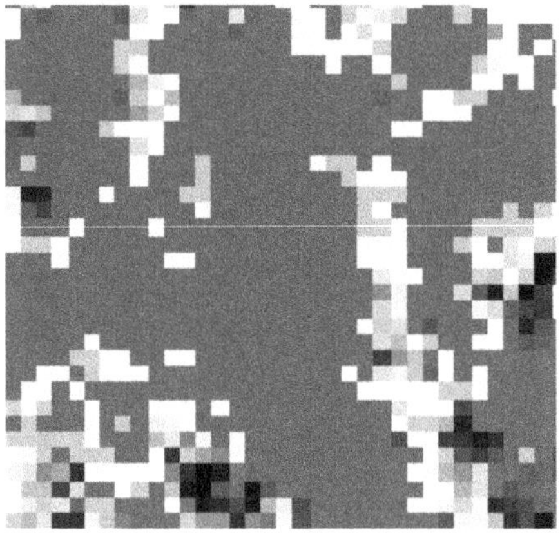

Figure 5. A raster representing the comparison between the LANDFIRE and HDCC data. White means LANDFIRE reported more canopy cover while black indicates the inverse.

Table 3 shows that the LFCC over reported for every orthomosaic.

Table 3. Mean difference and the standard deviation of that difference for each area.

	Mean	Standard Deviation
Belshazzar	17%	14
New Start	4%	13
Grimes Creek	6%	15
NW Placerville	1%	11
0530 Placerville	10%	9
Edna Creek	4%	1
South Experimental Forest	3%	12
North Experimental Forest	7%	12
East Placerville	7%	12
West Placerville	2%	13

3.4. Tree Mortality Mapping Assessment

The standard deviation and mean collected from the ten non-burnt orthomosaics show that there needs to be at least a change of 27% to know there has been a change with 95% confidence. If the difference is high enough to say there has been a change next the HDCC layer can be looked at to see whether it was a passive crown fire or an active crown fire. Pixels that did not contain adequate canopy reduction (more than 26%) from comparing pre-fire LFCC to post-fire HDCC are considered to be inconclusive. This would include unburned areas, surface fire, passive crown fire, and active crown fire. If an area had less than 27% canopy cover then we cannot say whether any canopy reduction occurred. This method was used on orthomosaics acquired over portions of the Mesa fire (a 16,000-hectare mixed severity fire located on the Payette National Forest in southern Idaho) to calculate burn intensity. This fire worked well as there were areas of active crown fire, passive crown fire, and surface fire. Figure 6 shows the Southwest area of the Mesa fire which had a lot of Active Crown fire, while Figure 7 the east section which had more passive crown fire and possible surface fire.

(a) (b)

Figure 6. (**a**) Hyperspatial data after a burn. (**b**) Types of fire: Red = Active, Yellow = Passive, Black = inconclusive crown fire activity.

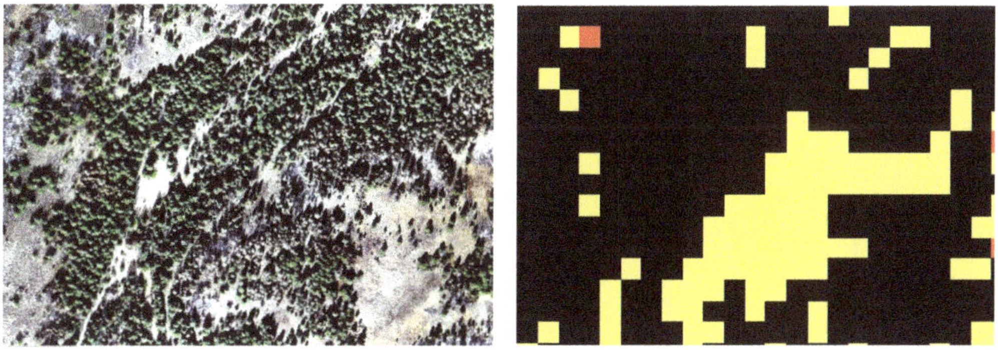

(a) (b)

Figure 7. (**a**) Hyperspatial data after a burn. (**b**) Types of fire: Red = Active, Yellow = Passive, Black = inconclusive crown fire activity.

4. Discussion

Post-fire mapping products are commonly used by land managers as they are developing post-fire management plans which specify strategies for mitigating the effects of the fire. In order for this data to be most useful, it needs to be readily available to managers. The advantage of using imagery acquired by sUAS is that it can be available as quickly as managers can get out to the site and conduct image acquisition flights. The timing of the flights can be tailored to optimize meteorological conditions that are conducive to acquiring image acquisition by enabling managers to acquire imagery as soon as the scene is not obscured by smoke and there is low enough cloud cover to ensure adequate illumination of the scene.

The MR-CNN was found to be very effective in identifying tree crowns. The classifier was found to have virtually no false positives with a specificity of 99%. The false negatives were still minimal, with a sensitivity of 84%.

Scott [18] showed that LANDFIRE Canopy Cover typically is over reported. As expected, when comparisons were made of study areas that had not experienced wildland fire, the HDCC was on average 2.5 percentage points lower than the LANDFIRE Canopy Cover. While the mean difference between the HDCC and LFCC was only a few percentage points, the higher standard deviation which typically ranged a bit more than 10 percentage points indicates that the canopy cover between the HDCC and LFCC products differed more significantly than indicated solely by the means.

The canopy reduction mapping over a burned area was very promising. When ran on the Mesa fire most pixels correctly identified surface fire as well as both active and passive crown fire. These methods work best on areas that had higher pre-fire canopy cover and canopy reduction. There needs to be at least a 27% difference for there to be a known change. Various theories to bring this number down are brought up in the future work Section 5.1.

In addition to higher accuracy mapping of canopy cover as well as tree mortality as measured by canopy reduction, the use of sUAS allows for much more current data acquisition than is possible with the LANDFIRE products. Using imagery acquired with a sUAS, it would be possible to have the HDCC product available within a day of flying a study area. By contrast, the current LANDFIRE 2.0.0 Remap effort is using Landsat imagery that was acquired in 2016. The initial product releases which covered the majority of the Contiguous US did not occur until 2019, meaning even the initial product releases were derived from imagery that was already three years out of date [8]. While canopy cover can be mapped with much higher accuracy using hyperspatial imagery acquired with a sUAS, this research team found that current technical and regulatory constraints on drone usage realistically will only allow for the acquisition of up to 600 hectares per day [16]. To effectively map large class F (>400 ha) and G (>5000 ha) fires [19], larger and more efficiently acquirable imagery satellite system still needs to be utilized to obtain a large enough analysis extent to map the whole fire. As the spatial resolution of satellite imagery increases with some of the new high-resolution sensors that are being put into orbit, it may be possible to leverage this methodology, applying it to new generations of imagery as they approach hyperspatial resolution.

5. Conclusions

Detection and mapping of trees from hyperspatial drone imagery enables more accurate mapping of canopy cover, a measure of forest density which is used to determine ecosystem departure from historic fire regimes, a primary measure of ecosystem resilience [20]. A comparison of a pre-fire LANDFIRE Canopy Cover product against post-fire HDCC resulted in a very effective measure of canopy fire reduction, which is a measure of tree mortality resulting as an effect of the wildland fire.

5.1. Future Work

While this effort was highly effective in establishing a timely, effective and cost efficient means of mapping canopy cover and tree mortality, there are many areas of future research which could result in an improvement in the effectiveness of this methodology.

An improvement in wildland fire extent mapping can be achieved by identifying unburned trees that are surrounded by surface vegetation. Current classification methods developed through this effort [1,3] which map burn extent are not capable of detecting that a surface fire burned underneath an unburned tree crown due to the view of the surface vegetation being obscured from the sensor on the sUAS by the tree crown. While it may not be possible to determine whether the biomass consumption under the tree crown is high or low due to the presence of white ash, it would be possible to identify that the surface vegetation was burned and update the burn extent and aerial mapping accordingly based on assumptions made based on whether surface vegetation visible next to the crown was burned.

While the specificity of the MR-CNN was quite high at 99%, the accuracy of the classifier could be improved by refining the training data for the MR-CNN, based on observations of false negatives during validation. This refinement would likely improve tree crown classification, reducing the number of false negatives, thereby increasing the classifier sensitivity, which was 84% in the validation dataset used for this effort.

An improvement in tree crown classification accuracy might be able to be achieved by using pixels' heights as additional input to the classifiers. Per-pixel height values were available from a digital surface model produced during the orthomosaic creation process but were not used as an input to the MR-CNN in this effort.

Due to the need to split each orthomosaic, the MR-CNN classifier can miss trees that are on the border of these splits since they look like half trees. This problem could be eliminated by modifying the snapping and density tool to run larger files than what they are currently limited to due to OpenCV's file size limits. This would allow entire orthomosaics to be processed at once.

A more major problem that arises with splitting up the image is when 30 m pixels overlap. This happens because the LANDFIRE and hyperspatial image cannot line up perfectly. Because of this the 5 cm data has an extra pixel on the bottom and right when it is clipped. The density tool creates a whole 30-m pixel from these few extra 5 cm pixels on the edges. When the 30-m rasters are recombined the data at the overlap is corrupted. This problem can be fixed either creating a more direct method to handle the extra pixel or by getting rid of the split. Getting rid of the split would require OpenCV to be able to handle larger orthomosaics. This problem is most likely a big factor in the high standard deviation. If fixed, the difference needed to be confident there has been a change should drop.

Author Contributions: Conceptualization, D.A.H.; Methodology, D.A.H., K. Brothers, S.D.J., J.W.; software, D.A.H., K.L.B., S.D.J., J.W.; validation, K.L.B.; formal analysis, J.C.; investigation, D.A.H., K.L.B., S.D.J.; data curation, D.A.H.; writing, D.A.H., J.C., K.L.B., S.D.J., J.W.; visualization, K.L.B.; supervision, D.A.H.; project administration, D.A.H.; funding acquisition, D.A.H. All authors have read and agreed to the published version of the manuscript.

Funding: This research was funded by the USDA Forest Service Boise National Forest, Forest Service Agreement No, 18-CS-11040200-0025.

Institutional Review Board Statement: Not applicable.

Informed Consent Statement: Not applicable.

Data Availability Statement: All data and source code from this project is available at the https://firemap.nnu.edu/tree-mortality.

Acknowledgments: We would like to acknowledge the students in the Northwest Nazarene University Department of Math and Computer Science who have helped with different aspects of this effort, including current and former students Nicholas Hamilton, Jonathan Hamilton, Braelyn Boerner, Gabriel Murphy, Enoch Levandovsky, Gabriel Johnson, Austin White, Aleesha Chavez and Andrew Welk. Additionally, we would like to acknowledge Northwest Nazarene University who funded the research efforts of many of the students mentioned.

Conflicts of Interest: The authors declare no conflict of interest.

Appendix A. Accuracy of Calculated Value for Canopy Cover

In this appendix, the accuracy of the value for canopy cover calculated from the hyperspatial orthomosaic is shown to be better than that calculated from LANDFIRE. This conclusion was stated in Section 2.2.2, but depends on mathematics beyond just routine statistical inference. The details are given here to justify the conclusion.

Setting and notation: Suppose that there is a 30 m pixel depicting a region of ground with canopy cover value $c \in [0,1]$. The actual value c would be unknown in practice. The same 30 m $\times$ 30 m region of ground is viewed as an $n \times n$ region of smaller pixels in the hyperspatial orthomosaic. In practice, n would be equal to 600, and the 600 $\times$ 600 orthomosaic would consist of 5 cm pixels, but the symbol n will be used for now for greater clarity.) The number of true "tree" pixels is n^2c, and the number of "non-tree" pixels is $n^2(1-c)$. Each of the n^2 pixels is assigned by the MR-CNN a binary value $A = 1$ (tree) or $A = 0$ (non-tree). Then the canopy cover is calculated from the hyperspatial orthomosaic as

$$\overline{A} = \frac{\sum A}{n^2}, \tag{A1}$$

where the sum is taken over all n^2 pixels.

The assignments are subject to error. Suppose the probability of correctly assigning $A = 1$ to a "tree" pixel (sensitivity) is known to be $p \in [0,1]$, and the probability of correctly assigning $A = 0$ to a "non-tree" pixel (specificity) is known to be $q \in [0,1]$.

Question: What level of accuracy can one expect for the value of canopy cover calculated from the $n \times n$ pixel region of the hyperspatial orthomosaic corresponding to a Landsat pixel? Precisely, what is the standard error of this value? Is this significantly smaller than what one might hope to obtain from LANDFIRE data?

Mean and standard deviation of calculated value: The expected value (mean) of the calculated value $\overline{A}$ is

$$E\left[\overline{A}\right] = cp + (1-c)(1-q) = 1 - q + c(p+q-1). \tag{A2}$$

The standard deviation is

$$\sqrt{\frac{n^2cp(1-p) + n^2(1-c)q(1-q)}{(n^2)^2}} = \frac{\sqrt{cp(1-p)+(1-c)q(1-q)}}{n} \tag{A3}$$

Adjustment of value for canopy cover: Recall that c is the proportion of small pixels representing ground that has canopy cover. It is the number wanted but not known. One solves for c in the expression for $E\left[\overline{A}\right]$ above to obtain

$$c = \frac{E\left[\overline{A}\right]+q-1}{p+q-1}, \tag{A4}$$

and thus

$$C = \frac{\overline{A}+q-1}{p+q-1} \tag{A5}$$

is an unbiased estimator of c. This will be called the "adjusted canopy cover". The standard error of C is just the standard deviation of $\overline{A}$ divided by $p+q-1$, which is

$$\frac{\sqrt{cp(1-p)+(1-c)q(1-q)}}{n(p+q-1)} \tag{A6}$$

Next, this expression will be examined more closely to obtain an upper bound.

Derivation of upper bound: The quantity $x(1-x)$ is decreasing for on the interval $[0.5, 1]$.

So that since $p > 0.70$ and $q > 0.95$,

$$p(1-p) < (0.70)(1-0.70) \text{ and } q(1-q) < (0.95)(1-0.95). \tag{A7}$$

It also follows that

$$(0.70)(1-0.70) > (0.95)(1-0.95), \quad \text{(A8)}$$

whereupon

$$c(0.70)(1-0.70) + (1-c)(0.95)(1-0.95) \quad \text{(A9)}$$

(which is a weighted mean of (0.70)(1 − 0.70) and (0.95)(1 − 0.95)) is an increasing function of c. Combining this fact with the first two inequalities, one obtains

$$\begin{aligned} &cp(1-p) + (1-c)q(1-q) \\ &< c(0.70)(1-0.70) + (1-c)(0.95)(1-0.95) \\ &< (0.80)(0.70)(1-0.70) + (1-0.80)(0.95)(1-0.95) \end{aligned} \quad \text{(A10)}$$

for

$$p > 0.70, \quad q > 0.95, \quad \text{and } c < 0.80. \quad \text{(A11)}$$

The actual canopy cover (c) in the training data used was less than 0.80, the sensitivity (p) was better than 0.70, and the specificity (q) better than 0.95. Using also $p+q-1 > 0.70+0.95-1$, one obtains

$$\frac{\sqrt{cp(1-p)+(1-c)q(1-q)}}{n(p+q-1)} < \frac{\sqrt{(0.80)(0.70)(1-0.70)+(1-0.80)0.95(1-0.95)}}{600(0.70+0.95-1)} \approx 0.001080, \quad \text{(A12)}$$

which compares favorably with results obtained with LANDFIRE 30 m pixels. If one assumes that $p > 0.80$ (which is true in most cases) rather than merely $p > 0.70$, the upper bound on the standard error of C becomes

$$\frac{\sqrt{(0.80)(0.80)(1-0.80)+(1-0.80)0.95(1-0.95)}}{600(0.80+0.95-1)} \approx 0.000824, \quad \text{(A13)}$$

which would be impossibly low to achieve with LANDFIRE 30 m pixels.

Conclusion: The adjusted value of canopy cover calculated from the hyperspatial orthomosaic has lower standard error than we could hope to obtain from LANDFIRE 30-m pixels.

References

1. Hamilton, D. Improving Mapping Accuracy of Wildland Fire Effects from Hyperspatial Imagery Using Machine Learning. Ph.D. Thesis, University of Idaho, Moscow, ID, USA, 2018.
2. Keeley, J.E. Fire intensity, fire severity and burn severity: A brief review and suggested usage. *Int. J. Wildl. Fire* **2009**, *18*, 116–126. [CrossRef]
3. Hamilton, D.; Pacheco, R.; Myers, B.; Peltzer, B. kNN vs. SVM: A Comparison of Algorithms. In *Proceedings of the Fire Continuum—Preparing for the Future of wildland Fire*; Missoula, MT, USA, 21–24 May 2018, p. 358. Available online: https://www.fs.usda.gov/treesearch/pubs/60581 (accessed on 1 December 2020).
4. Hamilton, D.; Hamilton, N.; Myers, B. Evaluation of Image Spatial Resolution for Machine Learning Mapping of Wildland Fire Effects. In *Proceedings of SAI Intelligent Systems Conference*; Springer: Cham, Switzerland, 2019; pp. 400–415.
5. Hamilton, D.; Myers, B.; Branham, J. Evaluation of Texture as an Input of Spatial Context for Machine Learning Mapping of Wildland Fire Effects. *Signal Image Process. Int. J.* **2017**, *8*, 146–158. [CrossRef]
6. Rothermel, R.C. *Predicting Behavior and Size of Crown Fires in the Northern Rocky Mountains*; U.S. Department of Agriculture, Forest Service, Intermountain Forest and Range Experiment Station: Ogden, UT, USA, 1991.
7. Cruz, M.G. Development and testing of models for predicting crown fire rate of spread in conifer forest stands. *Can. J. For. Res.* **2005**, *35*, 1626–1639. [CrossRef]
8. LANDFIRE. Landfire Program. 2020. Available online: https://landfire.gov/ (accessed on 25 November 2020).
9. Reeves, M.C.; Ryan, K.C.; Rollins, M.G.; Thompson, T.G. Spatial fuel data products of the LANDFIRE Project. *Int. J. Wildl. Fire* **2009**, *18*, 250. [CrossRef]
10. Hamilton, D.; Bowerman, M.; Collwel, J.; Donahoe, G.; Myers, B. A Spectroscopic Analysis for Mapping Wildland Fire Effects from Remotely Sensed Imagery. *J. Unmanned Veh. Syst.* **2017**, *5*, 146–158. [CrossRef] [PubMed]
11. Scott, J.H.; Reinhardt, E.D. Assessing crown fire potential by linking models of surface and crown fire behavior. In *USDA Forest Service Research Paper*; USDA: Fort Collins, CO, USA, 2001; p. 1.

12. He, K.; Gkioxari, G.; Dollar, P.; Girshick, R. Mask R-CNN. In Proceedings of the IEEE International Conference on Computer Vision, Venice, Italy, 22–29 October 2017.
13. Lin, T.Y.; Maire, M.; Belongie, S.; Hays, J.; Perona, P.; Ramanan, D.; Dollár, P.; Zitnick, C.L. Microsoft COCO: Common objects in context. In *European Conference on Computer Vision*; Springer: Cham, Switzerland, 2014; pp. 740–755.
14. Dutta, A.; Zisserman, A. *The VIA Annotation Software for Images, Audio and Video*; Springer: New York, NY, USA, 2019. [CrossRef]
15. Matese, A.; Toscano, P.; Di Gennaro, S.F.; Genesio, L.; Vaccari, F.P.; Primicerio, J.; Belli, C.; Zaldei, A.; Bianconi, R.; Gioli, B. Intercomparison of UAV, aircraft and satellite remote sensing platforms for precision viticulture. *Remote Sens.* **2015**, *7*, 2971–2990. [CrossRef]
16. Goodwin, J.; Hamilton, D. *Archaeological Imagery Acquisition and Mapping Analytics Development*; Boise National Forest: Boise, ID, USA, 2019.
17. Calkins, A.; Hamilton, D. *Archaeological Imagery Acquisition and Mapping Analytics Development*; Boise National Forest: Boise, ID, USA, 2018.
18. Scott, J. Review and Assessment of LANDFIRE Canopy Fuel Mapping Procedures. LANDFIRE, LANDFIRE Bulletin. 2008. Available online: https://landfire.cr.usgs.gov/documents/LANDFIRE_Canopyfuels_and_Seamlines_ReviewScott.pdf (accessed on 30 July 2020).
19. National Wildfire Coordinating Group (NWCG). Size Class of Fire. 2020. Available online: www.nwcg.gov/term/glossary/size-class-of-fire (accessed on 20 May 2020).
20. Barrett, S.; Havlina, D.; Jones, J.; Hann, W.; Frame, C.; Hamilton, D.; Schon, K.; Demeo, T.; Hutter, L.; Menakis, J.; et al. *Interagency Fire Regime Condition Class Guidebook*; Version 3.0; USDA Forest Service, US Department of Interior, and The Nature Conservancy: Rolla, MO, USA, 2010.

remote sensing

MDPI

Article

Mapping Burn Extent of Large Wildland Fires from Satellite Imagery Using Machine Learning Trained from Localized Hyperspatial Imagery

Dale Hamilton *, Enoch Levandovsky and Nicholas Hamilton

Department of Mathematics and Computer Science, Northwest Nazarene University, Nampa, ID 83686, USA; elevandovsky@nnu.edu (E.L.); nicholashamilton@nnu.edu (N.H.)
* Correspondence: dhamilton@nnu.edu; Tel.: +1-(208)-467-8769

Received: 28 October 2020; Accepted: 9 December 2020; Published: 15 December 2020

Abstract: Wildfires burn 4–10 million acres annually across the United States and wildland fire related damages and suppression costs have exceeded $13 billion for a single year. High-intensity wildfires contribute to post-fire erosion, degraded wildlife habitat, and loss of timber resources. Accurate and temporally adequate assessment of the effects of wildland fire on the environment is critical to improving the of wildland fire as a tool for restoring ecosystem resilience. Sensor miniaturization and small unmanned aircraft systems (sUAS) provide affordable, on-demand monitoring of wildland fire effects at a much finer spatial resolution than is possible with satellite imagery. The use of sUAS would allow researchers to obtain data with more detail at a much lower initial cost. Unfortunately, current regulatory and technical constraints prohibit the acquisition of imagery using sUAS for the entire extent of large fires. This research examined the use of sUAS imagery to train and validate burn severity and extent mapping of large wildland fires from various satellite images. Despite the lower resolution of the satellite image, the research utilized the advantages of satellite imagery such as global coverage, low cost, temporal stability, and spectral extent while leveraging the higher resolution of hyperspatial sUAS imagery for training and validating the mapping analytics.

Keywords: support vector machine; fuzzy logic; wildland fire extent; wildland fire severity; small unmanned aircraft systems; landsat

---

1. Introduction

This study examines the use of sub-decimeter hyperspatial imagery acquired with a small unmanned aircraft system (sUAS) to train machine learning algorithms to map wildland fire severity and extent from Landsat imagery. This effort increases the accuracy with which severity and extent can be mapped using hyperspatial imagery beyond the study area extent constraints observed due to the current technical and regulatory limitations resulting from the spatial extent of imagery that can be acquired using sUAS. Extending the usefulness of hyperspatial sUAS imagery beyond current flight extent restrictions will provide managers with increased actionable knowledge, leading to improved management decisions and increased ecosystem resilience.

A century of fire suppression and fire prevention communities has led to the current departure of wildlands from fire return intervals typically experienced under pre-European settlement conditions. Wildlands in the western United States (US) are experiencing a much higher incidence of catastrophic fires [1]. Millions of hectares of western United States wildlands are impacted by wildland fire annually, with wildlands burned in some fire seasons exceeding four million hectares [2], with suppression costs exceeding three billion dollars annually [3]. High-intensity wildland fires contribute to post-fire erosion, degraded wildlife habitat, and loss of timber resources. This loss results in negative impacts on ecosystem resilience as well as communities in the wildland-urban interface where 25,790 structures were burned in 2018 [2]. Additionally, wildland fires across the US claim more lives than any other type of natural disaster, resulting in the average loss of twenty wildland firefighters per year [4,5]. The 2018 Camp Fire in northern California alone resulted in 85 fatalities with estimates of insured losses running approximately ten billion dollars [6]. Effective management of wildland fire is a critical dimension of maintaining healthy and sustainable wildlands. Actionable knowledge of the relationships between fuel, fire behavior, and the effects on the ecosystem and human development can help land managers develop elegant solutions to wildfire problems. Remotely sensed imagery is commonly relied on in assessing the impact of fire on the ecosystem [7]. The knowledge gained from remotely sensed data enables land managers to better understand the effects fire has had on the landscape and develop a more effective management response facilitating ecosystem recovery and resiliency.

1.1. Background

Although researchers face many challenges in understanding fire behavior and its corresponding effect, we investigate the improvements of what changing data resolution creates when analyzing wildland fires. The improvement with which wildland fire severity and extent can be mapped from medium resolution satellite imagery using machine learning trained from hyperspatial sUAS imagery relied on previously published efforts, including identification of the ecological factors of interest, identification of current methods of mapping wildland fire extent, and utilization of sUAS as a remote sensing vehicle.

1.1.1. Wildland Fire Severity and Extent

The term "wildland fire severity" can refer to many different effects observed through a fire cycle, from determining how intense an active fire had burned to the ecosystem's response to the fire over the subsequent years. This study investigates the direct or immediate effects of a fire, such as biomass consumption, as observed in the days and weeks after the fire is contained [8]. Therefore, this study defines burn severity as the measurement of biomass (or fuel) consumption [9].

"Wildland fire extent" refers to the area in which a wildland fire has consumed organic material. Identification of burned area extent within an image can be achieved by exploiting the spectral separability between burned organic material and unburned vegetation [10,11]. Patchiness examines the fire's spatial completeness within the extent of the fire, examining how much biomass remains unburned within the fire perimeter [12]. This study defines burn extent as the area within which a fire has consumed organic materials, omitting the unburned islands of vegetation within the burned area perimeter.

Vector-based digitization of fire perimeters: Burn extent is often recorded as a polygon geospatial feature, recorded by a global navigation satellite system (GNSS) either from a helicopter flying along the perimeter of a fire or by ground-based mapping of the fire-edge with a handheld GNSS receiver. For the pilot, the burned area edge can be difficult to discern, especially when high vegetation canopy closure and shadows reduce visibility of surface vegetation from above. For ground-based personnel mapping the burned area either on foot or from a motorized vehicle, following the perimeter can be complicated by both rough terrain and the non-uniform manner in which wildland fire burns across the landscape. Most wildland fires contain islands of unburned vegetation dispersed throughout the burned area, ranging in size up to hundreds of hectares. These unburned islands are typically not mapped due to safety concerns as well as the impracticality of traversing the edge of each of these islands either aerially or on the ground. Even an extinguished fire can be a dangerous environment for humans to work in. Structurally unsound trees are known to fall upon and injure or kill unsuspecting people. Additionally, the patchy, convoluted edge of the fire can often be hard to delineate from the ground, let alone from a manned aircraft [13].

Raster-based mapping of burn extent and severity: The most common metric used for mapping wildland burn severity from medium satellite imagery is the normalized burn ratio (NBR) which is the normalized difference between the near-infrared (NIR) and shortwave infrared (SWIR) bands [9], calculated as:

$$\text{NBR} = \frac{(\text{SWIR} - \text{NIR})}{(\text{SWIR} + \text{NIR})} \quad (1)$$

While NBR is effectively used for burn severity mapping, differenced NBR (dNBR), calculated as pre-fire NBR (NBR_{pre}) minus post-fire NBR (NBR_{post}) [7], has been found to have a stronger indicator to burn severity than just NBR_{post} alone [14]. Satellite imagery is a useful dataset from which to calculate dNBR due to the ability to obtain pre-burn imagery corresponding to a study area containing an unplanned wildland fire.

The Landsat program has been provided continuous satellite coverage of the earth at 30 m resolution since 1982, providing 38 years of imagery from which to extract fire history data. The US Departments of Agriculture and Interior maintain the Monitoring Trends in Burn Severity (MTBS) program to map burn perimeters and severity across the US starting in 1984. The MTBS program maps wildland fires across the US using dNBR [7]. Due to a large number of fires across the US, the MTBS program only maps fires exceeding 400 hectares in the western US and exceeding 200 hectares in the eastern US [15]. With the offset in the orbits of Landsat 7 and 8, Landsat's 8-day flyover interval, easily allows analysts to access bi-temporal imagery preceding the fire for NBRpre and post-fire for NBRpost. In the event the fire study area was obscured by clouds or smoke during a pass-over, another scene can be used from either the preceding pass-over for NBRpre or in a succeeding pass-over for NBRpost as necessary. While MTBS is extensively used for mapping large fires across the US, it has been shown to overestimate the burn extent by four to sixteen percent due to oversimplification of burned area polygons and not mapping large unburned islands [16].

1.1.2. On Origins of Fires Used in This Study

As for the data used in this research, this study utilized a set of fire data from previous similar research efforts. The wildfire burn areas used in this study are a collection of medium to large size fires in southwestern Idaho ranging from xeric sagebrush steppe fires located within the United States Department of Interior Bureau of Land Management Boise District to the mesic upper Payette and Boise River watersheds in the United States Department of Agriculture Forest Service Boise National Forest. The naming schema for these fires, as displayed in Table 1, are usually determined by a nearby notable landmark such as a creek or a hill. For example, the Hoodoo fire burn area burned near and across the Hoodoo Creek near Idaho City, Idaho. Typically, these names were the official names of these fires assigned by the local jurisdiction and used by dispatchers. The locations of the burned areas of which acquisition flights were conducted are shown in Figure 1.

1.1.3. Utilization of sUAS for Mapping Wildland Fire

The proliferation of small unmanned aircraft system technology has made the procurement and use of remotely sensed imagery a viable possibility for many organizations that could not afford to obtain such data in the past. A small unmanned aircraft system (sUAS) is a designation given by the US Federal Aviation Administration to UAS that weigh between 0.25 kg and 25 kg [17]. Most commercially available sUAS come with an onboard digital camera, a multi-spectral sensor with three bands capturing visible light in the blue, green and, red spectrum ranging from 400 nm to 700 nm [18]. Spectral responses in the visible spectra can be used to differentiate between different image features such as white and black ash [10,11,19] and other features of interest to fire managers such as vegetation type [10,20].

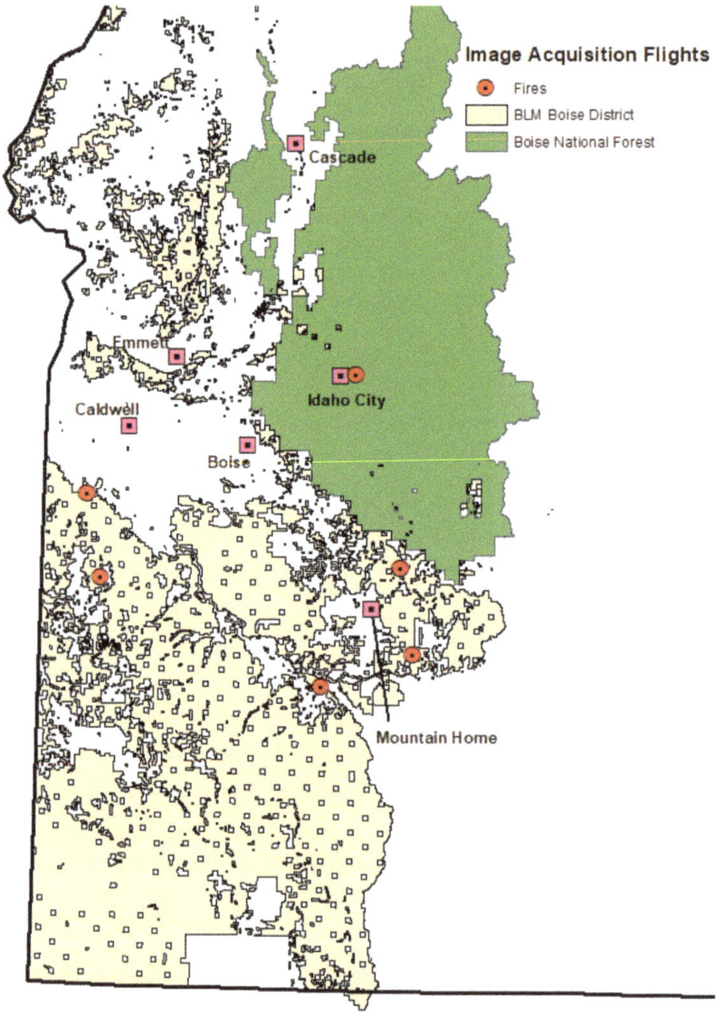

Figure 1. Geographic distribution of wildland fires over which post-suppression image acquisition flights were conducted in southwestern Idaho.

New advances in sUAS capabilities enable imagery acquisition with a spatial resolution of centimeters and temporal resolution of minutes [21]. The temporal responsiveness of acquiring imagery with a sUAS is significantly increased due to the increased availability of the sUAS being able to be flown at any desired time as opposed to Landsat imagery which can only be acquired when the satellite flies over the scene every 16 days, assuming the scene is not obscured by smoke or clouds during the flyover. However, although a manned aircraft can be a viable alternative, United States governmental agencies update their aerial manned photography through programs such as the Department of Agriculture National Agricultural Imagery Program every few years. However, this is nowhere near the update frequency of satellite imagery as needed in this experiment. In comparison to sUAS, on-demand manned aircraft aerial photography is far more expensive, with rental costs for manned aircraft running thousands of dollars per hour. High-resolution imagery was only needed on-demand and therefore sUAS imagery was the best temporally responsive and cost-effective solution to the US land management agency regulatory need to acquire post-fire data including mapping burn extent and fire effects within 14 days after fire containment.

Aerial imagery for this project was acquired with a DJI Phantom 4 with a 12-megapixel color camera. The imagery acquired with the sUAS was taken while flying at an altitude of 120 m above ground level (AGL), giving the photos a spatial resolution of 5 cm per pixel [9]. Objects that are wider than that pixel resolution will be discernible in the acquired hyperspatial imagery as shown in Figure 2a. The black rectangles in the image are burned areas. Small lines and patches of white within the burned area are white ash from sagebrush which was fully combusted by the fire. The unburned vegetation consists primarily of annual and perennial grasses and forbs, Wyoming big sagebrush (*Artemisia tridentata* spp. *Wyomingensis*), and yellow rabbitbrush (*Chrysothamnus viscidiflorus*). The scene shown in Figure 2 contains two western juniper trees (*Juniperus occidentalis* spp. *occidentalis*).

Features easily identified in hyperspatial (5 cm) imagery are lost in medium resolution (30 m) Landsat satellite imagery, being aggregated into more dominant neighboring features. Figure 2b shows the same scene as the preceding image but resampled to 30 m spatial resolution having 48 pixels aligned in six rows by eight columns. Mapping burn severity and extent was found to have significantly lower accuracy when using imagery with 30 m spatial resolution such as Landsat than was found with hyperspatial imagery acquired by flying the sUAS over the burned area [19]. While burn severity and extent can be mapped with much higher accuracy using hyperspatial imagery acquired with a sUAS, this research team found that current technical and regulatory constraints on drone usage realistically will only allow for the acquisition of up to 600 hectares per day [23]. That same flight extent would be a small part of a single Landsat scene. To effectively map large class F (>400 ha) and G (>5000 ha) fires [24], a larger and more efficiently acquirable imagery satellite system such as Landsat still needs to be utilized to obtain a large enough analysis extent to map the whole fire.

(a)

Figure 2. *Cont.*

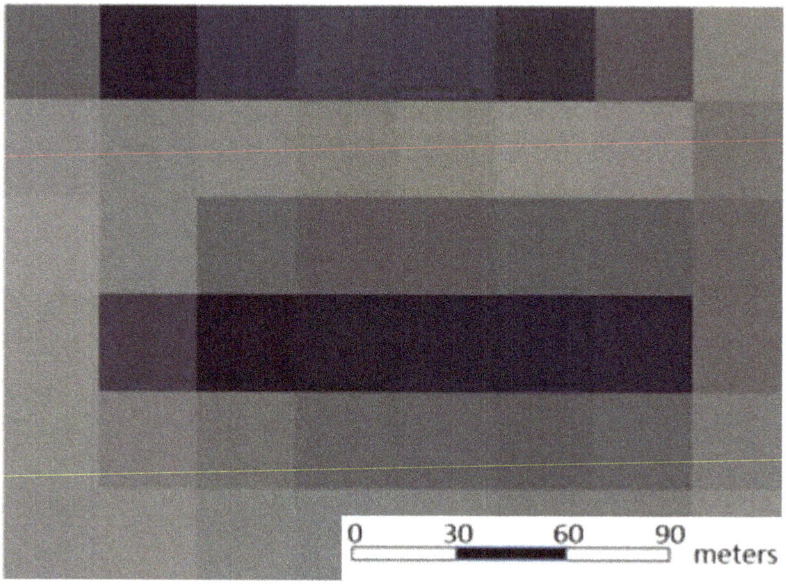

(**b**)

Figure 2. (**a**) Image of a rangeland study area acquired with a Phantom 4 sUAS flying at 120 m AGL with a spatial resolution of 6 cm per pixel. (**b**) Same scene resampled to 30 m resolution with six rows and eight columns of pixels [22].

1.1.4. Thirty Meter Burn Severity Mapping Analytics Trained with Hyperspatial Classification

In evaluating the effects of spatial resolution on burn severity and extent mapping accuracy, Hamilton [19] established the following methodology which enabled the use of hyperspatial burn severity classes of unburned vegetation, black ash and white ash as classified using a Support Vector Machine (SVM) [22] for training machine learning algorithms to map burn severity from medium resolution (30 m) imagery. This previous effort evaluated only the effect of spatial resolution on mapping accuracy, using 30 m imagery that was resampled from the hyperspatial imagery acquired with the sUAS, thereby removing other variables from consideration which could affect accuracy, such as sensor radiometric resolution, atmospheric influence, and temporal resolution. Machine learning classifiers mapped burn severity and extent from 30 m imagery separately, and were trained using hyperspatial burn severity and extent classifications using the following methods:

1. Hyperspatial orthomosaics (5 cm) were resampled to have medium resolution of 30 m (30 m), which is an equivalent spatial resolution to Landsat imagery, with the spatial reference for the 30 m imagery being calculated from the spatial reference from the 5 cm orthomosaic.
2. Labeling 30 m training pixels using fuzzy logic by:
    a. Calculating 5 cm burn severity class pixel density within each 30 m pixel, where density is the percentage of 5 cm pixels for a specific class that are found within the containing 30 m pixel.
    b. Applying fuzzification, the post-fire class density is used to establish where fuzzy set membership transitions from 0 to 1 over a range of values. Burn extent transitioned from unburned to burned between 35 and 65 percent of 5 cm pixels being unburned as shown in Figure 3. Burn severity transitioned between low biomass consumption and high biomass consumption between 33% and 50% [25] of burned pixels being classified as white ash as shown in Figure 4.

c. Activate fuzzy rules by applying fuzzy logic. When evaluating a series of fuzzy if statements, as shown in Figure 5, the data is defuzzified by selecting for activation the action associated with the if expression that has the highest value where a fuzzy AND is evaluated as taking the minimum value of either of the associated operands.

3. Each of the 30 m pixels was labeled with training data labels with the SVM training on 70 percent of the 30 m training pixels. The remaining labeled 30 m pixels withheld for validation of the SVM.

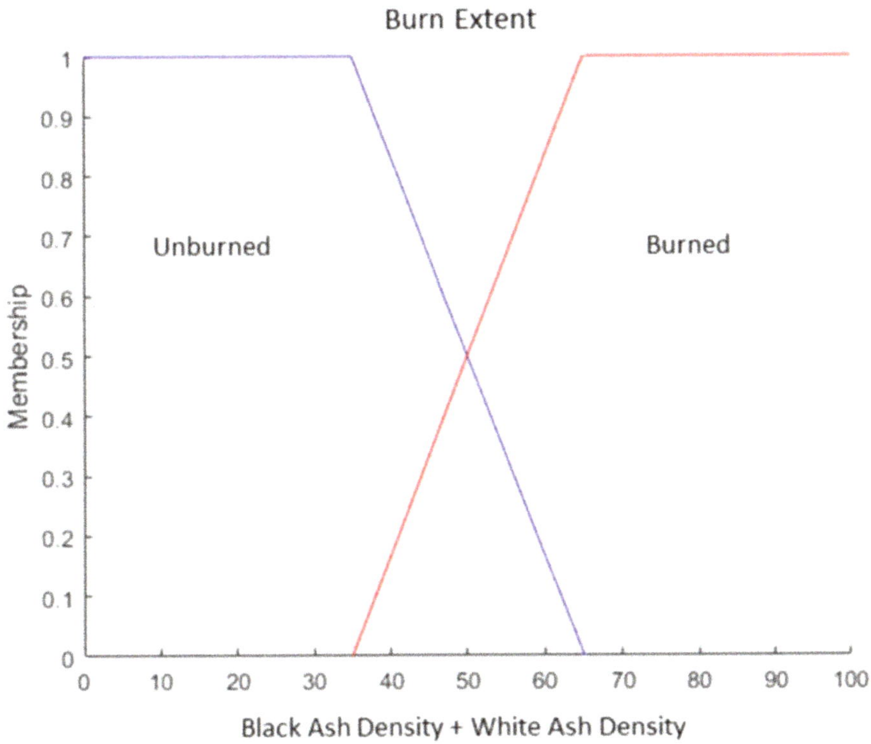

Figure 3. Burn extent fuzzy set showing fuzzy set membership of 30 m pixel transition between Unburned and Burned from 35 to 65 percent of 5 cm pixels being classified as burned.

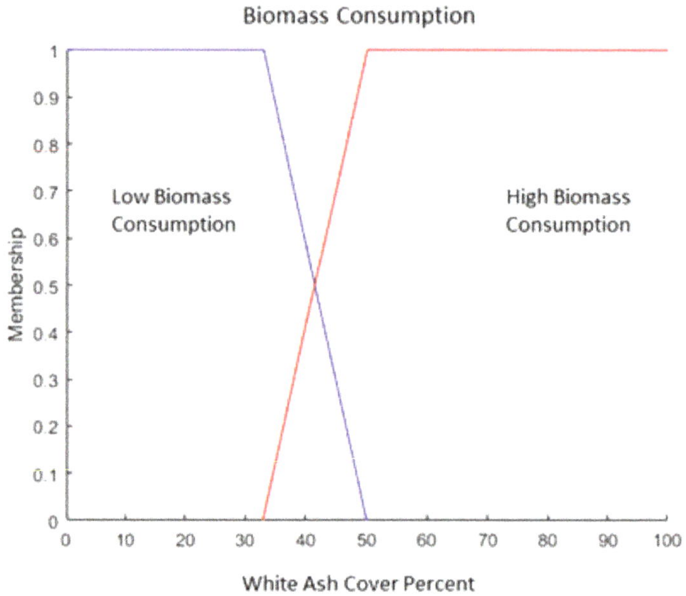

Figure 4. Biomass Consumption fuzzy set showing fuzzy set membership of 30 m pixel transition between Low and High Biomass Consumption from 33 to 50 percent of 5 cm burned pixels being classified as white ash.

If (extent:burned AND combust:high)
    training pixel = White Ash
If (extent:burned AND combust:low)
    training pixel = Black Ash
If (extent:unburned)
    training pixel = Unburned

Figure 5. Fuzzy logic algorithm for labeling 30 m pixels from 5 cm burn severity classes.

2. Materials and Methods

In order to complete the goal of this effort, a method was developed which mapped and analyzed wildland fire extent using spatial satellite imagery. This was done by converting high-resolution hyperspatial training data to a lower resolution training data using fuzzy logic. This created satellite resolution training data and enabled the burn extent to be determined and analyzed from satellite imagery using an SVM. Although this experiment will be using Landsat 8 imagery in particular, this method is not bound to any specific earth-observing satellite and can be applied to any spatial imagery provided the correct format. The fires used in the research were from the same set of fires used from a previous study [19,22] as mentioned in Section 1.1.2.

Hamilton [19] mapped wildland fire burn-extent from hyperspatial data with an accuracy of up to 98% on 5 cm pixels. For this experiment, hyperspatial training data will be assumed to be an adequate reference point when determining the accuracy of the SVM on the 30 m spatial imagery. This assumption enables the experiment to use hyperspatial training data resampled to satellite resolution training data using fuzzy logic. This experiment also investigated the effect different spectral bands have on burn extent and severity mapping accuracy.

2.1. Assembling Spatial Extent Burn Indicator

As mentioned before, one of the objectives of the application of this experiment was to provide an efficient and a more consistent method of running the experiment so that the results of the experiment were consistent and adequate to analyze using the scientific method. Thus, an application was assembled to evaluate the data in a consistent manner. The application for this experiment is divided into four sections. The first part of the application adjusts the geoposition alignment of the hyperspatial and satellite imagery. The second part utilizes an SVM to create hyperspatial training data from hyperspatial imagery. The third part converts hyperspatial training data into satellite resolution spatial training data using fuzzy logic. The final part of the application runs the SVM on the satellite scene using the newly created training data. This application will provide an effective and efficient way to analyze accuracy impact.

2.1.1. Coregistering Hyperspatial and Medium Resolution Images

Before resampling the resolution of the training image, acquired with an sUAS, from hyperspatial resolution to satellite medium resolution, such as is acquired by the Landsat Satellite project, preprocessing and imagery alignment is required to establish alignment with the smaller pixels and the larger medium resolution image pixels. The first alignment step reprojects the imagery so the hyperspatial image and the medium resolution image have the same spatial reference. In Figure 6, the projection for the medium resolution image (orange) and the hyperspatial sUAS image (blue) have different spatial references, making the pixel boundaries misaligned between the two images.

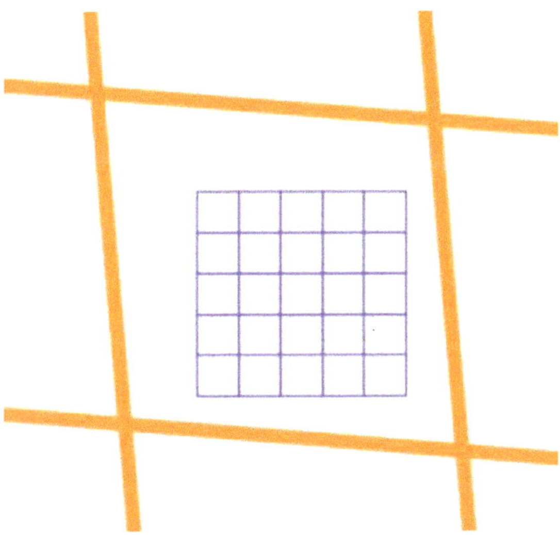

Figure 6. Medium resolution satellite image pixels (orange) and hyperspatial sUAS image pixels (blue) prior to reprojection.

Reprojection to the same spatial reference results in pixel boundaries that are aligned between the images as shown in Figure 7.

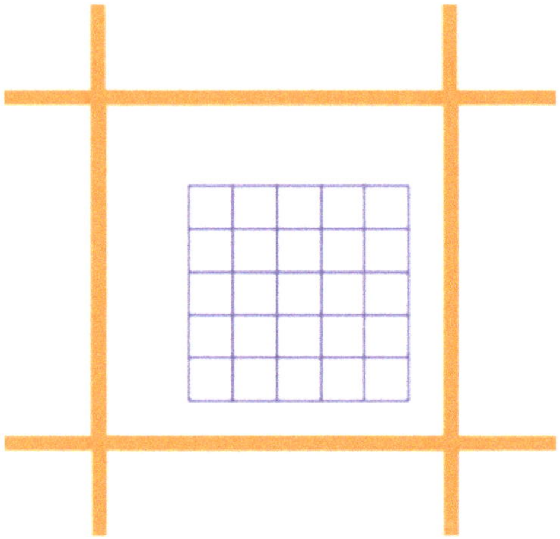

Figure 7. Example of projection before the border is added.

Once the pixel boundaries are aligned between the hyperspatial and medium resolution images, the medium resolution image is cropped so it only contains pixels that overlap the hyperspatial image. At that point, the extent of the hyperspatial image does not correspond to pixel boundaries, as is shown by the white space between the blue pixels and the boundary of the overlapping orange pixel in Figure 6. The extent of the hyperspatial image needs to be expanded out to the edge of the clipped medium resolution image. Additional rows and columns of null pixels are added around the hyperspatial image using OpenCV, expanding the extent of the hyperspatial image until it corresponds to the pixel outer pixel boundaries of the overlapping medium resolution pixels, which are shown in green in Figure 8.

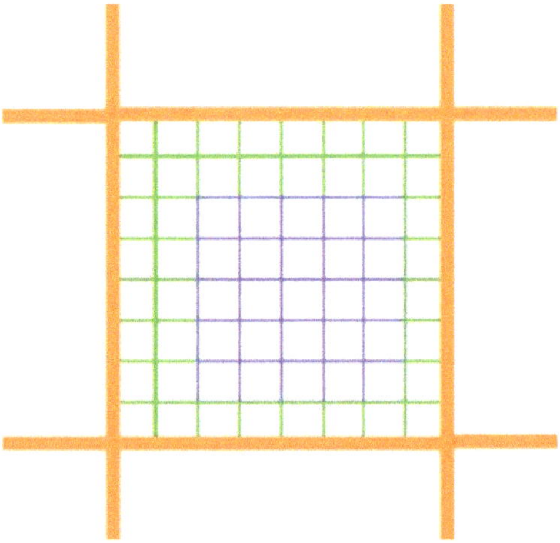

Figure 8. Example of projection after the border is added.

Once the additional rows and columns of null pixels are added to the hyperspatial image, the spatial resolution of the hyperspatial image is adjusted to account for the addition of the new null pixels around the perimeter of the orthomosaic as shown in Figure 9.

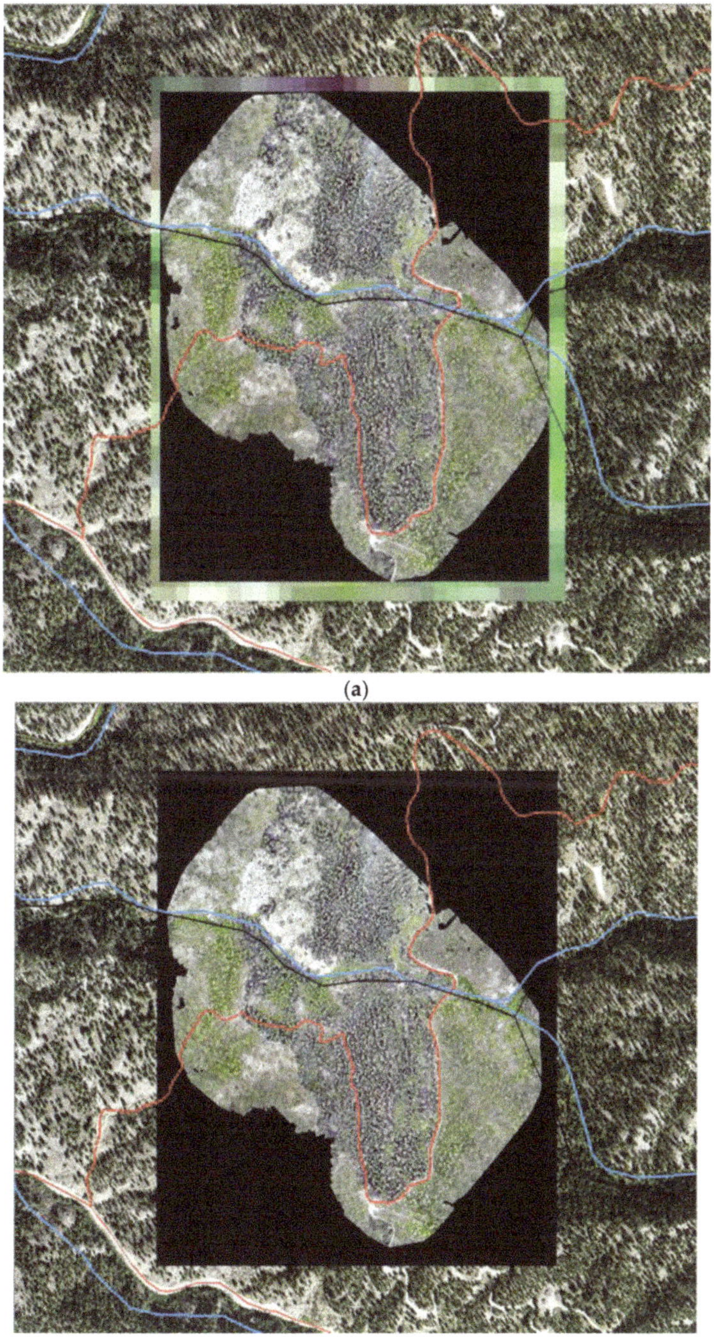

(a)

(b)

Figure 9. (**a**) sUAS orthomosaic rendered on top of clipped Landsat image; (**b**) sUAS orthomosaic with a buffer of null pixels giving it the same extent as the clipped Landsat image. Red polylines represent roads and trails. Blue polylines represent creeks. Black hatched polyline represents a historic railgrade.

2.1.2. Creating Hyperspatial Training Data

Once all the spatial imagery was aligned, the hyperspatial imagery burn extent and severity were mapped separately using a support vector machine. The training data used for the hyperspatial imagery was from a data set that had previously been processed while mapping burned areas over multiple years, with accuracy ranging up to 98% [19,21]. After burn severity and extent were mapped, the data was denoised using image morphology, reducing sub-object sized noise [26]. An example of this noise reduction on the burn extent is shown in Figure 10. The output would then be used as hyperspatial burn extant data resampled to satellite resolution burn extent data used as training data in the next section. A similar process was done with the severity case and the images were combined using the logic mentioned in Section 1.1.4.

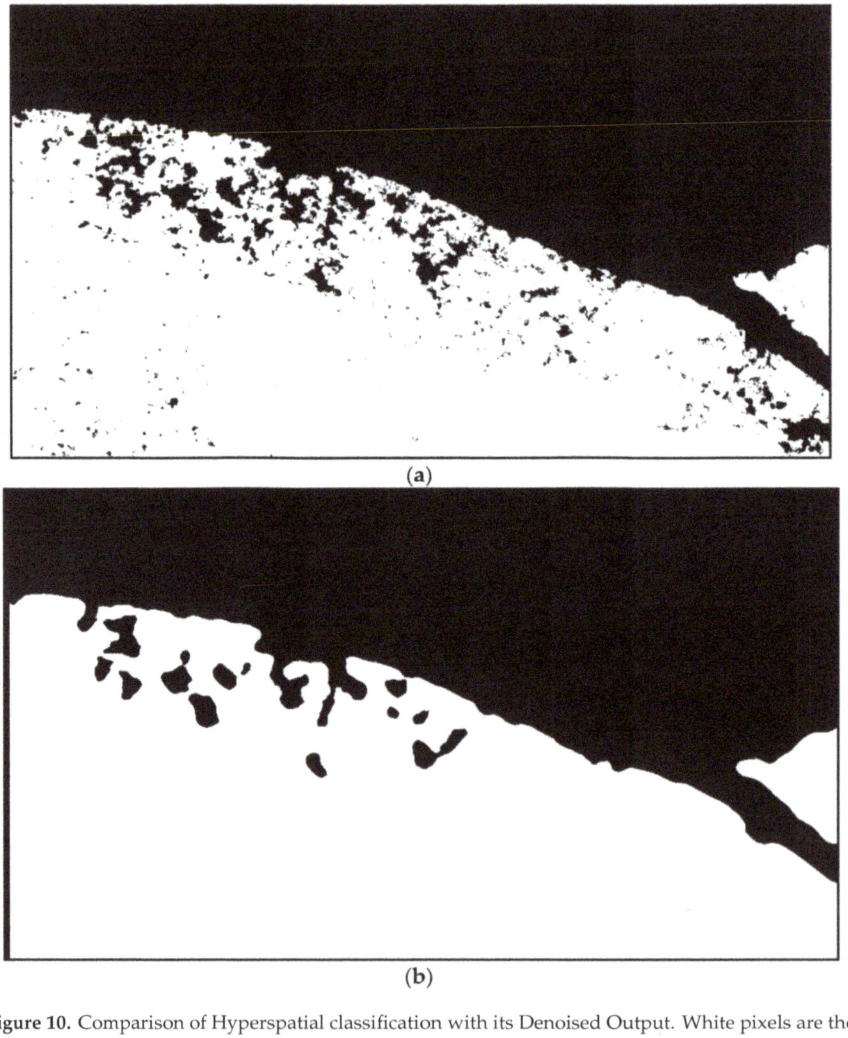

(a)

(b)

Figure 10. Comparison of Hyperspatial classification with its Denoised Output. White pixels are the burned data and the black pixels are the unburned data. (**a**) example of an image before denoise tool is applied; (**b**) example of an image after denoise tool is applied. These images is a post processing view of the bitmap output data with each pixel containing one class value Burned or Unburned.

2.1.3. Resampling Training Data to 30 m Using Fuzzy Logic

After generating the hyperspatial burn severity output, fuzzy control was used to label overlapping medium resolution pixels to be used for training an SVM to map burns from satellite imagery. Fuzzy logic allows decisions to be based on imprecise boundaries rather than relying on precise boundaries that are used by Boolean logic. This use of vagueness allows the expression of how much the data fits given criteria, transitioning from one class to another over a range of values. Fuzzy logic is often more applicable to ecological data than the crisp delineations resulting from Boolean logic, where data will transition from one class to another at a single threshold value.

The fuzzy set theory allows the specification of how well an object satisfies a vague criterion [27] with fuzzy logic providing a means for specifying that the transition from one class to another is not demarked at a single value but transitions from one class to another over a range of values [28]. For example, Hamilton [19] defined the transition from low to high biomass consumption as occurring between 33 and 50 percent of burned pixels being classified as white ash as shown in Figure 4. Rather than set membership being expressed as either zero or one, as is the case with Boolean logic, fuzzy logic allows set membership to be specified as a range of membership from 0.0 to 1.0. For example, using these thresholds, a plot with 40 percent white ash cover from the Biomass consumption fuzzy sets shown in Figure 3 would have 0.41 membership in the high biomass consumption set and 0.59 membership in the low biomass consumption set as shown in Figure 11.

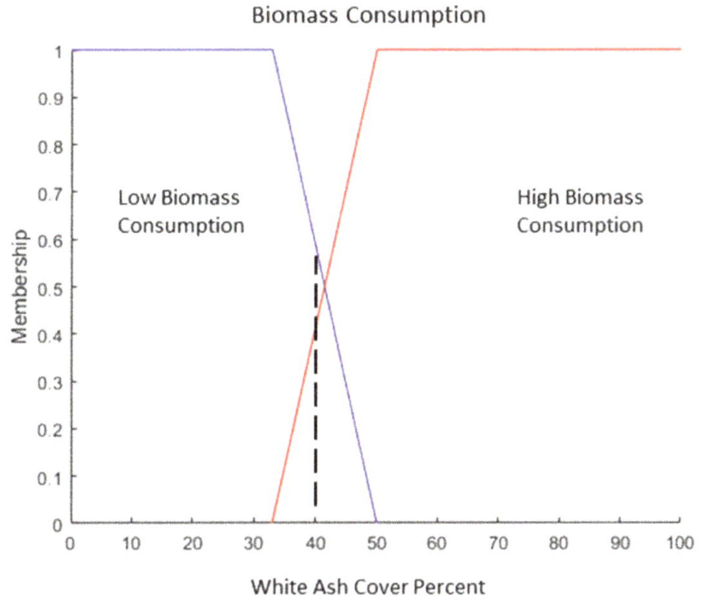

Figure 11. Biomass Consumption Fuzzy Set Membership. With 40 percent White Ash Cover denoted by the dotted line, there is 0.41 membership in High Biomass Consumption and 0.59 membership in Low Biomass Consumption.

Assignment of the post-fire burn extent and severity classes to a 30 m pixel is based on a combination of whether the pixel burned and white ash cover within the pixel. In neither case is Boolean logic appropriate for determining set membership. Increasing the density of a class by a handful of 5 cm pixels and as a result changing a Boolean expression from false to true does not adequately describe the state of the pixel. As mentioned in the background, the first set of data to be evaluated was between the burned and unburned sets, with the burned set consisting of the combination of pixels classified as either burned or unburned. The fuzzy logic transition for these criteria from 5 cm pixels to 30 m was from 35% to 65% [19], as previously shown in Figure 3.

An additional set membership was used that measured biomass consumption, which evaluated the relationship between the density of black ash and white ash. The white ash cover was expressed as a percentage of white ash to burned (white and black ash) pixels, transitioning between low and high biomass consumption over a transition from 33% to 50% [19,23], as previously shown in Figure 4.

Fire biomass consumption needed to be only evaluated on the burn extent, and thus any white ash detection outside of the burned boundaries was disregarded. This filter could be applied using fuzzy logic to fuzzy set membership, a fuzzy AND is expressed as taking the minimum value of the expressions on either side of the AND operator. Likewise, a fuzzy OR is expressed as taking the maximum value of the expressions on either side of the OR operator. When evaluating a series of fuzzy IF statements, as previously shown in Figure 5, the data are defuzzified by selecting for activation the action associated with the IF expression with the highest value.

After the fuzzy logic was applied, the data would be converted into 30 m training data by taking the most dominant set as determined by the fuzzy logic control and was labeled either white ash, black ash, or unburned accordingly. Since the snapping software aligned all the images, the newly created training data from the fuzzy logic training data came out aligned with the associated satellite scene and would be ready to train an SVM using Landsat imagery, which had been labeled using fuzzy logic on the hyperspatial burn extent and severity classifications as described above. After the training data were created, one third of the training data were set aside as validation data to test the classifier's accuracy.

2.1.4. Running the SVM on the Satellite Scene

After the satellite scene training data was created, the satellite scene and the newly created satellite resolution training data was used as the input into the SVM. However, before running the SVM the spatial training data was clipped to balance out the burned and unburned pixels for consistency across fires. This clipping was necessary to balance for the ratio between burned and unburned pixels is normalized between 45–55%. Balancing the burned and unburned pixels was done because most of the selected fires used in this experiment had an imbalance burned and unburned pixel ratio where the unburned pixels numerically dominated the burned pixels. Thus, a solution was developed through an algorithm that repeatedly stripped off each side until the ratio was met. This solution would be sufficient to the imbalance issue because most of the edge pixels were predominantly burned pixels in all the fires that were used. After preprocessing was complete, the SVM would create a TIF image of the fire's burn extent and severity through 30-m resolution.

2.1.5. Collecting Results

Once the SVM finished classifying, an output image of the entire satellite scene was produced. A simple testing program was developed to analyze the SVM's accuracy when it mapped burn extent and severity on a satellite. Once the data was normalized, all the pixels were compared with the training data, and the equation shown in Equation (2) was applied to determine the accuracy of the SVM's burn extent.

$$ExtentAccuracy = \frac{|TruePixels|}{|TotalPixels|} \tag{2}$$

However, to determine burn severity accuracy, only the white ash and black ash pixels were evaluated. Thus, all unburned pixels were omitted, and the accuracy calculation applied accuracy validity to white and black ash pixels, as shown in Equation (3).

$$SeverityAccuracy = \frac{|TrueBlackAsh| + |TrueWhiteAsh|}{|TotalBurnedPixels|} \tag{3}$$

3. Results

A set of fires previously used by Hamilton in [19,22] was selected for analysis. The estimated accuracy of mapping out burn severity using color satellite imagery (comprised of red, green, and blue spectral bands) in comparison to the high-resolution results from color imagery, as shown by Hamilton [19], is shown in Table 1.

Table 1. Experiment Results Compared with the 5 cm from [19,22].

	5 cm Extent	30 m Extent Results	5 cm Severity	30 m Severity Results
Elephant	98.5	58.52	98.67	95.30
MM106	86.58	71.5	98.68	100
Hoodoo	99.29	73.92	95.22	41.50
Immigrant	98.34	77.67	97.97	100
Jack	94.74	57.09	96.97	19.06
Owyhee	87.65	85.40	87.42	100
Mean	93.32	73.116	95.252	72.112

This experiment also investigated how the accuracy would change if color satellite imagery were replaced for a set of infrared (IR) bands. The IR imagery comprised the following Landsat bands: shortwave infrared 1 (0.156–1.660 μm), shortwave infrared 2 (2.100–2.300 μm), and near infrared (0.845–0.885μm) [29]. The results of this experiment are shown in Table 2.

Table 2. Color Experiment vs. IR Experiment.

	30 m Color Extent	30 m IR Extent	30 m Color Severity	30 m IR Severity
Elephant	58.52	55.95	95.30	97.83
MM106	71.5	69.1	100	100
Hoodoo	73.92	71.75	41.50	36.48
Immigrant	77.67	78.19	100	100
Jack	57.09	56.68	19.06	17.40
Owyhee	85.40	56.61	100	N/A [1]
Mean	70.68	64.71	75.98	70.34

[1] The fire was too small for the experiment to detect any white ash.

4. Discussion

This experiment's results did not converge to a specific accuracy, but instead the accuracy varied across fires. This was expected as many of the fires used in the experiment varied in severity, differed between forest and grass fire, had different atmospherical conditions, and varied in canopy cover. For example, the Elephant fire had a thin layer of cloud haze above it, which skewed off the color balance to be higher, which decreased the SVM's accuracy of detected darker burned pixels. Although each image may have one or multiple of these challenges, these problems were addressed through many various methods during the development of the experiment. However, all could not be met seamlessly. One of these challenges, such as the canopy cover, was addressed by applying the denoise tool, as previously shown in Figure 10. This tool would allow an unburned tree canopy and other smaller unburned objects with burned pixels around it to be marked as noise and replaced in with burnt pixels. Another challenge pertaining to the atmospherical condition was based on the fact that each satellite image, covering the same spatial area, had slightly different solar lighting, atmospheric density, and moisture. This color imbalance problem was solved by utilizing different hyperspatial training imagery for every fire in the experiment, although the separate training data were all created a consistent way. Ideally, the experiment should not need separate training data for each fire. Without creating separate training data for each fire, the experiments overall accuracy plummeted from color imbalance from atmospheric interference. Although bear in mind that each of the separate training was generated a consistent way as done by Hamilton [22]. However, creating separate training data is necessary because each satellite image has its own unique state. Future work could create a single set of training data that is cross-compatible with all fires, which will account for the ecological, atmospherical, and temporal setting of the fire.

As shown in Table 1, the experiment results for the 30 m burn extent shows that the accuracy of mapping burn extent from a satellite image using hyperspatial training data has a mean of 71% of 30 m extent results accuracy. The accuracy of the burn extent of the 30 m was expected to be lower than the accuracy of the 5 cm burn extent [19]. Many factors could impair the accuracy level. One of these factors include is the satellite image had a lot of atmospheric dissidence, which caused the sUAS training data to pick up far more unblemished data as previously discussed. Another of these factors includes that most of these fires were considered small fires and included many edge cases as shown in Figure 12. These edge cases would severely decrease the accuracy level of smaller fires. Therefore, accuracy is expected to increase when mapping larger wildland fires. To increase the overall accuracy of this experiment, future edge case analysis can be done to significantly increase the accuracy of mapping out wildland fires with satellite imagery.

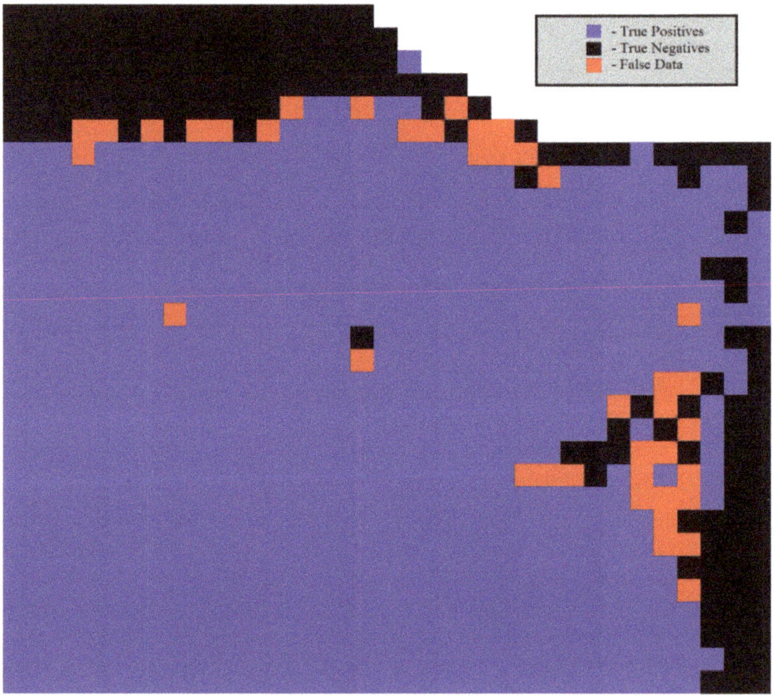

Figure 12. Some error analysis concluded that most False Positives or Negatives, indicated in red, are predominately located at the edge of the burn area.

Observing the IR experiment in compared to the RGB experiment, a decrease in accuracy is observed. However, this is contradictory to what was expected [9]. The accuracy decreases, for the most part, is slight, and after further analysis most of the error falls on the SVM edge cases similar to what is shown in Figure 12. A solution for this error is to use IR imagery hyperspatial and spatial imagery in this experiment, which theoretically will decrease error. However, this is left for future work. However, since the experiment was done on parts of the burn area where edge cases are predominately located, theoretically, the SVM would have better accuracy on a larger fire as the ratio between edge pixels and non-edge pixels is decreased.

5. Conclusions

This experiment successfully applied sUAS hyperspatial data to be used as training data to map out the burn extent on a satellite imagery. As shown in Table 2, the experiment went further than just using color imagery to determine accuracy. This experiment has also applied IR spectroscopy imagery in which has been previously proven to improve fire mapping; thus, results were expected to improve [9]. However, as previously discussed, the IR experiment results proved contradictory to what was expected and previous work [9]. Thus, it was assumed that the issue could be readdressed by using IR hyperspatial training data for the IR imagery as previously discussed.

One of the research goals was to determine the change in accuracy of burn extent mapping between the hyperspatial imagery and satellite imagery. The tool was created and tested on this experiment which showed improved accuracy from 5 cm to 30 m. However, for future work, one could take this research further by determining how the decrease in the resolution would decrease the accuracy of fire extent and severity mapping. This knowledge could enable future researchers to quickly and accurately determine which resolution would best fit their area of research, given their resources and budget constraints.

Source code (Supplementary Materials) developed as well as data used in this effort are available for download at https://firemap.nnu.edu/satellite-burn-mapping.

Implications for Local Management

Improved mapping of fire effects resulting from the development of methods, analytic tools, and metrics resulting in increased fire effects mapping accuracy will improve fire management. These improvements affect post-fire recovery planning and other management operations, which are driven by the extent of the fire and how much biomass was consumed by the fire. Development of post-fire recovery plans includes outlining the restoration activities that determine management response to the fire, facilitating ecosystem recovery and resiliency [7]. In order to reduce risks to affected resources, managers will also need increased capacity to determine potential effects on neighboring resources such as hydrologic features, infrastructure, and wildlife habitat.

Existing data about vegetation within the perimeter of a fire are rendered obsolete because of a disturbance such as a wildland fire. Increasingly accurate burn extent and biomass consumption mapping will improve efforts to update geospatial vegetation layers such as existing vegetation type, cover and height to accurately reflect fire fuel data following a wildland fire [7].

Improved mapping will also result in more complete fire history data, facilitating the inclusion of the spatial extent and biomass consumption of small fires which are commonly omitted from fire history data [19]. The inclusion of knowledge about small fires will reduce the omission of the most ecologically diverse areas burned within a landscape [30]. Additionally, calculations will be improved for the departure of current fire frequency from historical fire frequency, a key metric for determining ecosystem resilience [10].

6. Future Work

This research was unique in a way that the initial intent was to pave a path for future research. As discussed, prior, each satellite scene has its own unique solar lighting, atmospheric density, and moisture effect. This variance across each fire would cause a shift in the pixel values for each of the classes used in this experiment. For example, a burned pixel from a satellite image in early June may have a SWIR value of 1000, but in late August, the SWIR value of a burned pixel in a satellite image may have a value of 700. This could be caused by the solar position during the time of year, the lack of moisture towards the end of the year, and random atmospheric concentration. To solve this problem, one could calibrate the experiment by utilizing ground objects known to remain consistent throughout the year, such as pavement or lakes. Taking those base values to calibrate each satellite image could fix the problem and avoid utilizing different training data for each fire experiment. However, due to complexity, alternatively, a consistent set of hyperspatial data was used to create the satellite training data instead, as shown in Section 2.1.2.

Another challenge that occurred was the color images showing better accuracy than the IR images. As mentioned before, this is contradictory to previous work [9]. However, as mentioned previously, most of the inaccuracy occurred at the edge cases, as shown in Figure 12. Thus, to improve accuracy, more work on edge case analysis can be done, which could significantly improve the experiments accuracy. However, due to data and time constraints, this was left off for future work.

Supplementary Materials: The following are available online from the authors at https://firemap.nnu.edu/satellite-burn-mapping: source code and data.

Author Contributions: Conceptualization, D.H.; methodology, D.H. and E.L.; software, D.H., E.L., and N.H.; validation, E.L.; formal analysis, D.H. and E.L.; investigation, D.H. and E.L.; data curation, D.H., N.H., and E.L.; writing, D.H., E.L., and N.H.; visualization, D.H., E.L., and N.H.; supervision, D.H.; project administration, D.H.; funding acquisition, D.H. All authors have read and agreed to the published version of the manuscript.

Funding: This research was funded by the USDA Forest Service Boise National Forest, Forest Service Agreement No. 18-CS-11040200-0025.

Acknowledgments: We would like to acknowledge the students in the Northwest Nazarene University Department of Math and Computer Science who have helped with different aspects of this effort, including current and former students Nicholas Hamilton, Jonathan Hamilton, Gabriel Johnson, Aleesha Chavez and Andrew Welk. Additionally, we would like to acknowledge Northwest Nazarene University who funded the research efforts of many of the students mentioned.

Conflicts of Interest: The authors declare no conflict of interest.

References

1. Wildland Fire Leadership Council. Available online: https://www.forestsandrangelands.gov/documents/strategy/strategy/CSPhaseIIINationalStrategyApr2014.pdf (accessed on 30 July 2020).
2. Hoover, K.; Hanson, L.A. Wildfire Statistics. Congressional Research Service. October 2019. Available online: https://crsreports.congress.gov/product/pdf/IF/IF10244 (accessed on 14 May 2020).
3. National Interagency Fire Center (NIFC). Suppression Costs. 2020. Available online: https://www.nifc.gov/fireInfo/fireInfo_documents/SuppCosts.pdf (accessed on 18 May 2020).
4. National Interagency Fire Center (NIFC). Wildland Fire Fatalities by Year. 2020. Available online: https://www.nifc.gov/safety/safety_documents/Fatalities-by-Year.pdf (accessed on 18 May 2020).
5. Zhou, G.; Li, C.; Cheng, P. Unmanned aerial vehicle (UAV) real-time video registration for forest fire monitoring. In Proceedings of the 2005 IEEE International Geoscience and Remote Sensing Symposium, Seoul, Korea, 29 July 2005. [CrossRef]
6. Insurance Information Institute. Facts + Statistics: Wildfires|III. Available online: https://www.iii.org/fact-statistic/facts-statistics-wildfires (accessed on 18 May 2020).
7. Eidenshink, J.C.; Schwind, B.; Brewer, K.; Zhu, Z.-L.; Quayle, B.; Howard, S.M. A project for monitoring trends in burn severity. *Fire Ecol.* **2007**, *3*, 3–21. [CrossRef]
8. Keeley, J.E. Fire intensity, fire severity and burn severity: A brief review and suggested usage. *Int. J. Wildland Fire* **2009**, *18*, 116–126. [CrossRef]
9. Key, C.H.; Benson, N.C. Landscape Assessment (LA). In *FIREMON: Fire Effects Monitoring and Inventory System. Gen. Tech. Rep. RMRS-GTR-164-CD*; U.S. Department of Agriculture, Forest Service, Rocky Mountain Research Station: Fort Collins, CO, USA, 2016; Volume 164, p. 155.
10. Hamilton, D.; Bowerman, M.; Colwell, J.; Donahoe, G.; Myers, B. A Spectroscopic Analysis for Mapping Wildland Fire Effects from Remotely Sensed Imagery. *J. Unmanned Veh. Syst.* **2017**. [CrossRef] [PubMed]
11. Lentile, L.B.; Holden, Z.A.; Smith, A.M.; Falkowski, M.J.; Hudak, A.T.; Morgan, P.; Benson, N.C. Remote sensing techniques to assess active fire characteristics and post-fire effects. *Int. J. Wildland Fire* **2006**, 15. [CrossRef]
12. Morgan, P.; Hardy, C.; Swetnam, T.; Rollins, M.; Long, D. Mapping fire regimes across time and space: Understanding coarse and fine-scale fire patterns. *Int. J. Wildland Fire* **2001**, *10*, 329–342. [CrossRef]

13. Kolden, C.A.; Weisberg, P.J. Assessing Accuracy of Manually-mapped Wildfire Perimeters in Topographically Dissected Areas. *Fire Ecol.* **2007**, *3*, 22–31. [CrossRef]
14. Escuin, S.; Navarro, R.; Fernandez, P. Fire severity assessment by using NBR (Normalized Burn Ratio) and NDVI (Normalized Difference Vegetation Index) derived from LANDSAT TM/ETM images. *Int. J. Remote Sens.* **2008**, *29*, 1053–1073. [CrossRef]
15. USDA Forest Service Geospatial Technolgy and Applications Center (GTAC). Monitoring Trends in Burn Severity. *Monit. Trends Burn Sev.* **2020**. Available online: https://www.mtbs.gov/ (accessed on 20 May 2020).
16. Sparks, A.M.; Boschetti, L.; Smith, A.M.; Tinkham, W.T.; Lannom, K.O.; Newingham, B.A. An accuracy assessment of the MTBS burned area product for shrub–steppe fires in the northern Great Basin, United States. *Int. J. Wildland Fire* **2015**, *24*, 70–78. [CrossRef]
17. Federal Aviation Administration (FAA). Frequently Asked Questions. *Unmanned Aircr. Syst. Freq. Asked Quest.* **2020**. Available online: https://www.faa.gov/uas/resources/faqs/ (accessed on 20 May 2020).
18. Lebourgeois, V.; Bégué, A.; Labbé, S.; Mallavan, B.; Prévot, L.; Roux, B. Can commercial digital cameras be used as multispectral sensors? A crop monitoring test. *Sensors* **2008**, *8*, 7300–7322. [CrossRef] [PubMed]
19. Hamilton, D.; Hamilton, N.; Myers, B. Evaluation of Image Spatial Resolution for Machine Learning Mapping of Wildland Fire Effects. In Proceedings of the SAI Intelligent Systems Conference, London, UK, 5–6 September 2019; pp. 400–415.
20. Rango, A.; Laliberte, A.; Herrick, J.E.; Winters, C.; Havstad, K.; Steele, C.; Browning, D. Unmanned aerial vehicle-based remote sensing for rangeland assessment, monitoring, and management. *J. Appl. Remote Sens.* **2009**, *3*, 033542. [CrossRef]
21. Laliberte, A.S.; Herrick, J.E.; Rango, A.; Winters, C. Acquisition, orthorectification, and object-based classification of unmanned aerial vehicle (UAV) imagery for rangeland monitoring. *Photogramm. Eng. Remote Sens.* **2010**, *76*, 661–672. [CrossRef]
22. Hamilton, D. *Improving Mapping Accuracy of Wildland Fire Effects from Hyperspatial Imagery Using Machine Learning*; The University of Idaho: Moscow, ID, USA, 2018.
23. Goodwin, J.; Hamilton, D. *Archaeological Imagery Acquisition and Mapping Analytics Development*; Boise National Forest: Boise, ID, USA, 2019.
24. National Wildfire Coordinating Group (NWCG). Size Class of Fire. 2020. Available online: www.nwcg.gov/term/glossary/size-class-of-fire (accessed on 20 May 2020).
25. Lewis, S.; Robichaud, P.; Frazier, B.; Wu, J.; Laes, D. Using hyperspectral imagery to predict post-wildfire soil water repellency. *Geomorphology* **2008**, *95*, 192–205. [CrossRef]
26. Gonzalez, R.; Woods, R. *Digital Image Processing: Pearson Prentice Hall*; Pearson Prentice Hall: Upper New Jersey, NJ, USA, 2008.
27. Russell, S.; Norvig, P. *Artificial Intelligence: A Modern Approach*, 3rd ed.; Prentice Hall: Upper Saddle River, NJ, USA, 2010.
28. Han, J.; Kamber, M.; Pei, J. *Data Mining: Concepts and Techniques*, 3rd ed.; Morgan Kaufmann: Boston, MA, USA, 2012.
29. National Aeronautics and Space Administration (NASA). Landsat 8 Bands. 21 July 2020. Available online: https://landsat.gsfc.nasa.gov/landsat-8/landsat-8-bands/ (accessed on 22 July 2020).
30. Hamilton, D.; Hann, W. Mapping landscape fire frequency for fire regime condition class. In Proceedings of the Large Fire Conference, Missoula, MT, USA, 19–23 May 2014.

Publisher's Note: MDPI stays neutral with regard to jurisdictional claims in published maps and institutional affiliations.

remote sensing

MDPI

Article

Contextualizing the 2019–2020 Kangaroo Island Bushfires: Quantifying Landscape-Level Influences on Past Severity and Recovery with Landsat and Google Earth Engine

Mitchell T. Bonney [1,*], Yuhong He [1] and Soe W. Myint [2]

1 Department of Geography, Geomatics and Environment, University of Toronto Mississauga, 3359 Mississauga Rd, Mississauga, ON L5L 1C6, Canada; yuhong.he@utoronto.ca

2 School of Geographical Sciences and Urban Planning, Arizona State University, 975 S Myrtle Ave, Tempe, AZ 85281, USA; soe.myint@asu.edu

* Correspondence: mitchell.bonney@mail.utoronto.ca

Received: 27 October 2020; Accepted: 30 November 2020; Published: 2 December 2020

Abstract: The 2019–2020 Kangaroo Island bushfires in South Australia burned almost half of the island. To understand how to avoid future severe 'mega-fires' and how vegetation may recover from 2019–2020, we can utilize information from the bulk of historical fires in an area. Landsat time-series of vegetation change provide this opportunity, but there has been little analysis of large numbers of fires to build a landscape-level understanding and quantify drivers in an Australian context. In this study, we built a yearly cloud-free surface reflectance normalized burn ratio (NBR) time-series (1988–2020) using all available summer Landsat images over Kangaroo Island. Data were collected in Google Earth Engine and fitted with LandTrendr. Burn severity and post-fire recovery were quantified for 47 fires, with a new recovery metric facilitating comparison where fire frequency is high. Variables representing the current burn, fire history, vegetation structure, and topography were related to severity and yearly recovery with random forest and bivariate analysis. Results show that the 2019–2020 bushfires were the most widespread and severe, followed by 2007–2008. Vegetation recovers quickly, with NBR stabilizing ten years post-fire on average. Severity is most influenced by fire frequency, vegetation capacity and land use with more severe burns in nature conservation areas with dense vegetation and a history of frequent fires. Influence on recovery varied with time since fire, with initial (year 1–3) faster recovery observed in areas with less surviving vegetation. Later (year 6–10) recovery was most influenced by a variable representing burn year and further investigation indicates that precipitation increases in later post-fire years likely facilitated faster recovery. The relative abundance of eucalypt woodlands also has a positive influence on recovery in middle and later years. These results provide valuable information to land managers on Kangaroo Island and in similar environments, who should consider adjusting practices to limit future mega-fire risk and potential ecosystem shifts if severe fires become more frequent with climate change.

Keywords: Landsat; time-series; Google Earth Engine; LandTrendr; NBR; random forest; fire history

1. Introduction

The 2019–2020 Australia bushfires were among the largest and most devastating on record. It is estimated that close to 200,000 km^2 of land burned across the country, with projected losses of nearly three billion animals and potentially dozens of rare species [1,2]. There was also a major impact on society, with thousands of houses destroyed, 33 people killed and billions in economic losses. Kangaroo Island, off the coast of South Australia, was heavily impacted with about half of its land area burned. The 2020 fires have especially impacted the island's animal populations, with thousands of kangaroos, koalas, sheep, and other species killed and endangered species, including the dunnart and glossy black cockatoo, being pushed close to extinction [3,4].

Bushfires are becoming increasingly common in Australia as the climate warms [5]. Temperatures in Australia since 2010 are over 1 °C warmer than as recently as the mid-1970s [6] and combined with a period of drought creates a high fire risk scenario. The 2019–2020 fires occurred in the context of Australia's hottest and driest year on record, and a drought that had been ongoing for two years [2]. Despite increased mega-fire occurrence, including the 2009 Black Saturday fires that led to 173 deaths in Victoria, there is still policy debate about the effectiveness and implementation of forest management techniques such as fuel-reduction burning [7]. The occurrence of another devasting fire season a decade later has undoubtedly revived the debate over best practices in terms of land management policies and reducing fire risk. As fires are expected to become larger and more common [2], a better understanding of fire and recovery processes across the landscape is needed. If land managers knew landscape-level drivers of burn severity and recovery from past fires, they could then make more informed decisions about how and where to allocate limited resources across their landscape to mitigate future burns. Quantification of severity, recovery, and their drivers are also important at broad scales in terms of grasping their impact on forest carbon budgets [8].

A compelling way to obtain a landscape-level understanding of fire processes is to build a comprehensive dataset of fire history (e.g., when, where, severity, recovery) for an area. Remote sensing provides this opportunity, with the freely available and 30 m spatial resolution Landsat platform being used to observe burn severity and recovery since at least the 1990s [9,10]. These early studies primarily focused on observing single or small groups of fires with at least two images, but not enough to achieve a long-term yearly resolution of post-fire recovery. However, since the opening of the Landsat archive in 2008, the remote sensing community can support more dense Landsat time-series analysis [11]. More recent studies have made use of this to build longer time-series at near-annual temporal resolutions, but still often focus on a single fire [12]. Many of these types of studies use dozens of images downloaded from USGS to first calculate the difference between pre- and post-fire image spectral values with a given vegetation index and then quantify trends from subsequent post-fire images.

In the last few years, the development of segmentation algorithms such as LandTrendr that fit noisy yearly data to simplified trajectories and their release on Google Earth Engine (GEE), a cloud-based big data processing platform, has enabled more advanced analysis [13–15]. GEE allows for time-series to be built with hundreds or thousands of images since they do not have to be downloaded and can instead be accessed in the cloud. Studies have begun to make use of these platforms and algorithms to examine fires by building spectrally consistent surface reflectance time-series and quantifying severity and recovery as large negative trajectories followed by gradual positive trajectories [16,17]. However, the use of Landsat time-series to analyze the bulk of recent fires across a region and build a general understanding of fire processes related to these changes has focused on boreal and alpine forests [18–21].

Despite the long history of fire-focused remote sensing, there are still disagreements between fire studies that warrant more exploration. The normalized difference vegetation index (NDVI) is commonly used to quantify burn severity and recovery [12,19,22]. However, it is the normalized burn ratio (NBR) that has proven more effective in many different forested environments [16,20,23,24] because the shortwave infrared (SWIR) band is more sensitive to forest structure and moisture [25]. In Australian eucalypt forests, Hislop et al. [26] concluded that NBR is more reliable than NDVI and six other indices for estimating severity and recovery after using Glass's delta to compare pre- and post-fire data. Studies have reported spectral recovery periods differing from as low as three years to multiple decades with variations linked to location (e.g., vegetation type, terrain, climate etc.), the fire itself (e.g., severity) and the vegetation index used [12,16,23,27,28]. Hislop et al. [26] found that NDVI recovered to pre-fire levels after 3–5 years, while NBR took 8–10 years to recover.

There has been a wide range of studies concerned with identifying drivers behind differences in post-fire recovery. Burn severity, often quantified by dNBR or ΔNDVI, is commonly used as a predictor of recovery rate, with some studies indicating that more severely burned areas recover faster [16,23,29] and other studies linking more severe burns with slower recovery [24,27,30] or showing no significant relationship [12]. Other variables, such as topography [12,22,31], climate [16,19,24] and vegetation type [16,28] are used, with differing relationships to recovery. In south-east Australia, vegetation type analysis has shown differing spectral response to fire for forested and shrubland classes of various vegetation densities [32]. There has been less study on the influence of the historical fire regime, such as time since previous fire or longer-term frequency, on recovery [31,33]. On Kangaroo Island, where fires are common, fire frequency is an important variable controlling the distribution and survival of many fire-adapted vegetation groups, with recommended intervals from as low as six years to multiple decades [34,35]. Within the long-term Landsat record, there is thus an opportunity to monitor recovery in areas impacted by regular fires, with studies often focusing on single burn pixels and removing those burned multiple times from analysis [28]. Furthermore, there has been little study of how variables influencing recovery themselves change through time, with studies conducting comparisons for a single selected post-fire year [16]. By comparing at a yearly interval, we can build a full temporal understanding of influences on recovery from initial to late recovery stages. Although there has been extensive work on identifying drivers of recovery, less consideration has been given to influences on severity with fuel amount and time since last fire considered [36–39].

In this study, we use Landsat time-series, fire history spatial data, and GEE to understand landscape-level influences on burn severity and post-fire recovery in eucalypt-dominated forests across a region influenced by frequent fires. This study is based on remote sensing: vegetation, severity, and recovery through time are inferred from Landsat-based proxies of these variables and the LandTrendr algorithm in GEE [13–15]. We build a methodology that is useful for areas that undergo multiple burns within the Landsat record and use these methods to best understand the fire dynamics of the region. The primary objectives are to (1) monitor the fire history (i.e., extent, severity, recovery) of Kangaroo Island forested areas, especially in the context of the 2019–2020 bushfires; and (2) quantify the influence of landscape-level variability (e.g., in current burn context, fire history, vegetation structure and topography) on severity and yearly recovery. Identifying how influences on recovery change through time after fires is an important aspect. We use these objectives to gain information on why the 2019–2020 bushfires were so severe and how vegetation may recover from these bushfires in the coming years based on the historical data.

2. Materials and Methods

2.1. Study Area

Our focus is on Kangaroo Island (Figure 1). This island, located in South Australia southwest of Adelaide, consists of a mix of mostly eucalypt forest (nature conservation, residual cover, plantations) and agricultural (grazing pastures, cropping) landscapes [40,41]. Dominant tree species vary across the island, with *Eucalyptus diversifolia* being common along the coast, *E. remota* common in the dense western forests of Flinders Chase National Park (NP), *E. cladocalyx* common in riparian landscapes and *E. baxteri* fragmented across Flinders Chase NP and in central agricultural settings [35,42]. Species grow as either trees with a single stem or as mallees which have multiple stems and may take on a shrubby growth form [42]. Woodlands grow denser and are more widespread in the conserved western and extreme south of the island. However, residual native cover remains along riparian corridors, ridges and road networks in the more developed central and eastern parts of the island in notable quantities [41]. Growing development in the last half century has put some of these remnant and forest edge landscapes under threat [40,41]. The warmest months are December to March, and the wettest months are June to August [6]. In the summer, temperatures reach an average high of about 24 °C (19 °C annual average) and have been steadily warming in recent decades (Figure 2A). Annual precipitation fluctuates, with an average of 567 mm since 1988, and is consistently highest in the forested west and central parts (Figure 2B).

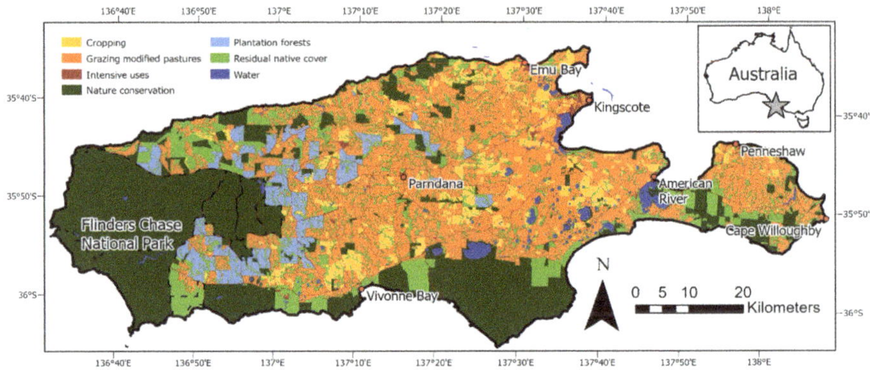

Figure 1. Kangaroo Island, Australia. Land use classes based on the Australian Land Use and Management Classification V8 [43]. Population centres and other points of interest shown. Flinders Chase NP (and the associated Ravine Des Casoars Wilderness Protection Area) shown with a dashed black outline [44].

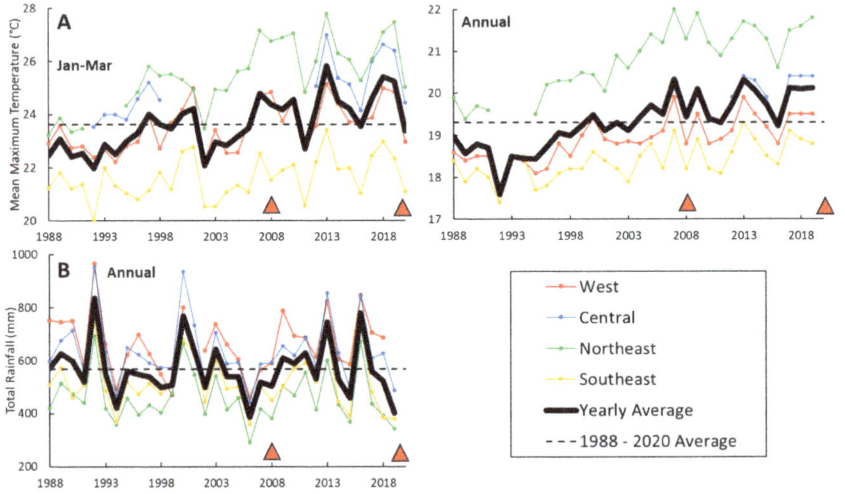

Figure 2. Mean maximum temperatures (**A**) and total annual rainfall (**B**) across Kangaroo Island from 1988-2020. Red triangles indicate the two most widespread fire seasons on record (2008, 2020). Climate data collected from on-island weather stations [6] and merged into regional clusters. West: stations within and near Flinders Chase NP (4 stations); Central: nearby Parndana and Vivonne Bay (9); Northeast: nearby Kingscote (5); Southeast: nearby American River and Cape Willoughby (5). Yearly is the average of these clusters.

Kangaroo Island's climate and landscape make it vulnerable to regular bushfires, especially during drier periods (Figure 2B) and in forested areas. Spatial records of fires have been kept since the early 1930s, although data collection is sparse until the mid-1950s [45]. The two most widespread fire seasons occurred in summer 2007–2008 and 2019–2020 (below as 2008 and 2020) where 950 km^2 and 2100 km^2 burned, respectively. Both fires occurred after multiple years of below average rainfall. The 2008 fires were initially ignited by lightning strikes and burned for two weeks, although it was mostly contained within Flinders Chase NP [46]. The 2020 fire season was devastating for Australia overall, but was especially widespread on the island. The fires were not contained to the park and burned throughout the agricultural countryside, with population centres across the island being evacuated [3]. Although the island's eucalypt vegetation is adapted to fire, the severity and close temporal proximity of these mega-fires over the same areas may impact their ability to recover. In the years before these fires, local analysis indicated that much of the area of all vegetation groups besides *E. remota* have not burned in over 40 years and recommend the introduction of prescribed burning [35].

2.2. Building Landsat Fire History Time-Series

Using GEE, a yearly time-series (1988–2020) of cloud-free surface reflectance medoid data from Landsat Thematic Mapper (TM), Enhanced Thematic Mapper Plus (ETM+), and Operational Land Imager (OLI) was constructed over Kangaroo Island. ETM+ imagery includes SLC-on and SLC-off data. Clouds and cloud shadows were automatically removed from input images with the CFMASK algorithm [47]. Georeferenced images were converted to surface reflectance using the LEDAPS algorithm for TM and ETM+ and LaSRC for OLI [48,49]. To minimize between-sensor discrepancies, OLI bands were converted to ETM+ spectral equivalents using the regression equations reported in Roy et al. [50]. When multiple cloud-free values were available for a given pixel in a given year, medoid compositing reduced them by finding the value that is closest to the center (i.e., median) for both the X (i.e., time of year) and Y (i.e., reflectance) axes. [51]. These processing steps were completed in GEE using the buildSRcollection code within LandTrendr to assure a consistent time-series and limit atmospheric and sensor influences to the highest extent possible [14].

For each year, satellite images were eligible for the time-series if collected between December 25 and March 25. This approximately corresponds to the Australian summer used in other Landsat time-series studies in the region [26,52] while also maximizing the amount of cloud-free data. However, three years in the study period (1988–2020) are heavily impacted by missing data (i.e., 1990, 1994, 1997), with 1997 having no available images (Figure 3). Some other years (e.g., 1989, 1992, 1998, 2011, 2012) also have small areas of missing data. Data availability for each year was checked in GEE using the buildClearPixelCountCollection code within LandTrendr [14].

To infer vegetation abundance in connection to moisture, the surface reflectance time-series was converted to the normalized burn ratio (*NBR*), which is calculated as follows:

$$NBR = \frac{NIR - SWIR}{NIR + SWIR} \tag{1}$$

where *NIR* represents the near-infrared Landsat band, and *SWIR* represents a shortwave infrared band [53]. In Landsat TM and ETM+, *NIR* is band 4 and *SWIR* is band 7. In Landsat OLI, *NIR* is band 5, and *SWIR* is band 7. We decided to use *NBR* because it has best quantified fire damage and recovery in other eucalypt forests [26]. A related index called the differenced *NBR (dNBR)* is used to quantify burn severity in remote sensing. It is defined as the difference between NBR in the year before a fire (*preNBR*) and the NBR in the year after the same fire (*postNBR*) based on available imagery (Figure 4A).

$$dNBR = preNBR - postNBR \tag{2}$$

If a *dNBR* year is missing imagery, LandTrendr (discussed below) will fit the decline over two years when filling in the missing value (Figure A1). In this case, *preNBR* is defined as the *NBR* value before the decline and *postNBR* is the lowest *NBR* value at the end of the decline. In this study, all *NBR*-based indices are multiplied by 1000. Based on visual interpretation of burns, the focus on forested land cover and the need to minimize more gradual declines being classified as burns, we set 100 *dNBR* as the 'burned' threshold.

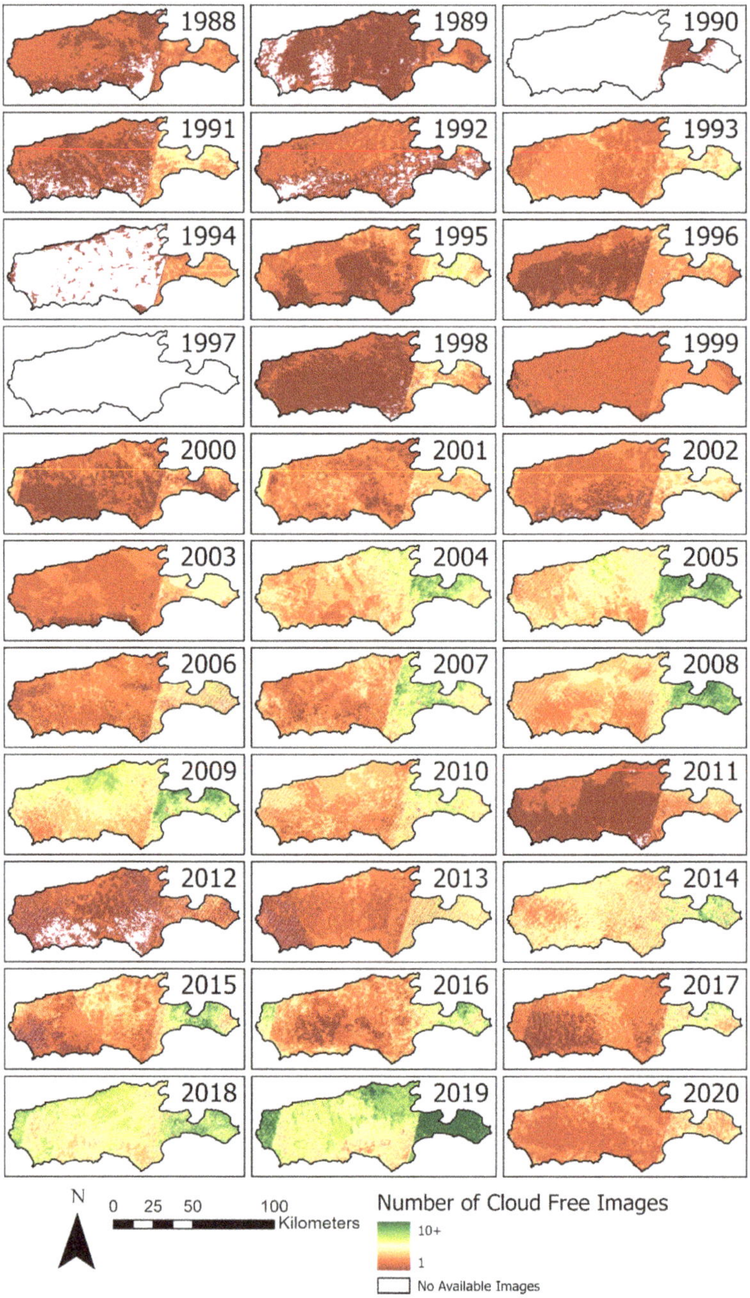

Figure 3. Number of cloud free images available for every pixel in each summer (December 25–March 25) across Kangaroo Island.

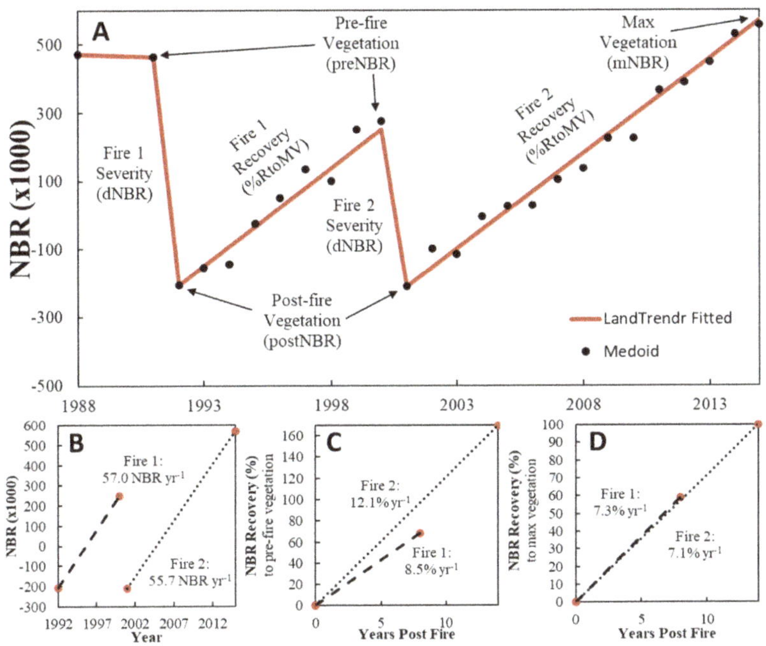

Figure 4. Conceptual model outlining LandTrendr and the quantification of burn severity (*dNBR*) and recovery (*%RtoMV*). (**A**) LandTrendr fitted and medoid *NBR* values for an example pixel that experienced fires in the 1992 and 2001 seasons. (**B–D**) Comparison of methods to quantify post-fire recovery for a multi-burn pixel. (**B**) Fitted *NBR* slope. (**C**) *NBR* recovery (%) to *preNBR* [16]. (**D**) *NBR* recovery (%) to the maximum amount of vegetation for that pixel in the time-series (*mNBR*).

The *NBR* time-series was input into the GEE implementation of LandTrendr, which is an algorithm that creates a time-series of spectral-temporal segments similar to a piecewise linear function [14]. LandTrendr functions on the pixel-level by first removing ephemeral spikes due to non-real change (e.g., un-masked clouds) and fitting a simple linear regression to the time-series [15]. Real disturbances are differentiated from non-real change based on how quickly the spectral index recovers to the pre-change values, with non-real changes returning to pre-change values after a single or very small number of years (the threshold of which is defined by recoveryThreshold). Breakpoint years are then identified based on their deviation from the regression line, with the number of final breakpoint years (and therefore the number of segments) defined by the maxSegments and vertexCountOvershoot parameters. Straight line segments are used to connect breakpoint years, with successively simpler models created from the maxSegments number down to one segment (i.e., simple linear regression). The simplest model with the best fit is chosen as the final fitted time-series for the pixel in question (Figure 4A).

LandTrendr provides a default set of parameters for general use [14] (Table 1). However, for more accurate quantification of spectral-temporal trends, it is recommended to find a set of parameters that best fits the process being observed over a study area [54]. To do this, we selected 100 random pixels across Kangaroo Island and built their fitted time-series using different variations of LandTrendr parameters. The LandTrendr developers provide a web-app that allows for efficient testing of different parameter sets (https://emaprlab.users.earthengine.app/view/lt-gee-pixel-time-series). For each of these pixels, we recorded any year that experienced a *dNBR* value considered a burn (i.e., 100+) in the fitted time-series. We then cross-referenced this information with a spatial dataset of fire polygons created by the Government of South Australia [45], who record historical burn scars using digitization from aerial photography, *NBR* calculation from satellite imagery and other sources and methods. A successful match occurred when the pixel in question experienced a recorded fire, and *dNBR* was at least 100 in the same year. Note that not all *dNBR* values above 100 are due to fires; plantation harvest, and other human and natural disturbances may also lead to high *dNBR* values.

Table 1. LandTrendr parameters. Default: generalized LandTrendr parameters. Selected: LandTrendr parameters that best characterize fire history on Kangaroo Island. Fire Identification: percent of sample pixels in a recorded fire polygon that exhibits an *NBR* loss of at least 100 during the correct year.

Parameters	Default	Selected
maxSegments	6	8
spikeThreshold	0.9	0.9
vertexCountOvershoot	3	3
preventOneYearRecovery	TRUE	TRUE
recoveryThreshold	0.25	0.75
pvalThreshold	0.05	0.05
bestModelProportion	0.75	0.75
minObservationsNeeded	6	6
Accuracy		
All	70%	81%
Forest	80%	93%

For our random sample of pixels, we determined which parameter set most accurately identified recorded fires (Table 1). *NBR* more consistently identified fires compared to other indices, such as the NDVI and Tasseled Cap Angle. The default parameters correctly identified 70% of recorded fires overall and 80% in forested land cover, which is where fires are expected to be more easily observed by Landsat and is the focus of this study. However, the most accurate LandTrendr parameters correctly identified 81% of recorded fires overall and 93% of forest fires. In this parameter set, maxSegments was increased from 6 to 8 to best fit the time-series of some multi-burn pixels. The recoveryThreshold was also increased from 0.25 to 0.75 because post-fire spectral vegetation indices recover quickly in this area, sometimes incorrectly activating this filter (which is meant to remove erroneous segments) at the default value. This parameter set was selected for use in subsequent analysis. A full description of all LandTrendr parameters is available in Kennedy et al. [14,15].

With the selected parameters, the yearly surface reflectance time-series was converted to a LandTrendr fitted time-series across Kangaroo Island in GEE. Pixel-level information was extracted from the fitted time-series. However, since our focus is on monitoring fire history, we used the spatial dataset of recorded fires for subsequent analysis [45]. From this dataset, we defined our unit of analysis as a fire area extent over one km^2 that burned between 1988 and 2020. There are 47 fires that fit these criteria (Table 2).

Another key aspect of this study focuses on monitoring post-fire recovery. LandTrendr is well suited to quantify this using the fitted *NBR* slope of the post-fire segment (Figure 4B). A more useful method, which quantifies percent *NBR* recovery for years post fire in terms of how much a burned pixel has recovered to pre-fire vegetation amounts (*%RtoPfV*), allows for comparison of different fires [16] (Figure 4C). It is calculated for a given year *X* as:

$$\%RtoPfV = \frac{(Year\ X\ NBR - postNBR)}{(preNBR - postNBR)} * 100 \quad (3)$$

This method is robust in many scenarios but may be misleading in areas that experience additional fires before the area in question has fully recovered from the previous fire. For example, in Figure 4C, *%RtoPfV* quantifies recovery for two fires that burn the same pixel nine years apart. Since Fire 2 occurred when the pixel was still recovering from Fire 1, it more quickly reaches *preNBR* and thus indicates a faster recovery rate even though their *NBR* slopes are very similar.

Table 2. Recorded fires on Kangaroo Island [45] from 1988 to 2020 with extent polygons larger than 1 km^2. Vegetation species determined by the most common National Vegetation Information System (NVIS) 5.1 [42] Level 5–6 vegetation species overlapping burned pixels for each fire. Fires listed in chronological order.

ID	Fire Season	Date	Extent (Burned) km^2	Elevation (m)	Vegetation Species
P1_5	1988–1989	20-October	2.4 (1.2)	195.4 ± 15.7	*Eucalyptus diversifolia*
P1_0	1989–1990	22-November	6.7 (1.9)	37.7 ± 9.6	*E. diversifolia*
P1_1	1989–1990	22-March	1.4 (0.9)	71.9 ± 6.9	*E. diversifolia*
P1_4	1990–1991	31-December	132.6 (90.9)	221.0 ± 55.9	*E. remota*
P1_2	1990–1991	09-February	1.6 (0.7)	16.0 ± 4.6	*E. remota*
P1_3	1991–1992	31-December	248.7 (194.6)	136.2 ± 67.1	*E. baxteri*
P2_1	1993–1994	09-December	1.5 (0.4)	63.7 ± 5.2	*E. diversifolia*
P2_2	1993–1994	24-March	2.3 (1.9)	93.4 ± 26.9	*E. diversifolia*
P2_0	1994–1995	11-December	22.0 (9.8)	32.9 ± 17.3	*Leucopogon parviflorus*
P2_3	1996–1997	15-December	242.7 (194.3)	36.7 ± 15.7	*E. diversifolia*
P2_4	1996–1997	27-March	44.2 (23.3)	152.2 ± 43.9	*E. remota*
P3_4	2000–2001	31-January	21.5 (19.7)	195.7 ± 34.9	*E. diversifolia*
P3_7	2000–2001	18-February	10.7 (8.1)	286.5 ± 11.1	*E. baxteri*
P3_6	2002–2003	02-November	63.7 (62.7)	208.4 ± 61.0	*E. diversifolia*
P3_9	2003–2004	14-January	2.5 (1.2)	26.2 ± 3.7	*E. cladocalyx*
P3_5	2004–2005	19-December	1.3 (0.9)	15.2 ± 6.4	*E. remota*
P3_2	2004–2005	28-April	17.6 (5.5)	81.7 ± 22.9	Undifferentiated
P3_3	2004–2005	28-April	1.6 (0.2)	137.5 ± 35.4	*E. remota*
P3_10	2005–2006	20-January	31.6 (31.5)	80.7 ± 10.4	*E. baxteri*
P3_1	2005–2006	21-January	1.1 (0.9)	268.8 ± 2.3	*E. remota*
P3_0	2006–2007	11-October	2.3 (2.0)	43.4 ± 7.2	*E. diversifolia*
P3_11	2006–2007	11-October	6.3 (5.4)	94.5 ± 10.3	*E. diversifolia*
P3_12	2006–2007	11-October	5.5 (4.4)	123.6 ± 32.2	*E. baxteri*
P3_8	2006–2007	11-November	23.6 (21.8)	258.0 ± 24	*E. cladocalyx*
P4_2	2007–2008	07-September	8.7 (8.4)	92.6 ± 32.5	*E. cladocalyx*
P4_0	2007–2008	06-December	2.3 (1.6)	237.8 ± 10.5	*E. cladocalyx*
P4_1	2007–2008	06-December	29.5 (24.5)	161.4 ± 44.5	*E. diversifolia*
P4_3	2007–2008	06-December	162.9 (161.0)	32.9 ± 11.0	*E. diversifolia*
P4_4	2007–2008	06-December	50.3 (18.9)	76.8 ± 36.8	*E. baxteri*
P4_5	2007–2008	06-December	604.7 (599.8)	154.2 ± 67.1	*E. remota*
P5_0	2010–2011	04-December	1.6 (0.2)	19.4 ± 7.2	*E. diversifolia*
P5_10	2010–2011	08-February	3.2 (3.0)	179.6 ± 14.4	Undifferentiated
P5_2	2011–2012	14-March	4.3 (3.7)	37.7 ± 12.7	*E. diversifolia*
P5_3	2011–2012	06-April	2.5 (2.0)	205.1 ± 36.5	*E. remota*
P5_5	2013–2014	22-April	4.5 (3.2)	34.7 ± 6.5	*E. diversifolia*
P5_6	2014–2015	31-May	3.7 (3.4)	240.3 ± 25.5	*E. diversifolia*
P5_8	2015–2016	25-October	1.5 (1.5)	277.6 ± 12.7	*E. remota*
P5_4	2015–2016	20-March	1.8 (1.1)	28.6 ± 3.2	Unclassifiable
P5_9	2016–2017	18-April	3.0 (2.6)	275.0 ± 18.0	*E. remota*
P5_12	2016–2017	18-April	2.0 (1.9)	263.9 ± 5.6	*E. remota*
P5_7	2017–2018	24-January	1.1 (0.3)	47.2 ± 7.3	*E. cosmophylla*
P5_1	2017–2018	08-February	3.7 (2.6)	84.0 ± 55.2	Undifferentiated
P5_11	2018–2019	06-December	11.3 (8.9)	103.0 ± 24.5	*E. baxteri*
P6_0	2019–2020	20-December	8.4 (2.9)	76.7 ± 18.2	*E. cladocalyx*
P6_1	2019–2020	21-December	165.5 (132.3)	183.1 ± 56.8	*E. cladocalyx*
P6_2	2019–2020	21-December	2.1 (1.6)	65.2 ± 7.1	*E. diversifolia*
P6_3	2019–2020	30-December	1918.8 (1630.8)	152.6 ± 72.7	*E. remota*

Therefore, we modified *%RtoPfV* to better account for multi-burn situations (Figure 4D). Rather than recovery to pre-fire vegetation, we instead quantify the percent recovery to the inferred maximum vegetation capacity for a given pixel (*%RtoMV*):

$$\%RtoMV = \frac{(Year\ X\ NBR - postNBR)}{(mNBR - postNBR)} * 100 \tag{4}$$

where *mNBR* is defined as the highest fitted *NBR* value in the time-series. With this approach, multiple recoveries in the same pixel are gauged against the same value, which allows for a more direct comparison of post-fire recovery. In Figure 4D, *%RtoMV* accurately quantifies that the recovery rate for both fires is very similar despite Fire 2 occurring amid recovery from Fire 1. Kangaroo Island experiences regular fires, with parts of Flinders Chase NP having burned at least nine times since 1930 (Figure 5), meaning that *%RtoMV* is more robust in this setting.

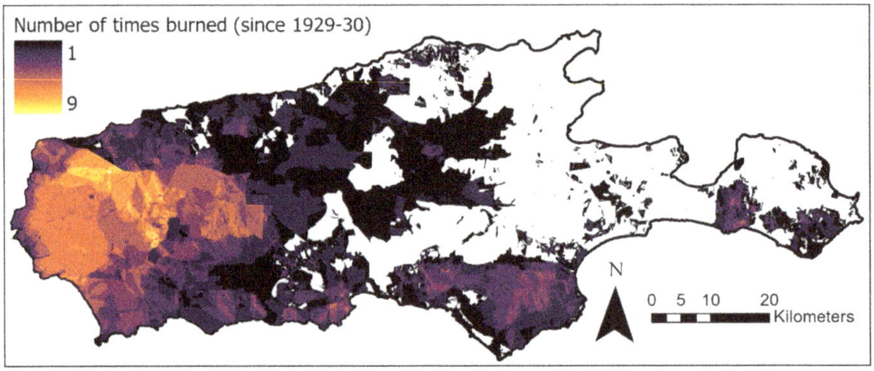

Figure 5. Number of times burned since 1930. Based on the size of fire history polygons [45] to be consistent across the entire record. Polygons from 1950 and onwards were used in the creation of fire history variables (BLX, SLB).

Burn severity (i.e., *dNBR*) and post-fire recovery (i.e., *%RtoMV*) were calculated at the pixel-level and averaged for each fire since that is the unit of analysis for this landscape-level study. Eligible fire polygons were used to extract *NBR* values from the fitted time-series. Only pixels with a *dNBR* value of at least 100 within these polygons were used in subsequent analysis. For *%RtoMV* calculation, the *postNBR* year is considered year 0, and %RtoMV is quantified for the following years. The recovery trajectory for each pixel ends when either the time-series ends (i.e., 2020), when another recorded fire occurs (i.e., spatial-temporal overlap with a fire polygon), or when some other notable damage occurs to vegetation (i.e., fitted *dNBR* of at least 50). When less than 25% of pixels are left recovering from a fire, the average recovery trajectory also ends.

2.3. *Assessing Landscape-Level Influences*

To quantify the influence of landscape-level variability on burn severity and recovery, 20 variables were built for comparison (Table 3). These variables mostly fall into four categories: the current burn (i.e., burn extent, burn year, time of year, *postNBR*, *dNBR*), fire history (i.e., years since last burn, longer-term fire frequency), vegetation structure (i.e., *preNBR*, *mNBR*, land use, vegetation group, species) and topography (i.e., elevation, slope, aspect, topographic position, topographic wetness). One additional variable assesses the influence of missing data in the Landsat time-series (i.e., part of burn area impacted by missing data in a *dNBR* year). These variables were built from the Landsat time-series data, the fire polygon data [45], Australian land use data [43], NVIS [42] and SRTM 1 Arc-Second Global DEM data [55]. Fire history variables are based on the spatial-temporal overlap of fire polygons (e.g., Figure 5). Final land use classes and vegetation groups were chosen based on how common they were across all fires. Information was extracted at the pixel and landscape level. If acquired at the pixel-level, mean values were calculated for each fire.

Table 3. Descriptions of variables tested for influence on burn severity (*dNBR*) and recovery (*%RtoMV* year 1–10). Pixel-level variables are based on burned pixels ($dNBR \geq 100$) only and averaged for each fire to match landscape-level analysis.

Variable	Level	Category	Description
Burned extent (BExt)	Landscape	Current burn	Area (km^2) burned in fire polygon, log10 transformed
Pre-fire vegetation (preNBR)	Pixel	Vegetation structure	NBR value the last year before burn x1000 (Figure 4)
Max vegetation (mNBR)	Pixel	Vegetation structure	Max NBR value across full time-series x1000 (Figure 4)
Land use percentage	Pixel	Vegetation structure	Land use class for each pixel, used to build fire-level composition
Vegetation group percentage	Pixel	Vegetation structure	Vegetation group for each pixel, used to build fire-level compsition
Most common land use	Landscape	Vegetation structure	Highest percentage land use across pixels
Most common vegetation group	Landscape	Vegetation structure	Highest percentage vegetation group across pixels
Most common species	Landscape	Vegetation structure	Highest percentage dominant species across pixels
Years since last burn (SLB)	Pixel	Fire history	Number of years since the last burn, pixels with no recorded burns given a value of 70 (i.e., at least 70 years since the last burn)
Times burned in last X years (BLX)	Pixel	Fire history	Number of times burned in the last X (70, 60, 50, 40, 30, 20, 10) years
Elevation (ELV)	Pixel	Topography	Elevation above sea level (m)
Slope (SLP)	Pixel	Topography	Slope (°) calculated from ELV
Aspect	Pixel	Topography	Aspect (°) calculated from ELV
Topographic position index (TPIX)	Pixel	Topography	Difference (m) between the center pixel and average ELV of an X by X (3, 5, 9, 15, 31, 65, 111) pixel moving window [56]
Topographic wetness index	Pixel	Topography	Ln((Flow Accumulation + 1)/Tan (((SLP) × π)/180))) [57]
Fire season (BYr)	Landscape	Current burn	Year corresponding with the end of fire season when burn occurred
Time of year	Landscape	Current burn	Number of days from December 31st when burn occurred
Missing data (MD)	Landscape		Indicates if a burn occurred when and where no cloud-free Landsat data were available (i.e., one of dNBR years impacted by missing data)
Post-fire vegetation (postNBR)	Pixel	Current burn	NBR value the first year after burn x1000, only tested in %RtoMV models (Figure 4)
Burn severity (dNBR)	Pixel	Current burn	NBR difference between preNBR and postNBR x1000, only tested in %RtoMV models (Figure 4)

Random forest (RF) modeling was used to quantify the influence of these landscape-level variables on burn severity and recovery. RF models create many high variance decision trees that are built using bootstrap sampling (i.e., with replacement) from the original dataset [58]. Each RF model included 500 decision trees (ntree). Due to the relatively small sample size of the analysis, we ran each RF model iteration 30 times with the goal of finding the smallest set of variables that explained the highest average variation in *dNBR* or *%RtoMV*. Separate average RF runs were conducted for each year of

%*RtoMV* from years 1 to 10. Only variables that improved overall model performance and were not highly correlated to each other were included in the final model runs. Modeling was conducted using the "randomForest" package in R [59].

Similar to Bright et al. [16], we evaluated overall model performance with percent root mean square error (%*RMSE*) and percent variance explained (%*VE*):

$$\%RMSE = \frac{\sqrt{MSE}}{\overline{X}} * 100 \tag{5}$$

$$\%VE = \frac{1 - MSE}{var(X)} * 100 \tag{6}$$

where X is either *dNBR* or %*RtoMV*, and *MSE* is the mean square error. %*VE* explained is the pseudo R^2 for RF models. We also evaluated the individual influence of each explanatory variable with the model improvement ratio (MIR). MIR is a standardized measure of variable importance (i.e., influence on *dNBR* or %*RtoMV*) between 0 and 1, where 0 is least important, and 1 is most important [60].

Further comparisons between those variables and both *dNBR* and %*RtoMV* were completed at the fire level. Scatterplots were created to relate these variables to *dNBR*, with strength (R^2) and statistical significance ($p < 0.05$) tested. Accompanying boxplots were produced by splitting average *dNBR* values into three groups based on natural breaks in each explanatory variable and testing for significant differences with the nonparametric Mann-Whitney (MW) and Kruskal-Wallis (KW) tests. Time-series plots relating the variables and %*RtoMV* from years 1 to 10 were also generated. Average %*RtoMV* trajectories were also split into three groups, and significant differences were tested for each year with the same tests.

3. Results

3.1. Fire History

Kangaroo Island experiences regular fires. The four most widespread fire seasons by area burned were 1954 (17.4% of the island burned), 1959 (16.4%), 2008 (19.5%) and 2020 (45.9%) (Figure 6A).

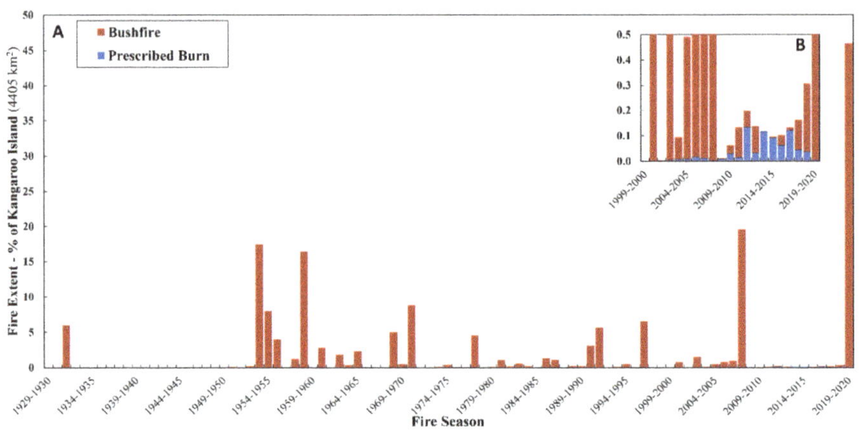

Figure 6. Percent of Kangaroo Island burned in each fire season from 1929–30 to 2019–20 (**A**), based on the size of fire history polygons [45] to be consistent across the entire record. Polygons from 1950 and onwards were used in the creation of fire history variables (BLX, SLB). Prescribed burns are planned fires on government land and are minor compared to bushfires. Inset graph (**B**) with all recorded burns since 2000 shows that prescribed burns never burned more than 0.2% of Kangaroo Island in a single season.

These fires have historically been most common in the densely forested west, south, and extreme east parts (Figure 7), with some parts of Flinders Chase NP burning every ten years on average (Figure 5). Landsat 30 m data have been collected over the island since 1988, where 47 fires over one km^2 have been recorded (Table 2). These fires have varied in location, size (BExt: 0.75 ± 0.90), and burn severity (*dNBR* 484 ± 173) over time (Figures 7 and A2). Fires in densely forested areas, especially Flinders Chase NP, are generally larger and burn more severely. Fires in residual forest cover, fragmented across agricultural areas, are usually smaller and burn less severely. The only fire season where large swaths of the agricultural landscape burned was 2020. Broadly, the 1990s fires burned less severely than more recent fires. The two largest fires, occurring in 2008 and 2020, both followed a roughly decade-long period of relatively little burning. This is especially true for the 2020 fires, where only 65 km^2 (1.5% of the island) burned between 2009 and 2019. These two fires were also the most severe, with 2008 having a *dNBR* of 797 ± 210 and 2020 having a *dNBR* of 820 ± 294.

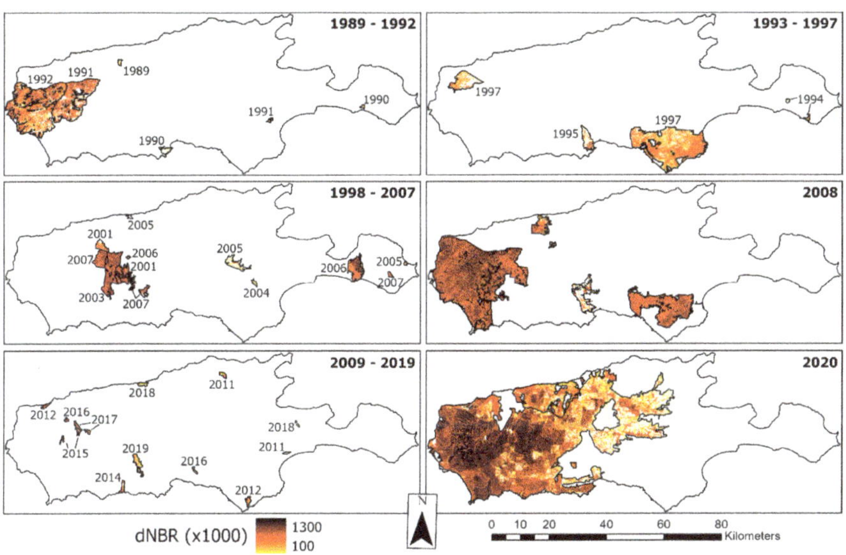

Figure 7. Burn severity (dNBR) for all recorded fires (>1 km^2) during the Landsat record (1988–1989 to 2019–20). Split into six periods to reduce overlap in multi-burn areas and highlight differences in burn extent through time. See Figure A2 for violin plot distributions of burn severity for each fire. See Figure A3 for a closer view of two smaller fires.

Fires have burned across Kangaroo Island in many different landscapes. However, when isolating burned pixels based on our *dNBR* analysis, it is clear that the vast majority of pixels burned in forested and shrubby areas (e.g., Figure A3). Based on land use (Figure 8A), most fires burned either primarily on nature conservation land (i.e., Flinders Chase NP, other large forested areas) or on residual native cover (i.e., fragmented forest patches located near agricultural land). One fire burned primarily on a plantation forest. Giving all fires equal weight, only 5% of burning occurred on land use classified as grazing modified pastures or cropping, and closer inspection of some of these areas indicates possible land use change (e.g., pasture to residual cover) between data date and fire date. Based on vegetation group (Figure 8B), most fires burned either primarily in denser eucalypt woodlands or mallee woodlands, which are also mostly eucalypts in a different growth form. Some fires burned primarily in other woodlands and shrublands, including *Casuarina* and *Acacia* dominated areas, but most of this class includes residual native cover that was not classified by NVIS [42] and plantation forests. Again, the influence of non-forest and shrubland is minimal, with cleared land covering similar areas as pastures and cropping land use. Eucalypts dominate the species composition of burned areas, with 42 of 47 fires consisting primarily of *Eucalyptus* spp. (Table 2).

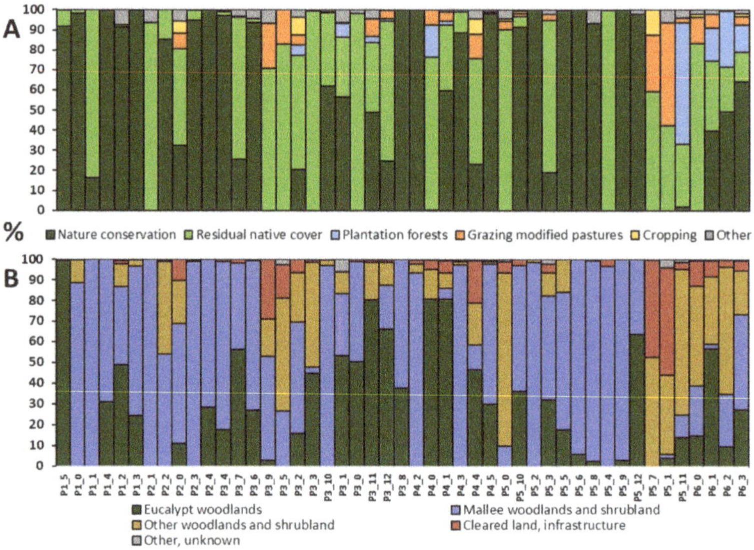

Figure 8. Land use and vegetation group composition of burned pixels (dNBR ≥ 100) for each fire. (**A**) Land use classes derived from the Australian Land Use and Management Classification V8 [43]. (**B**) Vegetation groups based on NVIS Major Vegetation Groups [42]. Visual inspection against ArcGIS Basemap imagery was used to confirm accurate land use and vegetation group classification. Fires listed chronologically as in Table 2.

The average recovery trajectories of the 40 eligible fires (i.e., not including recent fires with less than two years of recovery data), as quantified by *%RtoMV*, indicate that burned vegetation usually recovers quickly on Kangaroo Island (Figure 9).

In the first year, *NBR* recovers 22.3 ± 8.2% to its inferred maximum vegetation capacity on average. By year three, *NBR* recovers over half (51.3 ± 12.5%) of its capacity. Average *%RtoMV* follows a logarithmic pattern, reaching 80.3 ± 10.6% in year nine and stabilizing around 85% after ten years. Most fires follow a similar recovery pattern. The fastest recovering fire reached 97 *%RtoMV* in year six before the area was burned again in 2020. This fire (P5_3) occurred in a sloped riparian mallee *E. remota* woodland adjacent to pastures. The slowest recovering fire (P2_1) reached 48 *%RtoMV* in year 10, before the area was burned again in 2006. This fire burned in a mallee *E. diversifolia* woodland and shrubland area that is transitioning from a pasture land use. It also had the lowest burn severity (*dNBR*: 190 ± 66) of any recorded fire.

3.2. *Severity: Landscape-Level Influences*

RF modeling identifies seven landscape-level variables that together most influence burn severity (*dNBR*) (Figure 10). From most to least important, these variables are BL50, *mNBR*, percent nature conservation land use (%NC), BExt, BYr, TPI65, and MD (Table 3). Longer-term fire frequency, inferred by BL50, and vegetation capacity, inferred by *mNBR*, are practically identical in terms of their strong influence on *dNBR*. BL50 and *mNBR* have an average MIR of 0.964 ± 0.04 and 0.960 ± 0.06 respectively. Both BL50 and *mNBR* outcompeted similar variables, SLB and *preNBR*, that were also correlated to *dNBR* but not as strongly. %NC also has a strong influence on *dNBR*, with an MIR of 0.91 ± 0.07. A RF model of just BL50, *mNBR* and %NC can explain over 50% of the variation in *dNBR*. After these variables, there is a clear drop in influence, with BExt having an MIR of 0.44 ± 0.05, BYr 0.26 ± 0.04,

TPI65 0.20 ± 0.04 and MD 0.11 ± 0.02. However, each of these variables slightly increases model performance compared to them not being included. Overall, well over half (58.7 ± 1.3%) of the variance in *dNBR* is explained by these seven variables, and %RMSE is 22.9 ± 0.4%.

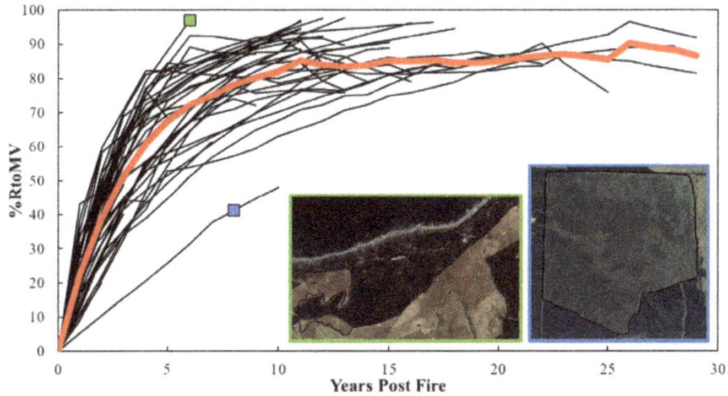

Figure 9. Average post-fire recovery (*%RtoMV*) trajectories for all recorded fires (≥1 km^2) during the Landsat record (1988–2020). Only burned pixels (*dNBR* ≥ 100) analyzed. Red: overall average recovery. Green: Google Earth image six years after the fastest recovering fire. Blue: eight years after slowest recovering fire. Black line indicates fire polygon area for both fires [45]. Note that a fire can only reach 100 *%RtoMV* if every burned pixel comes to its *mNBR* value in the same post-fire year.

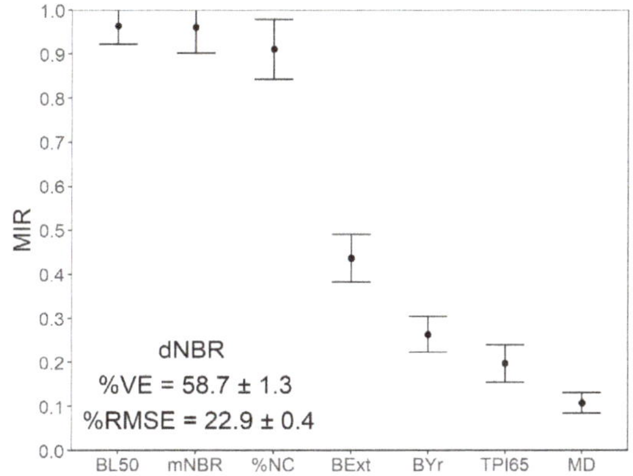

Figure 10. RF outputs for burn severity (*dNBR*) over 30 runs (Mean ± SD). Variable importance quantified by MIR. See Table 3 for full variable names and descriptions. %NC is the percentage of burned pixels in each fire comprised of nature conservation land use (Figures 1 and 8A).

The strength and direction of the relationships between each of these variables and *dNBR* vary at the fire scale. BL50 has a significant positive relationship with *dNBR*, with 30% variance explained (Figure 11A). This relationship is significant throughout the range of BL50; fires in areas that have burned more in the past burn more severely. When there are fires, areas that have not burned in the last 50 years burn less severely (*dNBR*: 367 ± 129) than areas that had burned once or only partially (*dNBR*: 497 ± 138), which in turn burn less severely than areas that have burned more than once (*dNBR*: 666 ± 115). *mNBR* has a significant positive relationship with *dNBR*, with 36% variance explained (Figure 11B). This relationship is most strong at lower levels of severity, with 68% variance in *dNBR* explained by *mNBR* when *mNBR* is below 450. Fires in areas with a greater vegetation capacity (*mNBR* > 600) burn significantly more severely (*dNBR*: 579 ± 139) than fires in low vegetation areas (*mNBR* < 450; *dNBR*: 348 ± 98), but this mostly levels off past 450 *mNBR* where there is no significant difference from 600 *mNBR*. %NC has a significant positive relationship with *dNBR*, with 23% variance explained (Figure 11C). %NC forms three distinct groups, where fires in areas with little to no nature conservation land use burn significantly less severely (*dNBR*: 327 ± 144) than the two groups with higher %NC (524 ± 117 and 496 ± 167). Interestingly, the middle group of %NC, representing fires with nature conservation cover between 16% and 64%, has a strong relationship with *dNBR* by itself (R^2 = 0.74***). This may be because fires that contain an increasingly mixed land use of nature conservation and mostly residual native cover are generally larger fires that spill over into both land use settings. Relatedly, BExt has a significant positive relationship with *dNBR*, with 35% variance explained (Figure 11D). Within the relationship, there is a significant increase between fires that burn an area less than 10 km^2 (BExt < 1; *dNBR*: 426 ± 152) and larger fires (621 ± 136), especially the most extensive and severe fires from 2008 and 2020. Although BYr, TPI65 and MD marginally improve RF model performance, they do not have a significant relationship with *dNBR* by themselves.

3.3. *Recovery: Landscape-Level Influences*

RF modeling identifies six landscape-level variables that together most influence post-fire recovery (*%RtoMV*) (Figure 12). In the initial years 1–3, surviving vegetation (*postNBR*) is most important. This importance relative to the other variables decreases over time, especially in year 3 where eucalypt woodland percentage (EucPer), dNBR, SLB, BL40 and BYr all increase in MIR. In year 4 SLB is the most important and in year 5 EucPer is the most important. In these middle years, all included variables have a stronger relative influence on *%RtoMV* with no variable having a MIR below 0.4. In the later years 6–10, BYr is the most important. This relative influence on *%RtoMV* becomes stronger towards year 10, with all other variables having a MIR below 0.5 by year 8. Generally, *postNBR* linearly decreases in importance, while BYr follows a sigmoid shape of increasing importance from years 1–10. The other variables (EucPer, dNBR, SLB, BL40) are less important in the initial (i.e., 1–2) and later (i.e., 8–10) years but peak in importance during the middle (i.e., 4–5) years as influence on *%RtoMV* transitions from *postNBR* to BYr. Overall, model performance increases from years 1–10 with %VE increasing from 41.3 ± 1.5% to 69.9 ± 0.9% and %RMSE remaining mostly stable from 7.6 ± 0.1% to 7.0 ± 0.1%.

The relationship between each of these variables and *%RtoMV* varies in strength, direction, and through time. Burned areas with little vegetation remaining (i.e., *postNBR* < 0) recover faster than areas with more vegetation remaining (i.e., *postNBR* > 0), and these differences are most significant in post-fire year 1–3 (Figure 13A). In year 1, *postNBR* < −200 fires have recovered almost double (*%RtoMV* 28.3 ± 7.4) compared to *postNBR* > 0 fires (15.7 ± 6.6%). Higher severity (*dNBR* ≥ 400) burn areas recover faster than lower severity (*dNBR* < 400) burns, and these differences are most significant in years 3–10 (Figure 13B). In year 3, when dNBR reaches its highest influence on *%RtoMV*, higher severity burns have recovered 55.7 ± 9.4% while lower severity burns have recovered 42.8 ± 13.2.%. Areas that burn where historical fires have not occurred (i.e., BL40 = 0) recover somewhat faster than areas with frequent fires (i.e., BL40 > 1), and this difference is significant in years 4–7 (Figure 13C). Areas that burn where there was no recent previous fire (i.e., SLB 30+) recover faster than areas that had already burned recently (i.e., SLB < 30), and these differences are most significant in years 4–6 (Figure 13D). In year 4, when SLB is the most influential variable, SLB < 30 fires have recovered 53.9 ± 15.0% while SLB 30+ fires have recovered 66.0 ± 11.1%. Areas that have a higher percentage of eucalypt woodlands (i.e., EucPer > 40%) recover faster than areas that have lower amounts of this vegetation group and these differences are most significant in years 3–10 (Figure 13E). In year 5, when EucPer is the most influential variable, EucPer > 40% fires have recovered 75.7 ± 8.2% while EucPer < 10% fires have recovered 59.5 ± 13.9%. This difference is among the largest and most significant of any variable at any year. More recent fires (i.e., BYr > 2000) have recovered faster than older fires (i.e., BYr < 2000), and this difference is most significant in years 4–10 (Figure 13F). The difference between 1990s fires and 2000s fires in years 8–10 represents the most significant difference between any two groups of fires in terms of *%RtoMV*, with 1990s fires at 70.8 ± 9.1% and 2000s fires at 87.9 ± 5.0% in year 10.

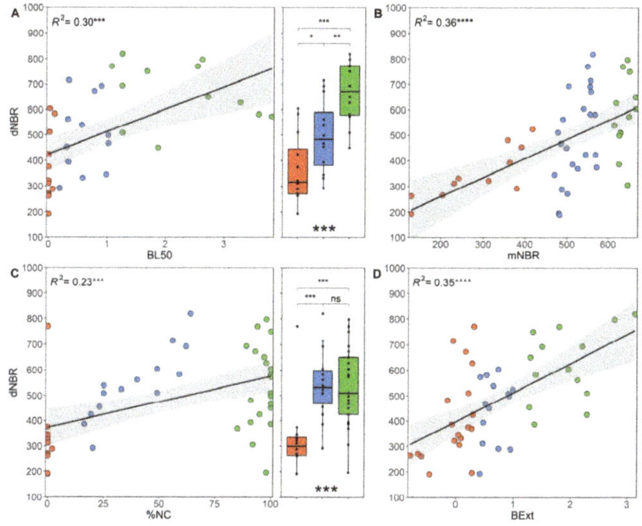

Figure 11. Relationship between variables that improved RF model performance and *dNBR*. Only variables that have a significant relationship with *dNBR* shown. From most to least important: (**A**) BL50 (**B**) *mNBR*, (**C**) %NC, (**D**) BExt. See Table 3 for full variable names and descriptions. Each scatterplot includes a boxplot with fires split into groups based on differences within each variable. MW between group test significance is shown above each boxplot, and KW overall test significance is shown below.

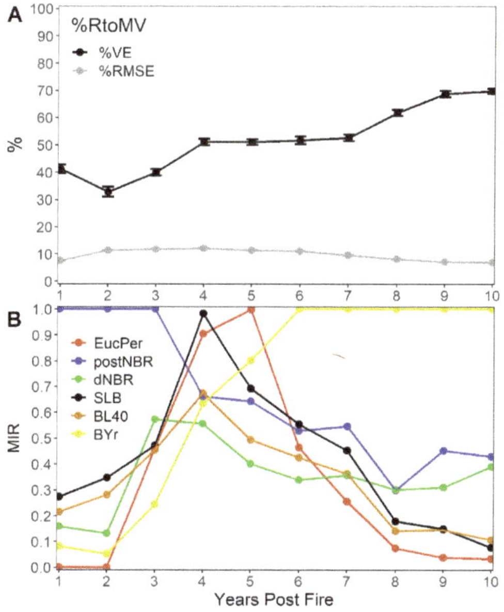

Figure 12. RF outputs for fire recovery (*%RtoMV* years 1–10) over 30 runs for each year. (**A**) Overall model results over time (Mean ± SD). (**B**) Variable importance over time as quantified by MIR. SD bars are not included here to make the figure legible. See Table 3 for full variable names and descriptions. EucPer is the percentage of burned pixels in each fire comprised of the eucalypt woodlands vegetation group (Figure 8B).

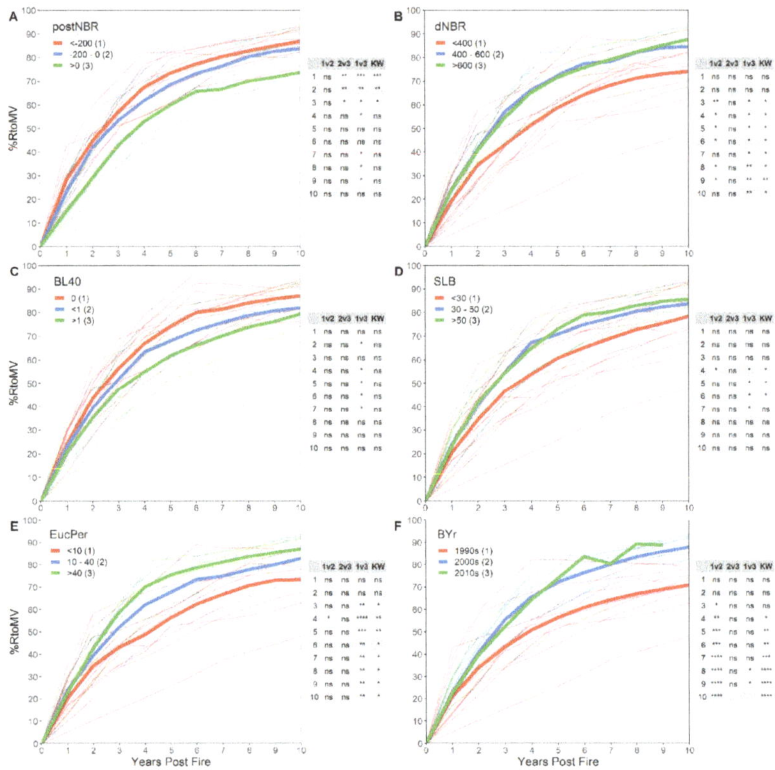

Figure 13. Relationship between variables that improved model performance and *%RtoMV* for the first 10 years post-fire. (**A**) *postNBR*, (**B**) *dNBR*, (**C**) BL40, (**D**) SLB, (**E**) EucPer (% Eycalypt woodlands vegetation group), (**F**) BYr. See Table 3 for full variable names and descriptions. Each time-series shows fires split into groups based on differences within each variable. MW between group test significance is shown in a table next to each plot and is based on numbers in each legend. KW overall test significance is shown in the far-right column.

4. Discussion

4.1. Fire History in the Context of the 2019–2020 Bushfires

Remote sensing time-series results confirm that the 2019–2020 bushfires were the most widespread and devastating to occur on Kangaroo Island since the late-1980s and likely since at least the early 1930s. They were the largest in extent, burned the most severely, and affected large areas that had not experienced burns in the modern record (Figures 5–7). That the second most widespread and severe fire season was only 12 years prior should be concerning if we hope to maintain the island's unique natural environment. Both the 2020 and 2008 fires occurred in the context of increasingly hot temperatures across the island (Figure 2A). Furthermore, both fire seasons occurred after at least three years of below-average annual precipitation (Figure 2B), which undoubtedly led to dry and fire-prone environments.

Although local and national authorities do not have direct control of temperature and precipitation, they can mitigate damage through the implementation of fuel-reduction burning, firebreaks, and other forest management techniques [7]. The relative lack of fire, especially in the decade before 2020, likely contributed to fuel build-up and higher burn severity (Figures 6 and 7). Although a local project on the island recommended fire intervals of 17–40 years [35], scientific studies on eucalypt forests elsewhere in southern Australia have found that shorter intervals of 6–10 years, through prescribed burns, are needed to effectively reduce fire risk [34,61]. Prescribed burns have been recorded on the island since the early 2000s, but only at small scales (32.6 km^2 recorded; Figure 6B) and as part of trial projects [62]. More involved forest management practices may be required to prevent future mega-fires. This is especially true since the four most common tree species on Kangaroo Island (*E. diversifolia, E. remota, E. cladocalyx* and *E. baxteri*) are fast growing (leading to large fuel buildup), highly flammable (from combustible oils and leaves and dense litter) and have the capacity to withstand and recover from fires (with lignotubers or epicormic buds). Prescribed burns can be used to limit fuel buildup, allow for controlled burning of flammable material and approximate a natural part of the life cycle for these fire-adapted species.

As in another Landsat time-series study of forest recovery in south-east Australia [26], post-fire NBR for Kangaroo Island fires recovers and stabilizes after about ten years (Figure 9). This is notably faster than in the western United States [16,28], Spain [12], boreal Canada [28] and elsewhere. These studies come from many different environments and employ different methods but our results indicate that Kangaroo Island eucalypt forests are relatively well adapted to fire and can quickly assume the regeneration process [35]. Depending on the specific eucalypt species impacted, this may occur with resprouting from lignotubers that protect buds from fire damage or from epicormic buds that sprout post-fire. However, even fire-tolerant eucalypts may not be able to survive higher severity fires, as shown from a study in south-eastern Australia [63]. These fires produce greater seedling regeneration but also a higher likelihood of eucalypt mortality, with the potential to eventually transition eucalypt forests to more open woodland or savannah environments. NBR recovery from the large 2008 fires (93 ± 6% after 10 years) indicates that eucalypt forests may be able to recover in the same manner after the 2020 fires. However, the unprecedented extent and severity of the 2020 fires should still be concerning because they may kill so many trees that recovery from re-sprouting becomes inadequate at the landscape-level, which means that eucalypt forests may not be able to recover their pre-fire density [63]. To end, it should be noted that we did not see any significant impact of specific eucalypt species (Table 2) on results at this scale, but did see significantly higher severity in nature conservation land use (which is mostly mallee and eucalypt woodlands) and faster recovery in the eucalypt woodland vegetation group.

4.2. Influences on Burn Severity

Included variables explain well over half of the variation in burn severity (i.e., *dNBR*). Severity is influenced mostly by fire frequency (i.e., BL50), vegetation capacity (i.e., *mNBR*) and land use (i.e., %NC) (Figure 10), with nature conservation areas of higher fuel loads and more historical fires burning more severely (Figure 11). Burn extent also has moderately strong and significant relationships with severity when compared individually, but RF results show that in a multivariate scenario, much of those relationships are explained by *mNBR*, BL50 and %NC. The influence of vegetation capacity on severity is not surprising, as ground-level and remote sensing analysis have come to similar conclusions elsewhere [37]. On Kangaroo Island, Peace and Mills [46] explain that Flinders Chase NP has both dense eucalypt canopies (esp. *E. remota*) and understory vegetation, especially in riparian areas, which led to high fuel loads and thus the severity of the 2008 fires.

Flinders Chase NP and other nature conservation land use areas (Figure 1) overlap with the mallee woodlands vegetation group (Figure 8). In this vegetation group, eucalypt species grow in multiple stems with a shorter and shrubby growth form [42]. The abundance of mallees in these areas is not surprising, as they burn often (Figure 5) are highly flammable and this type of growth form is common when lignotuber eucalypts are resprouting [64]. It is also possible that frequent burning leads to an increase in low canopy biomass for combustible mallees and understory shrub species, which would provide amble fuel for higher severity fires. Furthermore, the lack of modification and interference in nature conservation land use (esp. wilderness areas), means fuel build up is less managed [43]. It is thus likely that the importance of %NC in influencing dNBR is related in part to both fire frequency (i.e., BL50) and increased biomass fuel (i.e., *mNBR*).

In terms of fire history, studies in western North America have found that burn severity is lower when time since last fire (i.e., SLB) is lower, depressing severity even 20–30 years later [38,39]. We do not see this on Kangaroo Island, where *dNBR* is above average in areas that burned less than 10 years previously and have experienced multiple fires in the last 50 years. This difference can likely be explained by the relatively quick recovery of vegetation observed (Figure 9). However, our results do broadly correspond with Estes et al. [36], who found that severity was highest 30–50 years after the previous fire. The five highest severity fires on the island occurred in areas that burned between 1 and 3 times in the last 50 years and 24–36 years after the previous fire, with the somewhat shorter interval potentially related to faster regrowth of fire-adapted eucalypts. These studies do not consider a longer timeline in terms of the number of fires over multiple decades (i.e., BL50), which was a more important influencer than SLB.

The ecological reasoning for higher *mNBR* leading to more severe burns is clear and related to increased fuel loads in these areas, as discussed previously. This type of connection for BL50 is less clear, but it may be that areas subject to more frequent fires contain species (i.e., *Eucalyptus* spp.) that are better adapted to them and even require higher severity burns (although not too high) for most productive bud resprouting, seedling regeneration and subsequent regrowth [63]. Whatever the reasoning, these results provide valuable information to help mitigate future mega-fires like that of 2020. A key requirement for mitigation should be the limitation of fuel build-up through active pre-fire management methods [7,37]. However, it may not be feasible to manage all densely vegetated areas. Therefore, we recommended forest managers be especially conscious of fuel build up in those densely vegetated areas that have also burned many times in the past and to prioritize these types of settings if resources are limited. Furthermore, care should be given to limiting fire expansion from nature conservation areas (where managers may have less control on things like fuel buildup) to other settings. In 2020, bushfires were able to make the jump from Flinders Chase NP to forest plantations and residual native cover areas in the central part of the island, where corridors of remnant eucalypt forest likely facilitated fire expansion into large swaths of agricultural land.

4.3. Influences on Post-Fire Recovery

Included variables explain much of the variation in post-fire recovery (i.e., *%RtoMV*), with %VE increasing in later years. Recovery can be split into three periods with different landscape-level influences (Figures 12 and 13). In the initial years (i.e., 1–3), surviving vegetation (i.e., *postNBR*) most dictates recovery, with areas that have less remaining vegetation recovering more quickly. *postNBR* is significantly related to burn severity ($R^2 = 0.45$ ****), with higher severity burns usually leaving less remaining vegetation. Here, we find that higher *dNBR* burns recover more quickly than lower *dNBR* burns, supporting those studies that have come to similar conclusions [16,23,29]. However, *postNBR* outcompetes *dNBR* for importance in our RF analysis, especially in the initial years, indicating that more attention should be paid to recovery in different post-fire environments separate from severity.

In the middle years (i.e., 4–5), time since last burn and eucalypt woodlands percentage is most influential, but this appears to be a transitional period. After this, variables related to the past situation of the burn (i.e., *postNBR, dNBR*), previous fire history (i.e., SLB, BL40), and vegetation structure (i.e., EucPer) decline in relative influence on recovery. However, the strong relationship between eucalypt woodlands abundance and recovery rate for most post-fire years (Figure 13E) deserves more discussion. Eucalypt woodlands are common in the most densely vegetated interior areas of Flinders Chase NP and along riparian corridors here and across Kangaroo Island. We have previously established that eucalypts burn severely and that severe burns are associated with faster recovery. Eucalypt woodlands have a denser canopy than mallee woodlands [42] and their abundance in riparian areas provide both more opportunity for survival to recovery and better access to moisture [65], allowing for a relatively fast transition from burn scar to recovering mallees to denser eucalypt woodlands. That eucalypt woodlands are common in riparian corridors may explain why certain topographic variables like topographic position and wetness did not influence results.

It is in the later years (i.e., 6–10) that BYr becomes the clearly dominant influencer with 1990s fires recovering significantly slower than more recent fires. This indicates that the current (relative to burn year) context for recovering areas takes precedent over past disturbances. Bright et al. [16] found, nine years post-fire, that precipitation and temperature variables explained much of the variation in recovery. On Kangaroo Island, it also appears that climate change between fires in different time periods is responsible for variations in later year recovery. In the western portion, where most fires have occurred, summer temperatures 6–10 years after the faster recovering 2000s fires were 1.6 °C warmer on average, and annual temperatures 1° C warmer, than the later recovery years for 1990s fires (Figure 2A). Total annual precipitation 6–10 years after the 2000s fires (693 mm) averaged 76 mm more annually than the same period for 1990s fires (617 mm), a 12% increase (Figure 2B). It appears that BYr may represent a proxy of differences in climate through time, becoming most influential in later recovery years. Malak and Pausas [29] found that precipitation begins to directly influence post-fire vegetation only after seven years in a Spanish dryland setting. Eucalypt forests in this region are also moisture limited, meaning that short-term precipitation increase in the 2010s has likely contributed to the faster recovery of more recent fires, even in the broader context of precipitation decrease in the region due to global climate change [65].

As Kangaroo Island recovers from the 2019–2020 bushfires in the coming decade, historical recovery trends and drivers can provide some insights into how this may occur, with the caveat that their larger extent and higher severity are unprecedented in the modern record (Figures 6 and 7). Based on our findings, land managers could expect recovery to be fastest in the first few years post-fire, especially where it burned more severely, and there is little surviving vegetation. Severely burned eucalypt woodlands growing along riparian corridors are prime candidates for fast recovery based on our results. Parts of the severely burned Flinders Chase NP is already going through early recovery stages with bud resprouting, epicormic growth and fast-growing eucalypts and shrub species beginning to green up just weeks after the burn [66]. The impact of previous burn history should also be considered, with areas that have experienced more frequent and recent fires (i.e., higher BL40 and lower SLB) recovering slower than less burned areas. This should be especially concerning when you consider that climate change in south-east Australia is expected to further increase fire frequency in the coming decades [5], which may lead to heavily burned areas failing to recover their historical tree canopy density or transitioning to a new type of ecosystem [31]. In the long term, it is climate, especially precipitation, which appears to most influence the recovery rate.

4.4. Limitations

This study made use of Landsat imagery and other freely available data sources to monitor the recent fire history of Kangaroo Island. As such, we did not conduct field work, but did collect information on vegetation types and species composition [42]. As our and other results show, vegetation type in Australia does play a role in remotely sensed burn severity [32] and recovery [22]. However, it is not a static variable with factors like fire frequency potentially leading to lasting vegetation type change [31]. Even though vegetation type changes are likely minor over our 33-year study period, we should note that differences between fire dates and data dates for land use and vegetation groups may have some impact.

Climate was not directly incorporated into our analysis, which instead focused on landscape-level variables related to the current burn, fire history, vegetation structure, and topography. We did not anticipate that climate would play a major role since the analyzed fires occurred on an island with a mostly consistent climate. However, considering these fires occurred in the context of a changing climate (Figure 2), it becomes clear there is an influence on the probability of a severe fire occurring and on long-term post-fire recovery. A more robust examination of climatic influences on fire history in the region, especially in the context of a warming and drying climate, would be beneficial.

Topographical variables had only a minor influence on burn severity (i.e., TPI65) and no influence on post-fire recovery. Other studies, mostly concerned with recovery, have found topographical influence from strong [31] to weak [16] to none [12]. The lack of topographical influence in this study may be due to the landscape-scale of the analysis. When variables, including topography, are averaged at the fire level, the spatial variability within the burn area is lost. Having the unit of analysis being the fire itself is advantageous for building broader-scale understanding and monitoring large numbers of fires, but future research may consider combining this full fire history approach with finer-scale analysis within each fire. Recent studies have considered the spatial variability within multiple fires in their analysis, but with a smaller number of fires than analyzed here [16,24]. Future research may consider splitting individual fires into vegetation group, topography, and other subsets, while still analyzing a large number of fires to build more robust conclusions.

Missing data (i.e., MD), especially in the early years of analysis (Figure 3), was also a concern. If no Landsat imagery is available over a fire the year that it burns, LandTrendr will fit the decline in NBR based on the next available year (Figure A1). Some vegetation will have regrown in the intervening period, leading to the expectation that these fires will be artificially less severe than others even if adjusting *dNBR* years account for this. Despite this, MD only has a minor influence on severity in RF analysis and a comparison between those fires where part of the burn area had missing data, and other fires shows no significant difference in *dNBR* (Figure A4A). More vegetation may have also regrown depending on the fire date, compared to Landsat image collection date. Fires dated before December 25 (i.e., shorter time until Landsat date) were compared to fires dated after March 25 (i.e., longer time) and there is a minor difference in *dNBR* that is not significant (Figure 4B). A combined approach, highlighting fires that have either missing data or are dated after March 25 similarly show a minor, but not significant difference in *dNBR* (Figure A4C). MD does not influence post-fire recovery and fire date had no influence on either severity or recovery RF models. In future, near-real time (as opposed to yearly) fitting algorithms for Landsat such as Stochastic Continuous Change Detection can be used to limit the time gap between fire disturbance and post-fire detection [67].

5. Conclusions

Landsat images through time, supplemented with other spatial datasets available at no cost, provide useful observations about recent fire history and associated drivers. On Kangaroo Island, we found a recent (1988–200) history of frequent fires, with the 2008 and 2020 seasons being most widespread and severe. The fire-adapted eucalypt-dominant vegetation on the island shows a remarkable capacity to regrow, with %RtoMV stabilizing in only ten years on average. However, there is cause for concern, as those areas that have most frequently burned are also recovering more slowly. As climate change leads to warmer and dryer conditions, more frequent fires may curtail the recovery of these forests and potentially trigger transitions to more open woodlands or entirely new ecosystems. More generally, we found that burn severity was most influenced by variables representing fire frequency, vegetation capacity and land use (i.e., nature conservation). Land managers interested in reducing severity should consider fuel-reduction practices where allowed, especially in areas that have been subject to more frequent fires. Furthermore, we found varying influences on yearly post-fire recovery, with variables representing surviving vegetation important in early years and an inferred proxy of post-fire climate important in later years after a transitional period where the time since the previous burn and eucalypt woodland percentage was important. In other eucalypt environments, we might expect those most severely burned areas with less remaining vegetation to have faster initial recovery and in later years for growth to be guided by precipitation in moisture limited settings.

Results from this study suggest that even without access to plot-level information or other fine-scale data, a comprehensive understanding of fire history and influences on severity and recovery processes can be obtained. This type of analysis can be conducted efficiently, freely, and remotely, especially at present with travel restrictions due to COVID-19. The cloud computing and big data capabilities of GEE and the trajectory fitting approach of LandTrendr prove to be an effective combination for building landscape-level understandings of environmental change, especially when those changes are recent, and there is a need for timely analysis.

Author Contributions: Conceptualization, M.T.B., Y.H. and S.W.M.; methodology, M.T.B.; software, M.T.B.; validation, M.T.B.; formal analysis, M.T.B.; investigation, M.T.B. and Y.H.; resources, M.T.B. and Y.H.; data curation, M.T.B.; writing—original draft preparation, M.T.B.; writing—review and editing, M.T.B., Y.H. and S.W.M.; visualization, M.T.B.; supervision, Y.H.; project administration, Y.H.; funding acquisition, M.T.B. and Y.H. All authors have read and agreed to the published version of the manuscript.

Funding: This research was funded by the Natural Sciences and Engineering Research Council of Canada (NSERC) through the NSERC Discovery Program awarded to Dr. Yuhong He. It was also supported through the NSERC Postgraduate Scholarship, Centre for Urban Environments (University of Toronto Mississauga), and Department of Geography, Geomatics and Environment funding to Mitchell Bonney.

Acknowledgments: Analysis was conducted at the Remote Sensing and Spatial Ecosystem Modeling laboratory in Mississauga. We thank lab members for their advice and support throughout the project. We thank Justin Braaten for his correspondence related to LandTrendr and GEE. We thank Benjamin Bright for his correspondence related to NBR trajectories.

Conflicts of Interest: The authors declare no conflict of interest. The funders had no role in the design of the study; in the collection, analyses, or interpretation of data; in the writing of the manuscript, or in the decision to publish the results.

Appendix A

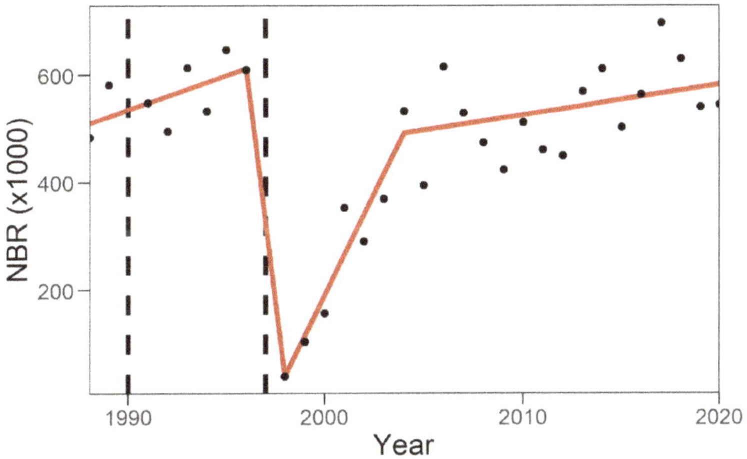

Figure A1. Example pixel that experienced a fire in December 1996. Black dots are medoid surface reflectance values, and the red line is the LandTrendr fitted model as in Figure 4. Vertical dashed black lines indicate years with no available imagery. The *postNBR* year 1997 is missing, so LandTrendr fits a two-year decline to 1998. In this example, *preNBR* is 1996 and *postNBR* is now 1998.

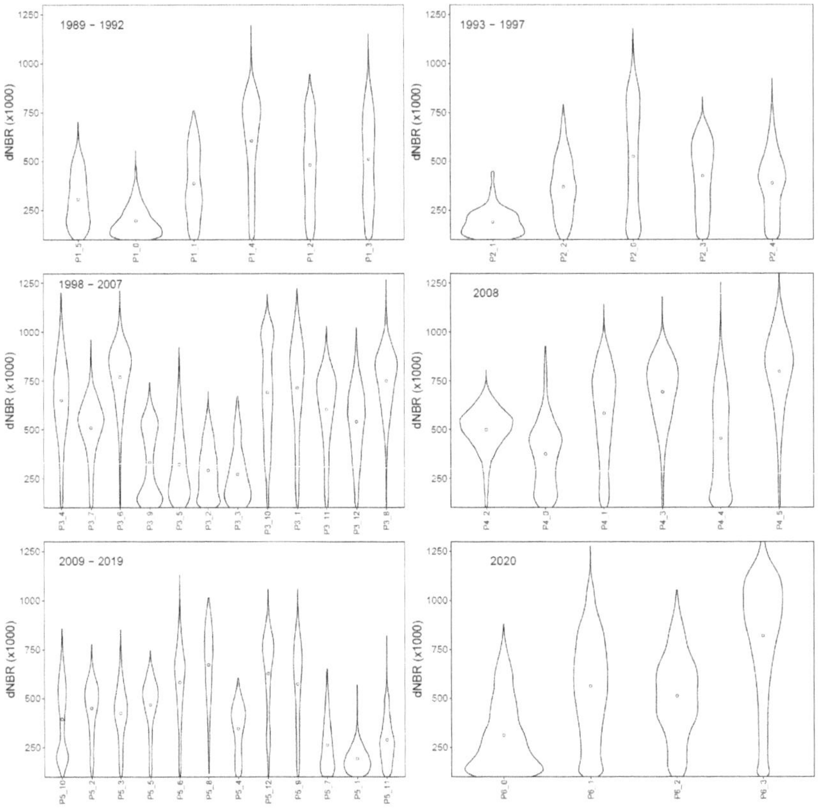

Figure A2. Violin plots showing the distribution of burn severity (*dNBR*) around the mean for all recorded fires. Split into six periods, as in Figure 7.

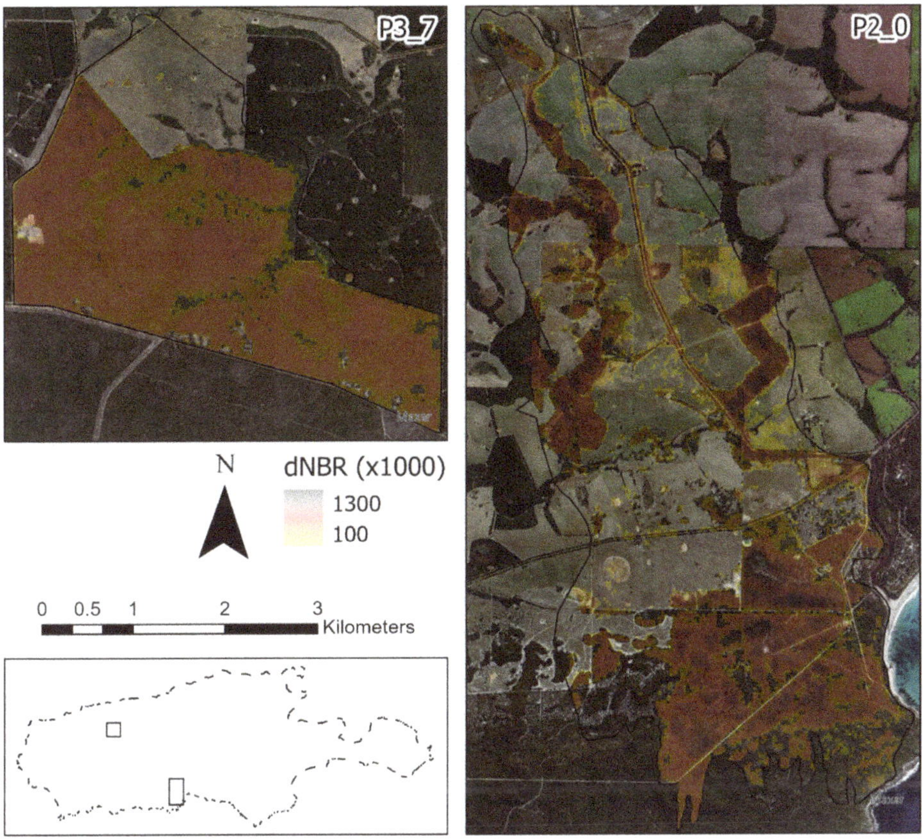

Figure A3. Closer view of two select fires (P3_7, P2_0), with *dNBR* set to 75% transparency. P3_7 burned in February 2001 at the northern edge of Flinders Chase NP. P2_0 burned in December 1994 just west of Vivonne Bay. In both cases (and for other fires that burn in these types of landscapes), *dNBR* values above 100 almost exclusively occur in treed and dense shrubby land cover. Grasslands, agriculture and other non-woody land cover, although within the fire polygons [45] have very limited influence. Note that the ArcGIS basemap imagery over these areas are dated to 2019 (P3_7) and 2018 (P2_0) so there may have been landscape change in intervening period.

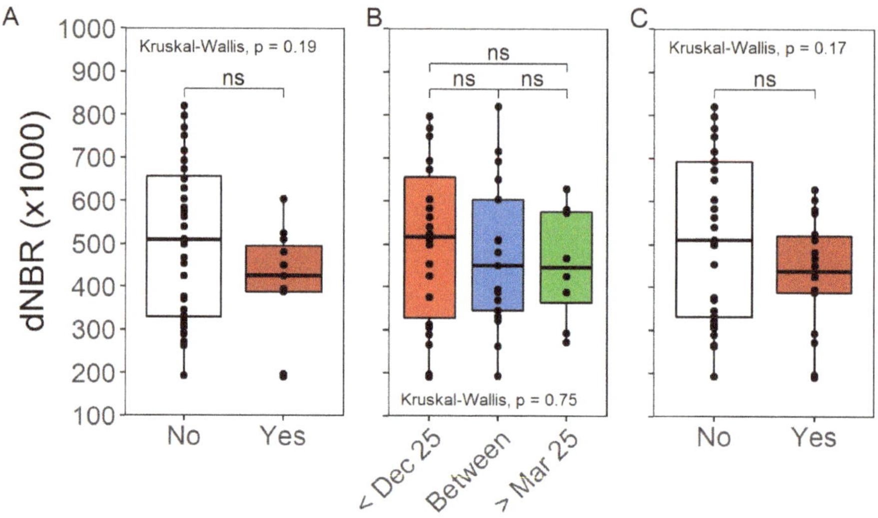

Figure A4. Influence of Landsat image time-series on *dNBR*. (**A**) Missing data: No = little to no influence of missing data on *dNBR* years. Yes = Notable portion of burn area impacted by missing data in *dNBR* years. (**B**) Fire dates [45]: < Dec 25 = fire dated before December 25 in fire season. > Mar 25 = fire dated after March 25 in fire season. Between = fire dated between December 25 and March 25 in fire season. Because image collection dates were December 25–March 25, the fires dated before these dates get a *postNBR* value soon after the burn while the fires dated after these dates get a post *NBR* value in the next year. The *postNBR* lag for fires dated during the image collection window depends on the medoid image that was selected over that area. (**C**) Combination: Yes = fires that either were impacted by missing data or are dated after March 25. These fires have the longest lag between burn date and *postNBR* date. No = all other fires.

References

1. Richards, L.; Brew, N.; Smith, L. 2019–20 Australian Bushfires—Frequently Asked Questions: A Quick Guide. Available online: https://www.aph.gov.au/About_Parliament/Parliamentary_Departments/Parliamentary_Library/pubs/rp/rp1920/Quick_Guides/AustralianBushfires (accessed on 1 August 2020).
2. Royal Commission into National Natural Disaster Arrangements. Interim Observations, 31 August 2020. Available online: https://naturaldisaster.royalcommission.gov.au/publications/interim-observations-1 (accessed on 15 November 2020).
3. Khalil, S. Australia Fires: 'Apocalypse' Comes to Kangaroo Island. Available online: https://www.bbc.com/news/world-australia-51102658 (accessed on 3 August 2020).
4. Tarabay, J. There's No Place Like Kangaroo Island. Can It Survive Australia's Fires? Available online: https://www.nytimes.com/2020/02/04/world/australia/kangaroo-island-fire.html (accessed on 24 June 2020).
5. Clarke, H.G.; Smith, P.L.; Pitman, A.J. Regional signatures of future fire weather over eastern Australia from global climate models. *Int. J. Wildland Fire* **2011**, *20*, 550–562. [CrossRef]
6. Australian Government: Bureau of Meteorology. Historical Weather Observations and Statistics. Available online: http://www.bom.gov.au/climate/data-services/station-data.shtml (accessed on 15 May 2020).
7. Attiwill, P.M.; Adams, M.A. Mega-fires, inquiries and politics in the eucalypt forests of Victoria, South-Eastern Australia. *For. Ecol. Manag.* **2013**, *294*, 45–53. [CrossRef]
8. Yang, J.; Tian, H.; Tao, B.; Ren, W.; Lu, C.; Pan, S.; Liu, Y. Century-scale patterns and trends of global pyrogenic carbon emissions and fire influences on terrestrial carbon balance. *Glob. Biogeochem. Cycles* **2015**, *29*, 1549–1566. [CrossRef]
9. Viedma, O.; Meliá, J.; Segarra, D.; García-Haro, J. Modeling rates of ecosystem recovery after fires by using Landsat TM data. *Remote Sens. Environ.* **1997**, *61*, 383–398. [CrossRef]

10. White, J.D.; Ryan, K.C.; Key, C.C.; Running, S.W. Remote sensing for forest fire severity and vegetation recovery. *Int. J. Wildland Fire* **1996**, *6*, 125–136. [CrossRef]
11. Wulder, M.A.; Masek, J.G.; Cohen, W.B.; Loveland, T.R.; Woodcock, C.E. Opening the archive: How free data has enabled the science and monitoring promise of Landsat. *Remote Sens. Environ.* **2012**, *122*, 2–10. [CrossRef]
12. Vicente-Serrano, S.M.; Pérez-Cabello, F.; Lasanta, T. *Pinus halepensis* regeneration after a wildfire in a semiarid environment: Assessment using multitemporal Landsat images. *Int. J. Wildland Fire* **2011**, *20*, 195–208. [CrossRef]
13. Gorelick, N.; Hancher, M.; Dixon, M.; Ilyushchenko, S.; Thau, D.; Moore, R. Google earth engine: Planetary-scale geospatial analysis for everyone. *Remote Sens. Environ.* **2017**, *202*, 18–27. [CrossRef]
14. Kennedy, R.E.; Yang, Z.; Gorelick, N.; Braaten, J.; Cavalcante, L.; Cohen, W.B.; Healey, S. Implementation of the LandTrendr algorithm on google earth engine. *Remote Sens.* **2018**, *10*, 691. [CrossRef]
15. Kennedy, R.E.; Yang, Z.; Cohen, W.B. Detecting trends in forest disturbance and recovery using yearly Landsat time series: 1. LandTrendr—Temporal segmentation algorithms. *Remote Sens. Environ.* **2010**, *114*, 2897–2910. [CrossRef]
16. Bright, B.C.; Hudak, A.T.; Kennedy, R.E.; Braaten, J.D.; Khalyani, A.H. Examining post-fire vegetation recovery with Landsat time series analysis in three Western North American forest types. *Fire Ecol.* **2019**, *15*. [CrossRef]
17. Quintero, N.; Viedma, O.; Urbieta, I.R.; Moreno, J.M. Assessing landscape fire hazard by multitemporal automatic classification of Landsat time series using the google earth engine in West-Central Spain. *Forests* **2019**, *10*, 518. [CrossRef]
18. Frazier, R.J.; Coops, N.C.; Wulder, M.A.; Hermosilla, T.; White, J.C. Analyzing spatial and temporal variability in short-term rates of post-fire vegetation return from Landsat time-series. *Remote Sens. Environ.* **2018**, *205*, 32–45. [CrossRef]
19. Meng, R.; Dennison, P.E.; Huang, C.; Moritz, M.A.; D'Antonio, C. Effects of fire severity and post-fire climate on short-term vegetation recovery of mixed-conifer and red fir forests in the Sierra Nevada mountains of California. *Remote Sens. Environ.* **2015**, *171*, 311–325. [CrossRef]
20. Pickell, P.D.; Hermosilla, T.; Frazier, R.J.; Coops, N.C.; Wulder, M.A. Forest recovery trends derived from Landsat time series for North American boreal forests. *Int. J. Remote Sens.* **2015**, *37*, 138–149. [CrossRef]
21. White, J.C.; Wulder, M.A.; Hermosilla, T.; Coops, N.C.; Hobart, G.W. A Nationwide annual characterization of 25 years of forest disturbance and recovery for Canada using Landsat time series. *Remote Sens. Environ.* **2017**, *194*, 303–321. [CrossRef]
22. Sever, L.; Leach, J.; Bren, L. Remote sensing of post-fire vegetation recovery; a study using Landsat 5 TM imagery and NDVI in North-East Victoria. *J. Spat. Sci.* **2012**, *57*, 175–191. [CrossRef]
23. Epting, J.; Verbyla, D. Landscape-level interactions of prefire vegetation, burn severity, and postfire vegetation over a 16-year period in interior Alaska. *Can. J. For. Res.* **2005**, *35*, 1367–1377. [CrossRef]
24. Liu, Z. Effects of climate and fire on short-term vegetation recovery in the boreal larch forests of Northeastern China. *Sci. Rep.* **2016**, *6*, 37572. [CrossRef]
25. Schroeder, T.A.; Wulder, M.A.; Healey, S.P.; Moisen, G.G. Mapping wildfire and clearcut harvest disturbances in boreal forests with Landsat time series data. *Remote Sens. Environ.* **2011**, *6*, 1421–1433. [CrossRef]
26. Hislop, S.; Jones, S.; Soto-Berelov, M.; Skidmore, A.; Haywood, A.; Nguyen, T.H. Using Landsat spectral indices in time-series to assess wildfire disturbance and recovery. *Remote Sens.* **2018**, *10*, 460. [CrossRef]
27. Fernandez-Manso, A.; Quintano, C.; Roberts, D.A. Burn severity influence on post-fire vegetation cover resilience from Landsat MESMA fraction images time series in Mediterranean forest ecosystems. *Remote Sens. Environ.* **2016**, *184*, 112–123. [CrossRef]
28. Yang, J.; Pan, S.; Dangal, S.; Zhang, B.; Wang, S.; Tian, H. Continental-scale quantification of post-fire vegetation greenness recovery in temperate and boreal North America. *Remote Sens. Environ.* **2017**, *199*, 277–290. [CrossRef]
29. Malak, D.A.; Pausas, J.G. Fire regime and post-fire normalized difference vegetation index changes in the Eastern Iberian peninsula (Mediterranean basin). *Int. J. Wildland Fire* **2006**, *15*, 407–413. [CrossRef]
30. Díaz-Delgado, R.; Lloret, F.; Pons, X. Influence of fire severity on plant regeneration by means of remote sensing imagery. *Int. J. Remote Sens.* **2003**, *24*, 1751–1763. [CrossRef]
31. Meng, R.; Dennison, P.E.; D'Antonio, C.M.; Moritz, M.A. Remote sensing analysis of vegetation recovery following short-interval fires in Southern California shrublands. *PLoS ONE* **2014**, *9*, 110637. [CrossRef]

32. Hammill, K.A.; Bradstock, R.A. Remote sensing of fire severity in the Blue Mountains: Influence of vegetation type and inferring fire intensity. *Int. J. Wildland Fire* **2006**, *15*, 213–226. [CrossRef]
33. Lippitt, C.L.; Stow, D.A.; O'Leary, J.F.; Franklin, J. Influence of short-interval fire occurrence on post-fire recovery of fire-prone shrublands in California, USA. *Int. J. Wildland Fire* **2013**, *22*, 184–193. [CrossRef]
34. Boer, M.M.; Sadler, R.J.; Wittkuhn, R.S.; McCaw, L.; Greirson, P.F. Long-term impacts of prescribed burning on regional extent and incidence of wildfires—Evidence from 50 years of active fire management in SW Australian forests. *Forest Ecol. Manag.* **2009**, *259*, 132–142. [CrossRef]
35. Dowie, D. *Age Class Distributions of Seven Vegetation Groups on Kangaroo Island. Fahrenheit 451—A Fire Management Program for Biodiversity and Asset Protection on Kangaroo Island*; Government of South Australia: Adelaide, Australia, 2006.
36. Estes, B.L.; Knapp, E.E.; Skinner, C.N.; Miller, J.D.; Preisler, H.K. Factors influencing fire severity under moderate burning conditions in the Klamath Mountains, northern California, USA. *Ecosphere* **2017**, *8*. [CrossRef]
37. Garcia-Llamas, P.; Suárez-Seoane, S.; Taboada, A.; Fernández-García, V.; Fernández-Guisuraga, J.M.; Fernández-Manso, A.; Quintano, C.; Marcos, E.; Calvo, L. Assessment of the influence of biophysical properties related to fuel conditions on fire severity using remote sensing techniques: A Case study on a large fire in NW Spain. *Int. J. Wildland Fire* **2019**, *28*, 512–520. [CrossRef]
38. Parks, S.A.; Miller, C.; Nelson, C.R.; Holden, Z.A. Previous fires moderate burn severity of subsequent wildland fires in two large western US wilderness areas. *Ecosystems* **2013**, *17*, 29–42. [CrossRef]
39. Stevens-Rumann, C.S.; Prichard, S.J.; Strang, E.K.; Morgan, P. Prior wildfires influence burn severity of subsequent large fires. *Can. J. For. Res.* **2016**, *46*, 1375–1385. [CrossRef]
40. Kangaroo Island Council. Kangaroo Island Development Plan. Available online: https://www.kangarooisland.sa.gov.au/council/plans/development/plan (accessed on 3 August 2020).
41. Paton, D.C.; Prescott, A.M.; Davies, R.J.P.; Heard, L.M. The distribution, status and threats to temperate woodlands in South Australia. In *Temperate Eucalypt Woodlands in Australia: Biology, Conservation and Restoration*; Hobbs, R.J., Yates, C.J., Eds.; Surrey Beatty & Sons: Chipping Norton, Australia, 2000; pp. 57–85.
42. Department of Agriculture, Water and the Environment. National Vegetation Information System (NVIS). Available online: https://www.environment.gov.au/land/native-vegetation/national-vegetation-information-system (accessed on 16 November 2020).
43. Australian Government: Department of Agriculture, Water and the Environment. Australian Land Use and Management Classification Version 8 (October 2016). Available online: https://www.agriculture.gov.au/abares/aclump/land-use/alum-classification (accessed on 3 February 2020).
44. Government of South Australia: Department for Environment and Water. Conservation Reserve Boundaries. Available online: https://data.sa.gov.au/data/dataset/conservation-reserve-boundaries (accessed on 13 November 2020).
45. Government of South Australia: Department for Environment and Water. Bushfires and Prescribed Burns History. Available online: https://data.sa.gov.au/data/dataset/e5434c77-9815-48e6-8ea7-fb35c78f6786 (accessed on 3 April 2020).
46. Peace, M.; Mills, G. *A Case Study of the 2007 Kangaroo Island Bushfires*; Technical Report No. 053; Australian Government: Bureau of Meteorology, The Centre for Australian Weather and Climate Research: Melbourne, Australia, 2012.
47. Zhu, Z.; Wang, S.; Woodcock, C.E. Improvement and Expansion of the FMASK algorithm: Cloud, cloud shadow, and snow detection for Landsats 4–7, 8, and sentinel 2 images. *Remote Sens. Environ.* **2015**, *159*, 269–277. [CrossRef]
48. Masek, J.G.; Vermote, E.F.; Saleous, N.E.; Wolfe, R.; Hall, F.G.; Huemmrich, K.F.; Gao, F.; Kutler, J.; Lim, T.K. A Landsat surface reflectance dataset for North America, 1990–2000. *IEEE Geosci. Remote Sens. Lett.* **2006**, *3*, 68–72. [CrossRef]
49. Vermote, E.; Justice, C.; Claverie, M.; Franch, B. Preliminary analysis of the performance of the Landsat 8/OLI land surface reflectance product. *Remote Sens. Environ.* **2016**, *185*, 46–56. [CrossRef]
50. Roy, D.P.; Kovalskyy, V.; Zhang, H.K.; Vermote, E.F.; Yan, L.; Kumar, S.S.; Egorov, A. Characterization of Landsat-7 to Landsat-8 reflective wavelength and normalized difference vegetation index continuity. *Remote Sens. Environ.* **2016**, *185*, 57–70. [CrossRef]

51. Flood, N. Seasonal composite Landsat TM/ETM+ images using the medoid (a multi-dimensional median). *Remote Sens.* **2013**, *5*, 6481–6500. [CrossRef]
52. Lehmann, E.A.; Wallace, J.F.; Caccetta, P.A.; Furby, S.L.; Zdunic, K. Forest cover trends from time series Landsat data for the Australian continent. *Int. J. Appl. Earth Obs. Geoinf.* **2013**, *21*, 453–462. [CrossRef]
53. Key, C.H.; Benson, N.C. Landscape assessment: Remote sensing of severity, the normalized burn ratio. In *FIREMON: Fire Effects Monitoring and Inventory System*; USDA Forest Service, Rocky Mountain Research Station, Fort Collins: Denver, CO, USA, 2006; pp. 305–325.
54. Yang, Y.; Erskine, P.D.; Lechner, A.M.; Mulligan, D.; Zhang, S.; Wang, Z. Detecting the dynamics of vegetation disturbance and recovery in surface mining area via Landsat imagery and LandTrendr algorithm. *J. Clean. Prod.* **2018**, *178*, 353–362. [CrossRef]
55. NASA Jet Propulsion Laboratory. U.S. Releases Enhanced Shuttle Land Elevation Data. Available online: https://www.jpl.nasa.gov/news/news.php?release=2014-321 (accessed on 21 March 2020).
56. Weiss, A.D. Topographic position and landforms analysis. In Proceedings of the Presented at ESRI Users Conference, San Diego, CA, USA, 9–13 July 2001.
57. Beven, K.J.; Kirkby, M.J. A physically based, variable contributing area model of basin hydrology. *Hydrol. Sci. Bull.* **1979**, *24*, 43–69. [CrossRef]
58. Breiman, L. Random Forests. *Mach. Learn.* **2001**, *45*, 5–32. [CrossRef]
59. Liaw, A.; Wiener, M. Classification and regression by random forest. *R News* **2002**, *2*, 18–22.
60. Murphy, M.A.; Evans, J.S.; Storfer, A.S. Quantify bufo boreas connectivity in Yellowstone national park with landscape genetics. *Ecology* **2010**, *91*, 252–261. [CrossRef] [PubMed]
61. Tolhurst, K.G.; McCarthy, G. Effect of prescribed burning on wildfire severity: A landscape-scale case study from the 2003 fires in Victoria. *Aust. For.* **2016**, *79*. [CrossRef]
62. Government of South Australia: Kangaroo Island Landscape Board. Bushfire Recovery. Available online: https://landscape.sa.gov.au/ki/land-and-water/Bushfire_recovery (accessed on 3 August 2020).
63. Bennett, L.T.; Bruce, M.J.; MacHunter, J.; Kohout, M.; Tanase, M.A.; Aponte, C. Mortality and recruitment of fire-tolerant eucalypts as influenced by wildfire severity and recent prescribed fire. *For. Ecol. Manag.* **2016**, *380*, 107–117. [CrossRef]
64. Haslem, A.; Kelly, L.T.; Nimmo, D.G.; Watson, S.J.; Kenny, S.A.; Taylor, R.S.; Bennett, A.F. Habitat or fuel? Implications of long-term, post-fire dynamics for the development of key resources for fauna and fire. *J. Appl. Ecol.* **2010**, *48*, 247–256. [CrossRef]
65. Butt, N.; Pollock, L.J.; McAlpine, C.A. Eucalypts face increasing climate stress. *Ecol. Evol.* **2013**, *3*, 5011–5022. [CrossRef]
66. Government of South Australia: Department of Environment and Water. Find out how South Australia's Flinders Chase National Park is recovering post-bushfire. Available online: https://www.environment.sa.gov.au/goodliving/posts/2020/04/flinders-chase-plants-regenerating (accessed on 24 June 2020).
67. Ye, S.; Rogan, J.; Zhu, Z.; Eastman, J.R. A near-real-time approach for monitoring forest disturbance using Landsat time series: Stochastic continuous change detection. *Remote Sens. Environ.* **2020**, *252*, 112167. [CrossRef]

remote sensing

MDPI

Article

Mapping Burned Areas of Mato Grosso State Brazilian Amazon Using Multisensor Datasets

Yosio Edemir Shimabukuro [1,*], Andeise Cerqueira Dutra [1], Egidio Arai [1], Valdete Duarte [1], Henrique Luís Godinho Cassol [1], Gabriel Pereira [2,3] and Francielle da Silva Cardozo [4]

1 Earth Observation and Geoinformatics Division, National Institute for Space Research, São José dos Campos 12227-010, SP, Brazil; andeise.dutra@inpe.br (A.C.D.); egidio.arai@inpe.br (E.A.); valdete.duarte@inpe.br (V.D.); henrique.cassol@inpe.br (H.L.G.C.)

2 Department of Geoscience, Federal University of São João del-Rei, São João del-Rei 36307-352, MG, Brazil; pereira@ufsj.edu.br

3 Graduate Program in Physical Geography, Faculty of Philosophy, Languages and Literature, and Human Sciences, University of São Paulo, São Paulo 05508-080, SP, Brazil

4 Graduate Program in Geography, Federal University of São João del-Rei, São João del-Rei 36307-352, MG, Brazil; franciellecardozo@ufsj.edu.br

* Correspondence: yosio.shimabukuro@inpe.br; Tel.: +55-12-3208-6483

Received: 1 October 2020; Accepted: 18 November 2020; Published: 21 November 2020

Abstract: Quantifying forest fires remain a challenging task for the implementation of public policies aimed to mitigate climate change. In this paper, we propose a new method to provide an annual burned area map of Mato Grosso State located in the Brazilian Amazon region, taking advantage of the high spatial and temporal resolution sensors. The method consists of generating the vegetation, soil, and shade fraction images by applying the Linear Spectral Mixing Model (LSMM) to the Landsat-8 OLI (Operational Land Imager), PROBA-V (Project for On-Board Autonomy–Vegetation), and Suomi NPP-VIIRS (National Polar-Orbiting Partnership-Visible Infrared Imaging Radiometer Suite) datasets. The shade fraction images highlight the burned areas, in which values are represented by low reflectance of ground targets, and the mapping was performed using an unsupervised classifier. Burned areas were evaluated in terms of land use and land cover classes over the Amazon, Cerrado and Pantanal biomes in the Mato Grosso State. Our results showed that most of the burned areas occurred in non-forested areas (66.57%) and old deforestation (21.54%). However, burned areas over forestlands (11.03%), causing forest degradation, reached more than double compared with burned areas identified in consolidated croplands (5.32%). The results obtained were validated using the Sentinel-2 data and compared with active fire data and existing global burned areas products, such as the MODIS (Moderate Resolution Imaging Spectroradiometer product) MCD64A1 and MCD45A1, and Fire CCI (ESA Climate Change Initiative) products. Although there is a good visual agreement among the analyzed products, the areas estimated were quite different. Our results presented correlation of 51% with Sentinel-2 and agreement of $r^2 = 0.31$, $r^2 = 0.29$, and $r^2 = 0.43$ with MCD64A1, MCD45A1, and Fire CCI products, respectively. However, considering the active fire data, it was achieved the better performance between active fire presence and burn mapping (92%). The proposed method provided a general perspective about the patterns of fire in various biomes of Mato Grosso State, Brazil, that are important for the environmental studies, specially related to fire severity, regeneration, and greenhouse gas emissions.

Keywords: burned areas detection; shade fraction image; linear spectral mixing model; VIIRS; PROBA-V; Landsat-8 OLI

1. Introduction

In Brazil, burning of vegetation biomass is a management practice used to create and to maintain cattle pasture and to expand agricultural frontier [1]. The burning practice is widely used in the production process in the Brazilian Amazon (rainforest) and Cerrado (savanna) biomes and is a factor that drives the agricultural expansion in these regions. Farmers burn their land to convert forests to cropland or pasture, to control the spread of weeds, pests, or diseases, and to stimulate pasture regrowth, as well [2–6].

Burns occur every year, and especially during the dry season, with most of the fires occurring towards the end of the dry season. During this time, risk of fire is highest due to factors, such as dry weather, the predominance of grass and other flammable materials, and increased exposure to ignition sources, such as flames, heated surfaces, sparks, lightning, and human action [7], facilitate the occurrence of the burning process [2,8–10]. Occasionally, non-anthropogenic factors, such as lightning, are also responsible for burning.

In the Legal Amazon, for example, it is estimated that at least 9% (1.99 million hectares) of the degraded primary forest areas were caused by fire during 2000–2013 time period [11]. In addition, prospects of forest degradation resulting from fires escaping from nearby agricultural areas into forests may be a growing risk in many regions in Brazil [12]. In this context, the intensity, duration, extent, and period of the year when fires occur will determine how a particular ecosystem will be resistant to this event and how it will recover to the previous state (resilient). In some cases, this phenomenon can be beneficial for vegetation, as is the case with pastures of the Brazilian savanna (Cerrado biome), where the low-intensity burning event accelerates the availability of nutrients for plants by mineralization [7]. However, successive burnings alter the rates of infiltration and evapotranspiration of water in the soil, increasing the soil's susceptibility to wind and water erosion. Moreover, unlike the Brazilian Cerrado, the Amazon rainforest is not adapted to fire; therefore, vegetation in most cases is almost entirely consumed by burning, leading to a gradual increase of tree mortality [7,13].

Forest burning brings numerous environmental impacts, such as soil depletion and loss of biodiversity of flora and fauna [7], as well as causing significant damage to private properties and society as a whole [14]. In addition, trace gases and aerosols released by burning biomass cause changes in the energy balance of the atmosphere, leading to local, regional, and global climate changes [9,15]. Biomass burning in tropical regions contributes to ~32% of global CO emissions to the atmosphere. In this context, Cerrado is one of the most critical sources of trace gases due to the frequency and extent of burning [16,17], where the interaction of human activities mold temporal and spatial patterns of fires emissions [18].

Analysis of the timing and distribution of forest fires is a crucial tool for efficient fire management across regions experiencing various fire events. It provides information for fire prevention planning and is also critical for assessing economic losses and ecological effects, monitoring changes in land use and land cover, and developing atmospheric and climate impact models due to vegetation biomass burning [19].

Remote sensing represents a particularly useful tool for mapping and quantifying burned areas in large and difficult to reach areas. These data allow for active fire detection, which registers the instantaneous temperature of objects in combustion using spectral bands between 4 and 11 μm [20,21]; burned area estimation, which uses the spectral bands in the visible, near-infrared, and mid-infrared regions of the electromagnetic spectrum to detect changes in the spectral characteristics of land cover before and after the occurrence of fires [22,23]; fire patterns and severity [24,25]; and studies of vegetation regeneration and risk analysis [26,27]. Moreover, data derived from remote sensing are the most efficient source of information to understand the fire dynamics as they allow the observation of large areas of the surface at high temporal resolutions [28,29].

Several studies using burned areas (BA) detection algorithms have been conducted. Reference [30] used an algorithm based on Normalized Burned Ratio differencing (dNBR) to map the BA from MODIS (Moderate Resolution Imaging Spectroradiometer) data in three different ecosystems, such as boreal forest of Central Siberia, Mediterranean-type ecosystem of California, and sagebrush steppe of the Great Basin. The results showed that in each ecosystem, the BA estimates were within 15% compared to the high-resolution base, presenting a strong potential for regional and ecosystem-specific burned area mapping applications. Reference [31] developed a multitemporal two-phase burned area (BA) algorithm using Sentinel-2 MSI (MultiSpectral Instrument) reflectance images and active fires from MODIS sensors in the whole Sub-Saharan Africa. The algorithm used the spectral indices NBR2 (Normalized Burn Ratio 2), MIRBI (Mid Infrared Burn Index), and the NIR (near infrared) spectral region. The validation was based on a two-stage stratified random sampling of Landsat images and also compared to MCD64A1. The results showed an omission error of 26.5% and commission error of 19.3%, and, when compared to MCD64A1, the product presented an 80% larger area than NASA (National Aeronautics and Space Administration) BA product. The limitations found relies on the lower temporal resolution, impacting in the mapping due to cloud/cloud shadows.

In this context, some limitations increase errors in burned area products. Significant source of errors include the presence of clouds, which significantly reduces the ability to detect a fire hotspot due to the attenuation of the spectral radiance emitted by the flaming and smoldering phases in the biomass burning process [32], the lack of data on the moment of fire occurrence and, especially, the incompatibility of the temporal and spatial resolutions of some sensors, making these sensors unsuitable for the identification of burns. One way to reduce the uncertainty in the detection of burned surfaces is by using sensors data that have geometric, radiometric, and temporal resolutions appropriate to map the location and the discrimination of burned areas.

There are several global burned areas maps available, such as MODIS MCD45A1, MODIS MCD64A1, and Fire CCI (ESA Climate Change Initiative). However, in general, they present different results due to data characteristics, image dates, and applied methods for mapping [33,34]. These differences in the estimates occur because burning events are dynamics and affect different sizes on the ground, requiring sensors data with different spatial and temporal resolutions. For example, Landsat sensors [35,36] with a medium spatial resolution (30 m) are adequate to estimate the burned areas, but their limited temporal resolution (16 days) reduce their utility in the savanna and pastureland areas which regenerate rapidly following fires. In addition, product derived from Sentinel-2 imagery would greatly contribute to better understanding the impacts of small fires in global fire events. In contrast, the MODIS sensor with daily acquisition and moderate spatial resolution (250 m) provides useful information for the global mapping of burned areas but underestimates small fire scars [37,38]. Other sensors with similar high temporal resolution as MODIS for burned areas assessment include NPP-VIIRS (National Polar-Orbiting Partnership-Visible Infrared Imaging Radiometer Suite) and PROBA-V (Project for On-Board Autonomy–Vegetation) (at each 2 days) [39–41].

Combining information from different sensors can fill the gap between high temporal resolution and medium spatial resolution in an operational and accurate burned area product. However, this high volume of data requires approaches that highlight specific information to facilitate further classification tasks. An alternative approach to reduce the dimensionality of image data and enhance specific information for digital interpretation is the use of fraction images [42,43]. These images constitute synthetic bands with information on endmember proportions and they are generated by Linear Spectral Mixing Model (LSMM). For example, vegetation fraction images highlight the vegetation cover, especially cropland areas; soil fraction images highlight the contrast between forest and non-forested areas, and shade fraction images highlight water bodies and burned areas. In this manner, many efforts have been reported using LSMM to map burned areas, especially in the Brazilian Amazon [44–46].

The main objective of this work was to propose a new method to map burned areas using multisensor fraction images by taking advantage of the high temporal resolution of Suomi NPP-VIIRS and PROBA-V, and the medium spatial resolution of Landsat-8 OLI (Operational Land Imager) images for the year 2015. In order to evaluate the method, we use as reference the burned area map obtained using Sentinel-2 data, and compare our results with active fire data and the available monthly burned area products of MODIS MCD64A1, MODIS MCD45A1, and Fire CCI.

2. Materials and Methods

2.1. Study Area

The study area is Mato Grosso State (Figure 1), the third largest state in Brazil located in the central-west region and covering an area approximately 904,984 km^2. Mato Grosso State consists of three biomes: Amazon (54%), Cerrado (40%), and Pantanal (6%), according to the official Brazilian national classification system [47]. Mato Grosso State presents complex biodiversity, which, due to variable climate and terrain relief, resulted in different vegetation types. The Amazon biome consists of closed vegetation, high canopy, and a continuous layer of crowns; the Cerrado biome, a savanna formation, is characterized by tree species with tortuous trunks and typical openness canopy; and the Pantanal biome, a seasonal wetland, is composed by savannas and marshes [48].

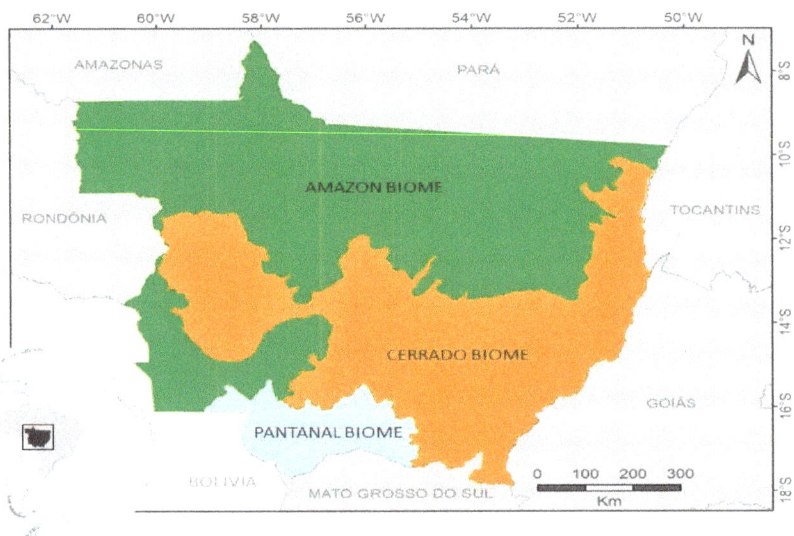

Figure 1. Location of the Mato Grosso State in the South American continent (inset figure). Mato Grosso State corresponds to the selected study area of this work, encompassing three biomes: Amazon, Cerrado, and Pantanal [47].

Mato Grosso Sate stands out as a center of Brazilian agribusiness tied to illegal deforestation. The state falls partly within the 'arc of deforestation' in the southern part of the Brazilian Legal Amazon, an area with one of the highest annual deforestation rates. According to the Monitoring Deforestation of the Brazilian Amazon Forest by Satellite (PRODES) data [49,50], the Mato Grosso State has an accumulated deforestation (1988–2019) corresponding to 146,140 km^2 area, which represents approximately 16% of its territory. In terms of total degraded area, the state is the second in the ranking among the states of the Legal Amazon. Deforestation activities cause, among others, habitat fragmentation, which leaves the remaining forest more vulnerable to edge effects and fire [51,52].

Mato Grosso State is among the states that lead in the number of active fires detected [53]. For example, in the 1988–2020 period, 40,000 fires were detected in the state, accounting for approximately 27% of the total active fires observed in the Legal Amazon [53]. Most of the active fires occurred between June and November, a period with the least precipitation received. In the dry season, it is expected the increasing risk of the fire spreading and escaping, affecting nearby forest areas and causing forest degradation, that is an important source for the REDD+ program (Reduction of Emissions from Deforestation and Forest Degradation) since these areas remain as forest areas but with high ecosystem loss.

2.2. Remote Sensing Datasets

2.2.1. PROBA-V

We acquired data from PROBA-V, a mini-satellite designed as a continuity to the 15 years of the SPOT-VEGETATION (Satellite Pour l'Observation de la Terre–Vegetation) mission, and for the Sentinel-3 mission of the European Space Agency (ESA). To meet the user's community, the spectral channels of the PROBA-V are similar to the SPOT VEGETATION instrument with four bands: blue (438–486 nm), red (615–696 nm), near-infrared (NIR) (772–914 nm), and medium infrared (MIR) (1564–1634 nm). The PROBA-V has a spatial resolution of 100 m at nadir and 300 m at the edge of scan.

In this study, we used 5-day Top-Of-Canopy PROBA-V image composite acquired at a spatial resolution of 100 m for the period between May and October 2015. The images were downloaded from the Vito Earth Observation portal (http://www.vito-eodata.be/PDF/portal/Application.html#Home), and all images had their clouds and respective shadows removed using the information from the Status Map layer in a program developed in the Interactive Data Language (IDL).

2.2.2. Suomi NPP-VIIRS

We downloaded 153 VIIRS daily reflectance (VNP09GA) images acquired between May and October 2015 from the U.S. Geological survey (USGS) at https://e4ftl01.cr.usgs.gov/VIIRS/VNP09GA.001/. The downloaded images consisted of three spectral bands, I1 (red, 600–680 nm), I2 (NIR, 846–885 nm), and I3 (MIR, 1580–1640 nm), and had a spatial resolution of 500 m. The 153 daily reflectance images were later composited into a cloud-free monthly VIIRS surface reflectance images using a script in Google Earth Engine and image quality flags 1 (QF1) layer.

2.2.3. Landsat-8 OLI

The Operational Land Imager (OLI) on Landsat-8 satellite has 11 spectral bands (panchromatic, 500–680 nm; multispectral, 430–450 nm; visible, 450–690 nm; NIR, 850–880 nm; MIR, 1570–2290 nm; cirrus, 1360–1380 nm; and thermal infrared, 10600–12510 nm). The sensor has a temporal resolution of 16 days and spatial resolution of 30 m except for the panchromatic band (15 m). The Landsat-8 OLI images were acquired during the dry season, from August to November, most acquired in October (41 images) and September (7 images), with August and November having only one image, due to the persistency of clouds (Appendix A). The images were downloaded from USGS archive: https://earthexplorer.usgs.gov/.

2.2.4. Sentinel-2 MSI

The Multispectral Instrument (MSI) on Sentinel-2A Level-1C dataset was used in this study as a reference to evaluate the burned areas classification by the proposed method. The Sentinel-2A MSI has 13 spectral bands from visible to SWIR (Short-wave Infrared) (B1, 432.2–453.2; B2, 459.4–525.4; B3, 541.8–577.8; B4, 649.1–680.1; B5, 696.6–711.6; B6, 733–748; B7772.8–792.8; B8, 779.8–885.8; B8A, 854.2–875.2; B9, 935.1–955.1; B10, 1358–1389; B11, 1568.2–1659.2; B12, 2114.9–2289.9). Each single satellite (2A and 2B) has a temporal resolution of 10 days and the combined constellation revisit is 5 days, with spatial resolution varying from 10 to 60 m. In this case, only bands with 10 m were used in this study. Due to the Sentinel-2A launch date (June 2015), only a segment of Mato Grosso State was available in 2015 in the period from August to November to compare the same period between the datasets used in this work. The images were downloaded from Joint Research Center website (https://jeodpp.jrc.ec.europa.eu/forobs/sentinel.py).

2.3. Global Burned Area Products

Burned area is considered as a non-permanent land cover change, and, in this manner, a variety of algorithms have been applied to detect and to map burned areas, leading to different results spatially and temporally [34]. In this work, the identified burned areas based on the proposed approach were spatially compared with three available global burned area products for an agreement analysis (Table 1).

Table 1. Overview of burned area products selected for agreement analysis.

Producer	Product	Collection	Sensor	Spatial Resolution	Period
European Space Agency	Fire CCI	5.1	MODIS	250 m	Monthly
NASA MODIS Land Science Team	MCD45A1	5.1	MODIS	500 m	Monthly
NASA MODIS Land Science Team	MCD64A1	6.0	MODIS	500 m	Monthly

The Fire CCI v5.1 products (https://geogra.uhttps://geogra.uah.es/fire_cci/ah.es/fire_cci/), managed by the European Space Agency, were obtained combining spectral information in the red and NIR bands from MODIS at 250 m resolution and thermal information from the MODIS active fire products. The daily-identified areas are aggregated in monthly compositions [54].

The National Aeronautics and Space Administration (NASA) produces two products of burned areas from the MODIS sensor onboard the Terra and Aqua Satellites, the MCD45A1 and the MCD64A1. The MCD45A1 product, monthly and global, uses a bidirectional reflectance distribution function (BRDF) model and analyzes the daily surface reflectance dynamics to locate rapid changes to identify burned areas [55]. The product MCD64A1 (https://e4ftl01.cr.usgs.gov/MOTA/MCD64A1.006/) is also monthly and global and uses images of surface reflectance, coupled with 1 km MODIS active fire data [56].

The monthly results provided by each product of MODIS MCD45A1, MODIS MCD64A1, and ESA Fire CCI are summed, individually, to have an annual product for 2015 to be compared with the results obtained by the proposed approach.

2.4. Fire Hotspots Products

Completing the agreement analysis, we performed a comparison with our results, global products and active fires detected by the MODIS Aqua and Terra Satellites from January to December 2015. Comparison with active fire observations indicates which products most reasonably capture the burning event by relating fire scars to thermal emissions [34,55,57]. The dataset was downloaded from the Queimadas Program managed by the Brazilian National Institute for Space Research (INPE) provided at the analyzed year for South America.

2.5. Land Cover Maps

2.5.1. Forest and Non-Forest Map

A thematic annual forest cover map available by the PRODES to 2015 year [50] was used as auxiliary data to estimate forest degradation by fire and burned areas in non-forested areas (Figure 2a). The PRODES forest definition cover areas under vegetation dominated or with the predominance of forest physiognomy, classified as Dense Ombrophilous Forest, Open Ombrophilous Forest, Seasonal Deciduous Forest, Pioneer Formation of Fluvial Influence Areas (Alluvial Vegetation), Wetland Woody Vegetation with Sandy accumulations (Campinarana), and ecotones (Forest and savanna contact).

2.5.2. Mato Grosso Cropland Map

We used the thematic 2015 annual croplands extension map from Arai et al. [58] as an auxiliary data to quantify the extension of burned areas in lands destined to agriculture (Figure 2b). The data were based on PROBA-V time series available at a 100 m spatial resolution. The annual cropland area estimation for the Mato Grosso State was 5.3 million hectares.

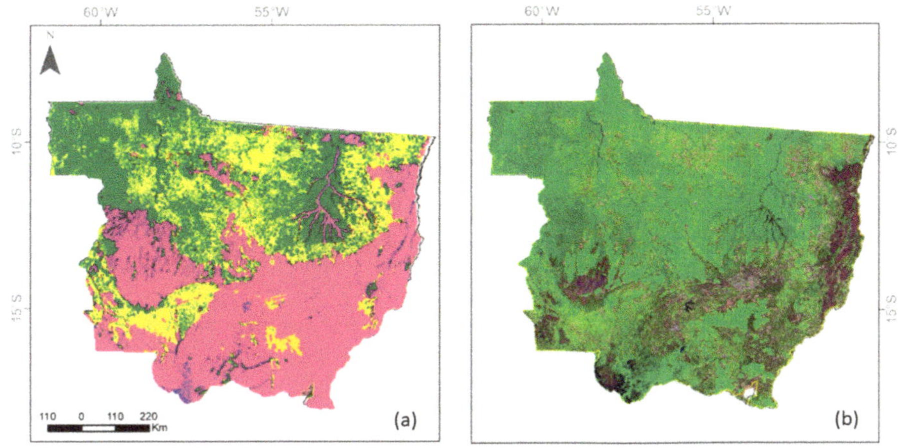

Figure 2. (**a**) PRODES 2015 showing the forest (in green) and non-forest (in magenta), deforestation (in yellow), and water (in blue) classes [47]; and (**b**) agriculture areas in 2015 (in light green) obtained from PROBA-V images [58].

2.6. Methodological Overview

The Linear Spectral Mixing Model (LSMM) was applied to those remote sensors datasets to generate the vegetation, soil, and shade fraction images based on the three pure pixels adopted by Arai et al. [58]. The LSMM assumes that pixel values are linear combinations of reflectance from different components, called endmembers (Equation (1)). The resulting images are the proportion or fraction of each endmember in each pixel [59] (Figure 3).

$$r_i = \sum_{j=1}^{m} x_j a_{i,j} + e_i, \tag{1}$$

where: r_i = the spectral reflectance in the ith spectral band; $a_{i,j}$ = the spectral reflectance of the jth component (endmember) in the ith spectral band; x_j = the proportion of the jth component within the pixel; $i = 1, \ldots, n$ (number of spectral bands used); $j = 1, \ldots, m = 3$ (number of considered components); and e_i = the residue for the ith spectral band.

The generation of these synthetic images is an alternative approach to reduce the dimensionality of image data and to enhance specific information (in this case, burned areas) for digital interpretation. Using the fraction images derived from the LSMM, a single image is composed containing the maximum proportion values of the shade fraction time series, comprising the burned areas information for PROBA-V and VIIRS datasets. This step was not applied to OLI images due to single images mosaic analyzed for the period.

Burned areas were obtained through digital classification of the shade fraction images using SPRING software [60] following a procedure based on image segmentation, unsupervised classification and assigning the corresponding classes to burned areas [61] for all sensor data (OLI, PROBA-V, and VIIRS) (Figure 4).

Image segmentation is a technique for grouping spatial information, whereby the algorithm assembles regions with and similar spectral characteristics. We used a region growing algorithm to perform the image segmentation. In such technique, two threshold parameters have to be set by the analyst to define segments (regions) at the subsequent classification procedure: (1) similarity threshold (the Euclidean distance between two regions with a different mean digital number of spectral clusters); and (2) an area threshold (the minimum area to be considered as a region, set by the number of pixels) [62]. In this work, the similarity and area thresholds were defined according to the sensors' characteristics, as (8, 100), (8, 10), and (8, 4) for Landsat-8 OLI, PROBA-V, and VIIRS, respectively.

Segmented images were classified using ISOSEG [62], a region classifier algorithm based on clustering techniques. This unsupervised algorithm uses the covariance matrix and the mean of the spectral values to estimate the centers of the classes in each region. The analyst defines an acceptance threshold, and the maximum allowed Mahalanobis distance that a mean digital number may be from the center of a class, considered as belonging to that class. The algorithm classifies the segments and separates them by classes based on the mentioned parameters. After the classification process and based on visual inspection, the labeling process is carried out only once, where all obtained classes from unsupervised classification were visually defined into classes with a real legend. This method was applied to the single shade fraction mosaic to obtain a final burned area classification for each sensor (PROBA-V, VIIRS, and OLI).

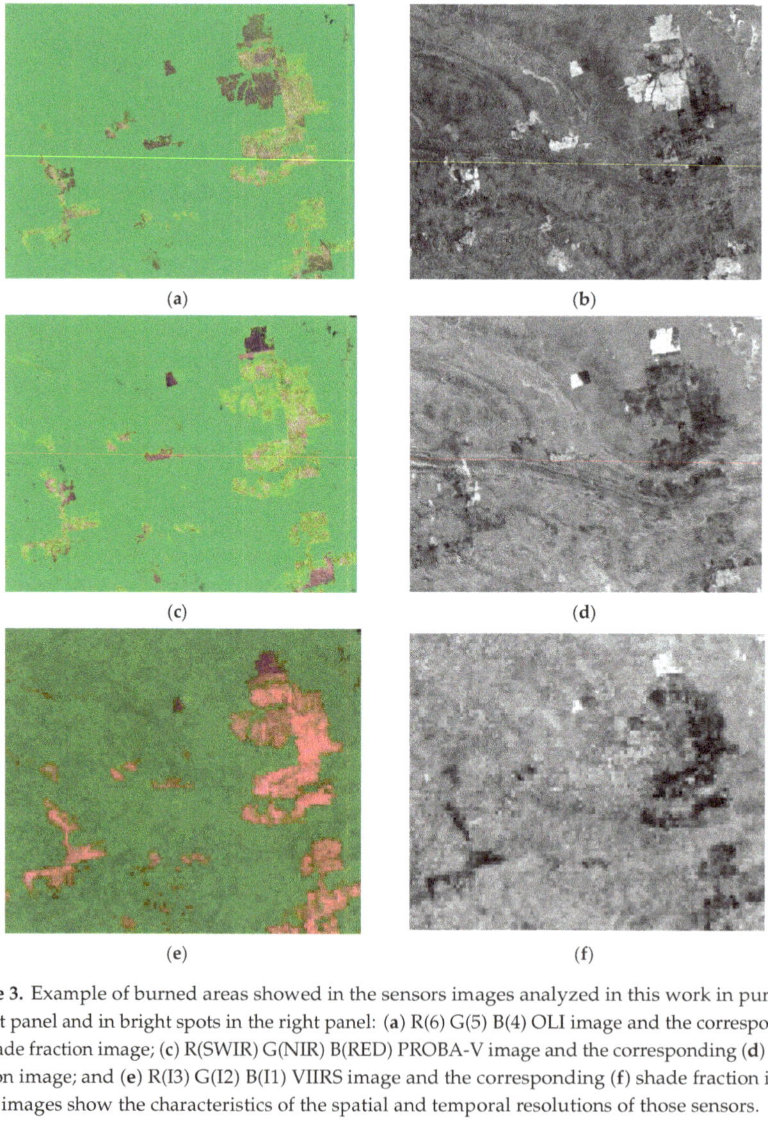

Figure 3. Example of burned areas showed in the sensors images analyzed in this work in purple in the left panel and in bright spots in the right panel: (**a**) R(6) G(5) B(4) OLI image and the corresponding (**b**) shade fraction image; (**c**) R(SWIR) G(NIR) B(RED) PROBA-V image and the corresponding (**d**) shade fraction image; and (**e**) R(I3) G(I2) B(I1) VIIRS image and the corresponding (**f**) shade fraction image. These images show the characteristics of the spatial and temporal resolutions of those sensors.

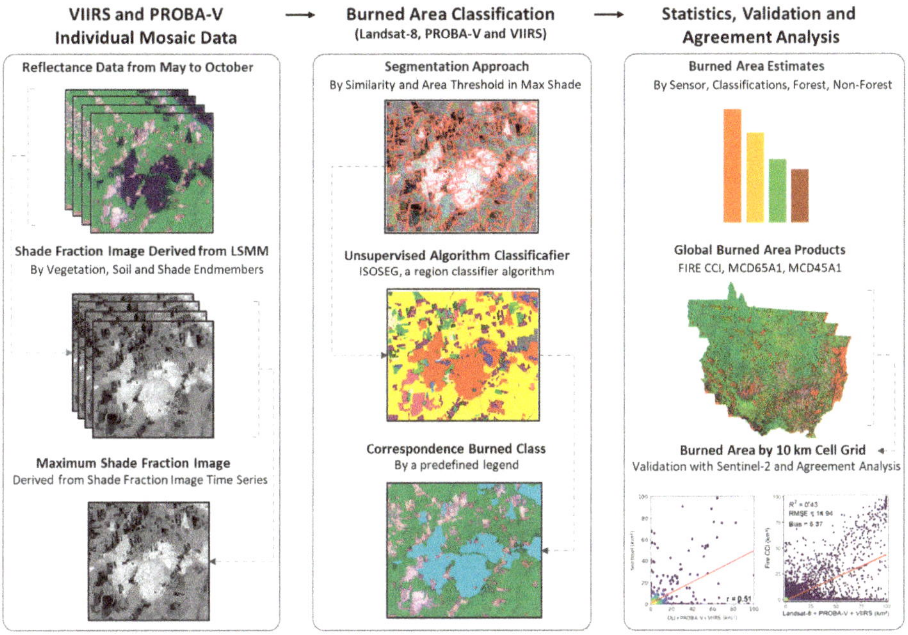

Figure 4. General methodological flowchart.

The final result is the mosaic of the burned area obtained from the three individual sensors. To show the visual consistency, a composite image was made highlighting the burned areas in the images of the three sensors; that is, the mapped burning pixels correspond to the original images of the sensors in which these areas were mapped. The burned areas were then calculated for the four results, including OLI, PROBA-V, NPP-VIIRS, and the final composite between the three products (OLI+PROBA-V+VIIRS). In addition, the burned area map obtained by the proposed method was superimposed on a PRODES and cropland datasets in order to analyze the spatial distribution and estimates of burned areas in the land use and land cover classes of the study area.

2.7. Evaluation and Agreement Analysis

In order to delineate burned areas in this study, 381 randomly distributed points in Sentinel-2 granules that included images from the period analyzed were used to estimate ground truth burned area, according to Equation (2):

$$Sample\ Size = \frac{Nz^2p(1-p)}{(N-1)e^2 + z^2p(1-p)}, \tag{2}$$

where: N is the population size, z is the confidence level (95%) from Z-score, e is the margin of error (5%), and p is the sample proportion (50%). Afterward, we used the Goodness of Variance Fit (GVF) to cluster the Sentinel-2 MSI granules statistics (Equation (3)) [63]:

$$GVF = \frac{\left(\sum_{i=1}^{O}(x_i-\mu)^2\right) - \left(\sum_{c=1}^{S}\sum_{i=1}^{O}(x_i-\varphi_c)^2\right)}{\sum_{i=1}^{O}(x_i-\mu)^2}, \tag{3}$$

where: O is the Sentinel-2 MSI granules, x_i is the sum of points inside the Sentinel-2 MSI granules, μ is the observations average, φ_c is the average of points over all granules, and S is the delimited classes in the optimization process. The clustering process separated the Sentinel-2 MSI granules in 2 groups, according to the density of random points (high and low), thus, burned areas were manually estimated in high density sampled granules (21LTC, 21LTE, 21LTG, 21LUC, 21LUD, and 21LUG).

Therefore, the burned areas derived from S2 MSI (considered the reference), Landsat-8 OLI, PROBA-V, Suomi NPP-VIIRS, and OLI+PROBA-V+VIIRS and the MCD45A1, MCD64A1, and Fire CCI global products were separated in several hexagonal grid cells with 10 km. This approach avoids the misregistration on the calculation of proportional errors for burned areas block sizes, and the bias in burned area estimation present in coarse and medium resolution burned area maps [34,64]. Then, a convolution mask was performed to estimate the agreement between products (Equation (4)).

$$BA_{(x,y.BAS,BAPi)} = \sum_{a=-\alpha}^{\propto} \sum_{b=-\beta}^{\beta} C(a,b).BAS(x+a,y+b) \cap C(a,b).BAPi(x+a,y+b), \quad (4)$$

where C(a,b) represents the convolution mask of M × N size (rows × columns); BAS is the burned areas estimated from Sentinel-2 data; BAPi is the burned areas estimated from burned areas products; and BA is the resulting grid defined for all points which the mask of M x N size completely overlaps the grid (x ϵ (α, M −α), y ϵ (β, N −β)).

The agreement analysis considered the same hexagonal grid cell of the evaluation analysis. We computed the proportion of burned areas of the auxiliary global burned area data products (MCD45A1, MCD64A1, and Fire CCI) and counted active fire observations by the presence or absence in each cell compared with the four results (OLI, PROBA-V, NPP-VIIRS, and OLI+PROBA-V+VIIRS). Based on these data, it was calculated the determination coefficient (r^2), root-mean-square error (RMSE), and bias.

3. Results

3.1. Burned Area Spatial Distribution and Estimates by Land Cover

The total burned area mapped from each sensor and by the final composite is showed in Figure 5b,d,f. It is important to highlight that burned areas results were obtained during the dry season and appear in dark color from the RGB mosaic in Figure 5a,c,e. A total of 55,694.32 km^2 of the burned area mapped by OLI sensor represents 6.15% of the study area, followed by VIIRS sensor with 29,768.72 km^2 of the burned area (3.28% of the study area); the PROBA-V sensor with 28,443.52 km^2 (3.14%); and then 93,027.92 km^2 (10.27%) mapped by the final composite (Figure 5h).

The estimates of burned areas superimposed on the PRODES and cropland datasets are shown in Table 2. The results pointed out that the most burned area estimate is located in non-Forest, reaching 66.57% (~61 thousand km^2) of the total area mapped by the final composite. Deforested areas, characterized by old (before 2015) and recent clear-cuts (2015 year), presented 21.54% (~19 thousand km^2) and 0.86% (~786 km^2), respectively. The results also showed less burn in croplands (5.32%, representing ~4 thousand km^2) than in forested areas (11.03%, ~10 thousand km^2).

Table 2. Estimated burned areas in deforested areas, forest, non-forest, and cropland areas using the final composite mapping.

Classes of Land Cover	Burned Area (km^2)	Total Burned Area (%)
2015 Deforestation	786.15	0.86%
Old Deforestation	19,745.43	21.54%
Non-Forest	61,009.31	66.57%
Forest	10,111.04	11.03%
Cropland	4873.32	5.32%

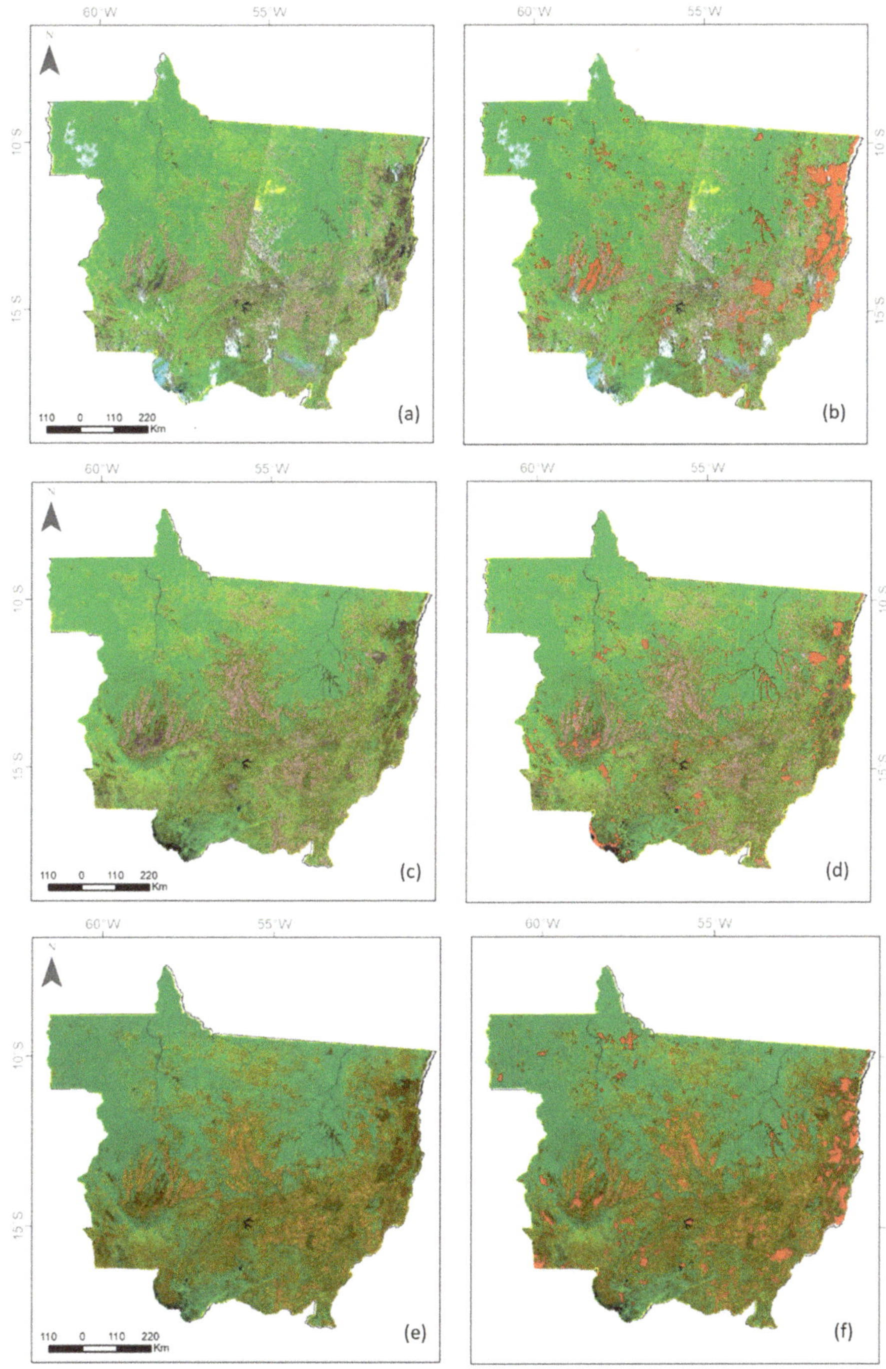

Figure 5. *Cont.*

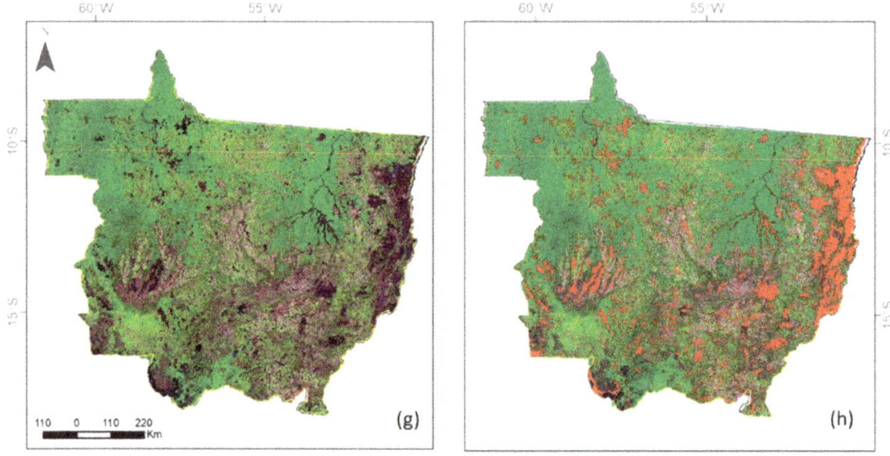

Figure 5. (**a**) Landsat OLI mosaic (August and November 2015); (**b**) burned areas over the mosaic showed in (**a**), showing the correspondence with dark areas; (**c**) PROBA-V mosaic composed of the pixel with high shade fraction value; (**d**) burned areas over the image in (**c**), showing the correspondence with dark areas; (**e**) VIIRS mosaic composed of the pixel with high shade fraction value; (**f**) burned areas over the image in (**e**), showing the correspondence with dark areas; (**g**) RGB Landsat OLI mosaic composite with the burned areas observed in the OLI, PROBA-V, and VIIRS images analyzed; and (**h**) burned areas over the mosaic showed in (**g**), showing the correspondence with dark areas.

3.2. Evaluation and Agreement Analysis

Figure 6 shows the evaluation graphics of burned areas maps for 6 granules (21LTC, 21LTE, 21LTG, 21LUC, 21LUD, and 21LUG) of S2 MSI reference data. It is noticed that the correlations vary according to the products and with burned areas size and location. The lower values ranged between 0.32 and 0.63 and the highest values ranged between 0.66 and 0.81 (all linear regressions were significant at $p < 0.05$, Student t-test). In the Sentinel-2 MSI burned areas, approximately 84% of grid cells had values lower than 5 km^2, with 41 hexagonal grid cells greater than 30 km^2. This indicates that burned area patterns correspond to small scars (as shown in density scatter plot of Figure 6a–g).

In analyzed products, we found several commission areas related mainly to dark red latosol, relief shading and water bodies that were not mapped in reference data. In these areas, where no burned scars occur, Fire CCI, MCD45A1, MCD64A1, Landsat-8 OLI, PROBA-V, NPP-VIIRS, and OLI+PROBA-V+VIIRS presented an inclusion of 150 km^2, 108 km^2, 488 km^2, 282 km^2, 930 km^2, 484 km^2, and 1712 km^2, respectively. Otherwise, we can find several hexagonal grid cells where burned areas were presented and fire products were unable to estimate (464 km^2, 550 km^2, 386 km^2, 547 km^2, 437 km^2, 1240 km^2, and 126 km^2 to Fire CCI, MCD45A1, MCD64A1, Landsat-8 OLI, PROBA-V, NPP-VIIRS, and OLI+PROBA-V+VIIRS, respectively).

Figure 7 compares only grid cells containing burned area between the results obtained by the proposed method (OLI+PROBA-V+VIIRS, OLI, PROBA-V, and NPP-VIIRS) with the available global datasets (MCD64A1, MCD45A1, and Fire CCI). There is a good visual agreement among the results from the four datasets, highlighted in a majority on the eastern Mato Grosso State, encompassing the Cerrado biome, and west, encompassing the Amazon biome. In the middle of the state, most cells in the grid contain only one or two global products with an agreement with the results of our method. The analysis based on active fire observation showed that most cells with the total agreement of burned detected by our method and global products present the highest number of active fire observations (from 51 to 133), being decreased when less agreement occurs.

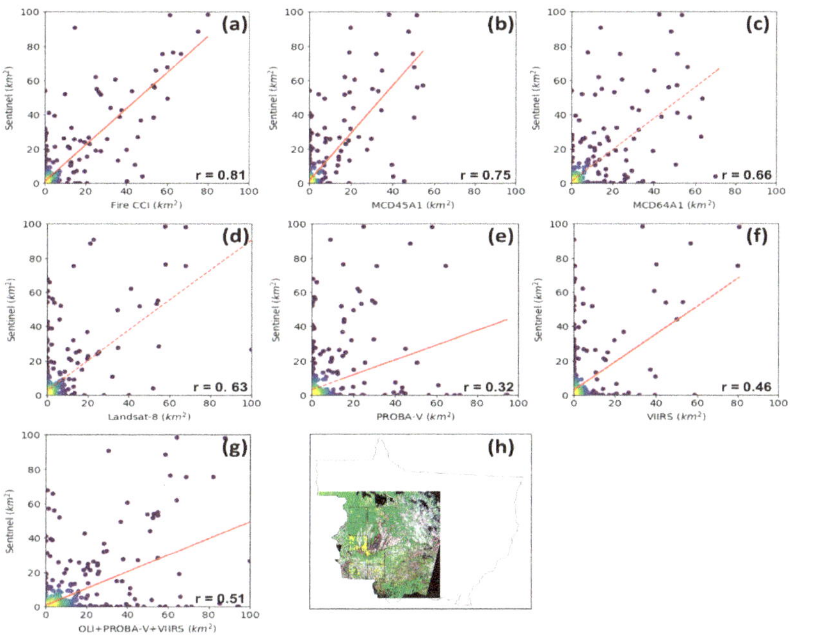

Figure 6. Assessment of burned areas mapped in S2 MSI (on y-axis) with Fire CCI (**a**), MCD45A1 (**b**), MCD64A1 (**c**), Landsat-8 OLI (**d**), PROBA-V (**e**), NPP-VIIRS (**f**), and OLI+ PROBA-V+VIIRS (**g**). Selected S2 MSI granules with burned areas in yellow (**h**).

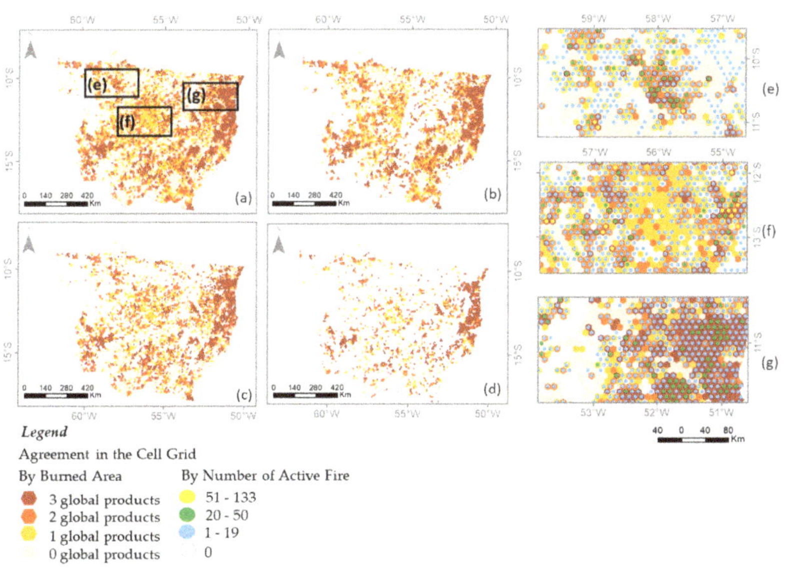

Figure 7. Comparison of results by the proposed approach in the Mato Grosso State with the three global burned area products in terms of an agreement in presence of burned area and final mapped composite (**a**); Landsat-8 OLI (**b**); PROBA-V (**c**); and NPP-VIIRS (**d**). In detail, specific areas are showing active fire spots classified by the number of observations in the cell (**e**,**f**,**g**). Blank areas inside the study area represents non-burned area by the classification results.

The analysis with active fire observations showed an agreement of 92.08% compared with the final composite burned area classification obtained using the proposed method in terms of the presence of active fire and burned area detection (Table 3). Compared with OLI, PROBA-V, and VIIRS, the agreement percentage decreases when lower the spatial resolution, with OLI sensor presenting the best result (71.97%) and VIIRS the lowest comparison (34.17%). Related to commission errors, where there is a burned area detected but no active fire observed, PROBA-V produced the highest commission error (51.27%), while the Fire CCI data had the lowest (8.79%).The omission errors (active fire detected and no burned area mapped) resulted in the best results for OLI sensor (28.03%) and the highest omission for VIIRS sensor (65.83%). Overall, the final composite classification (OLI+PROBA-V+VIIRS) results showed lower omission error (7.92%) but a higher commission error (72.73%) than all analyzed datasets.

Table 3. Percentages of grid cells compared by burned area detection and active fire observations.

Product	Active Fire and Mapped (%)	Non-Active Fire and Mapped (%)	Non-Active Fire and Non-Mapped (%)	Active Fire and Non-Mapped (%)
OLI+PROBA-V+VIIRS	92.08	72.73	27.27	7.92
OLI	71.97	44.44	55.56	28.03
PROBA-V	68.08	51.27	48.73	31.92
NPP-VIIRS	34.17	24.89	75.11	65.83
MCD64A1	64.57	16.43	83.57	35.43
MCD45A1	41.47	14.92	85.08	58.53
Fire CCI	51.89	8.79	91.21	48.11

There were substantial differences in mapped areal extent of burned areas across the various datasets (Table 4). Our results are overestimated compared with MCD64A1, MCD45A1, and Fire CCI datasets, respectively. However, these overestimates decrease when compared with individual sensor datasets.

Table 4. Burned area estimated by the proposed method and by the global burned area products.

Product	Area (km^2)
OLI+PROBA-V+VIIRS	96,461.24
OLI	55,694.32
PROBA-V	28,443.52
NPP-VIIRS	29,768.72
MODIS MCD64A1	40,364.75
MODIS MCD45A1	31,555.25
Fire CCI	34,991.13

Figure 8 shows the regression analysis of burned areas maps for each sensor and final composite compared with the global burned areas products by cell grid. Compared with the final composite, it was obtained that $r^2 = 0.30$ for MCD45A1, $r^2 = 0.32$ for MCD64A1, and $r^2 = 0.43$ for Fire CCI (Figure 8d,e,m, respectively). The RMSE varies from 16.94 to 14.94, and the bias between 7.78% to 5.84%. RMSE measures indicate a generally better spatial agreement between the products when the results are lower, being used as an indicator of the dispersion or tendency of classifiers to identify the same amount of burned areas across cells. However, the lack of bias in the regression model not necessarily has the same indicative [34]. In this case, the bias showed to be positive, and, as mentioned before, the final composite has a trend to overestimate the results compared with the global burned area products.

The regression results between the global burned area products and the burned areas mapped by each sensor showed Landsat-8 OLI with the best results (from $r^2 = 0.55$ to $r^2 = 0.36$), followed by PROBA-V (from $r^2 = 0.21$ to $r^2 = 0.12$) and NPP-VIIRS (from $r^2 = 0.18$ to $r^2 = 0.13$). The RMSE values were around 12.63, as observed with MCD45A1 and Landsat-8 OLI (Figure 8a), and 8.83 in MCD45A1 and PROBA-V (Figure 8b). The bias also showed to be positive for the most analysis, around

3.60% in MCD45A1 and Landsat-8 OLI (Figure 8a) and 0.74% in MCD45A1 and PROBA-V (Figure 8b), representing an overestimation. On the other hand, as observed previously, the mapped burned areas of PROBA-V and NPP-VIIRS sensors were underestimated when compared with MCD64A1 and Fire CCI products, representing a negative bias around to 1.20% in MCD64A1 and PROBA-V (Figure 8f) to 0.52% in Fire CCI and NPP-VIIRS (Figure 8l).

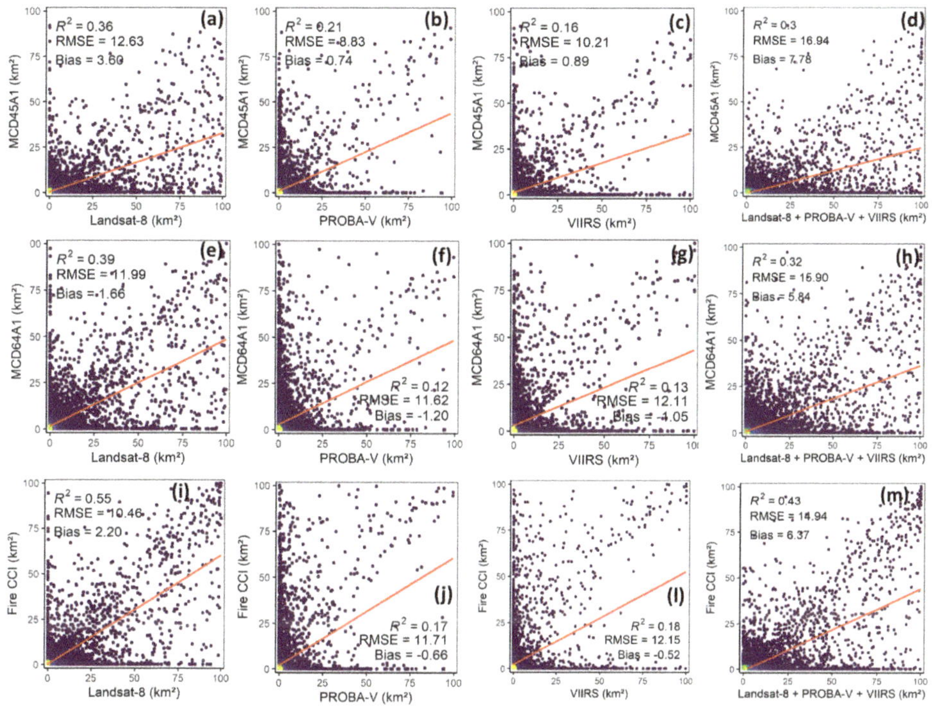

Figure 8. Scatter plots of regression by grid cells in comparison with MCD45A1 (**a–d**); MCD64A1 (**e–h**); and Fire CCI (**i–m**) with the results, considering grid cells with side dimensions of 10 km, to obtain the r^2, root-mean-square error (RMSE), and bias, based on the maximum number of cells.

4. Discussion

4.1. Burned Area by Land Cover

Historically, fire has been the main tool in the land use and land cover changes, and the impacts of fires on the environment, especially on the climate, require the analysis and monitoring of large territorial extensions in the long term. In a total of approximately of 96,461.24 km^2 of the burned areas identified in the Mato Grosso State, our results indicate that 66% of the total burned area occurs in non-forested areas, including savanna formations and grasslands, mainly founded in the Cerrado and Pantanal biomes. Twenty-one percent and 0.86% of burned areas were identified in older deforestation (before the 2015 year) and in deforested areas identified in the same year by the PRODES dataset, respectively. Critically, forested areas experienced 11% of the total burned area, and the occurrence of fires in forest formations can have negative social, economic, and environmental consequences, also favoring the conversion to deforestation. Additionally, the most effective policy instruments in reducing deforestation are not the most effective in reducing forest fires and vice-versa, suggesting that both objectives cannot be achieved with a single measure of political intervention [65]. In this manner,

Brazilian government policies still need to be rethought to contemplate the reduction of forest fires and the mitigation of their negative effects.

4.2. Overall Accuracy and Agreement Analysis

Although using the same MODIS sensor datasets, the existing global burned areas (Fire CCI, MCD45A1, and MCD64A1) products present different results. These differences are due to the methodologies and image dates used to generate such products. Our proposed method used image datasets from OLI, PROBA-V, and VIIRS sensors with different temporal and spatial resolutions. Then, we generated different results using these sensors datasets individually. We also generated a composite result by adding those individual results taking the advantages of higher temporal resolution of PROBA-V and VIIRS sensors and the medium spatial resolution of Landsat-8 OLI sensor. The results showed a good visual agreement among all burned areas maps produced by different sensors datasets and methodologies. However, the areal extents of burned areas were very different.

The proposed method used to generate the burned areas maps in this study showed different results according to the sensor analyzed. When evaluated with the reference data (Sentinel-2 MSI), the OLI sensor presented the best correlation (r = 0.63), followed by VIIRS (r = 0.46) and PROBA-V (0.32). In this context, the spatial resolution of OLI sensor (30 m) could explain the highest correlation. This analysis is corroborated by Reference [64], where the authors cite that a key determinant on the accuracy of burned area detection is the spatial mapping scale of dataset, with evidence that burned area maps produced from higher resolution observations presented reduced omission errors.

Although PROBA-V has better spatial resolution than VIIRS (100 m and 500 m, respectively), both presented similar results. The clouds and shadows were removed in the two products before the mapping, and the temporal analysis was the same (from May to October months). Therefore, the number of images utilized may explain the difference in results since VIIRS sensor is a daily product, compared to the composite of 5-days for PROBA-V, totalizing 153 and 36 images, respectively. Thus, some fires could be missed in PROBA-V, impacting in the evaluation process.

In relation to the evaluation using Sentinel-2 MSI data to delimited the burned area by visual interpretation, it is important to notice that the period analyzed (August to November) not coincide with PROBA-V and VIIRS dataset used in this study (May to November). Moreover, only a segment of Mato Grosso State was observed. Both limitations are related to the number of available images in 2015, due to the launch date (June 2015) of Sentinel-2A, which may cause uncertainties in the evaluation.

When comparing the reference data with the global burned area products, the Fire CCI, MCD45A1, and MCD64A1 presented a better correlation than OLI sensor, with a better agreement to Fire CCI (0.81), followed by MCD45A1 (0.75) and MCD64A1 (0.66). The number of scenes used to map the burned areas in OLI sensor may explain these results. Due to the persistency of clouds, some scenes could not be used and, in general, only one date was mapped for each of the 47 path/rows that comprise the state of Mato Grosso, most of them in October, missing some burned areas. In addition, the slightly better correlation of MCD45A1 compared to the MODIS improved product MCD64A1 may be related to the lower commission error presented by MCD45A1 (108 km^2) compared to MCD64A1 (488 km^2). Still, the final product OLI+PROBA-V+VIIRS not presented a good correlation (0.51), even when compared to OLI.

4.3. Specific Product of Burned Area by Each Dataset

The OLI sensor presented the highest values of the total burned area, with 55,694 km^2, followed by VIIRS and PROBA-V (29,768 km^2 and 28,443 km^2, respectively). As mentioned before, the better spatial resolution of OLI sensor could be a factor in detecting more burned areas, as well as the largest number of scenes used of VIIRS sensor. The OLI sensor resulted in the most total burned areas mapped comparing these results with global burned areas products, followed by Fire CCI product (34,991 km^2) and MCD64A1 (31,555 km^2). The better spatial resolution of Fire CCI (250 m) could be related to more burned areas detected and, according to Reference [34], Fire CCI and MCD64A1 are efficient in identifying burned area in cloudy regions due to their reliance on active fire detections as training data samples. PROBA-V presented the lowest burned area mapped, which may be linked to the number of scenes used when compared to the OLI+PROBA-V+VIIRS final product.

The spatial resolution of OLI could also explain the best results of this sensor related to commission errors (282 km^2), followed by VIIRS and PROBA-V (484 km^2 and 930 km^2, respectively). In Mato Grosso State, some elements may be confused with burned areas, like the presence of dark red latosol, especially in agricultural areas where they are most exposed. Cloud shadows, topographic shadowing, and water bodies have similar spectral characteristics as burned areas. According to Reference [66], other features that can be confused with the burned areas include clear cuts, land conversion, and non-fire forests mortality. Compared with the global burned area products, PROBA-V presented the highest commission error value, followed by MCD64A1 (488), VIIRS (484), OLI (282), Fire CCI (150), and MCD45A1 (108). The lowest commission error of burned areas by MCD45A1 could be explained by high omission error from this product, as will be mentioned later.

Related to omission errors, OLI and PROBA-V presented similar results, with 547 km^2 and 437 km^2, respectively. The number of scenes used to map the burned areas may explain the slight increase in the omission error of OLI sensor. The VIIRS sensor presented the highest omission error of burned area (1240 km^2), and, when compared to OLI and PROBA-V, the 500 m spatial resolution of the product may explain the differences, since approximately 84% of the burned areas found in reference data is composed by scars with less than 5 km^2. According to Reference [64], sensors with moderate and coarse spatial resolution presents a low degree of fidelity in the detection of small or highly fragmented fires. Besides, omissions can also occur when fires leave little residual heat and spread rapidly between satellite overpasses and when the differences in the spectral characteristics between pre- and post-fire are minimal [67]; and, to Reference [68], fires obscured by clouds, with short-lived thermal signatures and small fires also may not be identified.

Compared with the global burned area products, the VIIRS sensor resulted in higher omission error of burned area, followed by OLI (547 km^2), MCD45A1 (550 km^2), Fire CCI (464 km^2), PROBA-V (437 km^2), MCD64A1 (386 km^2), and OLI+PROBA-V+VIIRS final product (126 km^2). The OLI product showed a high rate of fires omission (ranking second). Thus, the temporal resolution is also crucial in the analysis and identification of burned areas, avoiding non-identification throughout the season. Related to global burned area products, the MCD64A1 presented the best performance, mostly due to the improvement in the algorithm, that integrates reflectance changes with active fire observations. Several validation studies for global products have identified the good performance of MCD64A1 in the identification of burned areas with low errors of omission [34,69,70]. Finally, the combined results of the proposed method presented the lowest omission of burned areas, indicating that the combination of higher temporal and medium spatial resolution could be a better approach to identifying burned areas.

4.4. Comparison with Active Fires

When comparing the final classification of burned areas with the active fire observations, it is noticed that related to omissions (active fires detected but not-mapped), the final composite product (OLI+PROBA-V+VIIRS) presented the best performance (7.92%), followed by OLI (28.03%), PROBA-V (31.92%), MCD64A1 (35.43%), Fire CCI (48.11%), MCD45A1 (58.53%), and VIIRS (65.83%). Therefore, sensors that present better spatial resolution and products with algorithms that use active fires present the best results. In this context, MCD45A1, that uses only reflectance images, and VIIRS, both with 500 m of spatial resolution, did not detect more than half of the burned areas even with active fires occurring.

Related to commission (burned area mapped with no active fire detected), Fire CCI presented the best performance (8.79%), indicating that the 250 m spatial resolution with the use of active fire product is efficient in not including burns that do not exist, or that may exist but are characterized by small scars where active fires are absent. The next following products presented the best results: MCD45A1 (14.92%), MCD64A1 (16.43%), VIIRS (24.89%), OLI (44.44%), PROBA-V (51.27%), and OLI

+ PROBA-V + VIIRS (72.73%). In this context, MCD45A1 and VIIRS that presented high values of omission errors related to active fires also presented low values of commission errors, the same pattern demonstrated by OLI + PROBA-V + VIIRS, that presented the best performance related to omission error and lowest performance related to commission errors.

The last analysis is related to the burned areas mapped that presented active fires. The OLI+PROBA-V+VIIRS product presented the best performance (92.08%), followed by OLI (71.97%), PROBA-V (68.08%), and MCD64A1 (64.57%). Besides, Fire CCI presents better spatial distribution compared to MCD64A1, and this product presented 51.89% of burned areas mapped in areas with active fires, followed by MCD45A1 (45.47%) that do not use active fires in the algorithm to map the burned areas. The sensor with the worst performance was VIIRS, with 34.17% of burned areas mapped, indicating that the 500 m spatial resolution does not perform well in areas with small burn scars, even with active fires.

In this context, the algorithm accuracies can be influenced by fire patch characteristics, since small and irregular burned area patches may lead to high omission errors because the spatial resolution of the pixels is too coarse to detect the smaller fires [71]. On the other hand, small and irregular unburned areas within large burned patches may lead to an increase in commission errors, particularly for high sensitivity algorithms [57].

In the Amazon biome, the burning activity is performed to clean newly deforested areas for agriculture expansion, and sometimes the fires escape to forest areas causing forest degradation. Nevertheless, these remaining areas contain less biomass and are not considered in the carbon balance yet. In non-forested areas, burnings are caused by the use of fire to improve the productivity of grassland and pastureland covers. In this context, previous validation studies have indicated that global satellite-derived burned area products typically perform best in regions where fire activity is more prevalent [69], and, for South America, Reference [70] found that the smallest relative uncertainties occur in the frequently burning savannas and grasslands.

5. Conclusions

The use of remote sensing burned areas products requires confident estimates uncertainties. Therefore, in this paper, a method combining products with higher spatial and temporal resolutions to estimate the burned areas, by using OLI, PROBA-V, and NPP-VIIRS datasets, was proposed. The shade fraction images showed the advantages to reduce the data volume to be analyzed and highlighting the burned areas facilitating the image classification process, even with a high volume of data. In the study area, 11% (10,111 km^2) of burnings occurred in forested areas, reaching the double of burned area extension compared with areas destined for agriculture (5%, 4873 km^2). This indicates that fire occurrences are less in areas intended for agriculture using mechanized practices, although it is still a common practice due to its low cost and cultural habits. On the other hand, Brazilian government policies still need to be rethought to contemplate the reduction of forest fires and the mitigation of their negative effects.

Regarding to the total burned area, the 30 m high spatial resolution of OLI sensor resulted in the best results and the less commission errors; however, when evaluated with reference data (Sentinel-2 MSI), the Fire CCI global burned area product presented the best correlation, and this fact can be explained by the persistency of clouds in Mato Grosso State of 2015 year, resulting in less available images of OLI sensor, indicating the crucial importance of temporal resolution in burned areas analysis.

The final product OLI+PROBA-V+VIIRS used by the proposed method overestimated the burned areas and presented a moderate correlation compared to the reference data; however, it presented the lowest omission of burned areas and the best performance related to the mapping of burnings in areas with active fire observations. Therefore, these results indicate that the combination of higher temporal and medium spatial resolutions sensor data could be a better approach to identifying burned areas, since the best spatial resolution is indicated to small scars identification and high temporal resolution is used to avoid omission errors. In this context, the temporal resolution is also extremely important, being necessary the use of all scenes available, and a cloud mask could assist in the use of images that have a longer revisit period, improving the results obtained. For future studies, the recommendation is to use a machine learning algorithm and cloud processing for monitoring burned areas using different datasets and a high volume of data for larger areas. In addition, using ancillary information, e.g., water bodies map, may avoid the confusion of burned areas and water bodies and, consequently, improve the classification process.

The proposed method provided a general perspective about the patterns of fire in the Amazon, Cerrado, and Pantanal biomes of Mato Grosso State, Brazil, discriminating by forestland, cropland, and deforested and non-forested areas. Such information is important for environmental studies, especially related to fire severity, regeneration, and greenhouse gas emissions.

Author Contributions: All authors made significant contributions to the work. Specific contributions include conceptualization and experimental design (Y.E.S., E.A., and A.C.D.), data processing and formal analysis (E.A., Y.E.S., A.C.D., F.d.S.C., G.P., and V.D.), draft preparation (Y.E.S., A.C.D., F.d.S.C., H.L.G.C., and G.P.), review and editing (Y.E.S., A.C.D., F.d.S.C., G.P., E.A., H.L.G.C., and V.D.). All authors have read and agreed to the published version of the manuscript.

Funding: This study was supported by the São Paulo Research Council (FAPESP-Grant 19806-3/2016) and by the National Council for Scientific and Technological Development (CNPq—Grant 380716/2019-4). H.L.G.C. thanks to FAPESP grant 2018/14423-4. G.P thanks to CNPq grant 441934/2018-8. F.S.C thanks to CNPq grant 407519/2018-1.

Acknowledgments: The authors would like to thank the anonymous reviewers for providing comments and suggestions that helped to improve the quality of the original manuscript.

Conflicts of Interest: The authors declare no conflict of interest.

Appendix A

Table A1. Landsat OLI images (path/row and dates) used in this study.

Path/Row	Date	Path/Row	Date
223/067	21/Oct/2015	227/067	17/Oct/2015
223/068	21/Oct/2015	227/068	17/Oct/2015
223/069	21/Oct/2015	227/069	17/Oct/2015
223/070	21/Oct/2015	227/070	17/Oct/2015
223/071	21/Oct/2015	227/071	17/Oct/2015
224/067	12/Oct/2015	227/072	17/Oct/2015
224/068	12/Oct/2015	228/066	08/Oct/2015
224/069	12/Oct/2015	228/067	08/Oct/2015
224/070	12/Oct/2015	228/068	08/Oct/2015
224/071	12/Oct/2015	228/069	08/Oct/2015
224/072	29/Nov/2015	228/070	08/Oct/2015
225/067	19/Oct/2015	228/071	08/Oct/2015
225/068	19/Oct/2015	229/065	15/Oct/2015
225/069	19/Oct/2015	229/066	15/Oct/2015
225/070	19/Oct/2015	229/067	15/Oct/2015
225/071	19/Oct/2015	229/068	15/Oct/2015
225/072	19/Oct/2015	229/069	15/Oct/2015
226/067	08/Sep/2015 10/Oct/2015	229/070	15/Oct/2015
226/068	08/Sep/2015 10/Oct/2015	229/071	15/Oct/2015
226/069	08/Sep/2015	230/066	06/Oct/2015
226/070	08/Sep/2015	230/067	06/Oct/2015
226/071	08/Sep/2015	230/068	06/Oct/2015
226/072	23/Aug/2015	231/066	11/Sep/2015
		231/067	11/Sep/2015 13/Oct/2015

References

1. Belcher, C.M. *Fire Phenomena and the Earth System: An Interdisciplinary Guide to Fire Science*; College of Life and Environmental Sciences, University of Exeter: Exeter, UK, 2013.
2. Coutinho, L.M. Fire in the ecology of the Brazilian cerrado. In *Fire in the Tropical Biota*; Goldammer, J.G., Ed.; Springer-Velarg: New York, NY, USA, 1990; pp. 82–105.
3. Alencar, A.; Nepstad, D.; Silva, E.; Brown, F.; Lefebvre, P.; Mendosa, E.; Almeida, D.; Carvalho, O., Jr. *Uso do Fogo na Amazônia: Estudos de Caso ao Longo do Arco de Desmatamento*; World Bank Report; World Bank: Brasilia, Brazil, 1997.
4. Sampaio, G.; Nobre, C.; Costa, M.H.; Satyamurty, P.; Soares Filho, B.S.; Cardoso, M.F. Regional climate change over eastern Amazonia caused by pasture and soybean cropland expansion. *Geophys. Res. Lett.* **2007**, *34*, 1–7. [CrossRef]
5. Marengo, J.A.; Nobre, C.; Salazar, L.F. Regional climate change cenarios in South America in the Late XXI Century: Projections and expected impacts. *Nova Acta Leopoldina* **2010**, *112*, 251–265.
6. Bustamante, M.M.C.; Roitman, I.; Aide, T.M.; Alencar, A.; Anderson, L.; Aragão, L.; Asner, G.P.; Barlow, J.; Berenguer, E.; Chambers, J.; et al. Towards an integrated monitoring framework to assess the effects of tropical forest degradation and recovery on carbon stocks and biodiversity. *Global Change Biology* **2016**, *1*, 92–109. [CrossRef] [PubMed]
7. Cochrane, M.A.; Ryan, K.C. Fire and fire ecology: Concepts and principles. In *Tropical Fire Ecology, Climate Change, Land Use and Ecosystem Dynamics*; Cochrane, M.A., Ed.; Springer: Chichester, UK, 2009; pp. 60–97. [CrossRef]
8. Freitas, S.R.; Longo, K.M.; Dias, M.A.F.S.; Dias, P.L.S.; Chatfield, R.; Prins, E.; Artaxo, P.; Grell, G.A.; Recuero, F.S. Monitoring the transport of biomass burning emissions in South America. *Environ. Fluid Mechanics* **2005**, *5*, 135–167. [CrossRef]
9. Fearnside, P.M.; Righi, C.A.; Graça, P.M.L.A.; Keizer, E.W.H.; Cerri, C.C.; Nogueira, E.M.; Barbosa, R.I. Biomass and greenhouse gas emissions from land-use change in Brazil's Amazonian "arc of deforestation": The states of Mato Grosso and Rondônia. *Forest Ecology Manag.* **2009**, *258*, 1968–1978. [CrossRef]
10. Leblon, B.; Bourgeau-Chavez, L.; San-Miguel-Ayanz, J. Use of Remote Sensing in wildfire management. In *Sustainable Development–Authoritative and Leading Edge Content for Environmental Management*; Curkovic, S., Ed.; InTech Press: Rijeka, Croatia, 2012; pp. 55–82.
11. Tyukavina, A.; Hansen, M.C.; Potapov, P.V.; Stehman, S.V.; Smith-Rodriguez, K.; Okpa, C.; Aguilar, R. Types and rates of forest disturbance in Brazilian Legal Amazon, 2000–2013. *Science Advances* **2017**, *3*, e1601047. [CrossRef]
12. Davidson, E.A.; De Araüjo, A.C.; Artaxo, P.; Balch, J.K.; Brown, I.F.; Mercedes, M.M.; Coe, M.T.; Defries, R.S.; Keller, M.; Longo, M.; et al. The amazon basin in transition. *Nature* **2012**, *481*, 321–328. [CrossRef]
13. Cardozo, F.S.; Pereira, G.; Shimabukuro, Y.E.; Moraes, E.C. Analysis and Assessment of the Spatial and Temporal Distribution of Burned Areas in the Amazon Forest. *Remote Sensing* **2014**, *6*, 8002–8025. [CrossRef]
14. Stephenson, C.; Handmer, J.; Betts, R. Estimating the economic, social and environmental impacts of wildfires in Australia. *Environ. Hazards* **2013**, *12*, 93–111. [CrossRef]
15. Van der Werf, G.R.; Randerson, J.T.; Giglio, L.; Collatz, G.J.; Kasibhatla, P.S.; Arellano, A.F., Jr. Interannual variability in global biomass burning emissions from 1997 to 2004. *Atmos. Chem. Phys.* **2006**, *6*, 3423–3441. [CrossRef]
16. Andreae, M.O. Biomass burning: Its history, use, and distribution and its impact on environmental quality and global climate. In *Global Biomass Burning: Atmospheric, Climatic, and Biospheric Implications*; Levine, J.S., Ed.; The MIT Press: Cambridge, MA, USA, 1991; pp. 2–21.
17. Durigan, G.; Ratter, J.A. The need for a consistent fire policy for Cerrado conservation. *J. Appl. Ecol.* **2016**, *53*, 11–15. [CrossRef]
18. Mataveli, G.A.V.; Silva, M.E.S.; Pereira, G.; Cardozo, F.S.; Kawakúbo, F.S.; Bertani, G.; Costa, J.C.; Ramos, R.C.; Silva, V.V. Satellite observations for describing fire patterns and climate-related fire drivers in the Brazilian savannas. *Nat. Hazards Earth Syst. Sci.* **2018**, *18*, 125–144. [CrossRef]
19. Ichoku, C.; Kahn, R.; Chin, M. Satellite contributions to the quantitative characterization of biomass burning for climate modeling. *Atmos. Res.* **2012**, *111*, 1–28. [CrossRef]

20. Justice, C.O.; Giglio, L.; Korontzi, S.; Owens, J.; Morisette, J.; Roy, D.; Descloitres, J.; Alleaume, S.; Petitcolin, F.; Kaufman, Y. The MODIS fire product. *Remote Sens. Environ.* **2002**, *83*, 244–262. [CrossRef]
21. Pereira, G.; Siqueira, R.; Rosário, N.E.; Longo, K.L.; Freitas, S.R.; Cardozo, F.S.; Kaiser, J.W.; Martin, J.; Wooster, M.J. Assessment of fire emission inventories during the South American Biomass Burning Analysis (SAMBBA) experiment. *Atmos. Chem. Phys.* **2016**, *16*, 6961–6975. [CrossRef]
22. Roy, D.P.; Lewis, P.E.; Justice, C.O. Burned area mapping using multi-temporal moderate spatial resolution data—a bi-directional reflectance model-based expectation approach. *Remote Sens. Environ.* **2002**, *83*, 263–286. [CrossRef]
23. Chuvieco, E.; Mouillot, F.; van der Werf, G.R.; Miguel, J.S.; Tanase, M.; Koutsias, N.; García, M.; Yebra, M.; Padilla, M.; Gitas, I.; et al. Historical background and current developments for mapping burned area from satellite Earth observation. *Remote Sens. Environ.* **2019**, *225*, 45–64. [CrossRef]
24. Coppoletta, M.; Merriam, K.E.; Collins, B.M. Post-fire vegetation and fuel development influences fire severity patterns in reburns. *Ecol. Appl.* **2016**, *26*, 686–699. [CrossRef]
25. Andrés, R.G.; Davies, G.M.; Waldron, S.; Scott, E.M.; Gray, A. Increased fire severity alters initial vegetation regeneration across Calluna-dominated ecosystems. *J. Environ. Manag.* **2019**, *231*, 1004–1011. [CrossRef]
26. Casady, G.M.; Marsh, S.E. Broad-scale environmental conditions responsible for post-fire vegetation dynamics. *Remote Sens.* **2010**, *2*, 2643–2664. [CrossRef]
27. Chen, Z. A literature Survey: The Effects of Forest Fire on Ecology and Regeneration. Ph.D. Thesis, Faculty of Forestry and the Forest Environment, Lakehead University, Thunder Bay, ON, USA, 2019; 43p.
28. Hantson, S.; Padilla, M.; Corti, D.; Chuvieco, E. Strengths and weaknesses of MODIS hotspots to characterize global fire occurrence. *Remote Sens. Environ.* **2013**, *131*, 152–159. [CrossRef]
29. Andela, N.; van der Werf, G.R.; Kaiser, J.W.; van Leeuwen, T.T.; Wooster, M.J.; Lehmann, C.E.R. Biomass burning fuel consumption dynamics in the tropics and subtropics assessed from satellite. *Biogeosciences* **2016**, *13*, 3717–3734. [CrossRef]
30. Loboda, T.; O'Neal, K.J.; Csiszar, I. Regionally adaptable dNBR-based algorithm for burned area mapping from MODIS data. *Remote Sens. Environ.* **2007**, *109*, 429–442. [CrossRef]
31. Roteta, E.; Bastarrika, A.; Padilla, M.; Storm, T.; Chuvieco, E. Development of a Sentinel-2 burned area algorithm: Generation of a small fire database for sub-Saharan Africa. *Remote Sens. Environ.* **2019**, *222*, 1–17. [CrossRef]
32. Schroeder, W.; Morisette, J.T.; Csiszar, I.; Giglio, L.; Morton, D.; Justice, C.O. Characterizing vegetation fire dynamics in Brazil through multisatellite Data: Common trends and practical issues. *Earth Interact.* **2005**, *9*, 1–26. [CrossRef]
33. Giglio, G.; Boschetti, L.; Roy, D.P.; Humber, M.L.; Justice, C.O. The Collection 6 MODIS Burned Area Mapping Algorithm and Product. *Remote Sens. Environ.* **2018**, *217*, 72–85. [CrossRef]
34. Humber, M.L.; Boschetti, L.; Giglio, L.; Justice, C.O. Spatial and temporal intercomparison of four global burned area products. *Int. J. Digital Earth* **2019**, *12*, 460–484. [CrossRef]
35. Hawbaker, T.J.; Vanderhoof, M.K.; Beal, Y.J.; Takacs, J.D.; Schmidt, G.L.; Falgout, J.T.; Williams, B.; Fairaux, N.M.; Caldwell, M.K.; Picotte, J.J.; et al. Mapping burned areas using dense time-series of Landsat data. *Remote Sens. Environ.* **2017**, *198*, 504–522. [CrossRef]
36. Cabral, A.I.R.; Silva, S.; Silva, P.C.; Vanneschi, L.; Vasconcelos, M.J. Burned area estimations derived from Landsat ETM+ and OLI data: Comparing Genetic Programming with Maximum Likelihood and Classification and Regression Trees. *ISPRS J. Photogramm. Remote Sens.* **2018**, *142*, 94–105. [CrossRef]
37. Chen, D.; Pereira, J.M.C.; Masiero, A.; Pirotti, F. Mapping fire regimes in China using MODIS active fire and burned area data. *Appl. Geogr.* **2017**, *85*, 14–26. [CrossRef]
38. Melchiorre, A.; Boschetti, L. Global Analysis of Burned Area Persistence Time with MODIS Data. *Remote Sens.* **2018**, *10*, 750. [CrossRef]
39. Urbanski, S.; Nordgren, B.; Albury, C.; Schwert, B.; Peterson, D.; Quayle, B.; Hao, W.M. A VIIRS direct broadcast algorithm for rapid response mapping of wildfire burned area in the western United States. *Remote Sens. Environ.* **2018**, *219*, 271–283. [CrossRef]
40. Fernández-Manso, A.; Quintano, C. A Synergetic Approach to Burned Area Mapping Using Maximum Entropy Modeling Trained with Hyperspectral Data and VIIRS Hotspots. *Remote Sens.* **2020**, *12*, 858. [CrossRef]

41. Shimabukuro, Y.E.; Arai, E.; Duarte, V.; Dutra, A.C. Assessment of Land Use Land Cover in Brazil, South America, Using Fraction Images Derived from Proba-V Datasets. In Proceedings of the IGARSS 2019–2019 IEEE International Geoscience and Remote Sensing Symposium, Yokohama, Japan, 28 July–2 August 2019; pp. 6401–6404.
42. Aguiar, A.P.D.; Shimabukuro, Y.E.; Mascarenhas, N.D.A. Use of synthetic bands derived from mixing models in the multispectral classification of remote sensing images. *Int. J. Remote Sens.* **1999**, *20*, 647–657. [CrossRef]
43. Shimabukuro, Y.E.; Ponzoni, F.J. *Spectral Mixture for Remote Sensing. Linear Model and Applications*; Springer Nature: Basel, Switzerland, 2017; p. 77. [CrossRef]
44. Anderson, L.O.; Aragão, L.E.O.C.; Lima, A.; Shimabukuro, Y.E. Detecção de Cicatrizes de Áreas Queimadas Baseada No Modelo Linear de Mistura Espectral e Imagens Índice de Vegetação Utilizando Dados Multitemporais Do Sensor MODIS/TERRA No Estado Do Mato Grosso, Amazônia Brasileira. *Acta Amaz.* **2005**, *35*, 445–456. [CrossRef]
45. Quintano, C.; Fernández-Manso, A.; Fernández-Manso, O.; Shimabukuro, Y.E. Mapping Burned Areas in Mediterranean Countries Using Spectral Mixture Analysis from a Uni-Temporal Perspective. *Int. J. Remote Sens.* **2006**, *27*, 645–662. [CrossRef]
46. Shimabukuro, Y.E.; Arai, E.; Duarte, V.; Jorge, A.; Santos, E.G.D.; Gasparini, K.A.C.; Dutra, A.C. Monitoring deforestation and forest degradation using multi-temporal fraction images derived from Landsat sensor data in the Brazilian Amazon. *Int. J. Remote Sens.* **2019**, *40*, 5475–5496. [CrossRef]
47. IBGE, Estados. 2016. Available online: Ftp://geoftp.ibge.gov.br/cartas_e_mapas/mapas_estaduais_e_distrito_federal/informacoes_ambientais/ (accessed on 7 April 2020).
48. Rossi, F.S.; Santos, G.A.A. Fire dynamics in Mato Grosso State, Brazil: The relative roles of gross primary productivity. *Big Earth Data* **2020**, 1–22. [CrossRef]
49. Assis, L.F.F.G.; Ferreira, K.R.; Vinhas, L.; Maurano, L.; Almeida, C.; Carvalho, A.; Rodrigues, J.; Maciel, A.; Camargo, C. TerraBrasilis: A Spatial Data Analytics Infrastructure for Large-Scale Thematic Mapping. *ISPRS Int. J. Geo-Inf.* **2019**, *8*, 513. [CrossRef]
50. INPE—Instituto Nacional de Pesquisas Espaciais. Coordenação Geral de Observação da Terra. Programa de Monitoramento da Amazônia e Demais Biomas. Desmatamento—Amazônia Legal. 2020. Available online: http://terrabrasilis.dpi.inpe.br/downloads/ (accessed on 15 April 2020).
51. Cochrane, M.A.; Laurance, W.F. Fire as a large-scale edge effect in Amazonian forests. *J. Trop. Ecol.* **2002**, *18*, 311–325. [CrossRef]
52. Silva Junior, C.H.L.; Aragão, L.E.O.C.; Fonseca, M.G.; Almeida, C.T.; Vedovato, L.B.; Anderson, L.O. Deforestation-Induced Fragmentation Increases Forest Fire Occurrence in Central Brazilian Amazonia. *Forests* **2018**, *9*, 305. [CrossRef]
53. INPE—Instituto Nacional de Pesquisas Espaciais. Portal do Monitoramento de Queimadas e Incêndios. 2020. Available online: http://www.inpe.br/queimadas (accessed on 15 April 2020).
54. Pettinari, M.L.; Lizundia-Loiola, J.; Chuvieco, E. ESA CCI ECV Fire Disturbance: D4.2 Product User Guide—MODIS, Version 1.0. 2020. Available online: https://www.esa-fire-cci.org/documents (accessed on 1 May 2020).
55. Roy, D.P.; Jin, Y.; Lewis, P.E.; Justice, C.O. Prototyping a global algorithm for systematic fire-affected area mapping using MODIS time series data. *Remote Sens. Environ.* **2005**, *97*, 137–162. [CrossRef]
56. Giglio, L.; Justice, C.; Boschetti, L.; Roy, D. MCD64A1 MODIS/Terra+Aqua Burned Area Monthly L3 Global 500 m SIN Grid V006. 2015. Available online: https://doi.org/10.5067/MODIS/MCD64A1.006 (accessed on 10 March 2020).
57. Boschetti, L.; Eva, H.D.; Brivio, P.A.; Grégoire, J.M. Lessons to be learned from the comparison of three satellite-derived biomass burning products. *Geophys. Res. Lett.* **2004**, *31*. [CrossRef]
58. Arai, E.; Eyji Sano, E.; Dutra, A.C.; Cassol, H.L.G.; Hoffmann, T.B.; Shimabukuro, Y.E. Vegetation Fraction Images Derived from PROBA-V Data for Rapid Assessment of Annual Croplands in Brazil. *Remote Sens.* **2020**, *12*, 1152. [CrossRef]
59. Shimabukuro, Y.E.; Smith, J.A. The least squares mixing models to generate fraction images derived from remote sensing multispectral data. *IEEE Trans. Geosci. Remote Sens.* **1991**, *29*, 16–20. [CrossRef]
60. Camara, G.; Souza, R.C.M.; Freitas, U.; Garrido, J. SPRING: Integrating remote sensing and GIS by object-oriented data model. *Comput. Graph.* **1996**, *20*, 395–403. [CrossRef]

61. Shimabukuro, Y.E.; Batista, G.T.; Mello, E.M.K.; Moreira, J.C.; Duarte, V. Using shade fraction image segmentation to evaluate deforestation in Landsat Thematic Mapper images of the Amazon region. *Int. J. Remote Sens.* **1998**, *19*, 535–541. [CrossRef]
62. Bins, L.S.; Fonseca, L.M.G.; Earthal, G.J. Satellite imagery segmentation: A region growing approach. In Proceedings of the VIII Simpósio Brasileiro de Sensoriamento Remoto, Salvador, BA, Brazil, 14–19 April 1996; pp. 677–680.
63. Wilks, D.S. *Statistical Methods in the Atmospheric Sciences*, 3rd ed.; Academic Press: Oxford, UK; Waltham, MA, USA, 2011.
64. Roy, D.P.; Boschetti, L. Southern Africa Validation of the MODIS, L3JRC, and GlobCarbon Burned-Area Products. *IEEE Trans. Geosci. Remote Sens.* **2009**, *47*, 1032–1044. [CrossRef]
65. Morello, T.F.; Parry, L.; Markusson, N.; Barlow, J. Policy instruments to control Amazon fires: A simulation approach. *Ecol. Econ.* **2017**, *138*, 199–222. [CrossRef]
66. Schroeder, T.A.; Schleeweis, K.G.; Moisen, G.G.; Toney, C.; Cohen, W.B.; Freeman, E.A.; Yang, Z.; Huang, C. Testing a Landsat-based approach for mapping disturbance causality in U.S. forests. *Remote Sens. Environ.* **2017**, *195*, 230–243. [CrossRef]
67. Hawbaker, T.J.; Radeloff, V.C.; Syphard, A.D.; Zhu, Z.; Stewart, S.I. Detection rates of the MODIS active fire product in the United States. *Remote Sens. Environ.* **2008**, *112*, 2656–2664. [CrossRef]
68. Schroeder, W.; Prins, E.; Giglio, L.; Csiszar, I.; Schmidt, C.; Morisette, J.; Morton, D. Validation of GOES and MODIS active fire detection products using ASTER and ETM+ data. *Remote Sens. Environ.* **2008**, *112*, 2711–2726. [CrossRef]
69. Padilla, M.; Stehman, S.V.; Ramo, R.; Corti, D.; Hantson, S.; Oliva, P.; Alonso-Canas, I.; Bradley, A.V.; Tansey, K.; Mota, B.; et al. Comparing the accuracies of remote sensing global burned area products using stratified random sampling and estimation. *Remote Sens. Environ.* **2015**, *160*, 114–121. [CrossRef]
70. Brennan, J.; Gómez-Dans, J.L.; Disney, M.; Lewis, P. Theoretical uncertainties for global satellite-derived burned area estimates. *Biogeosciences* **2019**, *16*, 3147–3164. [CrossRef]
71. Rodrigues, J.A.; Libonati, R.; Pereira, A.A.; Nogueira, J.M.; Santos, F.L.; Peres, L.F.; Santa Rosa, A.; Schroeder, W.; Pereira, J.M.; Giglio, L.; et al. How well do global burned area products represent fire patterns in the Brazilian Savannas biome? An accuracy assessment of the MCD64 collections. *Int. J. Appl. Earth Obs. Geoinf.* **2019**, *78*, 318–331. [CrossRef]

remote sensing

MDPI

Article

Remote Sensing of Burn Severity Using Coupled Radiative Transfer Model: A Case Study on Chinese Qinyuan Pine Fires

Changming Yin [1], Binbin He [1,*], Xingwen Quan [1], Marta Yebra [2,3,4] and Gengke Lai [1]

1. School of Resources and Environment, University of Electronic Science and Technology of China, Chengdu 611731, China; uestcycm@163.com (C.Y.); xingwen.quan@uestc.edu.cn (X.Q.); laigengke@gmail.com (G.L.)
2. Fenner School of Environment and Society, Australian National University, Canberra, ACT 2601, Australia; marta.yebra@anu.edu.au
3. Bushfire and Natural Hazards Cooperative Research Centre, East Melbourne, VIC 3002, Australia
4. Research School of Aerospace, Mechanical, and Environmental Engineering, Australian National University, Canberra, ACT 2601, Australia

* Correspondence: binbinhe@uestc.edu.cn; Tel.: +86-1343-812-1356

Received: 29 August 2020; Accepted: 23 October 2020; Published: 2 November 2020

Abstract: Burn severity mapping is critical to quantifying fire impact on key ecological processes and post-fire forest management. Satellite remote sensing has the advantages of high spatial-temporal resolution and large-scale monitoring and provides a more efficient way to evaluate forest fire burn severity than traditional field or aerial surveys. However, the proportion of tree canopy cover (TCC) affects the spectral signal received by remote sensing sensors from the background charcoal and ash. Consequently, not considering this factor normally leads a spectral confusion in burn severity retrieval. In this study, the burn severity of two Qinyuan forest fires was estimated using a coupled Radiative Transfer Model (RTM) and Sentinel-2A Multi-Spectral Instrument (MSI) reflectance data. A two-layer Canopy Reflectance Model (ACRM) RTM was coupled with the GeoSail RTM by replacing the spectra of the background input of GeoSail RTM to simulate the spectra of the three-layered forests for burn severity retrieval measured as the Composite Burn Index (CBI). The TCC data was then served to RTM parameterization and constrain the backward inversion procedure of the coupled RTM to alleviate spectral confusion. Finally, the inversion retrievals were evaluated using 163 field measured CBI. The coupled RTM can simulate the radiative transfer characteristics of three-layer vegetation and has greater potential to accurately estimate burn severity worldwide. To evaluate the merit of our proposed method, the CBI was estimated through coupled RTM inversion with TCC constraint (CP_RTM+TCC), coupled RTM inversion with global optimal search (CP-RTM+GOS), Forest Reflectance and Transmittance (FRT) RTM inversion with TCC constraint (FRT+TCC), and random forest (RF) algorithm. The results showed that the method proposed in this study (CP_RTM+TCC) yielded the highest estimation accuracy (R^2 = 0.92, RMSE = 0.2) among the four methods used as benchmark, indicating its reasonable ability to assist forest managers to better understand post-fire vegetation regeneration and forest management.

Keywords: burn severity; composite burn index; Tree canopy cover; RTM; Sentinel-2A

1. Introduction

Wildfire is a pervasive natural disturbance, which affects the global forest climate services by regulating the spatial distribution of forests, as well as the exchange of carbon, water, and energy between the land and atmosphere [1–3]. Burn severity has been defined as the impact of fire on post-fire environments, including vegetation and soil [4]. Accurate, robust, and timely estimations of burn severity are very important for the quantification of forest fire impact and post-fire forest management [5,6].

To quantify the burn severity and make it easier for forest managers to quantitatively assess the impact of fire on forest ecosystems, Key and Benson [7] proposed the composite burn index (CBI). The CBI ranges from 0 to 3, where "0" stands for unburned and "3" stands for all biomass was completely consumed. CBI divides the forest vegetation structure into five layers from top to bottom and visually evaluates the burning degree of each layer of vegetation to obtain the CBI_i of a specific i layer. Here, i, represents the number of layers, and the CBI of all layers is calculated from the average CBI of the five layers in the forest. Subsequently, De Santis and Chuvieco (2009) [8] modified the CBI into the GeoCBI that calculates the field burn severity by considering the fraction of cover (*FCOV*) of each vegetation layer. GeoCBI has better spectral consistency with satellite remotely sensed data but there is a risk of increasing the area of moderate and high burn severity in the area covered by sparse trees [9]. Because of its fine evaluation, CBI has high accuracy as a field measurement index to evaluate the burn severity [7]. However, the collection of CBIs is time-consuming and costly, especially in remote regions. To cover the remote regions, only a simplified assessment can be carried out from a helicopter in most cases [5,10]. Consequently, CBI is usually used as the ground truth to verify the accuracy of estimated burn severity. Remote sensing techniques provide a cost-effective approach for large-scale spatial post-fire burn severity assessment as it offers high temporal and spatial observation for landscapes [11].

Four kinds of optical remotely sensed data-based methods have been used for burn severity estimation: empirical spectral index-based methods, Spectral Mixture Analysis (SMA), machine learning methods, and Radiative transfer model (RTM)-based methods. Empirical spectral index-based methods make use of statistical models calibrated using field measured CBI and vegetation indices such as the differenced Normalized Burn Ratio (dNBR), the relative differenced Normalized Burn Ratio (RdNBR), and the differenced Land Surface Temperature and Enhanced Vegetation Index ratio (d(LST/EVI)) derived from satellite data [12–14]. These empirical models are straightforward and performed well for sensor-specific and site-dependent burn severity estimates, but they lack physical meaning and require site-specific calibration for promising accuracy [9]. Spectral Mixture Analysis (SMA)-based methods are also among the most commonly used methods. In these methods, burn severity is estimated by solving the sub-pixel mixing problem because the post-fire environments are usually composed of a mixture of ash, vegetation, and soil. This kind of method normally uses all reflective bands of satellite data and has achieved promising accuracy in Mediterranean regions [15–17]. However, it is limited by the need for an extensive, site-specific end-member library of different burn severities [5]. Machine learning methods for burn severity estimation typically use the Random Forest (RF) classifier [18,19]. RF classifier is well suited to addressing fire mapping problems [20,21] as it can account for multiple environment environmental factors at the same time [22]. However, the accuracy of burn severity classification accuracy largely depends on the selection of the training data, and the classification results are usually only applicable to specific locations [18].

RTM-based methods have been proposed to solve the limitation in accuracy and precision of previous burn severity studies and improve the estimation of burn severity from remotely sensed data as RTM are based on physical laws that provide an explicit connection between the biophysical and biochemical variables of vegetation and canopy reflectance [9,23–25]. The RTM selected to estimate burn severity should be sensitive to the spectral characteristic variation of as many vegetation layers as possible as the evaluation of CBI consider five vegetation layers [26]. RTMs are first "forward" applied to simulate reflectance and transmittance corresponding to specific burn severity values, expressed as CBI or GeoCBI. In the backward inversion procedure, the surface reflectance extracted from remote sensing data is used to retrieve the biophysical and biochemical parameters of the vegetation that corresponding to a certain CBI or GeoCBI value [24,25]. However, the RTMs which are often used to retrieve burn severity, e.g., Markov Chain Canopy Reflectance Model (MCRM) and GeoSail model, cannot directly simulate the radiative transfer characteristics of three-layer vegetation structure except for the Forest Reflectance and Transmittance (FRT) model [27–29]. The two-layer vegetation structure generally includes tree canopy and lower grassland on the soil surface, while the three-layer vegetation structure includes upper tree canopy, middle shrub, and lower grassland. Quan et al. [30] has proved that the coupled RTM has better performance than the single RTM as the coupled RTM can better simulate the radiative transfer characteristics of multi-layer vegetation structure forest.

All of the previously presented methods that estimate burn severity by use of a single optical post-fire remotely sensed image have a defect in that there will spectral confusion between different burn severity levels with the variation of tree canopy cover (TCC). This is because the post-fire dead leaf litter and charcoal on the soil surface will have different contributions to the spectral signal received by remote sensing sensors with the variation of TCC [12]. The charcoal and dead litter on the soil surface will have a stronger contribution to the observed spectral signal for the sparse tree-covered area, while for the dense tree-covered area, the spectral signal of remote sensing sensors received from the soil surface will be shielded by the green tree canopy. The variation of TCC will lead the same burn severity will produce different spectral characteristics and the same spectrum will correspond to different burn severity values. Yin et al. [9] significantly improved the CBI estimation accuracy by using prior TCC information into the forward simulations and backward inversions of FRT model. Because there is no TCC parameter in FRT model, the stand density is used as a proxy parameter to constrain TCC. However, TCC is not only determined by stand density but also canopy diameter. Moreover, the correlation between stand density and TCC has not been validated other than the Yatir forest in Israel [31], and therefore the established empirical relationship cannot be generalized, limiting the application of FRT RTM-based inversion for forest burn severity somewhere else.

Within this context, our objectives are to (1) develop an improved RTM-based method to retrieve burn severity by coupling A two-layer Canopy Reflectance Model (ACRM) and GeoSail models at canopy level, and (2) achieve more accurately burn severity mapping. ACRM is used to model the spectra of vegetation in the middle and low stratum, while GeoSail is used to model the spectrum of the discontinuous upper tree canopy layer. The tree coverage parameter *FCOV* is one of the input parameters of GeoSail model, therefore it can explicitly account for variations in TCC without a proxy parameter. Our hypothesis is the coupled RTM can simulate the radiative transfer characteristics of three-layer vegetation and has greater potential to accurately estimate burn severity worldwide given it directly accounts for variation in TCC to avoid misclassify burn severity without the need of determining a site-specific stand density–TCC relationship. More accurate estimation of burn severity will contribute to better guide post-fire forest management.

2. Materials and Methods

2.1. Study Area

The study area is located at Qinyuan County, central China (111°58′30″E–112°32′30″E, 36°20′20″N–37°00′42″N) (Figure 1). The study area is dominated by overstory vegetation of *Pinus tabuliformis* while the middle vegetation layer is dominated by low *Quercus acutissima* and *Rosa xanthine* and the grass over the soil surface are dominated by Gramineous weeds. Under the management of the local government's policy of closing hills for reforestation, the vegetation growth in the study area is very dense, and the TCC in most areas is more than 50%. The height of *Pinus tabulaeformis* in the study area ranged from 3 to 5 m, which has a positive correlation with fire behavior [32]. *Pinus tabulaeformis* will become very flammable in dry and hot weather for a long time. At the same time, pine balls will help the fire spread faster with the help of wind. The topography of the study area is rugged, and the altitudes range from 1020 to 1639 m. The average annual rainfall in this region is 656.7 mm. The average annual relative humidity is 65% and the average annual temperature is 8.7 °C. Maximum and minimum precipitations were recorded in July–September and November–March, respectively. From November to March there is little precipitation, and the temperature will continue to rise from about −10 °C to 15 °C. During this period, sparse precipitation and temperature rise will help forest ecosystems accumulate large amounts of dry and flammable fuels. Generally, the fire season begins in March and ends in May.

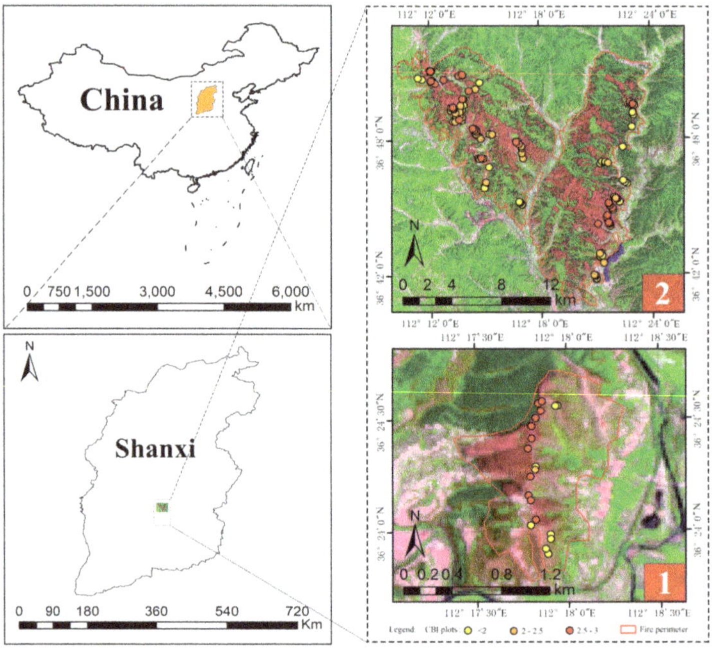

Figure 1. Location of the study area and surveyed field plots (dots in the figure) for fire 1 (bottom right panel) ignited on the 14 March (20 Composite Burn Indexes (CBIs)) and 2 (top right panel) ignited on the 29 March (143 CBIs). The background images used in the right two maps are the false-color composites (short wave infrared 2 (SWIR2), near-infrared (NIR), and red bands) of post-fire Sentinel-2A Multi-Spectral Instrument (MSI) data for the 30th June 2019. The fire perimeters are drawn manually based on differenced Normalized Burn Ratio (dNBR) data of the two fires.

2.2. Data Collection

In March 2019, two fires that caught the attention nationwide broke out in south (fire 1, 14 March) and north of (fire 2, 29 March) Qinyuan County (Figure 1). Fire 1 was small in scale (approximately burnt 48 ha), but six firefighters died during fire suppression because of sudden changes in wind direction. There were no casualties in fire 2 even though the burn area was larger (approximately 16,788 ha,). The post-fire field survey was conducted from 18 to 23 July 2019. The CBI was measured based on the field criterion proposed by Key and Benson (2006) and used as the ground truth to validate the accuracy of burn severity estimated by the methodology proposed in this study. There was a total of 20 and 143 field CBIs recorded for fire 1 and fire 2, respectively (Figure 1). Field plots were selected within areas with homogeneous burn severity levels, as recommended by Key and Benson [7].

2.3. Remote Sensing Data

2.3.1. Sentinel-2A MSI Data

In this study, the Sentinel-2A Multi-Spectral Instrument (MSI) data were used as the satellite data source to estimate burn severity. The Sentinel-2A MSI data has 13 spectral channels in the visible (VIS), near-infrared (NIR), and shortwave infrared (SWIR) spectral ranges. To keep the spatial resolution (20 m) consistency of surface reflectance data, nine out of the 13 spectral bands (bands

2–7, 8a, 11, and 12) were used to estimate burn severity (Table 1). Considering that NBR can provide more additional spectral information, the post-fire NBR, normalized of bands 8a and 12 were also used for burn severity estimation. To ensure that it is closest to the field survey time, the post-fire imagery was acquired on 30th June 2019. This image is the closest to the sampling date and has the best imaging quality. The Sentinel-2A MSI Level-1C data used in this study were downloaded from the Copernicus Open Access Hub (https://scihub.copernicus.eu/). The atmospheric correction of Sentinel-2A MSI data was conducted by use of the Sen2Cor Tool (version 2.3.1) [33,34]. The reflectance of the Level-1C Bottom-of-Atmosphere (BOA) was converted to Level-2A by using the default parameters. Sen2Cor consists of two main modules: the Scene Classification (SCL) module and the Atmospheric Correction (AC) module [35]. The AC is executed using a set of Look Up Tables (LUTs) generated by libRadtran [36]. AC includes three configuration parameters: aerosol type, atmosphere type, and ozone content. The aerosol type can be set as rural or maritime (the default setting is rural). There are two atmosphere types (mid-latitude summer or mid-latitude winter) that can be set, which will be automatically selected by Sen2Cor according to the geographic location and climatology of the scene [35]. The Ozone content is also automatically selected by Sen2Cor according to the geographic location and climatology.

Table 1. Spectral bands for the Sentinel 2A sensor.

Band Number	Central Wavelength (nm)	Bandwidth (nm)	Spatial Resolution (m)
1	442.7	21	60
2	492.4	66	10
3	559.8	36	10
4	664.6	31	10
5	704.1	15	20
6	740.5	15	20
7	782.8	20	20
8	832.8	106	10
8a	864.7	21	20
9	945.1	20	60
10	1373.5	31	60
11	1613.7	91	20
12	2202.4	175	20

2.3.2. Landsat Vegetation Continuous Fields Product

The Landsat Vegetation Continuous Fields (VCF) tree cover product contains estimates of the TCC in each 30 m pixel covered by woody vegetation greater than 5 m in height [37]. The product extracts all seven bands of Landsat 5 Thematic Mapper (TM) and/or Landsat 7 Enhanced Thematic Mapper Plus (ETM+) by rescaling the 250 m Moderate-resolution Imaging Spectroradiometer (MODIS) Vegetation Continuous Fields (VCF) Tree Cover layer. On 5 March 2019, the NASA Land Cover/Land Use Change Program released version 4 of the Global 30 m Landsat tree canopy cover product. The fourth edition of the TCC product replaced the previous version of global tree canopy cover estimates for 2000, 2005, 2010, and 2015. Previous versions of TCC were collected using the Landsat Global Land Survey, but version 4 was created by mining and processing entire Landsat archives, providing improved and accurate representation for specific years (https://landsat.gsfc.nasa.gov/global-30m-landsat-tree-canopy-version-4-released/). In this study, the TCC layer of 2015 was used to parameterize and constrain the RTM. The forests in the study area are natural forests with little human intervention, and there has been no forest fire in the study area since 2015, thus the TCC layer in 2015 can be used to represent the situation before the fire. For more information about this dataset please visit http://landcover.org/data/landsatTreecover. The TCC layer of Landsat VCF data was resampled to 20 m matching the spatial resolution of Sentinel-2A MSI data using nearest-neighbor interpolation.

2.4. Burn Severity Retrieval Using Coupled RTM

The methodology to retrieve burn severity using coupled RTM is carried out in four steps (Figure 2): (i) remote sensing data preprocessing, including the atmospheric correction of the Sentinel 2A MSI data and resampling of the Landsat VCF product; (ii) couple RTMs of leaf and canopy level for burn severity retrieval; (iii) forward modeling, including the parameterization of the coupled RTM based on the sensitivity analysis results of previous studies and construction of the LUT [24]; and (iv) Coupled RTM backward inversion procedure by finding the simulated surface reflectance (saved in the LUT) that most similar to the extracted spectrum from satellite remotely sensed data by use of a cost function [38]. The coupled RTM can simulate the radiative transfer characteristics of three-layer vegetation and has greater potential to accurately estimate burn severity worldwide given it directly accounts for variation in TCC to avoid misclassify burn severity without the need of determining a site-specific stand density–TCC relationship.

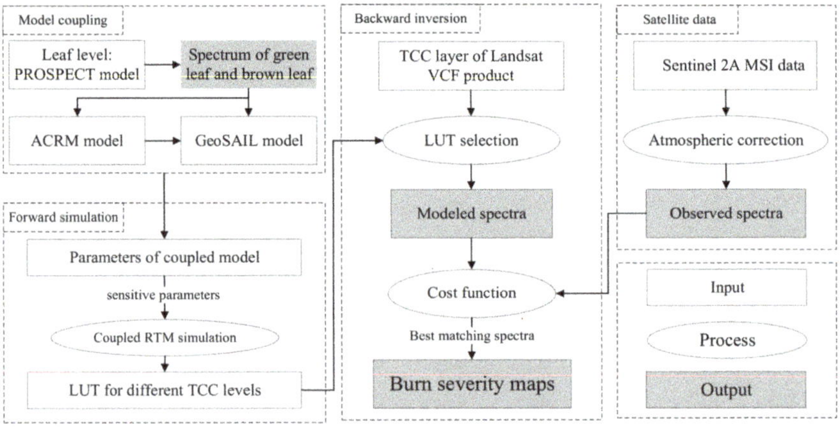

Figure 2. Methodological flowchart of this study. The four parts in the flowchart correspond to the four steps of burn severity inversion.

2.4.1. Model Selection and Coupling

Several vegetation layers are considered for CBI evaluation. Therefore, the selected RTM should have the capacity to simulate the spectrum of multi-layered forests. In this study, the leaf-level RTM PROSPECT [39], canopy-level RTM ACRM [40,41], and GeoSail RTM [42] were coupled to simulate the spectra of the multi-layered forests for burn severity retrieval. The coupled RTM has the capacity of simulating the radiative transfer characteristics of three-layer vegetation and has greater potential to accurately estimate burn severity worldwide given it directly accounts for variation in TCC to avoid misclassifying burn severity without the need of determining a site-specific stand density–TCC relationship.

The PROSPECT model was used to simulate the spectrum of a green and brown needle leaf. Although other RTMs may be more suitable for simulating the reflectance and transmittance of needles than PROSPECT (e.g., LIBERTY model proposed by Dawson et al. [43]), they are not as operational as PROSPECT because it requires more input variables with more complex parameters. Besides, PROSPECT has been successfully applied to the simulation of needle leaves' spectra in many studies [24,44,45]. The PROSPECT model is a leaf optical model, which calculates the reflectance and transmittance of the leaf hemisphere in the range of 400 to 2500 nm in steps of 1 nm. The calculated reflectance and transmittance are expressed as a function of six input parameters: leaf structure parameter, N (unit-less); C_{ab} (μg cm^{-2}); Equivalent Water Thickness, C_w (g cm^{-2}); Dry Matter Content, C_m (g cm^{-2}); carotenoid content, C_{ar} (μg cm^{-2}); and leaf brown pigment, C_s (unit-less).

ACRM assumes that the canopy consists of a homogeneous vegetation layer and a thin layer of vegetation on the soil surface. The model is an extension of the homogeneous multispectral canopy reflectance model MSRM [46] and Markov chain canopy reflectance model MCRM [47]. The model has a step size of 1 nm in the 400 to 2400 nm spectral range. This RTM is thus suitable for simulating the spectra of middle and lower canopies. Here, the ACRM was coupled with the GeoSail model by replacing the spectra of the bare soil input of GeoSail and considering the vertical stratification characteristics of vegetation in the study area. The ACRM was used to simulate the spectrum of middle vegetation layers and grass over the soil surface.

The GeoSail model was used to simulate the spectrum of the upper tree canopy layer since it is a geometric model that considers the geometry and shadowing effects of forest stands. The Geosail model described the canopy reflectance of heterogeneous and discontinuous vegetation types with relatively simple input parameters [42] and has been successfully used for burn severity retrieval [24]. The tree coverage parameter *FCOV* in the GeoSail model can be directly constrained by satellite-derived TCC without proxy parameters such as stand density, therefore the coupled RTM has greater potential than FRT RTM in burn severity retrieval. All the canopy vegetation parameters required by the coupled RTM are listed in Table 2. For the range and default values of model parameters, refer to the user's guide and previous studies [24,27,30].

Table 2. The input canopy vegetation parameters are used for the parameterization of the coupled Radiative Transfer Model (RTM). The simulation of radiative transfer characteristics of upper vegetation was based on GeoSsail model, while that of middle and lower vegetation layers was based on the A two-layer Canopy Reflectance Model (ACRM) model.

Parameters	Units	Abbreviation	Range	Defaults
Upper tree canopy GeoSail				
Fractional cover		*FCOV*	0–1	0.7
leaf area index	(m^2 m^{-2})	*LAI*	0.1–2.5	2
height to width ratio of the crown		*CHW*	1–8	2.36
sun zenith angle	(°)	*tts*	0–90	42.27
Middle vegetation layer ACRM				
*LAI2*_ground	(m^2 m^{-2})	*LAI2*	0.01–6	0.208
HS-parameter		*sl2*	0.02–0.4	0.15
foliage clumping parameter		*clmp2*	0.4–1	1.0
displacement parameter		*szz*	0–2	1.2
eccentricity parameter of LAD		*eln2*	0–4.5	3.99
modal leaf angle	(°)	*thm2*	0–90	53.37
n_ratio2		*n_ratio2*	0.6–1.3	0.991
leaf weight per area	g m^{-2}	*SLW2*	30–180	81.7
Lower vegetation layer ACRM				
*LAI1*_ground	(m^2 m^{-2})	*LAI1*	0.01–1.1	1.064
HS-parameter		*sl1*	0.02–0.4	0.15
foliage clumping parameter		*clmp1*	0.4–1	1
eccentricity parameter of LAD		*eln1*	0–4.5	3
modal leaf angle	(°)	*thm1*	0–90	75.469
n_ratio1		*n_ratio1*	0.6–1.3	1.224
leaf weight per area	g m^{-2}	*SLW1*	30–180	78.54

For the coupled RTM, the diffuse radiation was set as the dominant radiation for the grassland layer by the hypothesis that most of the direct radiation intercepted by the tree canopy. Therefore, the bi-hemispherical reflectance rather than the BRDF of the understory was modeled in this case. Besides, the spectral reflectance of the background originally in the GeoSail is assumed to be the reflectance of Lambertian while the spectra modeled by the PROSAIL (PROSPECT and Scattering by Arbitrarily Inclined Leaves) is non-Lambertian.

2.4.2. Model Parameterization and Forward Modeling

The parameterization of the coupled RTM was based on prior information obtained from field quantitative measurements, qualitative survey, previous studies [24,30], and Sentinel 2A MSI metadata. Only the parameters most sensitive to the change of the target bands according to previous analysis from De Santis et al. [24] and Yin et al. [9] were used, i.e., the fractional cover (*FCOV*), leaf area index (*LAI*), and water content (C_w) are most sensitive to the short-wave infrared (SWIR) band. In detail, the *LAI* and *FCOV* are the most sensitive parameters for the NIR band. The C_{ab}, *LAI*, and brown pigments content (C_s) are the most sensitive parameters for VIS bands. Therefore, the sensitive parameters selected for the GeoSail model were *FCOV* and *LAI* at canopy level and C_w, C_{ab}, and C_s at the leaf level, as illustrated in Table 3. In the CBI field criterion, burn severity is evaluated using a set of variables for each vegetation vertical strata, and the CBI increases with the proportion of brown over green leaves. Therefore, the PROSPECT model was used at the leaf level to simulate only one representative spectrum for green and brown reference leaf types. The parameterization for PROSPECT model were N = 2.5, C_{ab} = 70 µg cm^{-2}, C_w = 0.048 g cm^{-2}, and C_s = 0.2 for green leaf [48] while to N = 2.5, C_{ab} = 20 µg cm^{-2}, C_w = 0.008 g cm^{-2}, and C_s = 1.5 for brown leaf [49]. The simulated green and brown leaf spectrum based on the PROSPECT model was shown in Figure 3. The ACRM model was calibrated following Yin et al. [9]. For the middle vegetation layer, the sensitive parameters are *LAI* of middle vegetation and a series of biochemical parameters determining leaf color (N, C_{ab}, C_w, and C_s); for the lower vegetation layer, the sensitive parameters are *LAI* of lower vegetation and biochemical parameters determining leaf color (N, C_{ab}, C_w, and C_s) (Table 3).

Table 3. Combination of parameters used as inputs in the coupled RTM for simulating the reflectance corresponding to different CBI values. To improve the efficiency of model execution and alleviate the ill-posed problem, only the parameters sensitive to burn severity inversion were used to construct Look Up Tables (LUTs).

CBI	0.5	1	1.5	1.8	2	2.3	2.5	2.8	3
FCOV					0–0.8				
LAI3	0.8	0.8	0.8	0.78	0.75	0.72	0.7	0.5	0.05
leaf color	Green	Green	Green	Green	Green	Green	Green	Brown	Brown
LAI2	0.5	0.4	0.3	0.15	0.1	0.05	0.05	NAN	NAN
leaf color	Green	Brown	Brown	Brown	Brown	Brown	Brown	NAN	NAN
LAI1	3	0.3	NAN	NAN	NAN	NAN	NAN	NAN	NAN
leaf color	Brown	Brown	NAN	NAN	NAN	NAN	NAN	NAN	NAN
sub-stratum	0.6soil+ 0.4DCH	0.5soil+ 0.5DCH	DCH	DCH	DCH	DCH	DCH/ LCH	DCH/ LCH	DCH/ LCH

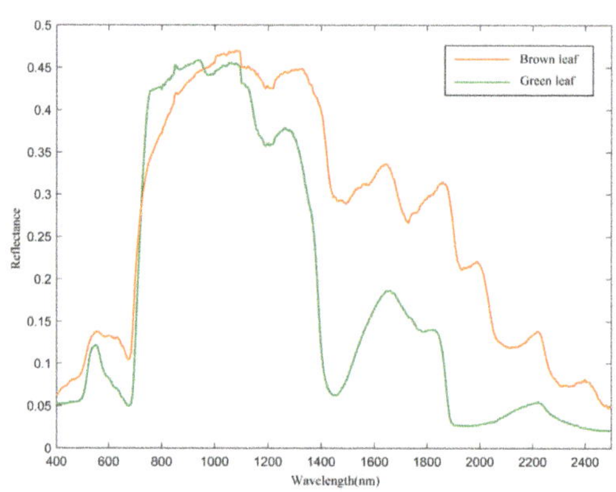

Figure 3. Modeled spectra of brown (brown curve) and green leaf (green curve) used as input for the simulation at canopy level based on the PROSPECT model.

Considering the uncertainty caused by unrealistic combinations of the input model variables, a supervised simulation scheme was executed to reduce the error because it is better than the full range variations [24,26]. The parameter *FCOV* (range from 0 to 0.8) in the GeoSail model was set as the only free variable to generate the LUT. When the coupling model is forward simulated, the spectral curves corresponding to different TCC conditions can be generated with the variation of *FCOV*. A combination example of the model input parameters used to generate LUT is listed in Table 3. The background shortly after wildfires usually consists mainly of soil and charcoal [24]. Therefore, three reference spectra were considered as the background layer for the model input: soil, dark charcoal (DCH), and light charcoal (LCH; a mixture of charcoal and ash) (Figure 4). The soil spectra were measured on sandy soil with moderate moisture, while the spectrum of dark and light charcoal was measured with GER 2600 field spectro-radiometer (Geophysical & Environmental Research Corporation, Millbrook, NY, USA) by De Santis et al. [24]. The simulated CBI value variation from 0.5–3 with steps of 0.1. The burn severity can be divided into low, moderate, and high level by use of CBI threshold values of 2 and 2.5 [9]. For low burn severity level, only the grassland and shrub on the soil surface were burned. For moderate burn severity level, the underlying grass and shrubs were completely burned and the understory of the tree was slightly burned, but the trees still retained the green canopy. For high burn severity level, the low and middle vegetation strata were completely burned and the crown was severely burned. For the part of CBI lower than 2.5, only the DCH scenario was simulated. For the part of CBI higher than 2.5, the organic matter in the soil and all vegetation strata is burned, the substratum can be charcoal or ash, therefore both DCH and LCH scenarios are simulated. Finally, there are 352 combinations of input parameters used to construct the LUT for burn severity retrieval.

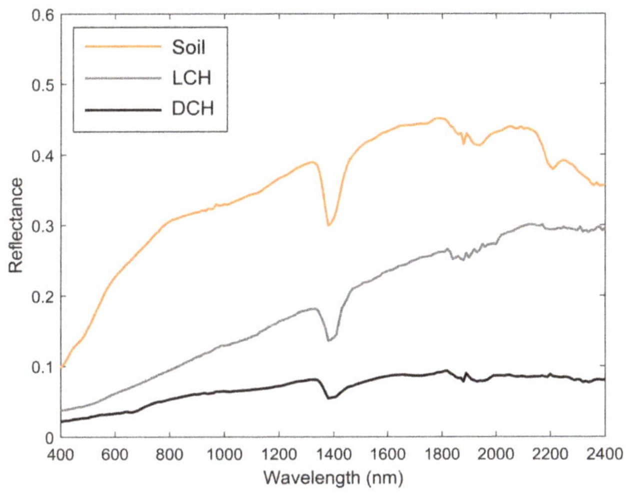

Figure 4. The reference spectrum used as post-fire background layer: soil (brown curve), dark charcoal (DCH; black curve), and light charcoal (LCH; a mixture of charcoal and ash, grey curve). The reference spectrum was provided by De Santis.

2.4.3. Coupled RTM Inversion

In this study, the LUT was used as the inversion algorithm to retrieve burn severity using Sentinel-2A MSI satellite remotely sensed data. We derived burn severity using the information on TCC to constrain the backward inversion procedure. For each pixel in the burnt area, the Landsat VCF product was served to select the corresponding LUT. The model simulated spectral reflectance in the LUT corresponding to a specific TCC range was subsequently compared to the extracted surface reflectance of the pixel by use of a cost function. The Spectral Angle Mapper (SAM) was used as the cost function to find the model simulated spectral reflectance that was most similar to the spectra reflectance extracted from remotely sensed data. SAM is a supervised classification technology based on pixels. SAM solves spectral similarity by calculating the spectral angle (SA) between two spectral vectors [50,51]. This angle is independent of the length of vectors and is therefore insensitive to the illumination or albedo effects (Equation (1)). Consequently, SAM can eliminate most of the errors introduced by terrain relief [23].

$$\mathrm{SA} = \cos^{-1}\left[\frac{\sum_{i=1}^{nb} t_i r_i}{\left(\sum_{i=1}^{nb} t_i^2\right)^{\frac{1}{2}}\left(\sum_{i=1}^{nb} r_i^2\right)^{\frac{1}{2}}}\right] \quad (1)$$

The i represents the band number, nb represents the number of the bands, and t and r represent the simulated and observed spectrum, respectively. SA between the simulated and observed reflectance was calculated as the cost function to estimate the biophysical and biochemical variables of the vegetation that corresponding to certain CBI values. The R square (R^2), root mean square error (RMSE), and slope between the measured and estimated CBI were used as the accuracy evaluation index. The backward inversion process in this study was executed by use of all nine Sentinel-2A MSI bands and the coupled RTM with TCC constraint (labeled as CP-RTM+TCC). In addition to the nine spectral bands of Sentinel-2A MSI data mentioned above, the post-fire NBR was also used as an additional vector to provide more spectral information and improve the accuracy of burn severity retrieval.

2.4.4. Methodological Comparison

To compare and evaluate the merit of our proposed method (labeled as CP-RTM+TCC), we derived burn severity using three other existing methods. In the first method, the coupled RTM is inverted using a global optimal search and without TCC constraint (labeled as CP-RTM+GOS). The second one is the method used by Yin et al. [9] (labeled as FRT+TCC). In this method, the burn severity was estimated using the FRT model with TCC constraint. The third method uses a random forest (RF) algorithm. The RF classifier is particularly useful for burn severity mapping with promising accuracy [18,19,52]. In the RF method, the independent variables are the nine spectral bands and the post-fire NBR index of Sentinel 2A MSI data, and the dependent variable is CBI. First, the field measured plots were randomly divided into three samples (each sample comprised about 54 CBI measurements). Second, two samples (i.e., the training sample) were used for RF model calibration, and the other sample (i.e., the testing sample) was used for the validation. Third, the process was repeated three times to adequately test the performance of the RF method [53]. To minimize the statistical error caused by each run of the RT model, we set the number of trees to 1000 and run 100 times to train each RF model.

3. Results

3.1. Influence of TCC on the Spectral Response of Burn Severity

The surface reflectance and post-fire NBR of all the field plots were extracted from Sentinel-2A MSI data from locations corresponding to our sampling locations and averaged by burn severity level and TCC (<20%, 20%–30%, 30%–50%, and >50%) to prove the influence of TCC on the spectral response and the confusion between different burn severity levels (Figure 5). Under this circumstance, the burn severity was divided into low, moderate, and high level by use of CBI threshold values of 2 and 2.5 [9]. Figure 5(a1,b1), and Figure 5(c1) represent the extracted mean reflectance from all field plots without constraining TCC. The averaged reflectance of all field plots obscured the reflectance variation between different TCC ranges. By analyzing the spectrum of three different levels of burn severity under different TCC, it can be found that the high-level burn severity is not sensitive to the change of tree coverage, while the medium level burn severity shows little sensitivity to the change of tree coverage, and the low-level burn severity is most sensitive to the change of tree coverage. This was also observed from the spectral angle (SA) between the spectra of lowest and highest TCC within the same burn severity. The SA was 0.68 (Figure 5(a2,a5)), 0.33 (Figure 5(b2,b5)), and 0.27 (Figure 5(c1,c5)) for low, moderate, and high burn severity, respectively. For TCC over 50%, the spectral characteristic of the high burn severity exhibited similar characteristics to the moderate and low burn severity with the lower TCC. For a detailed analysis of the influence of TCC on the inversion of burn severity, please refer to Yin et al. [9]. The results given in Figure 5 indicate that if the TCC is not considered in the inversion of burn severity, spectral confusion will be caused by the change of TCC between different burn severity levels.

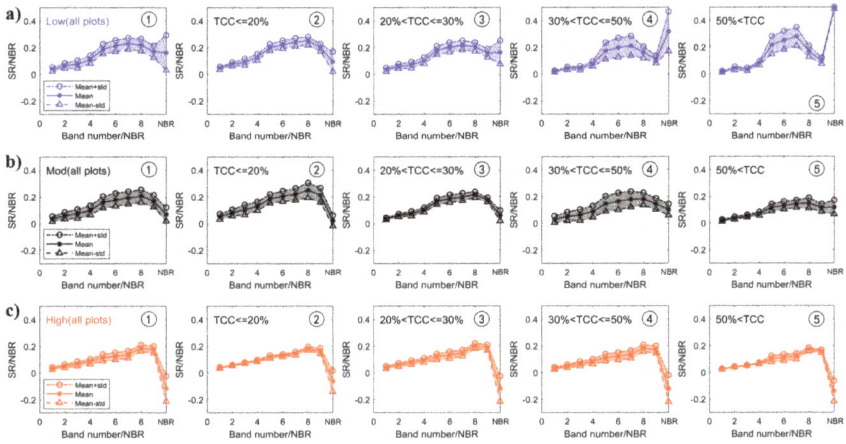

Figure 5. Mean and standard deviations (std) of the Sentinel-2A MSI spectral signatures of plots with (**a**) low, (**b**) moderate, and (**c**) high burn severities and TCC ranges. The x-axis represents the nine spectral bands and post-fire NBR, and the y-axis represents the value of the reflectance (SR) and post-fire NBR. The first column represents the reflectance from all field plots without constraining TCC; other columns represent the reflectance statistics in different TCC levels.

3.2. Evaluation of Burn Severity Estimates

A total of 163 CBI values recorded at two sample sites were used to validate the retrieved CBI. The CBI was estimated based on the coupled RTM constrained by TCC information provided by satellite estimates (CP_RTM+TCC). The validation results at both sites showed a high correlation between the observed and simulated CBI (R^2 equal to 0.96 for study site 1 (Figure 6a), 0.91 for study site 2 (Figure 6b), and 0.92 for two study sites (Figure 6c)) and an almost 1:1 linear fit (slope: ~1; constant: ~0). The RMSE is also very low (0.15 for study site 1, 0.21 for study site 2, and 0.20 for two study sites).

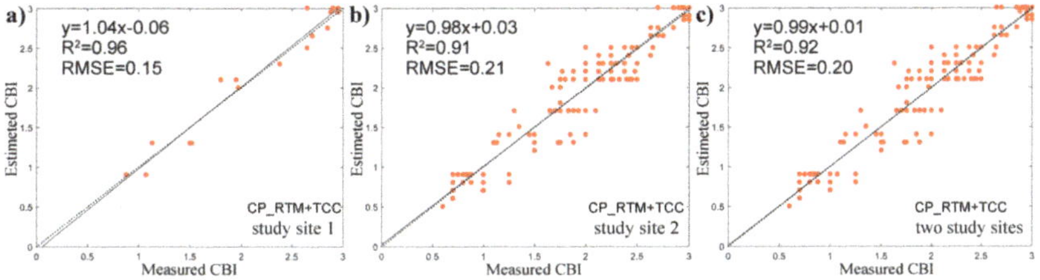

Figure 6. Field measured versus estimated burn severity (expressed as CBI) based on the method proposed in this study (CP_RTM+TCC) for (**a**) study site 1, (**b**) study site 2, and (**c**) two study sites. The dotted line and solid line represent the 1:1 line and fitting line, respectively.

When comparing the method proposed in this study with the burn severity estimations obtained with CP_RTM+GOS, FRT+TCC, and RF methods (Table 4 and Figure 7), the method proposed in this study but without TCC constrains (CP_ RTM+GOS method) presents slightly worse performance than FRT+TCC and RF. Figure 7b also shows that the estimated results of the FRT+TCC method yield a relatively higher accuracy (R^2 equal to 0.82, RMSE equal to 0.32, and slope equal to 0.81). As can be indicated from Figure 7c that the validation accuracy (R^2 = 0.76, RMSE = o 0.35 and slop = 0.7) based on the RF method is higher than CP_RTM+GOS but lower that of the FRT+TCC method. Overall, the CP_RTM+TCC method (Figure 6) performed best (R^2 equal to 0.92, RMSE equal to 0.2 and slop equal to 0.99) among these four methods.

Table 4. The accuracy summary of the four methods used in this study. The method proposed in this study is CP_RTM+TCC, which has the highest accuracy of burn severity estimation.

	CP_RTM+TCC (Proposed Method)	**CP_RTM+GOS**	**FRT+TCC**	**RF**
R^2	0.92	0.57	0.82	0.76
RMSE	0.2	0.57	0.32	0.35
Slop	0.99	0.74	0.81	0.7

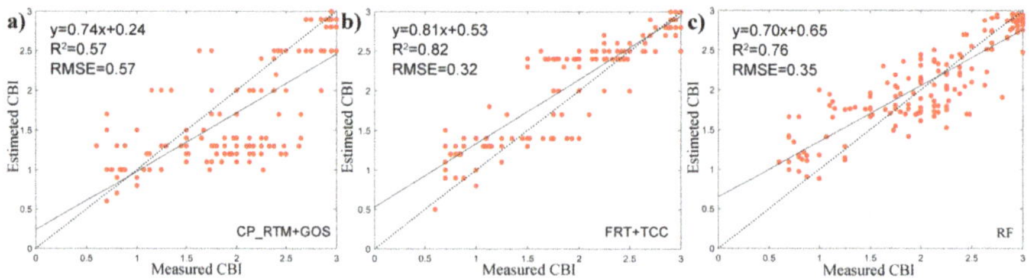

Figure 7. Field measured versus estimated burn severity (expressed as CBI) based on (**a**) CP_RTM+GOS, (**b**) FRT+TCC, and (**c**) RF method for two study sites. The dotted line and solid line represent the 1:1 line and fitting line, respectively.

3.3. Burn Severity Mapping

The focus of this study is burn severity estimation, so the dNBR as well as visual interpretation were used to extract burn area. Burn severity maps for the two sample sites of Qinyuan country were produced based on the burn area extraction and the CP_RTM+TCC method (Figure 8). The high-level burn severity range (CBI $\geq$ 2.5) in site 1 is mainly distributed in the western area because the elevation of the western area of the site is relatively high and its main vegetation type is dense and dry *Pinus tabulaeformis* forest, which is quite flammable. The eastern part corresponds to the foot and hillside with relatively low elevation, the vegetation types are mainly yellow rose, low bush, and dense herbaceous vegetation, and the burn severity type is mainly of medium (2 $\leq$ CBI < 2.5) and low level (CBI < 2). For study site 2, the main vegetation type is also *Pinus tabulaeformis.* The middle and high-level burn severity are mainly distributed at the top of the mountain with relatively high altitude, while the low-level burn severity is mainly distributed at the foot of the mountain with relatively low and flat terrain. From the spatial distribution map of burn severity in study area 2, we can also see that there are more areas with CBI over 2.8 and below 2.0, but fewer areas with CBI between 2.0 and 2.8 than in study site 1.

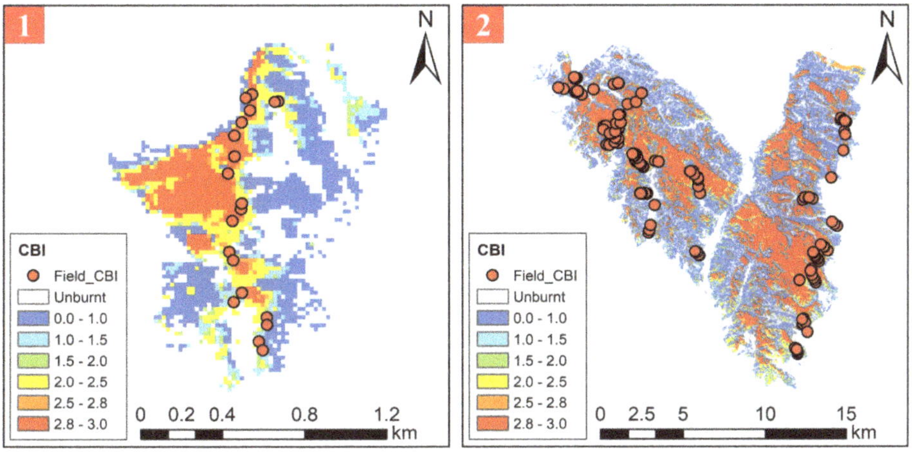

Figure 8. Burn severity maps for study sites 1 (left) and 2 (right) of Qinyuan fires achieved using the CP_RTM+TCC method proposed in this study. In order to show the spatial distribution characteristics of burn severity in the two study sites more finely, CBI was divided into six grades which vary from 0 to 3.

In the field measurement, in addition to the photos used to estimate the CBI value of the sample plots, we also selected some areas with a wide view to take panoramic photos for quantitative verification of burn severity inversion results. The qualitative validation was carried out by comparing the field landscape panoramic photos and the spatial distribution map of burn severity (Figure 9). A total of four validation points were selected: two (Figure 9b,c) of them point to the valley, and the other two (Figure 9a,d) point to the hillside. The qualitative accuracy validation results indicated that the burn severity map and the landscape panoramic photos from the field survey showed good spatial distribution consistency. This provided further evidence for the reliability of our burn severity estimates.

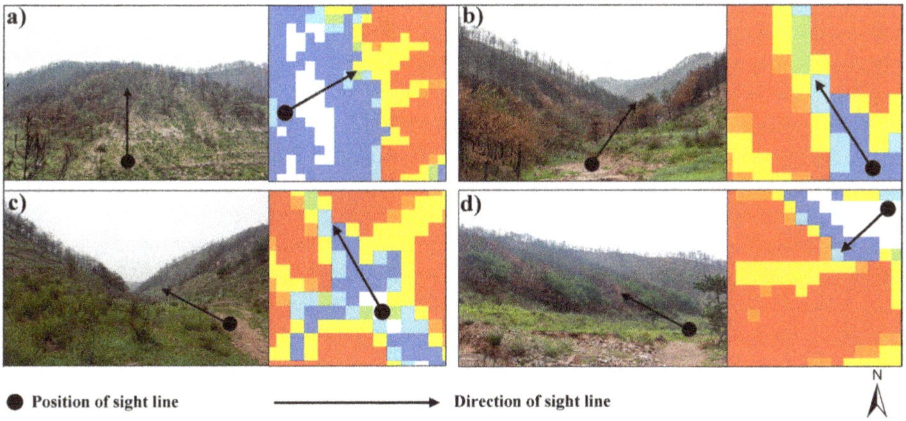

Figure 9. Examples of panoramic landscape photos (left of **a**–**d**) and burn severity map (right of **a**–**d**) to demonstrate the validity of the method proposed in this study. The black point represents the position of sight, the arrows represent the direction of the sight. The qualitative validation indicated that the burn severity map and the panoramic landscape photos showed good spatial distribution consistency.

4. Discussion

In this study, the burn severity of two forest fires was estimated using a coupled RTM and Sentinel-2A MSI reflectance data. ACRM assumes that the canopy consists of a homogeneous vegetation layer and a thin layer of vegetation on the soil surface, which is thus suitable for simulating the spectra of middle and lower canopies. The GeoSail model is a geometric model that considers the geometry and shadowing effects of forest stands and was used to simulate the spectrum of the upper tree canopy layer. The ACRM was coupled with the GeoSail model by replacing the spectra of the bare soil input of GeoSail and considering the vertical stratification characteristics of vegetation in the study area. The coupled RTM can simulate the radiative transfer characteristics of three-layer vegetation and has greater potential to accurately estimate burn severity worldwide given it directly accounts for variation in TCC to avoid misclassify burn severity without the need of determining a site-specific stand density–TCC relationship.

Compared with the three existing methods in other studies, the proposed method (CP_ RTM+TCC) in this study performed best. The CP_ RTM+GOS method performed worst because if TCC is not constrained, there is serious spectral confusion between the different burn severity levels. This is similar to the conclusion of Yin et al. [9], who found spectral confusion between different burn severity levels when the influence of TCC was not considered in the RTM forward and backward inversion procedure. Accuracy comparison results also showed that the FRT+TCC method yields a relatively higher accuracy. This is because the influence of tree canopy cover is considered in the FRT+TCC method. However, the accuracy was still lower than that of this study. This is because the stand density was used as a proxy parameter to constrain the TCC in the FRT+TCC method, but stand density is not the only determining parameter of tree coverage. Furthermore, the statistical correlation between stand density and the lack of universality in TCC occurs because the empirical correlation has not been validated other than the Yatir forest in Israel. Therefore, the indirect constraint of tree canopy cover through stand density will produce matching errors, which will affect the estimation accuracy of CBI. In essence, the RF algorithm estimates the burn severity according to the characteristics of spectral reflectance, so it cannot get rid of the influence of TCC. Therefore, the estimated accuracy based on the RF method is still lower than the method proposed in this study.

The spatial distribution of burn severity in the two study sites shows obvious boundary characteristics. Most regions have CBI values above 2.8 or below 2.0, but few regions between 2.0 and 2.8. Combined with field sampling, we analyzed two main reasons for this phenomenon. First, the interval between the fire and the field sampling was nearly 4 months and there was rainfall during this period. The vegetation recovery ability in the low and medium burn severity area was stronger than that in the high burn severity area. Most of the trees are burned completely under the high burn severity level, which hinders quick recovery, and the low and medium burn severity is only affected by the medium degree, supporting recovery faster speeds, which is consistent with the conclusion of Meng et al. [54]. On the other hand, the relationship between vegetation recovery rate after the fire and burn severity is not linear, but is a saddle curve [54]. This curve shows that the forest has certain fire resistance. For example, some trees under fire stress have a short-term loss of canopy leaves but do not die. Therefore, their canopies will recover in a short time after the fire. However, with the increase of forest burn severity, tree survival will decrease as canopies cannot recover from the damage. The areas affected by a moderate burn severity did not cause fatal damage to trees, therefore the recovery rate was the fastest, with quick recovery to the state of lower burn severity level after the fire. In most cases, the damage to trees caused by high burn severity is irreversible, implying slow recovery. Second, field sampling saw that the high burn severity level was mainly distributed in areas with high altitude and high slope. During the period from the fire (29 March) to the field sampling (18 July), there were several precipitation events (>83 mm), and the temperature continued to rise (from 6.9 to 26.3 °C) (Figure 10); this created good conditions for the post-fire recovery of vegetation in the study area. However, because of the fire-related decrease of water storage capacity in mountain forests, there was a slower recovery rate of trees at the top of mountains and in steep areas compared to valleys and relatively flat terrain. In the process of field survey, we also found that the spatial distribution of burn severity has a high correlation with terrain characteristics. For example, the areas with high burn severity are mostly distributed on the top of the mountain and steep hillside, which indicates that terrain factor is the important drive factor of burn severity. Future work will be focus on analyzing the relationship between post-fire burn severity and pre-fire environmental drive factors.

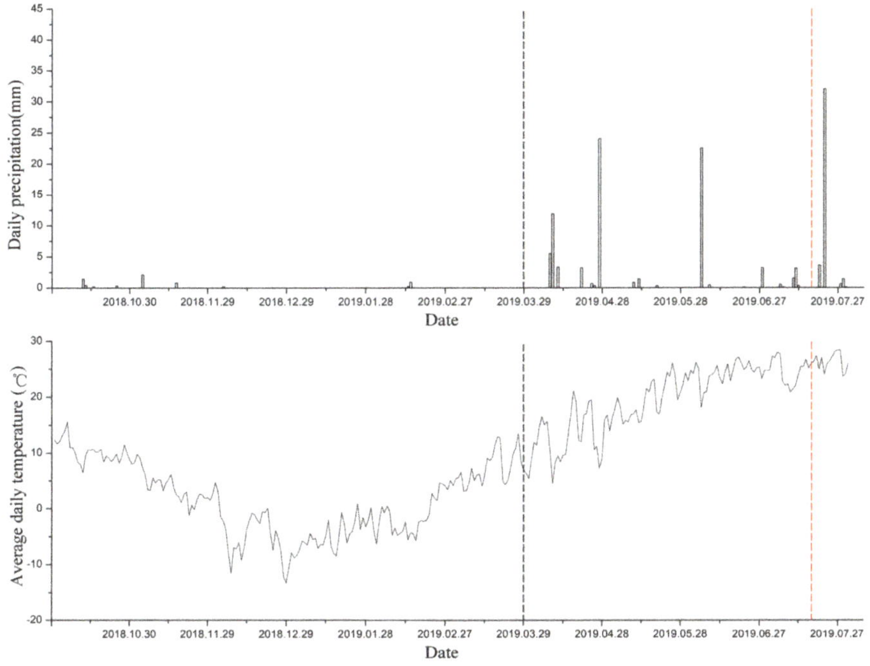

Figure 10. Average daily temperature and daily precipitation data for study site 2 obtained from Taiyuan weather station (longitude: 112.35°E, latitude: 37.37°N) during the period from October 2018 to July 2019. There is almost no precipitation from October 2018 to March 2019, and the temperature has continued to rise since January 2019. The location of the black dotted line represents the ignition date of the fire, and the location of the red dotted line represents the time of field sampling.

The meteorological data before the forest fire 2 in Qinyuan county were also analyzed. Figure 10 gives the daily average temperature and precipitation data of Taiyuan meteorological station nearest to Qinyuan county (longitude: 112.35°E, latitude: 37.37°N) from 1 October 2018 to 31 July 2019. There was almost no precipitation in the six months before the fire and the temperature continued to rise in the three months before the fire, which provided sufficient external conditions for the fire. This is also evident in the driving effect of meteorological factors on fire. Furthermore, due to the local policy of closing the mountain for forest cultivation, many areas in sample site 2 were completely closed, resulting in the accumulation of a large number of fuel load, including live fuel, such as *Pinus tabulaeformis* and other shrubs; herbaceous vegetation; and dead fuel represented by litter and humus on the ground. The accumulation of fuel load is also one of the important causes of fire ignition. Meteorological variables are one of the most important factors affecting post-fire vegetation regeneration. Vegetation regeneration and growth are generally highly depending on favorable postfire meteorological conditions, such as conditions with higher moisture and proper temperature [55]. The several precipitation events and the average daily temperature continued to rise in the study area after the fire created good conditions for vegetation regeneration.

5. Conclusions

Burn severity is of vital importance for post-fire forest management and will assist in post-fire vegetation regeneration. In this study, the burn severity of two forest fires in Qinyuan was estimated using coupled RTM and Sentinel 2A MSI data. Traditional optical satellite remotely sensed data-based methods that do not take the variation in TCC into account will lead to spectral confusion in burn severity retrieval. Yin et al. [9] significantly improved the CBI estimation accuracy by using prior TCC information into the forward simulations and backward inversions via the stand density parameter of the FRT model. However, TCC is not only determined by stand density but also canopy diameter. Moreover, the correlation between stand density and TCC is not universal, limiting the application of FRT RTM-based inversion for forest burn severity somewhere else. In this study, ACRM was coupled with the GeoSail model, by replacing the spectra of the bare soil input of GeoSail and considering the vertical stratification characteristics of vegetation in the study area. The tree coverage parameter *FCOV* in the GeoSail model was directly constrained by TCC without proxy parameters, and therefore has greater potential to accurately estimate burn severity worldwide. The verification of the accuracy of results using four methods (CP_RTM+TCC, CP_RTM+GOS, FRT+TCC, and RF) showed that the method proposed in this study (CP_RTM+TCC) has the highest burn severity estimation accuracy. The more accurate estimation of the burn severity obtained from the approach presented in this study will help land managers to better understand post-fire vegetation regeneration and help in post-fire forest management. Now that a new method for accurately estimate fire severity has been found, future work will be focus on analyzing the relationship between post-fire burn severity and pre-fire environmental drive factors.

Author Contributions: B.H. and X.Q. supervised and fund this research; C.Y., X.Q., and B.H. conceived and designed the experiments; C.Y. performed the experiments and analyzed the data; C.Y. and X.Q. wrote the first version of the manuscript; M.Y. refined the methodology and contributed to the final version of the manuscript; G.L. edited the manuscript. All authors have read and agreed to the published version of the manuscript.

Funding: This work was supported by the National Natural Science Foundation of China (Contract No. 41671361, 41601373, and 41801272) and the Fundamental Research Fund for the Central Universities (Contract No. ZYGX2017KYQD195).

Acknowledgments: We would like to especially thank De Santis who provided the field-measured reflectance of soil, dark charcoal, and light charcoal. The authors are grateful to Long Wang and Lin Chen for their assistance during the field campaigns.

Conflicts of Interest: The authors declare no conflict of interest.

References

1. Bonan, G.B. Forests and Climate Change: Forcings, Feedbacks, and the Climate Benefits of Forests. *Science* **2008**, *320*, 1444–1449. [CrossRef] [PubMed]
2. Liu, Z.; Ballantyne, A.P.; Cooper, L.A. Biophysical feedback of global forest fires on surface temperature. *Nat. Commun.* **2019**, *10*, 1–9. [CrossRef] [PubMed]
3. Trumbore, S.; Brando, P.; Hartmann, H. Forest health and global change. *Science* **2015**, *349*, 814–818. [CrossRef] [PubMed]
4. Chuvieco, E.; Riaño, D.; Danson, F.M.; Martín, M.P. Use of a radiative transfer model to simulate the postfire spectral response to burn severity. *J. Geophys. Res. Space Phys.* **2006**, *111*, 111. [CrossRef]
5. Edwards, A.C.; Russell-Smith, J.; Maier, S.W. A comparison and validation of satellite-derived fire severity mapping techniques in fire prone north Australian savannas: Extreme fires and tree stem mortality. *Remote Sens. Environ.* **2018**, *206*, 287–299. [CrossRef]
6. Frolking, S.; Palace, M.W.; Clark, D.B.; Chambers, J.Q.; Shugart, H.H.; Hurtt, G.C. Forest disturbance and recovery: A general review in the context of spaceborne remote sensing of impacts on aboveground biomass and canopy structure. *J. Geophys. Res. Space Phys.* **2009**, *114*, 114. [CrossRef]
7. Key, C.; Benson, N. Landscape assessment: Ground measure of severity, the Composite Burn Index. In *FIREMON: Fire Effects Monitoring and Inventory System*; Lutes, D.C., Ed.; USDA Forest Service, Rocky Mountain Research Station: Fort Collins, CO, USA, 2006; pp. LA8–LA15.

8. De Santis, A.; Chuvieco, E. GeoCBI: A modified version of the Composite Burn Index for the initial assessment of the short-term burn severity from remotely sensed data. *Remote Sens. Environ.* **2009**, *113*, 554–562. [CrossRef]
9. Yin, C.; He, B.; Yebra, M.; Quan, X.; Edwards, A.C.; Liu, X.; Liao, Z. Improving burn severity retrieval by integrating tree canopy cover into radiative transfer model simulation. *Remote Sens. Environ.* **2020**, *236*, 111454. [CrossRef]
10. Edwards, A.; Maier, S.W.; Hutley, L.B.; Williams, R.J.; Russell-Smith, J. Spectral analysis of fire severity in north Australian tropical savannas. *Remote Sens. Environ.* **2013**, *136*, 56–65. [CrossRef]
11. Chuvieco, E. *Earth Observation of Wildland Fires in Mediterranean Ecosystems*; Springer: Berlin/Heidelberg, Germany, 2009.
12. Miller, J.D.; Knapp, E.E.; Key, C.H.; Skinner, C.; Isbell, C.J.; Creasy, R.M.; Sherlock, J.W. Calibration and validation of the relative differenced Normalized Burn Ratio (RdNBR) to three measures of fire severity in the Sierra Nevada and Klamath Mountains, California, USA. *Remote Sens. Environ.* **2009**, *113*, 645–656. [CrossRef]
13. Miller, J.D.; Thode, A.E. Quantifying burn severity in a heterogeneous landscape with a relative version of the delta Normalized Burn Ratio (dNBR). *Remote Sens. Environ.* **2007**, *109*, 66–80. [CrossRef]
14. Roy, D.; Boschetti, L.; Trigg, S.N. Remote Sensing of Fire Severity: Assessing the Performance of the Normalized Burn Ratio. *IEEE Geosci. Remote Sens. Lett.* **2006**, *3*, 112–116. [CrossRef]
15. Fernández-Manso, O.; Quintano, C.; Fernández-Manso, A. Combining spectral mixture analysis and object-based classification for fire severity mapping. *For. Syst.* **2009**, *18*, 296. [CrossRef]
16. Quintano, C.; Fernández-Manso, A.; Roberts, D.A. Multiple Endmember Spectral Mixture Analysis (MESMA) to map burn severity levels from Landsat images in Mediterranean countries. *Remote Sens. Environ.* **2013**, *136*, 76–88. [CrossRef]
17. Quintano, C.; Fernández-Manso, A.; Roberts, D.A. Burn severity mapping from Landsat MESMA fraction images and Land Surface Temperature. *Remote Sens. Environ.* **2017**, *190*, 83–95. [CrossRef]
18. Collins, L.; Griffioen, P.; Newell, G.; Mellor, A. The utility of Random Forests for wildfire severity mapping. *Remote Sens. Environ.* **2018**, *216*, 374–384. [CrossRef]
19. Collins, L.; McCarthy, G.; Mellor, A.; Newell, G.; Smith, L. Training data requirements for fire severity mapping using Landsat imagery and random forest. *Remote Sens. Environ.* **2020**, *245*, 111839. [CrossRef]
20. Meddens, A.J.; Kolden, C.A.; Lutz, J.A. Detecting unburned areas within wildfire perimeters using Landsat and ancillary data across the northwestern United States. *Remote Sens. Environ.* **2016**, *186*, 275–285. [CrossRef]
21. Ramo, R.; Chuvieco, E. Developing a Random Forest Algorithm for MODIS Global Burned Area Classification. *Remote Sens.* **2017**, *9*, 1193. [CrossRef]
22. Hultquist, C.; Chen, G.; Zhao, K. A comparison of Gaussian process regression, random forests and support vector regression for burn severity assessment in diseased forests. *Remote Sens. Lett.* **2014**, *5*, 723–732. [CrossRef]
23. De Santis, A.; Chuvieco, E. Burn severity estimation from remotely sensed data: Performance of simulation versus empirical models. *Remote Sens. Environ.* **2007**, *108*, 422–435. [CrossRef]
24. De Santis, A.; Chuvieco, E.; Vaughan, P.J. Short-term assessment of burn severity using the inversion of PROSPECT and GeoSail models. *Remote Sens. Environ.* **2009**, *113*, 126–136. [CrossRef]
25. De Santis, A.; Asner, G.P.; Vaughan, P.J.; Knapp, D.E. Mapping burn severity and burning efficiency in California using simulation models and Landsat imagery. *Remote Sens. Environ.* **2010**, *114*, 1535–1545. [CrossRef]
26. Chuvieco, E.; De Santis, A.; Riaño, D.; Halligan, K. Simulation Approaches for Burn Severity Estimation Using Remotely Sensed Images. *Fire Ecol.* **2007**, *3*, 129–150. [CrossRef]
27. Kuusk, A.; Nilson, T. Forest reflectance and transmittance FRT user guide. *Sci. Chin. D* **2002**, *41*, 580–586.
28. Kuusk, A.; Kuusk, J.; Lang, M. Modeling directional forest reflectance with the hybrid type forest reflectance model FRT. *Remote Sens. Environ.* **2014**, *149*, 196–204. [CrossRef]
29. Kuusk, A.; Nilson, T.; Paas, M.; Lang, M.; Kuusk, J. Validation of the forest radiative transfer model FRT. *Remote Sens. Environ.* **2008**, *112*, 51–58. [CrossRef]
30. Quan, X.; He, B.; Yebra, M.; Yin, C.; Liao, Z.; Li, X. Retrieval of forest fuel moisture content using a coupled radiative transfer model. *Environ. Model. Softw.* **2017**, *95*, 290–302. [CrossRef]

31. Sprintsin, M.; Karnieli, A.; Sprintsin, S.; Cohen, S.; Berliner, P. Relationships between stand density and canopy structure in a dryland forest as estimated by ground-based measurements and multi-spectral spaceborne images. *J. Arid. Environ.* **2009**, *73*, 955–962. [CrossRef]
32. Lianqiang, L.; Shukui, N.; Changsen, T.; Ling, C.; Feng, C. Correlations between stand structure and surface potential fire behavior of *Pinus tabuliformis* forests in Miaofeng Mountain of Beijing. *J. Beijing For. Univ.* **2019**, *1*, 73–81. [CrossRef]
33. Gascon, F.; Bouzinac, C.; Thépaut, O.; Jung, M.; Francesconi, B.; Louis, J.; Lonjou, V.; Lafrance, B.; Massera, S.; Gaudel-Vacaresse, A.; et al. Copernicus Sentinel-2A Calibration and Products Validation Status. *Remote Sens.* **2017**, *9*, 584. [CrossRef]
34. Louis, J.; Debaecker, V.; Pflug, B.; Main-Knorn, M.; Bieniarz, J.; Mueller-Wilm, U.; Cadau, E.; Gascon, F. Sentinel-2 Sen2Cor: L2A processor for users. In Proceedings of the Living Planet Symposium 2016, Prague, Czech Republic, 9–13 May 2016; pp. 1–8.
35. Main-Knorn, M.; Pflug, B.; Louis, J.; Debaecker, V.; Müller-Wilms, U.; Gascon, F. Sen2Cor for Sentinel-2. In Proceedings of the Image and Signal Processing for Remote Sensing XXIII, Warsaw, Poland, 11–14 September 2017; p. 1042704.
36. Mayer, B.; Kylling, A. Technical note: The libRadtran software package for radiative transfer calculations—Description and examples of use. *Atmos. Chem. Phys. Discuss.* **2005**, *5*, 1855–1877. [CrossRef]
37. Sexton, J.O.; Song, X.-P.; Feng, M.; Noojipady, P.; Anand, A.; Huang, C.; Kim, D.-H.; Collins, K.M.; Channan, S.; DiMiceli, C.; et al. Global, 30-m resolution continuous fields of tree cover: Landsat-based rescaling of MODIS vegetation continuous fields with lidar-based estimates of error. *Int. J. Digit. Earth* **2013**, *6*, 427–448. [CrossRef]
38. Yebra, M.; Quan, X.; Riaño, D.; Larraondo, P.R.; Van Dijk, A.I.; Cary, G.J. A fuel moisture content and flammability monitoring methodology for continental Australia based on optical remote sensing. *Remote Sens. Environ.* **2018**, *212*, 260–272. [CrossRef]
39. Jacquemoud, S.; Baret, F. PROSPECT: A model of leaf optical properties spectra. *Remote Sens. Environ.* **1990**, *34*, 75–91. [CrossRef]
40. Kuusk, A. A Directional Multispectral Forest Reflectance Model. *Remote Sens. Environ.* **2000**, *72*, 244–252. [CrossRef]
41. Kuusk, A. A two-layer canopy reflectance model. *J. Quant. Spectrosc. Radiat. Transf.* **2001**, *71*, 1–9. [CrossRef]
42. Huemmrich, K.F. The GeoSail model: A simple addition to the SAIL model to describe discontinuous canopy reflectance. *Remote Sens. Environ.* **2001**, *75*, 423–431. [CrossRef]
43. Dawson, T.P.; Curran, P.J.; Plummer, S.E. LIBERTY—Modeling the Effects of Leaf Biochemical Concentration on Reflectance Spectra. *Remote Sens. Environ.* **1998**, *65*, 50–60. [CrossRef]
44. Cheng, Y.-B.; Zarco-Tejada, P.J.; Riaño, D.; Rueda, C.A.; Ustin, S. Estimating vegetation water content with hyperspectral data for different canopy scenarios: Relationships between AVIRIS and MODIS indexes. *Remote Sens. Environ.* **2006**, *105*, 354–366. [CrossRef]
45. Zarco-Tejada, P.J.; Miller, J.R.; Harron, J.; Hu, B.; Noland, T.L.; Goel, N.; Mohammed, G.H.; Sampson, P. Needle chlorophyll content estimation through model inversion using hyperspectral data from boreal conifer forest canopies. *Remote Sens. Environ.* **2004**, *89*, 189–199. [CrossRef]
46. Kuusk, A. A multispectral canopy reflectance model. *Remote Sens. Environ.* **1994**, *50*, 75–82. [CrossRef]
47. Kuusk, A. A Markov chain model of canopy reflectance. *Agric. For. Meteorol.* **1995**, *76*, 221–236. [CrossRef]
48. Kötz, B.; Schaepman, M.; Morsdorf, F.; Bowyer, P.; Itten, K.; Allgöwer, B. Radiative transfer modeling within a heterogeneous canopy for estimation of forest fire fuel properties. *Remote Sens. Environ.* **2004**, *92*, 332–344. [CrossRef]
49. Lang, M.; Nilson, T.; Kuusk, A.; Kiviste, A.; Hordo, M. The performance of different leaf mass and crown diameter models in forming the input of a forest reflectance model: A test on forest growth sampleplots and Landsat ETM images. In Proceedings of the ForestSat 2005, Boras, Switzerland, 31 May–3 June 2005.
50. Dennison, P.E.; Halligan, K.Q.; Roberts, D.A. A comparison of error metrics and constraints for multiple endmember spectral mixture analysis and spectral angle mapper. *Remote Sens. Environ.* **2004**, *93*, 359–367. [CrossRef]
51. Kruse, F.; Lefkoff, A.; Boardman, J.; Heidebrecht, K.; Shapiro, A.; Barloon, P.; Goetz, A. The spectral image processing system (SIPS)—Interactive visualization and analysis of imaging spectrometer data. *Remote Sens. Environ.* **1993**, *44*, 145–163. [CrossRef]

52. Parks, S.A.; Holsinger, L.M.; Koontz, M.J.; Collins, L.; Whitman, E.; Parisien, M.-A.; Loehman, R.; Barnes, J.L.; Bourdon, J.-F.; Boucher, J.; et al. Giving Ecological Meaning to Satellite-Derived Fire Severity Metrics across North American Forests. *Remote Sens.* **2019**, *11*, 1735. [CrossRef]
53. Wang, L.; Quan, X.; He, B.; Yebra, M.; Xing, M.; Liu, X. Assessment of the Dual Polarimetric Sentinel-1A Data for Forest Fuel Moisture Content Estimation. *Remote Sens.* **2019**, *11*, 1568. [CrossRef]
54. Meng, R.; Wu, J.; Zhao, F.; Cook, B.D.; Hanavan, R.P.; Serbin, S.P. Measuring short-term post-fire forest recovery across a burn severity gradient in a mixed pine-oak forest using multi-sensor remote sensing techniques. *Remote Sens. Environ.* **2018**, *210*, 282–296. [CrossRef]
55. Liu, Z. Effects of climate and fire on short-term vegetation recovery in the boreal larch forests of Northeastern China. *Sci. Rep.* **2016**, *6*, 37572. [CrossRef]

remote sensing

MDPI

Article

Repeat Oblique Photography Shows Terrain and Fire-Exposure Controls on Century-Scale Canopy Cover Change in the Alpine Treeline Ecotone

David McCaffrey * **and Chris Hopkinson**

Geography and Environment Department, University of Lethbridge, Lethbridge, AB T1K 3M4, Canada; c.hopkinson@uleth.ca

* Correspondence: david.mccaffrey@alumni.uleth.ca

Received: 25 March 2020; Accepted: 14 May 2020; Published: 15 May 2020

Abstract: Alpine Treeline Ecotone (ATE), the typically gradual transition zone between closed canopy forest and alpine tundra vegetation in mountain regions, displays an elevational range that is generally constrained by thermal deficits. At landscape scales, precipitation and moisture regimes can suppress ATE elevation below thermal limits, causing variability in ATE position. Recent studies have investigated the relative effects of hydroclimatic variables on ATE position at multiple scales, but less attention has been given to interactions between hydroclimatic variables and disturbance agents, such as fire. Advances in monoplotting have enabled the extraction of canopy cover information from oblique photography. Using airborne lidar, and repeat photography from the Mountain Legacy Project, we observed canopy cover change in West Castle Watershed (Alberta, Canada; ~103 km^2; 49.3° N, 114.4° W) over a 92-year period (1914–2006). Two wildfires, occurring 1934 and 1936, provided an opportunity to compare topographic patterns of mortality and succession in the ATE, while factoring by exposure to fire. Aspect was a strong predictor of mortality and succession. Fire-exposed areas accounted for 83.6% of all mortality, with 72.1% of mortality occurring on south- and east-facing slope aspects. Succession was balanced between fire-exposed and unburned areas, with 62.0% of all succession occurring on north- and east-facing slope aspects. The mean elevation increase in closed canopy forest (i.e., the lower boundary of ATE) on north- and east-facing undisturbed slopes was estimated to be 0.44 m per year, or ~44 m per century. The observed retardation of treeline advance on south-facing slopes is likely due to moisture limitation.

Keywords: alpine treeline ecotone; repeat photography; monoplotting; lidar; fire

---

1. Introduction

Alpine treeline ecotone (ATE), the transition zone between closed canopy forest and alpine tundra, occurs where tree growth is limited by thermal thresholds at high elevations [1,2]. At continental scales, the approximate elevation of ATE is predicted by the growth limitation hypothesis, which suggests an inverse relationship exists between latitude and ATE elevation [1] as a function of temperature in the growing season [2].

However, numerous modulating factors have been described, which suppress treeline below this thermal limit (for detailed list, see [3]). The relative effects of these modulating factors on ATE elevation vary by spatial scale [4–6]. At landscape scales, ATE is suppressed below thermal limits by geomorphic conditions [7,8], physiological stressors (e.g., moisture limitation [9], wind exposure [10]), and disturbances such as rockslides [7], avalanches [11], and fires [12]).

A meta-analysis [13] identifies 103 publications which describe long term records of treeline position (both ATE and latitudinal treeline) at 166 sites around the world, between 1900–2008. Approximately 52% of these sites have some form of treeline advance. Two studies report treeline recession and both cases are associated with disturbances. Linear regression with climate records reveals that many of the advances in ATE position in [13] are associated with increases in atmospheric temperature.

Modulating factors, which suppress ATE below thermal limits regardless of rising atmospheric temperature, are often studied in isolation, and the potential for interaction between modulating factors is rarely considered. Our research investigated the interaction between two of these modulating factors, fire disturbance and moisture limitation, and their combined influence on ATE position in a spruce-fir forest in the Rocky Mountains. Using a novel monoplotting technique to extract quantitative spatial data from a historic photographic record, we investigated whether postfire moisture limitation caused topographic patterns of vegetative regeneration, thereby selectively suppressing ATE in dry, fire-exposed environments.

1.1. Fire and Topography in the ATE

In the Rocky Mountains, fires are less frequent in the ATE than at lower elevations, and fire return intervals (FRIs) are positively correlated with elevation [14,15]. While relatively infrequent, fire remains an important disturbance agent in the ATE [16,17], and the effects of fire on subalpine regions are potentially increasing with climate change [18].

Topographic factors like aspect and slope influence many of the processes that affect fire behavior in mountain landscapes. Topography can alter precipitation regimes, species composition, lighting strike frequency, and drying caused by both solar insolation and wind [14]. For example, in North America, south-facing mountain aspects receive more solar insolation and are generally drier than north-facing aspects. This impacts fire density (i.e., number of fires per unit area), with greater density observed on south-facing aspects [19–21]. However, the increase in fire density on warm aspects does not translate to an increase in area burned on south-facing aspects [22,23]. Thus, small fires are ignited more frequently on south-facing aspects, but a large fire will burn a range of topographic conditions, without aspect bias.

Patterns of moisture limitation also have the potential to impact postfire regrowth, potentially suppressing ATE elevation below levels that would be expected if ATE was only limited by atmospheric temperature. At high elevations in the Rocky Mountains, postfire areas are predominantly colonized by spruce or fir. The initial postfire colonization in these areas can take 20–100+ years [24–27], leaving alpine meadows that persist for a century or more. Postfire rates of succession in subalpine forest can be twice as high on mesic slopes as on xeric slopes [28], and growth of subalpine fir is limited on south-facing aspects [29,30]. If delayed regrowth can inhibit colonization for almost a century, then the present position of fire-exposed treelines on south-facing aspects may be temporarily suppressed, appearing lower than would be expected if ATE were solely limited by atmospheric temperature. While stand-age reconstructions are useful at identifying the processes involved in postfire regrowth, few direct observations of an aspect effect on postfire regrowth have been made using imagery.

1.2. ATE Observation Techniques

Determining the interaction between aspect and regrowth in postfire environments using imagery requires high resolution observations that extend more than a century. These are uncommon. A trade-off exists between spatial resolution and temporal extent in ATE observation methods [31]. Multitemporal remote sensing can monitor vegetative land cover change at spatial resolutions required to correlate with topographic variables (10–100 m) [17], but observations of ATE using passive optical satellite imagery, like Landsat TM [11,32,33] only extend back to the 1970–80s. Aerial photographs are similarly useful for treeline change monitoring [17,34], but aerial photography records only extend to the late 1940s and early 1950s for much of North America.

Conversely, ATE observation records with very long temporal extents often have poor spatial resolution, prohibiting topographic analysis. For example, palynological records extend thousands of years before present, with some exceeding 10,000 years [35], but these observations can only provide a spatial resolution equivalent to the pollen distribution distance (<100 m). Dendrochronological records have offered the best combination of high spatial resolution (<10 m) and high temporal extent (~1000 yr). Tree-ring records are often used in ATE studies [9,36–38]. The spatial resolution and temporal extent of dendrochronological studies are both high, but the spatial extent is limited by onerous collection methods and the large sample size required to cover enough sites for topographic analysis [9,15].

Repeat photography can address the need for ATE observations that provide high spatial extent and resolution, and long temporal extent. This method is widely used to assess land cover change in the ATE. Patterns of ATE advance in eastern slope regions of the Canadian Rockies have been described using repeat photography [39–41]. High resolution photographs were used to identify shrub advance in the Swedish Scandes [42]. Repeat photography has also been used for a description of land cover change in the Ural Mountains [43]. While the temporal extent of repeat photographs, 100 years or more in cases, is useful for studies of change in the ATE, the method has been mostly restricted to qualitative analysis, given the previous inability to standardize spatial scale across oblique imagery. Quantitative spatial analysis of ATE has been attempted by draping a polygon *fishnet* over repeat photographs [44], in order to standardize the observation units in both historical and contemporary photographs, but spatial scale still varied between polygons, limiting the usefulness of this technique for spatial analysis.

Recent technological advances have enabled the fully quantitative spatial analysis of repeat photographs, using the WSL Monoplotting Tool (WSL-MT, note: WSL is the German acronym for "Eidgenössische Forschungsanstalt für Wald, Schnee und Landschaft", the Swiss Federal Institute for Forest, Snow and Landscape Research – see Supplementary Material for further information on WSL-MT), which produces georeferenced vector data from oblique photographs using tie points between oblique and aerial photographs, and high resolution topographic data [45]. The WSL-MT tool has been used to monitor land cover change by extracting spatial data from oblique repeat photographs [46] and to observe century-scale vegetation transitions in the Rocky Mountains of southern Alberta, Canada [47]. These experiments were conducted on photo pairs from the Mountain Legacy Project, a collection of over 120,000 historic survey images of the Canadian Rockies (1888–1958), of which over 6000 have 21st-century repeat photographs available under creative commons license [48]. Previous research, validated using airborne lidar, established that fractional cover classes could be assessed from oblique imagery at a spatial resolution of 20 m, using the WSL-MT [49].

1.3. Research Goals

We used a technique to assess fractional cover from oblique photographs, described in [49], to observe postfire vegetative change in repeat photographs of a watershed in the Canadian Rockies. We sought to determine: (1) if there were observable differences in canopy cover change between fire-exposed and non-fire-exposed regions of the watershed, (2) if changes in canopy cover correlate to topographic patterns that are consistent with expected edaphic controls on regrowth in the region (e.g., less regrowth on south-facing aspects), and (3) if it was possible to detect the rate of ATE advance in non-fire-exposed areas, where ATE position is regulated by climate.

2. Materials and Methods

2.1. Study Area

The study was conducted over the West Castle Watershed (WCW), Alberta, Canada (49.3° N, 114.4° W). WCW is in the headwaters of the Oldman River, on the eastern slopes of the Canadian Rockies, with an area of ~103 km^2, and an elevation range of ~1400–2600 m a.s.l. Aside from a small ski resort and village near the downstream end of the WCW (2.9% of the watershed area), and trails in the valley bottoms, the forested slopes demonstrate limited anthropogenic disturbance. Consequently, the WCW is an ideal study area for assessing recent natural shifts in the ATE of this part of the Canadian Rockies. Dominant species at ATE elevation in WCW are subalpine fir and Engelmann spruce.

2.2. Repeat Photographs

Repeat photographs of WCW were provided by the Mountain Legacy Project [48] (Figure 1). The historical images were taken in the summer of 1914 by Morrison Parsons Bridgeland. The repeat photographs were taken by the Mountain Legacy Project in July of 2006 (Table 1). Photograph pairs were selected to maximize the spatial coverage of the observed area of WCW. Using the same set of photographs, it was determined that 38.5 km^2 (37.7%) of the watershed was observable in these images [49] (Figure 2).

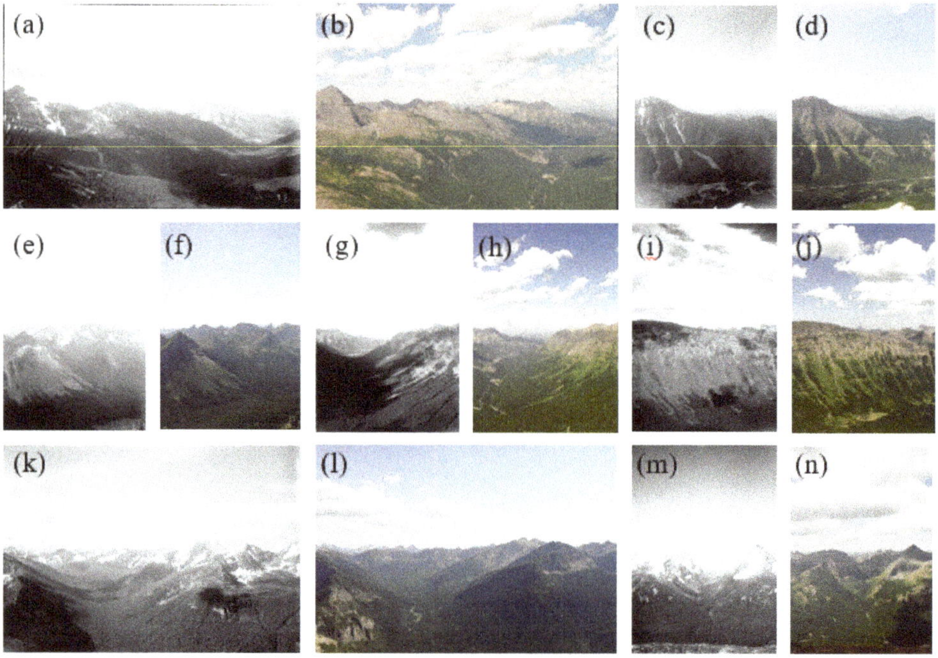

Figure 1. Seven repeat photograph pairs of WCW; the black and white series by M.P. Bridgeland, 1914, courtesy of Library and Archives Canada; color series copyright 2006, Mountain Legacy Project. Photo 1; (**a**) 1914, (**b**) 2006, Photo 2; (**c**) 1914, (**d**) 2006, Photo 3; (**e**) 1914, (**f**) 2006, Photo 4; (**g**) 1914, (**h**) 2006, Photo 5; (**i**) 1914, (**j**) 2006, Photo 6; (**k**) 1914, (**l**) 2006, Photo 7; (**m**) 1914, (**n**) 2006. See Table 1 for further details.

Table 1. Collection data for Mountain Legacy Project 2006 photographs. Information for the corresponding 1914 data can be found at www.mountainlegacy.ca.

Photograph Number	Elevation (m a.s.l.)	Latitude (N)	Longitude (W)	Photograph Date	MLP Photo ID (2006)
1	2295	49°16′3.62″	114°22′58.97″	28 July 2006	B0032P0048
2	2216	49°20′14.00″	114°23′22.81″	29 July 2006	B0033P0005
3	2390	49°17′7.01″	114°20′44.21″	28 July 2006	B0032P0073
4	2283	49°16′4.94″	114°22′56.77″	28 July 2006	B0032P0056
5	2283	49°16′4.94″	114°22′56.77″	28 July 2006	B0032P0055
6	2390	49°17′7.01″	114°20′44.21″	28 July 2006	B0032P0075
7	2462	49°18′7.26″	114°22′18.63″	30 July 2006	B0034P0005

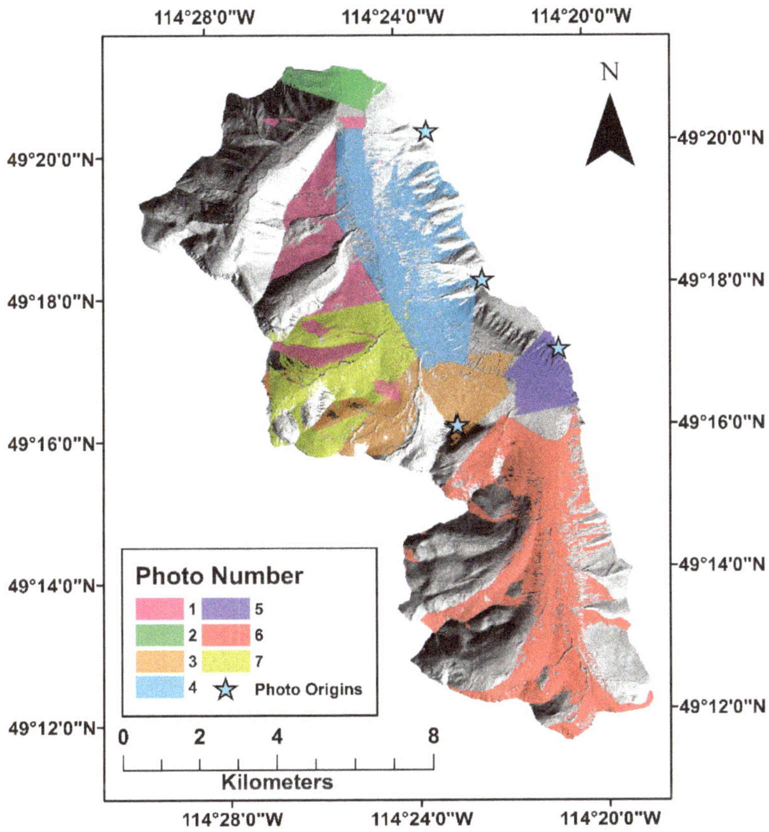

Figure 2. Visible region of WCW in each of the seven MLP photographs. In cases where the same region is visible in multiple photographs, the photograph used for analysis was determined by criteria described in [49].

2.3. Canopy Classification

Canopy cover was extracted from oblique photography with the WSL monoplotting tool [45], using a method described in [49]. The WSL-MT performs oblique to orthogonal vector transformation with a camera calibration, which generated by pairing control points between oblique photographs and a high resolution DEM, co-registered with aerial imagery. For the DEM requirement, a 1 m DEM was used, which was collected from an airborne lidar survey in October 2014. For the aerial imagery requirement, a SPOT 6 image of WCW was used (1.5 m, acquired July 31, 2014, © 2014 CNES – see Supplementary Material). In the absence of high resolution topographic and aerial data from 1914, the analysis relied on the camera calibration from the 2014 lidar DEM and 2006 SPOT image being applied to the 1914 oblique photographs. This approach was deemed reasonable, as no extreme changes in topography (e.g., rock slides, slumping) are apparent between the 1914 and 2006 datasets.

The monoplotting procedure results in a *fishnet* vector layer, where each cell represents a 20 × 20 m area (viewed orthogonally). This vector layer is draped over the oblique raster image, such that cells are scaled according to the perspective in the image (i.e., cells in the foreground appear larger, and cells in the background appear smaller) (Figure 3). This method circumvents the problem of varying spatial scale in the oblique image, allowing quantitative spatial analysis to be performed on the resulting raster.

Figure 3. Fishnet vector layer, developed using the WSL monoplotting tool, draped on an oblique image. Each cell in the fishnet represents a 20 × 20 m area, viewed orthogonally.

Calibrated fishnets were applied to each of the seven oblique images. Any fishnet grid cell that intersected with less than 75% of the viewshed (i.e., < 300 m^2) was omitted from the analysis (note: because these data were eventually rasterized, and each raster represented an area between 300–400 m^2, canopy cover and change is reported by grid cell count rather than surface area).

Once in an oblique projection, grid cells were manually assigned into one of six canopy cover classes, based on observed canopy openness and texture: (1) *No Cover*—grid cells devoid of vegetation; (2) *Low Vegetation*—grid cells appear vegetated, but context and texture suggest shrubs or krummholz, not upright trees > 2 m; (3) *Partial Canopy*—trees are present, but ground is visible in > 50% of the grid cell; (4) *Full Canopy*—trees cover > 50% of a grid cell; (5) *Snow*—snow covers > 50% of a grid cell; (6) *Structure*—any anthropogenic structure or non-vegetated land cover is present in the grid cell (e.g., buildings, roads, and trails) (Figure 4).

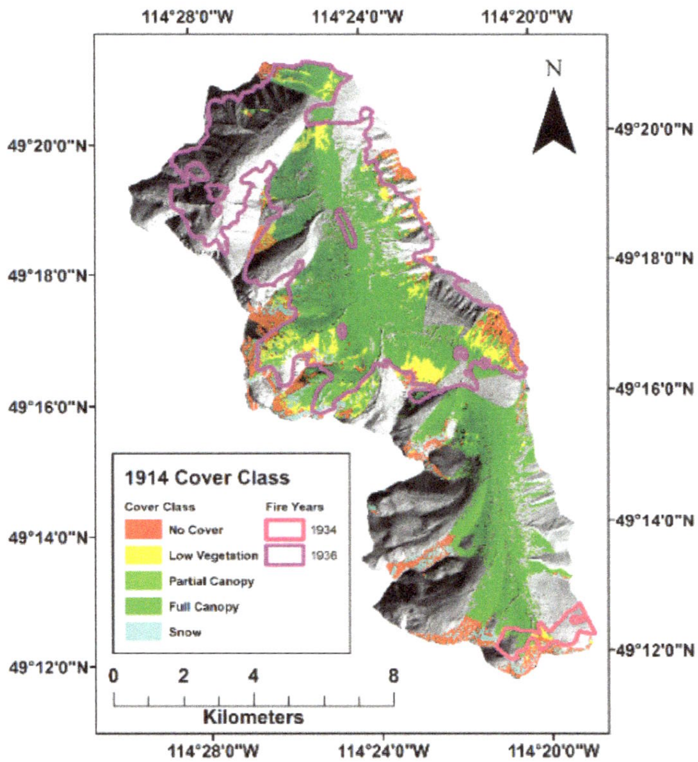

(a)

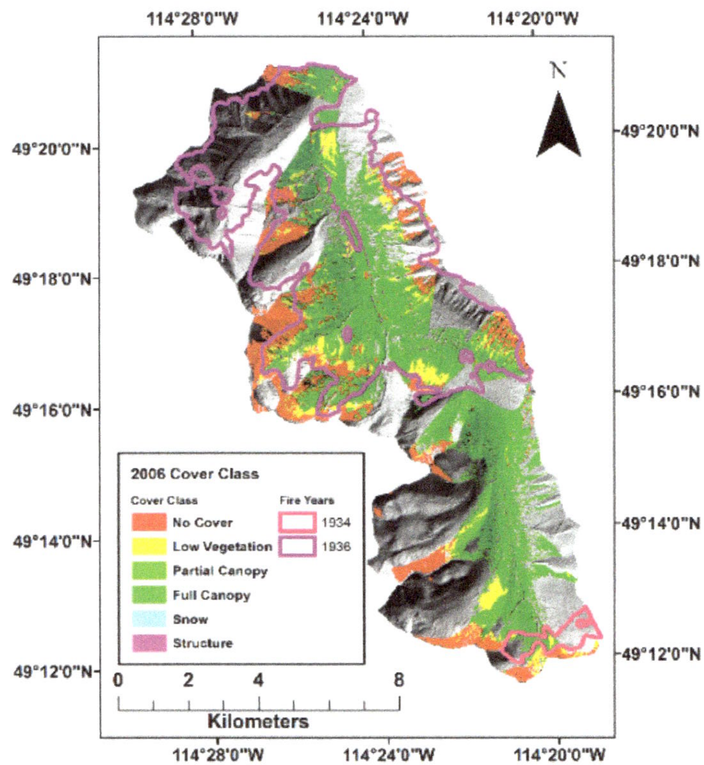

(b)

Figure 4. Canopy cover classification for WCW using oblique photography from (**a**) 1914 and (**b**) 2006. Each cell is a 20×20 m area, viewed orthogonally, where canopy cover has been assessed, ranging from *No Cover* to *Full Canopy*, and including cells with snow cover or modern structures. The extents of the 1934 and 1936 fires are provided for reference [50].

Given the focus on vegetative land cover change, further analysis focused on the first four, ordinal classes of vegetation (*no cover* through *full canopy*). Any grid cells having *snow* or *structure* in either 1914 or 2006 were omitted from analysis. The *snow* class was removed due to uncertainty of vegetation beneath the snow. The anthropogenic *structure* class was omitted because the processes causing vegetative land cover to change were distinct from the processes being studied. Of the total observed area, 7.2% was affected by the removal of snow cover, and 2.9% by the removal of anthropogenic structures.

2.4. Anthropogenic Disturbance

The Alberta Vegetation Inventory [51] was used to identify and omit areas of anthropogenic disturbance (see Supplementary Material). It was produced from interpretation of aerial photographs, and identifies several types of former and current disturbance, including former oil well pads, historical cut-blocks, settlements (including the Castle Mountain Ski Resort), and roads. These areas were aggregated into a single layer of anthropogenic disturbance so that topographic and other factors which affected land cover change could be considered independently of human activity.

2.5. Fire Disturbance

Two major historical fires occurred in the interval between the oblique photograph observations, one in 1934 and one in 1936. The extents of the historical fires were delineated in 2005, using a combination of visual evidence from a 1949 aerial photograph survey and descriptions from historical fire reports [50] (Figure 4). More detailed attributes of these fires, such as intensity and cause, are unknown, but previous research has demonstrated that fire intensity does not co-vary with topographic features such as aspect and slope in subalpine fir-spruce forests of the Rockies, as fires that reach these elevations are typically high intensity, crown replacing fires [22,23]. Thus, if we observe topographic differences in patterns of postfire regrowth, they are likely related environmental factors that affect regrowth, and not likely caused by topographic differences in fire intensity.

Fire suppression in early 20th-century was common practice, yet it is known that the Government of Alberta did not engage in fire suppression in the 1930s and 1940s, for reasons related to the Great Depression and the Second World War [52]. Fire suppression techniques did not affect the fire return interval of subalpine forests in the region [52]. Mean fire return intervals in the WCW range from 19.3 years near the mouth of the valley, to 70.8 years in the southern headwaters [53].

2.6. Topographic Analysis of Change

The four ordinal classes of canopy cover (1—*no cover*, 2—*low vegetation*, 3—*partial canopy*, 4—*full canopy*) were used to categorize vegetative change between 1914 and 2006. For each 20 x 20 m grid cell, the difference between the 2006 cover class and the 1914 cover class was used to generate a change class (Figure 5). For example, if canopy cover in 2006 was *no cover* (1), and canopy cover in 1914 was *full canopy* (4), then the change class for that area was $1 - 4 = -3$. The seven possible change classes were: −3 (high mortality), −2 (medium mortality), −1 (low mortality), 0 (no change), +1 (low succession), +2 (medium succession), +3 (high succession).

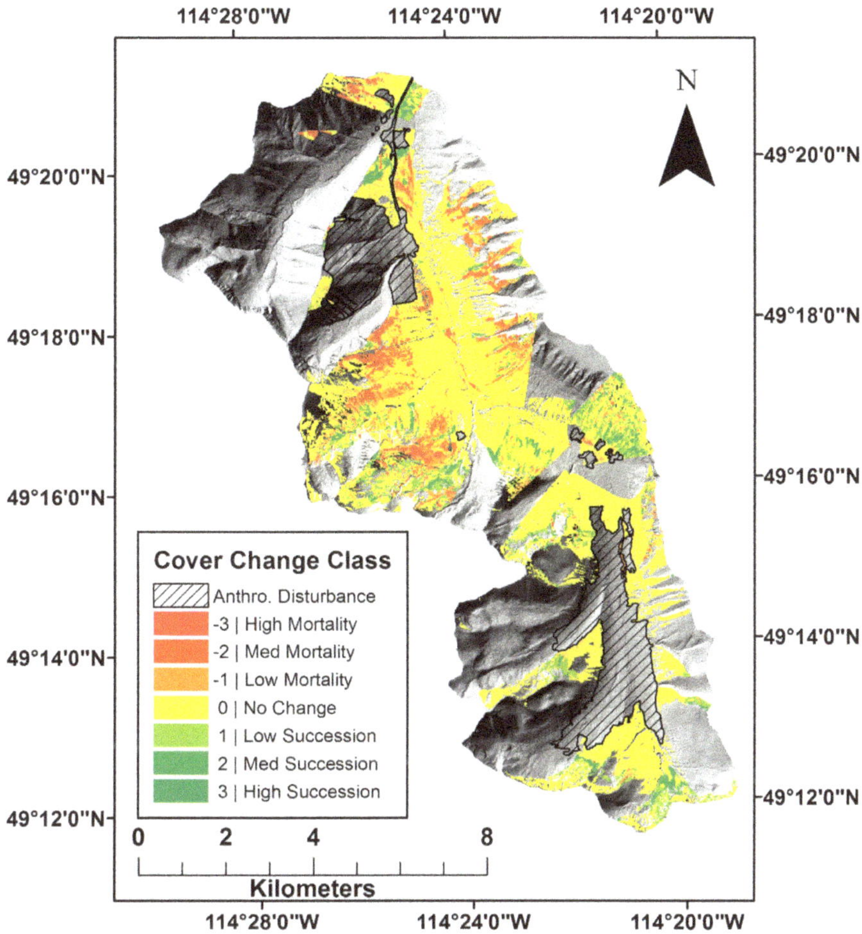

Figure 5. Cover change classification for period 1914–2006. Large areas of anthropogenic disturbance, such as a ski hill in the North of the watershed and a historic cut block in the South of the watershed, were omitted from analysis.

In preparation for the topographic analysis, a 1 m DEM was aggregated to 20 m and coregistered to the change classes: for elevation, the mean value of the 1 m cells was used for its corresponding 20 m raster; for slope, the degree angle of slope was calculated from the 1 m DEM, and the mean of degree values was used for the 20 m raster; for aspect, azimuth was calculated on the aggregated 20 m DEM, and discretized in eight intercardinal direction bins of 45° each: N = 337.5°–22.5°, NE = 22.5°–67.5°, E = 67.5°–112.5°, SE = 112.5°–157.5°, S = 157.5°–202.5°, SW = 202.5°–247.5°, W = 247.5°–292.5°, NW = 292.5°–373.5°.

Data were factored by fire exposure, with the 1934 and 1936 fire extents aggregated into a single layer. The distribution of topographic variables within each change cover class was contrasted between fire-exposed and non-fire-exposed areas in the time interval 1914–2006.

For aspect, we tested if the proportional distribution of aspects within each change class differed from expected aspect distribution in a way that suggested aspect influence on canopy cover. The 20 m DEM was sampled to establish the expected proportional distribution of aspect classes in the WCW. The observed proportional distribution of aspect within each change class was compared to the expected distribution with a chi-square goodness-of-fit test. To test the significance to each comparison, a Monte Carlo analysis using 100 repetitions was used to build a sampling distribution.

For elevation and slope, we tested if the distribution of these variables differed in between comparisons of change class. One-way ANOVA tested whether any of the seven classes' mean values differed from the others. For both elevation and slope comparisons, Welch's t-test was used on each of the 20 pairwise combinations of seven change classes to establish which classes differed from each other. Note that Welch's t-test corrects for both unequal variance between distributions and unequal sample size [54]. A Bonferroni correction was applied to the t-tests to account for multiple comparisons. All topographic analyses were completed solely outside of the anthropogenic disturbance area.

Finally, the rate of upslope ATE advance in undisturbed areas was estimated along five transects of canopy change cells. Along each transect, the boundary of closed canopy forest was identified in the 1914 and 2006 scenes. The closed canopy forest boundary is the lower extent of ATE (sometimes referred to as timberline). Given that the upper boundary (species limit) could vary within a 20 m grid cell, and may have been obstructed by snow, the closed canopy forest boundary was used as a proxy for total ATE change. Transects were selected where a distinct timberline (lower ATE boundary) could be observed across at least five grid cells (100 m), in areas without orographic boundaries to treeline advance (i.e., climatic treelines), and where there was no exposure to fire between observations. Five of these transects were identified. Elevation was sampled from each of the 20 m grid cells, and the mean elevations of the closed canopy fronts in the 1914 and 2006 scenes were compared.

3. Results

3.1. Aspect

Observed distribution of aspect significantly differed from expected distribution, with the majority of mortality seen on fire-exposed, warm aspects (Figure 6). In total, fire affected 63.2% of the observed area in WCW and 83.6% of all mortality occurred in fire-exposed areas. A strong interaction between aspect and fire was observed, with 72.0% of all mortality occurring on SW, S, SE, and E aspects. There was a high occurrence of succession classes in N and NE aspects, in both fire-exposed and non-fire-exposed areas. Succession was also notably high in fire-exposed SW aspects, and non-fire-exposed NE aspects. In total, north- and east-facing slope aspects accounted for 62.0% of all succession.

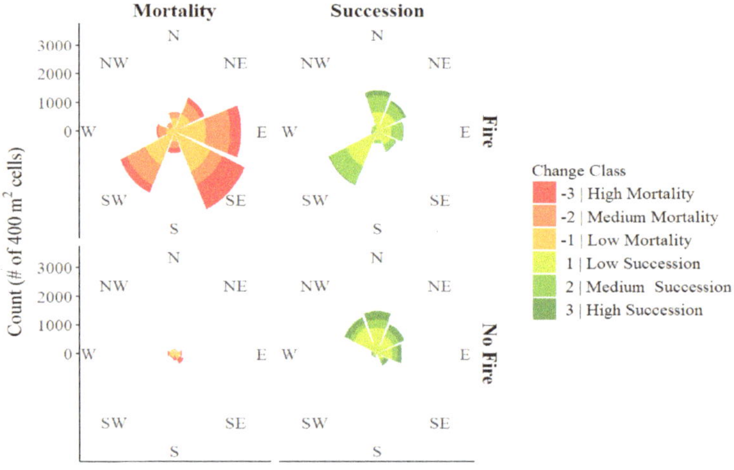

Figure 6. Mortality and succession in WCW over the period 1914–2006, factored by exposure to fire. Mortality is largely restricted to fire exposed areas, and is predominantly in the SW, SE, and E aspects.

The expected proportion distribution of aspect in WCW varied, as a function of the orientation of the valley; SW, E, and NE aspect had higher occurrence, while S, SE, and NW aspects had lower occurrence (Figure 7). We compared this expected proportion of aspects for each of the seven change classes (8 aspects x 7 change classes = 56 comparisons). The results of the chi-squared test show that significant differences between expected and observed aspect proportion occurred frequently; 46 out of 56 (82.1%) of the change class/aspect cases showed significant differences (Table 2). The greatest departures from expected proportions were seen in high occurrences of mortality on S and SE aspects, and high occurrences of succession on N and NE aspects.

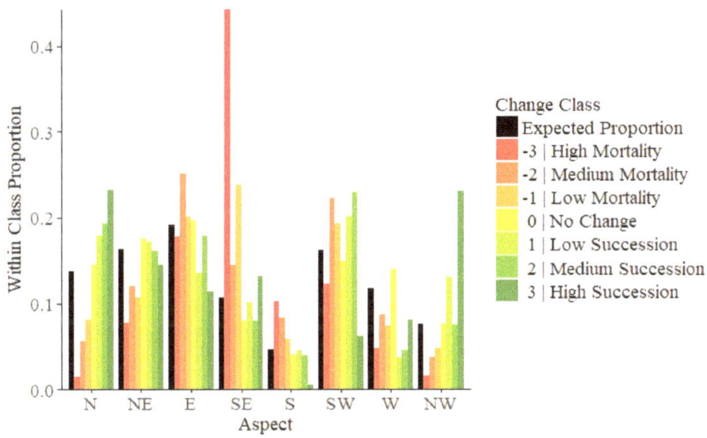

Figure 7. Expected proportion of aspects within the WCW (in black) compared to the observed proportion of aspect for each change class. For example, roughly 0.1 (or ~10%) of the land in WCW is on a SE aspect (black bar). If there were no aspect effect, we would expect the same proportion of high mortality cases to be on SE aspects. Instead, > 0.4 (or >40%) of high mortality occurs on SE aspects (red bar).

Table 2. Results of chi-squared test of aspect proportions. Significant values in bold. Significant differences in aspect proportion were seen in 43 out of 56 of the tests.

Aspect		Change Class						
		High Mortality	Medium Mortality	Low Mortality	No Change	Low Succession	Medium Succession	High Succession
N	χ^2	307.5	212.1	196.8	28.6	101.4	90.0	59.2
	P	<0.001 *	<0.001 *	<0.001 *	<0.001 *	<0.001 *	<0.001 *	<0.001 *
NE	χ^2	130.1	52.6	171.7	64.5	3.4	0.1	1.9
	P	<0.001 *	<0.001 *	<0.001 *	<0.001 *	0.066	0.728	0.163
E	χ^2	3.3	85.9	4.4	6.1	147.0	4.2	30.7
	P	0.069	<0.001 *	0.037	0.013	<0.001 *	0.041	<0.001 *
SE	χ^2	2863.3	55.9	1327.0	530.7	2.6	27.5	5.1
	P	<0.001 *	<0.001 *	<0.001 *	<0.001 *	0.109	<0.001 *	0.024
S	χ^2	177.1	121.3	26.7	34.4	0.1	3.3	29.6
	P	<0.001 *	<0.001 *	<0.001 *	<0.001 *	0.702	0.070	<0.001 *
SW	χ^2	27.8	101.1	52.0	71.2	81.5	114.1	58.0
	P	<0.001 *	<0.001 *	<0.001 *	<0.001 *	<0.001 *	<0.001 *	<0.001 *
W	χ^2	114.1	34.8	136.6	311.9	434.0	169.8	10.2
	P	<0.001 *	<0.001 *	<0.001 *	<0.001 *	<0.001 *	<0.001 *	<0.001 *
NW	χ^2	127.0	81.2	81.9	0.0	301.4	0.1	268.0
	P	<0.001 *	<0.001 *	<0.001 *	0.963	<0.001 *	0.806	<0.001 *

* Significant values.

3.2. Elevation

The distribution of elevation within change classes shows that the greatest canopy changes were seen in high elevation. *High mortality* and *high succession* classes had the highest mean elevation, while the lowest mean elevation was in the *no change* class (Figure 8). This result suggests that the degree of change, for both mortality and succession, covaried with elevation. To test this hypothesis, change classes were pooled by absolute value (e.g., −3 combined with 3, −2 combined with 2), and a Spearman rank test was performed. The results showed that degree of canopy change was significantly correlated with elevation ($r = 0.174$, $p < 0.001$), though the correlation was weak.

Table 3. Results of elevation t-test between change classes. Significant values in bold. Significant differences were seen in 20 out of 21 tests.

		Change Class					
		High Mortality	**Medium Mortality**	**Low Mortality**	**No Change**	**Low Succession**	**Medium Succession**
Medium Mortality	t	−24.042	-	-	-	-	-
	p	<0.001 *	-	-	-	-	-
Low Mortality	t	−20.153	6.643	-	-	-	-
	p	<0.001 *	<0.001 *	-	-	-	-
No Change	t	−42.218	−10.400	−22.928	-	-	-
	p	<0.001 *	<0.001 *	<0.001 *	-	-	-
Low Succession	t	−7.627	18.686	13.900	39.454	-	-
	p	<0.001 *	<0.001 *	<0.001 *	<0.001 *	-	-
Medium Succession	t	−24.636	−2.525	−8.754	5.790	−19.643	-
	p	<0.001 *	0.244	<0.001 *	<0.001 *	<0.001 *	-
High Succession	t	13.569	30.010	27.219	39.397	19.105	30.610
	p	<0.001 *	<0.001 *	<0.001 *	<0.001 *	<0.001 *	<0.001 *

* Significant values.

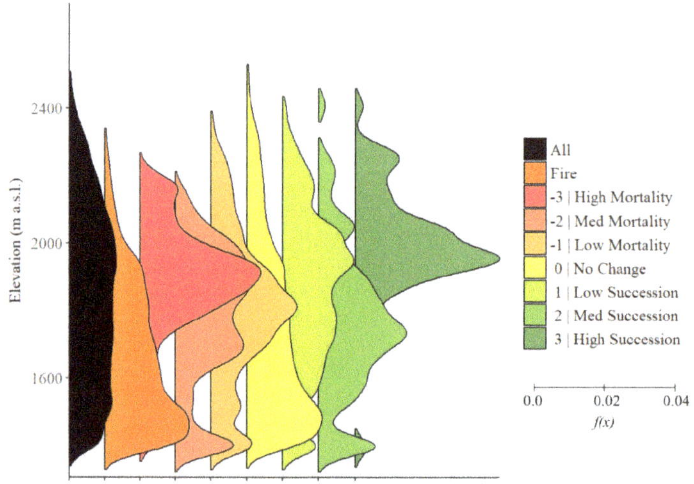

Figure 8. Elevation distributions in WCW, normalized with a probability density function. For each category, the hypsometric distribution of elevation within that category is displayed. The further right a curve extends from each baseline, the more likely it is to find values at that elevation within each category. Across all WCW, in black, elevation is evenly distributed, with elevation above 2200 m a.s.l. being less common. Fire, in orange, was more likely below 1600 m a.s.l., and high mortality and high succession were more likely at ATE elevations, between 1800–2100 m a.s.l. Corresponding t-tests can be seen in Table 3.

Between change class differences in elevation distribution were generally significant. One-way ANOVA showed that the mean elevation of at least one of the change classes was significantly different from the remaining groups ($F = 743.16$, $df = 6$, $p < 0.001$). The result of the pairwise t-tests showed that, of the 21 possible pairwise combinations, 20 showed significant differences (Table 3).

Mean elevation of fire-exposed areas was lower than mean elevation of non-fire-exposed areas in every change class (Figure 9). This is as expected, given the decreasing probability of fire with increasing elevation [15]. While factoring by fire showed between group differences in elevations, the significant correlation between degree of canopy change and elevation was preserved in both fire-exposed cases ($r = 0.340$, $p < 0.001$), and non-fire-exposed cases ($r = 0.110$, $p < 0.001$).

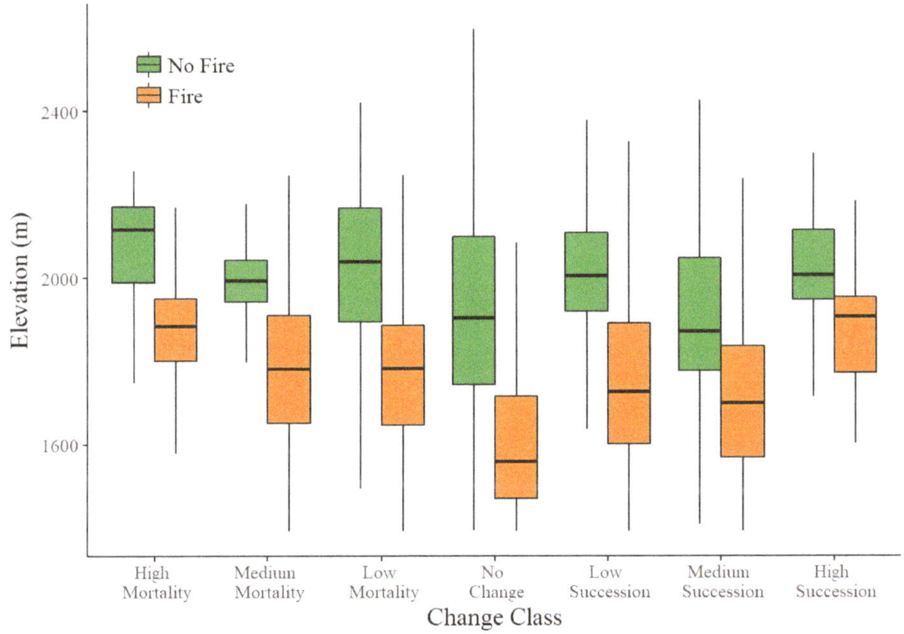

Figure 9. Distribution of elevation within change class, factored by fire exposure. In each canopy change class, median elevation is higher in areas that were not exposed to fire.

3.3. Slope

Significant differences in slope distribution were observed among change classes. The highest mean slope values were in the *high mortality* and *high succession* classes. The lowest mean slope value was in the *no change* category (Figure 10). As with elevation, the degree of change correlated positively with slope. Change classes were pooled by absolute value, and a Spearman rank test was performed. A weak correlation was found between slope and the degree of canopy change ($r = 0.126$, $p < 0.001$).

Table 4. Results of slope t-test between change classes. Significant values in bold. Significant differences were seen in 18 out of 21 tests.

		Change Class					
		High Mortality	**Medium Mortality**	**Low Mortality**	**No Change**	**Low Succession**	**Medium Succession**
Medium	t	−8.609	-	-	-	-	-
Mortality	p	<0.001 *	-	-	-	-	-
Low	t	−8.727	1.097	-	-	-	-
Mortality	p	<0.001 *	1.000	-	-	-	-
No	t	−24.947	−14.348	−21.092	-	-	-
Change	p	<0.001 *	<0.001 *	<0.001 *	-	-	-
Low	t	−12.396	−2.926	−4.793	13.790	-	-
Succession	p	<0.001 *	0.072	<0.001 *	<0.001 *	-	-
Medium	t	−14.204	−6.250	−8.061	4.742	−4.204	-
Succession	p	<0.001 *	<0.001 *	<0.001 *	<0.001 *	<0.001 *	-
High	t	−0.064	4.710	4.350	11.106	6.297	8.120
Succession	p	1.000	<0.001 *	<0.001 *	<0.001 *	<0.001 *	<0.001 *

* Significant values.

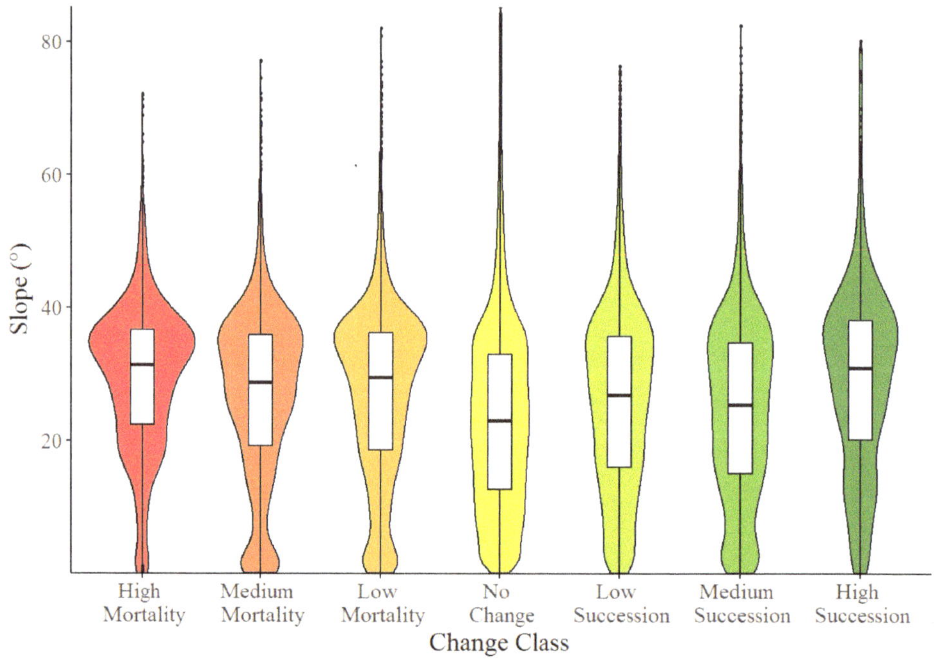

Figure 10. Distribution of slope within change class. The colored curves behind each box plot show a probability density function of the values in each distribution. Corresponding t-tests can be seen in Table 4.

The differences in slope distribution between change classes were generally significant. The slope distribution of at least one of the change classes was significantly different from the remaining groups (One-way ANOVA, F = 213.69, df = 6, $p < 0.001$). The result of the pairwise t-tests showed that, of the 20 possible pairwise combinations, 17 had significant differences (Table 4).

In a majority of change classes (−3 through 1), mean slope was lower in fire-exposed area than in non-fire-exposed areas (Figure 11). The hypsometry of this watershed shows a positive correlation between elevation and slope. Given that fires predominantly occurred at lower elevations, lower slope in fire-exposed areas was expected. In *medium* and *maximum succession* classes, fire-exposed areas had slightly higher mean slope than non-fire-exposed areas. Degree of canopy change and slope were significantly correlated in both fire-exposed cases ($r = 0.244$, $p < 0.001$), and non-fire-exposed cases ($r = 0.027$, $p < 0.001$).

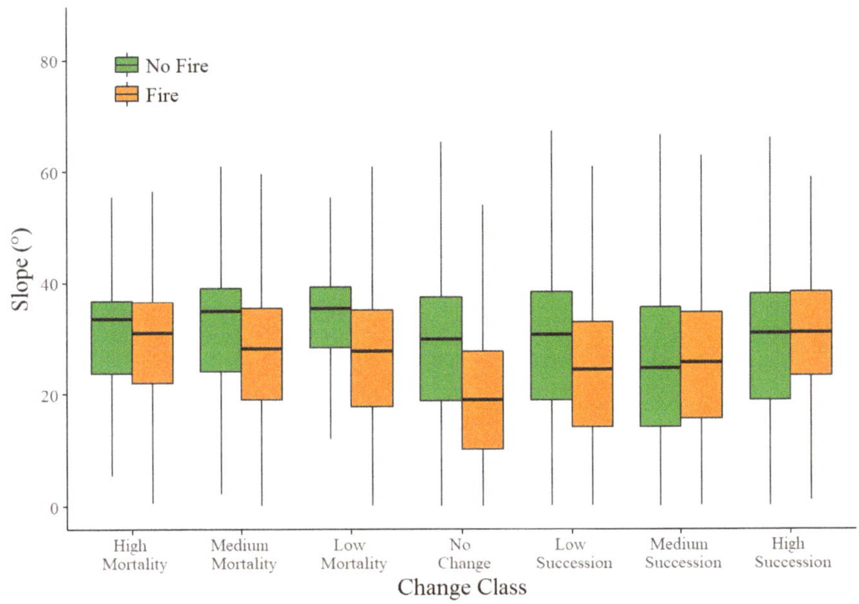

Figure 11. Distribution of slope within change class, factored by fire exposure. Median slope was generally higher in regions that were not exposed to fire.

3.4. Upslope Advance

Of the five closed canopy transects used to determine upslope elevation advance, all occurred in areas where Engelmann spruce was the dominant species in 2006 (Table 5). In each of these transects, mean elevation was higher in 2006 than in 1914. The mean elevation of closed canopy forest in these unburned areas increased by 40.5 m over 92 years, which translates to an annual increase of 0.44 m.

Table 5. Differences in mean elevation, between 1914 and 2006, across transects of cells in unburned areas, which represent the advanced of timberline, the lower boundary of ATE.

Transect	Number of Cells 1914	Mean Elevation 1914 (m a.s.l.)	Number of Cells 2006	Mean Elevation 2006 (m a.s.l.)	Elevation Difference (m a.s.l.)
1	6	2046.0	8	2053.5	7.5
2	7	1894.6	9	1897.0	2.4
3	35	2002.9	28	2028.5	25.6
4	24	1945.1	24	2033.5	88.4
5	29	1876.1	31	1954.7	78.6
Mean	20.2	1952.9	20.0	1993.4	40.5

4. Discussion

Our study demonstrated an example of topographic patterns of postfire colonization that conforms to the edaphic process model of postfire colonization, using direct quantitative observation with monoplotting on repeat photography. In fir-spruce forests, postfire regrowth in high elevation regions, like the ATE, likely has edaphic controls. As a result, colonization in xeric areas, like south-facing slopes, can lag several decades or even centuries behind regrowth in mesic areas, like north-facing slopes. These edaphic controls on colonization were generally suggested based on research using stand level age-reconstructions, but given the long gap in colonization between mesic and xeric colonization, few direct observations of this effect have been made in imagery. By increasing the temporal range of observation using monoplotting on historic repeat photography, the topographic differences in postfire colonization are now visible in imagery.

4.1. Aspect

Greater than expected proportions of succession occurred on N and NW aspects, while higher than expected proportion of mortality occurred on S and SE aspects. Correlation between aspect and canopy cover change was associated with fire-exposure, suggesting soil moisture stress levels may have been a significant barrier to postfire recruitment.

It is known that seedling recruitment of subalpine fir and Engelmann spruce is inhibited on warm-aspects in the ATE, primarily as a function of soil moisture stress caused by high temperature and solar radiation [55]. This process has been demonstrated experimentally: heating experiments in trees at ATE elevations show that moisture limitation both inhibits conifer seedling recruitment [56] and reduces carbon gain [57]. In a postfire environment, insolation would increase due to reduced interception and increased gap fraction in the burned canopy. This increased insolation could have caused aspect specific mortality on warm aspects in WCW, particularly the SW aspect. Postfire erosion may also contribute to mortality. A study in a watershed adjacent to WCW [58] shows that total suspended sediment levels are eight times higher in fire-exposed areas, an effect that is exaggerated by steep terrain at high elevations. High levels of erosion in fire-exposed, subalpine areas may suppress the moisture retention capacity of the soil and exacerbate other obstacles to recruitment.

On the north- and east-facing slopes in the WCW, environmental factors may have mitigated soil moisture stress. Reduced energy inputs and wind exposure diminish the moisture losses to snowpack sublimation, soil and vegetation evapotranspiration. For example, it is known that the end of winter mean snowpack depth on north-facing slopes can be ~40% greater than south-facing slopes in the southern Canadian Rockies [59]. North-facing snow packs also persist longer into the growing season, thus ensuring melt water is available to support processes of germination and plant maintenance. Aspect specific patterns of snow accumulation can affect the functional length of the growing season, in turn altering ATE position [9]. Additionally, aspect can predict postfire recruitment of subalpine fir [17], with cool, north-facing aspects having rates of establishment twice as high as on south-facing aspects. Interactions among solar radiation (aspect), fire history, and elevation (in addition to anthropogenic activity) predict broad patterns of succession in a large region of the Rocky Mountains of southern Alberta [47]; this includes the encroachment on closed canopy forest on alpine meadow, or ATE advance. Together, these findings suggest postfire colonization is highly dependent on soil moisture, and that this limitation is exacerbated at high elevations.

Aspect effects unrelated to fire disturbance are reported in a number of studies of ATE dynamics, globally [9,60–62]. However, the proposed mechanisms of these aspect effects are often site-specific, and vary by study region. These include differences in lapse rates among aspects [61] and diurnal cycles of cloud cover causing preferential aspects in equatorial rainforests [60].

The suggestion that soil moisture limitation, caused by warm and dry conditions, prevented postfire recruitment is consistent with known climatic history of the region. The early 20th-century is noted as being a historically warm period in the eastern slopes of the Canadian Rockies, with dendroclimatic records showing the periods from 1917–1936 and 1939–1958 as having the highest temperature anomalies since the 1400s [63]. In fact, one of these reconstructions listed 1936, the year of the Pass Creek fire in WCW, as the second warmest year in the ~1000 year record [63]. This observation suggests the possibility that abnormally high atmospheric temperatures may have a two-fold effect on subalpine forests, first causing the desiccated conditions required for a burn at high elevation, then limiting expected regrowth.

4.2. Elevation

High degree change was almost entirely restricted to elevations greater than 1800 m, which coincide with the lower extent of ATE at the study site. This finding is consistent with the expectation that, as a transitional boundary that occurs as a function of elevation, canopy cover at ATE would be subject to a high level of variability [13,64,65]. Fire-exposed areas had lower mean elevations in all change classes, but the majority of canopy cover change was seen in these areas. Elevations greater than 1800 m accounted for a proportionally small region of fire-exposed areas, and yet the highest degree of change was seen at those elevations. The positive correlation between change degree and elevation is maintained in fire exposed areas suggesting that postfire mortality and re-establishment are potentially affected by factors that covary with elevation.

In the Canadian Rockies, there is evidence of both upslope advance and downslope recession of ATE elevation over the past 300 years. Our estimate of ATE advance in areas undisturbed by fire is consistent with other observations in the region; reported a subalpine fir advance rate of 0.28–0.62 m yr^{-1} along an upslope transect [36].

4.3. Slope

Treelines in high slope areas are known to be subject to mechanical disturbances by avalanches [11,66], or enhanced soil erosion following disturbance [17,67], which introduces the prospect that treelines in WCW may be orographically regulated [6]. Cyclical patterns of disturbance and regrowth may explain the high degree of both mortality and succession seen in high slope areas. However, in a nearby site in Glacier National Park (GNP), avalanche-regulated treelines were stable over a >70 year repeat photograph record, showing no cycle of disturbance and advance [40]. If the observed slope pattern in WCW is a result of cyclical processes, then this discrepancy between WCW and GNP warrants further investigation.

The slope of fire-exposed areas was generally lower than non-fire-exposed areas. High terrain slopes can accelerate the spread of fire [68], but this effect was outweighed by the extent of the WCW fire that occurred on the low slopes of the valley bottom. In two cases, *medium* and *high succession* classes, slopes were greater in fire-exposed areas than in non-fire-exposed areas, which agrees with research showing that slope was one of the factors that predicted re-establishment of subalpine fir following disturbance by fire [17]. In that study, 75% of establishment occurs on slopes with angles between 40°–60°. The authors reason that this intermediate slope range is steep enough to avoid competition by shrubs during re-establishment, but shallow enough to avoid frequent disturbance by rockslide and avalanche. That explanation is inconsistent with the present finding, as areas with slopes above 30° saw a high proportion of both mortality and succession.

5. Conclusions

The WSL-MT monoplotting technique provided the ability to perform quantitative spatial analysis of mountain landcover change on oblique, historic repeat photography from the MLP. Canopy cover estimates from these images agree with canopy cover measurements using airborne lidar [49]. The method provides the ability to observe ATE at a spatial resolution and temporal extent which is sufficient to identify topographic correlation to change, which is an important step in understanding interactions between modulating factors. The MLP dataset is ideal for multi-scale investigation of change in the ATE; the spatial extent of MLP photographs ranges from individual meadows (~100 m), to valleys (~10 km), at sites which can be several hundred kilometers apart [47].

In ATE in the WCW, patterns of postfire colonization related to edaphic controls have hindered regrowth and locally suppressed ATE on south-facing aspects, while considerable postfire colonization was seen on north-facing aspects. This observation appears to directly support hypothesis of edaphic controls on colonization postfire. While fire is not a direct control on ATE position, it is in this case a modulating factor, and acts to locally suppress ATE below the thermal limits. While much has been learned about these edaphic controls using stand reconstructions, these can be limited in their understanding of microsite-processes and larger regional trends. Repeat photography provides the opportunity to observe the patterns at a high spatial resolution, over broad regions and diverse landscapes. Future research in the region would benefit from investigating the interactions between fire, moisture limitation, and other modulating factors, using the multi-scale approach to understanding ATE dynamics [4,5].

In unburned areas on north-facing slopes, where moisture is not limiting, the ATE has systematically advanced. These observations of successful regeneration and ATE advance are consistent with increasing alpine isotherms and energy availability, as is expected under many climatic change scenarios, provided sufficient moisture or nutrient availability. More research is needed to further refine these hypotheses, particularly with respect to snowpack quantities and persistence at and above ATE.

Supplementary Materials: Data for the analysis presented in this research are maintained in the following repository: https://osf.io/ha579/?view_only=9cdda5a7fd47413386851fc91ba1282b. The WSL Monoplotting software is available free of charge (https://www.wsl.ch/en/services-and-products/software-websites-and-apps/monoplotting-tool.html). Historic and repeat photography from the mountain legacy project are open (http://mountainlegacy.ca). Spot 6 imagery is commercially available (https://www.intelligence-airbusds.com/en/8693-spot-67). The Alberta Vegetation Inventory are Fire Disturbance layers are open, by request (https://www.alberta.ca/forest-and-vegetation-inventories-data.aspx).

Author Contributions: Conceptualization, D.M. and C.H.; methodology, D.M. and C.H.; software, D.M.; validation, D.M.; formal analysis, D.M.; data curation, D.M.; writing—original draft preparation, D.M.; writing—review and editing, C.H.; visualization, D.M.; supervision, C.H.; project administration, C.H.; funding acquisition, D.M. and C.H. All authors have read and agreed to the published version of the manuscript.

Funding: McCaffrey acknowledges funding from the NSERC AMYTHEST program and the University of Lethbridge Graduate Students' Association. Hopkinson acknowledges funding from the Alberta Innovates Water Innovation Program, the NSERC Discovery Grant Program, the Campus Alberta Innovates Program, and Alberta Environmental Protection.

Acknowledgments: Repeat photography was graciously provided by the Mountain Legacy Project. Claudio Bozzini is thanked for developing the WSL monoplotting tool and instructing our team on its use. Marie-Pierre Rogeau was instrumental in accessing the fire history data of the West Castle area.

Conflicts of Interest: The authors are unaware of any potential conflicts of interest.

References

1. Körner, C. A re-assessment of high elevation treeline positions and their explanation. *Oecologia* **1998**, *115*, 445–459. [CrossRef] [PubMed]
2. Körner, C.; Paulsen, J. A world-wide study of high altitude treeline temperatures. *J. Biogeogr.* **2004**, *31*, 713–732. [CrossRef]
3. Holtmeier, F.-K. *Mountain Timberlines: Ecology, Patchiness and Dynamics*; Springer Science & Business Media: Berlin/Heidelberg, Germany, 2009; Volume 36.
4. Case, B.; Duncan, R.P. A novel framework for disentangling the scale-dependent influences of abiotic factors on alpine treeline position. *Ecography* **2014**, *37*, 838–851. [CrossRef]
5. Weiss, D.J.; Malanson, G.; Walsh, S. Multiscale Relationships Between Alpine Treeline Elevation and Hypothesized Environmental Controls in the Western United States. *Ann. Assoc. Am. Geogr.* **2015**, *105*, 437–453. [CrossRef]
6. Holtmeier, F.-K.; Broll, G. Sensitivity and response of northern hemisphere altitudinal and polar treelines to environmental change at landscape and local scales. *Glob. Ecol. Biogeogr.* **2005**, *14*, 395–410. [CrossRef]
7. Butler, D.R.; Malanson, G.; Walsh, S.; Fagre, D.B. Influences of Geomorphology and Geology on Alpine Treeline in the American West—More Important than Climatic Influences? *Phys. Geogr.* **2007**, *28*, 434–450. [CrossRef]
8. Resler, L.M. Geomorphic Controls of Spatial Pattern and Process at Alpine Treeline. *Prof. Geogr.* **2006**, *58*, 124–138. [CrossRef]
9. Elliott, G.P.; Cowell, C.M. Slope Aspect Mediates Fine-Scale Tree Establishment Patterns at Upper Treeline during Wet and Dry Periods of the 20th Century. *Arct. Antarct. Alp. Res.* **2015**, *47*, 681–692. [CrossRef]

10. Holtmeier, F.-K.; Broll, G. Wind as an Ecological Agent at Treelines in North America, the Alps, and the European Subarctic. *Phys. Geogr.* **2010**, *31*, 203–233. [CrossRef]
11. Walsh, S.J.; Butler, D.R.; Allen, T.R.; Malanson, G. Influence of snow patterns and snow avalanches on the alpine treeline ecotone. *J. Veg. Sci.* **1994**, *5*, 657–672. [CrossRef]
12. Stine, M.B.; Butler, D.R. Effects of fire on geomorphic factors and seedling site conditions within the alpine treeline ecotone, Glacier National Park, MT. *Catena* **2015**, *132*, 37–44. [CrossRef]
13. Harsch, M.A.; Hulme, P.E.; McGlone, M.S.; Duncan, R.P. Are treelines advancing? A global meta-analysis of treeline response to climate warming. *Ecol. Lett.* **2009**, *12*, 1040–1049. [CrossRef] [PubMed]
14. Baker, W.L. *Fire Ecology in Rocky Mountain Landscapes*; Island Press: Washington, DC, USA, 2009.
15. Rogeau, M.-P.; Armstrong, G.W. Quantifying the effect of elevation and aspect on fire return intervals in the Canadian Rocky Mountains. *For. Ecol. Manag.* **2017**, *384*, 248–261. [CrossRef]
16. Colombaroli, D.; Henne, P.D.; Kaltenrieder, P.; Gobet, E.; Tinner, W. Species responses to fire, climate and human impact at tree line in the Alps as evidenced by palaeo-environmental records and a dynamic simulation model. *J. Ecol.* **2010**, *98*, 1346–1357. [CrossRef]
17. Stueve, K.M.; Cerney, D.L.; Rochefort, R.M.; Kurth, L.L. Post-fire tree establishment patterns at the alpine treeline ecotone: Mount Rainier National Park, Washington, USA. *J. Veg. Sci.* **2009**, *20*, 107–120. [CrossRef]
18. Cansler, C.A.; McKenzie, N.; Halpern, C.B. Area burned in alpine treeline ecotones reflects region-wide trends. *Int. J. Wildland Fire* **2016**, *25*, 1209. [CrossRef]
19. Barrows, J. Fire behavior in northern Rocky Mountain forests. In *USDA Forest Service Station Paper 29*; Northern Rocky Mountain Forest and Range Experiment Station: Missoula, MT, USA, 1951.
20. Fowler, P.M.; Asleson, D.O. The Location of Lightning-Caused Wildland Fires, Northern Idaho. *Phys. Geogr.* **1984**, *5*, 240–252. [CrossRef]
21. Howe, E.; Baker, W.L. Landscape Heterogeneity and Disturbance Interactions in a Subalpine Watershed in Northern Colorado, USA. *Ann. Assoc. Am. Geogr.* **2003**, *93*, 797–813. [CrossRef]
22. Baker, W.L.; Kipfmueller, K.F. Spatial Ecology of Pre–Euro-American Fires in a Southern Rocky Mountain Subalpine Forest Landscape. *Prof. Geogr.* **2001**, *53*, 248–262. [CrossRef]
23. Buechling, A.; Baker, W.L. A fire history from tree rings in a high-elevation forest of Rocky Mountain National Park. *Can. J. For. Res.* **2004**, *34*, 1259–1273. [CrossRef]
24. Aplet, G.H.; Laven, R.D.; Smith, F.W. Patterns of Community Dynamics in Colorado Engelmann Spruce-Subalpine Fir Forests. *Ecology* **1988**, *69*, 312–319. [CrossRef]
25. Peet, R.K. Forest vegetation of the Colorado Front Range. *Vegetatio* **1981**, *45*, 3–75. [CrossRef]
26. Romme, W.H.; Knight, D.H. Fire Frequency and Subalpine Forest Succession Along a Topographic Gradient in Wyoming. *Ecology* **1981**, *62*, 319–326. [CrossRef]
27. Veblen, T.T. Age and Size Structure of Subalpine Forests in the Colorado Front Range. *Bull. Torrey Bot. Club* **1986**, *113*, 225. [CrossRef]
28. Donnegan, J.A.; Rebertus, A.J. Rates and mechanisms of subalpine forest succession along an environmental gradient. *Ecology* **1999**, *80*, 1370–1384. [CrossRef]
29. Kelsey, K.C.; Redmond, M.D.; Barger, N.N.; Neff, J. Species, Climate and Landscape Physiography Drive Variable Growth Trends in Subalpine Forests. *Ecosystems* **2017**, *21*, 125–140. [CrossRef]
30. Redmond, M.D.; Kelsey, K.C. Topography and overstory mortality interact to control tree regeneration in spruce-fir forests of the southern Rocky Mountains. *For. Ecol. Manag.* **2018**, *427*, 106–113. [CrossRef]
31. Danby, R.K. Monitoring Forest-Tundra Ecotones at Multiple Scales. *Geogr. Compass* **2011**, *5*, 623–640. [CrossRef]
32. Allen, T.R.; Walsh, S.J. Spatial and compositional pattern of alpine treeline, Glacier National Park, Montana. *Photogramm. Eng. Remote Sens.* **1996**, *62*, 1261–1268.
33. Bolton, D.K.; Coops, N.C.; Hermosilla, T.; Wulder, M.A.; White, J.C. Evidence of vegetation greening at alpine treeline ecotones: Three decades of Landsat spectral trends informed by lidar-derived vertical structure. *Environ. Res. Lett.* **2018**, *13*, 084022. [CrossRef]
34. Danby, R.K.; Hik, D.S. Evidence of Recent Treeline Dynamics in Southwest Yukon from Aerial Photographs. *Arctic* **2009**, *60*, 60. [CrossRef]
35. Tinner, W.; Theurillat, J. Institute of Arctic and Alpine Research (INSTAAR) University of Colorado Uppermost Limit, Extent, and Fluctuations of the Timberline and Treeline Ecocline in the Swiss Central Alps during the Past 11,500 Years. *Arct. Antarct. Alp. Res.* **2003**, *35*, 158–169. [CrossRef]

36. Bekker, M.F. Positive Feedback Between Tree Establishment and Patterns of Subalpine Forest Advancement, Glacier National Park, Montana, USA. *Arct. Antarct. Alp. Res.* **2005**, *37*, 97–107. [CrossRef]
37. Mamet, S.D.; Kershaw, G.P. Subarctic and alpine tree line dynamics during the last 400 years in north-western and central Canada. *J. Biogeogr.* **2011**, *39*, 855–868. [CrossRef]
38. Sakulich, J. Reconstruction and spatial analysis of alpine treeline in the Elk Mountains, Colorado, USA. *Phys. Geogr.* **2015**, *36*, 471–488. [CrossRef]
39. Butler, D.R.; DeChano, L.M. Environmental Change in Glacier National Park, Montana: An Assessment through Repeat Photography from Fire Lookouts. *Phys. Geogr.* **2001**, *22*, 291–304. [CrossRef]
40. Butler, D.R.; Malanson, G.P.; Cairns, D. Stability of alpine treeline in Glacier National Park, Montana, USA. *Phytocoenologia* **1994**, *22*, 485–500. [CrossRef]
41. Klasner, F.L.; Fagre, D.B. A Half Century of Change in Alpine Treeline Patterns at Glacier National Park, Montana, USA. *Arct. Antarct. Alp. Res.* **2002**, *34*, 49–56. [CrossRef]
42. Kullman, L.; Öberg, L. Post-Little Ice Age tree line rise and climate warming in the Swedish Scandes: A landscape ecological perspective. *J. Ecol.* **2009**, *97*, 415–429. [CrossRef]
43. Moiseev, P.A.; Shiyatov, S.G. Vegetation Dynamics at the Tree-Line Ecotone in the Ural Highlands, Russia. In *Alpine Biodiversity in Europe*; Springer Science and Business Media: Berlin, Germany, 2003; Volume 167, pp. 423–435.
44. Roush, W.; Munroe, J.S.; Fagre, D.B. Development of a spatial analysis metho using ground-based repeat photography to detect changes in the alpine treeline ecotone, Glacier National Park, Montana, USA. *Arct. Antarct. Alp. Res.* **2007**, *39*, 297–308. [CrossRef]
45. Claudio, B.; Conedera, M.; Krebs, P. A New Monoplotting Tool to Extract Georeferenced Vector Data and Orthorectified Raster Data from Oblique Non-Metric Photographs. *Int. J. Herit. Digit. Era* **2012**, *1*, 499–518. [CrossRef]
46. Stockdale, C.A.; Bozzini, C.; Macdonald, S.E.; Higgs, E.; Macdonald, S.E. Extracting ecological information from oblique angle terrestrial landscape photographs: Performance evaluation of the WSL Monoplotting Tool. *Appl. Geogr.* **2015**, *63*, 315–325. [CrossRef]
47. Stockdale, C.A.; Macdonald, S.E.; Higgs, E. Forest closure and encroachment at the grassland interface: A century-scale analysis using oblique repeat photography. *Ecosphere* **2019**, *10*, e02774. [CrossRef]
48. Trant, A.J.; Starzomski, B.M.; Higgs, E. A publically available database for studying ecological change in mountain ecosystems. *Front. Ecol. Environ.* **2015**, *13*, 187. [CrossRef]
49. McCaffrey, D.R.; Hopkinson, C. Assessing Fractional Cover in the Alpine Treeline Ecotone Using the WSL Monoplotting Tool and Airborne Lidar. *Can. J. Remote Sens.* **2017**, *43*, 504–512. [CrossRef]
50. Wildfire Management Branch—Alberta Agriculture and Forestry. Fire History Polygons C5 FMU. 2017. Available online: https://wildfire.alberta.ca/resources/historical-data/spatial-wildfire-data.aspx (accessed on 14 May 2020).
51. Alberta Environment and Parks. *Alberta Vegetation Inventory*; Branch, R.I.M., Ed.; Alberta Environment and Parks: Edmonton, AB, Canada, 2012.
52. Rogeau, M.-P.; Flannigan, M.; Hawkes, B.C.; Parisien, M.-A.; Arthur, R. Spatial and temporal variations of fire regimes in the Canadian Rocky Mountains and Foothills of southern Alberta. *Int. J. Wildland Fire* **2016**, *25*, 1117–1130. [CrossRef]
53. Rogeau, M.P. *Fire History Study Castle River Watershed*; Alberta. Technical report prepared for Alberta Environment and Sustainable Resource Development; Forest Protection Branch, Southern Rockies Wildfire Management Area: Calgary, AB, Canada, 2012; 54p.
54. Lumley, T.; Diehr, P.; Emerson, S.; Chen, L. The Importance of the Normality Assumption in Large Public Health Data Sets. *Annu. Rev. Public Heal.* **2002**, *23*, 151–169. [CrossRef]
55. Germino, M.J.; Smith, W.K.; Resor, A.C. Conifer seedling distribution and survival in an alpine-treeline ecotone. *Plant Ecol.* **2002**, *162*, 157–168. [CrossRef]
56. Kueppers, L.M.; Conlisk, E.; Castanha, C.; Moyes, A.; Germino, M.J.; De Valpine, P.; Torn, M.S.; Mitton, J.B. Warming and provenance limit tree recruitment across and beyond the elevation range of subalpine forest. *Glob. Chang. Boil.* **2016**, *23*, 2383–2395. [CrossRef]
57. Moyes, A.; Germino, M.J.; Kueppers, L.M. Moisture rivals temperature in limiting photosynthesis by trees establishing beyond their cold-edge range limit under ambient and warmed conditions. *New Phytol.* **2015**, *207*, 1005–1014. [CrossRef]
58. Silins, U.; Stone, M.; Emelko, M.B.; Bladon, K.D. Sediment production following severe wildfire and post-fire salvage logging in the Rocky Mountain headwaters of the Oldman River Basin, Alberta. *Catena* **2009**, *79*, 189–197. [CrossRef]

59. Hopkinson, C.; Collins, T.; Anderson, A.; Pomeroy, J.; Spooner, I. Spatial Snow Depth Assessment Using LiDAR Transect Samples and Public GIS Data Layers in the Elbow River Watershed, Alberta. *Can. Water Resour. J.* **2012**, *37*, 69–87. [CrossRef]
60. Bader, M.; Ruijten, J.J. A topography-based model of forest cover at the alpine tree line in the tropical Andes. *J. Biogeogr.* **2008**, *35*, 711–723. [CrossRef]
61. Dang, H.; Zhang, Y.; Zhang, Y.; Zhang, K.; Zhang, Q. Variability and rapid response of subalpine fir (Abies fargesii) to climate warming at upper altitudinal limits in north-central China. *Trees* **2015**, *29*, 785–795. [CrossRef]
62. Greenwood, S.; Chen, J.-C.; Chen, C.-T.; Jump, A.S. Strong topographic sheltering effects lead to spatially complex treeline advance and increased forest density in a subtropical mountain region. *Glob. Chang. Boil.* **2014**, *20*, 3756–3766. [CrossRef] [PubMed]
63. Luckman, B.H.; Wilson, R. Summer temperatures in the Canadian Rockies during the last millennium: A revised record. *Clim. Dyn.* **2005**, *24*, 131–144. [CrossRef]
64. Körner, C. *Alpine Treelines: Functional Ecology of the Global High Elevation Tree Limits*; Springer Science & Business Media: Berlin/Heidelberg, Germany, 2012.
65. Tranquillini, W. *Physiological Ecology of the Alpine Timberline: Tree Existence at High Altitudes with Special Reference to the European Alps*; Springer: Berlin, Germany, 1979.
66. Butler, D.R.; Walsh, S. Site Characteristics of Debris Flows and their Relationship to Alpine Treeline. *Phys. Geogr.* **1994**, *15*, 181–199. [CrossRef]
67. Shakesby, R.; Doerr, S. Wildfire as a hydrological and geomorphological agent. *Earth Sci. Rev.* **2006**, *74*, 269–307. [CrossRef]
68. Werth, P.A.; Potter, B.E.; Clements, C.; Finney, M.A.; Goodrick, S.L.; Alexander, M.E.; Cruz, M.G.; Forthofer, J.; McAllister, S. *Synthesis of Knowledge of Extreme Fire Behavior: Volume I for Fire Managers*; USDA: Washington, DC, USA, 2011.

remote sensing

MDPI

Article

Identifying Post-Fire Recovery Trajectories and Driving Factors Using Landsat Time Series in Fire-Prone Mediterranean Pine Forests

Alba Viana-Soto *, Inmaculada Aguado, Javier Salas and Mariano García

Environmental Remote Sensing Research Group, Department of Geology, Geography and Environment, Universidad de Alcalá, Calle Colegios 2, 28801 Alcalá de Henares, Spain; inmaculada.aguado@uah.es (I.A.); javier.salas@uah.es (J.S.); mariano.garcia@uah.es (M.G.)

* Correspondence: alba.viana@uah.es; Tel.: +34-(918)-854-454

Received: 30 March 2020; Accepted: 4 May 2020; Published: 8 May 2020

Abstract: Wildfires constitute the most important natural disturbance of Mediterranean forests, driving vegetation dynamics. Although Mediterranean species have developed ecological post-fire recovery strategies, the impacts of climate change and changes in fire regimes may endanger their resilience capacity. This study aims at assessing post-fire recovery dynamics at different stages in two large fires that occurred in Mediterranean pine forests (Spain) using temporal segmentation of the Landsat time series (1994–2018). Landsat-based detection of Trends in Disturbance and Recovery (LandTrendr) was used to derive trajectory metrics from Tasseled Cap Wetness (TCW), sensitive to canopy moisture and structure, and Tasseled Cap Angle (TCA), related to vegetation cover gradients. Different groups of post-fire trajectories were identified through K-means clustering of the Recovery Ratios (RR) from fitted trajectories: continuous recovery, continuous recovery with slope changes, continuous recovery stabilized and non-continuous recovery. The influence of pre-fire conditions, fire severity, topographic variables and post-fire climate on recovery rates for each recovery category at successional stages was analyzed through Geographically Weighted Regression (GWR). The modeling results indicated that pine forest recovery rates were highly sensitive to post-fire climate in the mid and long-term and to fire severity in the short-term, but less influenced by topographic conditions (adjusted R-squared ranged from 0.58 to 0.88 and from 0.54 to 0.93 for TCA and TCW, respectively). Recovery estimation was assessed through orthophotos, showing a high accuracy (Dice Coefficient ranged from 0.81 to 0.97 and from 0.74 to 0.96 for TCA and TCW, respectively). This study provides new insights into the post-fire recovery dynamics at successional stages and driving factors. The proposed method could be an approach to model the recovery for the Mediterranean areas and help managers in determining which areas may not be able to recover naturally.

Keywords: post-fire recovery; Landsat; time series; LandTrendr; K-means; driving factors; Mediterranean; pine forests

1. Introduction

Wildfires constitute one of the most widespread and important natural disturbances of forest ecosystems, playing a paramount role in the dynamics of the terrestrial system [1]. Forest fires impact at a wide range of scales causing ecological, economic and human health impacts [2,3]. Specifically in Europe, the Mediterranean region registers the highest number of fires and burned areas [4], with around 85% of the total burnt area [5].

Notwithstanding, Mediterranean ecosystems are adapted to fire recurrence as it constitutes the most important natural disturbance, driving vegetation dynamics [6]. Mediterranean species have developed post-fire ecological strategies including resprouting capacity, seed bank persistence and increased dispersal capacity [7,8]. Nevertheless, land use changes and the impacts of climate change may affect the dynamics of post-fire ecological succession in the immediate future [3,9]. Although large fire (i.e., ≥500 ha) occurrence for the European Mediterranean region does not show a strong increasing trend in the recent decades [5], climate change projections indicate an increase in the frequency and intensity of megafires, as a result of more extended and severe seasonal droughts [10], which will impact ecosystems' species composition and functioning [3]. Forest ecosystems must adapt not only to changes in average climatic variables, but also to a wide variability with higher risk of extreme climatic events, such as prolonged droughts. Thus, forest management in European Mediterranean countries is challenging due to the vulnerability of natural regrowth capability of these ecosystems [11,12].

Time series of satellite data have long been used for retrospectively generating information on forest disturbance and recovery dynamics [13]. The opening of the Landsat archive in 2008, now available geometrically and radiometrically corrected, provided new opportunities for improved understanding of the mechanisms of forest changes [14,15]. Several studies have addressed the spatial and temporal analysis of post-fire vegetation dynamics through different forest ecosystems: Mediterranean [16,17], boreal [18,19], Siberian [20,21], temperate [22], tropical [23], savannah [24] or across different ecozones at the regional or national scale [25–28].

The use of the Landsat time series (LTS) for change detection has increased substantially in recent years as new methodological approaches have emerged [29]. Early approaches characterized post-fire recovery dynamics by applying linear regression functions to spectral trajectories obtained from the Landsat time series [30,31]. More recently, several change detection algorithms have been developed and widely used in analyzing forest changes, such as Landsat-based detection of Trends in Disturbance and Recovery (LandTrendr) [32] and Vegetation Change Tracker (VCT) [33], to provide change information on an annual time-scale [29]. Others include, Breaks For Additive Seasonal and Trend (BFAST) [34] and Continuous Monitoring of Forest Disturbance Algorithm (CMFDA) [35], which use a high frequency of time series. The trajectory-based segmentation algorithm LandTrendr enables the characterization of distinct subtrends within a simplified representation of the spectral trajectory, which provides the essential information needed to identify abrupt disturbances in forests (e.g., fire and harvest), as well as slowly evolving processes (e.g., regrowth and defoliation). The utility of LandTrendr has been demonstrated in different regions for assessing disturbance and recovery dynamics [25,28,36].

Several spectral measures can be derived from LTS and used as inputs for segmentation algorithms such as spectral indices or Tasseled Cap Transformations (TCT). Some spectral indices focused more on the red and near-infrared bands, making them sensitive to canopy greenness and photosynthetic activity, such as the Normalized Difference Vegetation Index (NDVI) employed for characterizing post-fire recovery [16,18,37], whereas other indices using the SWIR bands are more sensitive to vegetation moisture and forest structure [38], such as the Normalized Burn Ratio (NBR) [39], commonly used for recovery assessment [19,28,40]. TCT are created via linear transformations using defined coefficients [41] and have been widely used for studying forest changes [36,42–44]. TCT components correspond to the physical characteristics of vegetation: Brightness (TCB) is related to the pixel albedo of the land surface and values are typically high after a stand replacing disturbance; Greenness (TCG) is a contrast between the visible and near-infrared bands, being sensitive to green vegetation [19], and Wetness (TCW) is a contrast of the visible and near-infrared with the SWIR bands, making it sensitive to canopy moisture and structure [45]. Several metrics can be derived from TCT such as the TC Angle (TCA), which is related to the vegetation cover within the TCB-TCG spectral plane [46]. Considering that spectral indices are sensitive to different vegetation conditions, the use of TC components and derived metrics enables the characterization of different forest conditions [19,38,44,47].

Although the dynamic of post-fire vegetation recovery has been studied through different forest ecosystems, few studies have investigated the recovery driving factors in Mediterranean ecosystems [16,30,37,48] and fewer have focused on characterizing successional recovery stages [49,50], which are key to understanding forest changes for sustainable forest management. This study assesses the post-fire recovery dynamics at different stages in fire-prone Mediterranean pine forests. The specific objectives of this study were: (1) to identify the different post-fire recovery trajectories using temporal segmentation of LTS; (2) to analyze the recovery patterns for each trajectory group through stages and (3) to appraise the environmental and contextual drivers of the recovery process.

2. Materials and Methods

2.1. Study Area

This research was based on two large fires that occurred in the summer of 1994 (Figure 1): the Yeste Fire (7 August), which burned 11,685 ha of wooded area, and the Requena Fire (5 July), which burned 16,373 ha of wooded area. For this study, we selected sections that had neither burned in subsequent fires nor been reforested after the main disturbance of 1994 in order to ensure the analysis of natural recovery only.

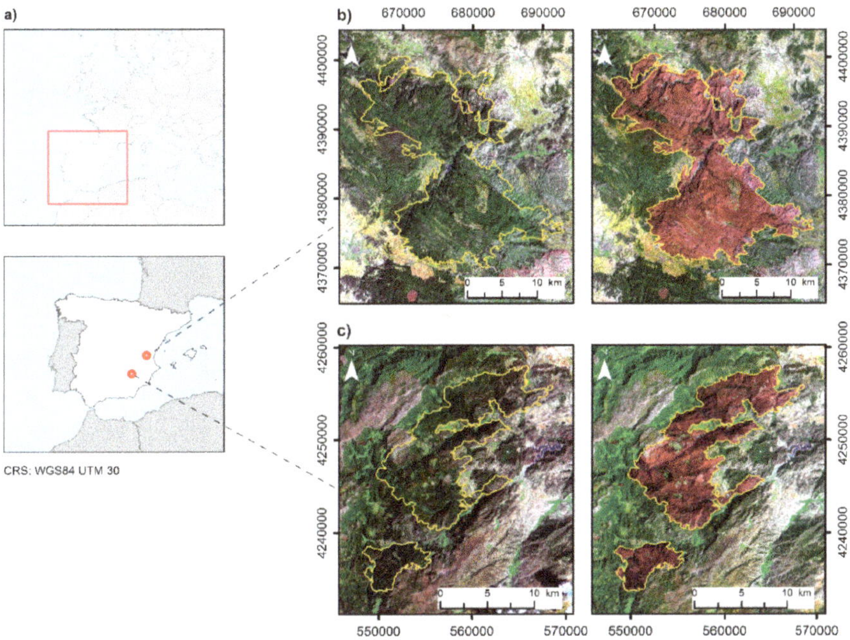

Figure 1. Study areas located in the Iberian Peninsula: Requena above, Yeste below; (**a**) location of the study areas; (**b**) pre- and post-fire Landsat composition for Requena RGB (SWIR2, NIR, Blue) and (**c**) pre- and post-fire Landsat composition for Yeste RGB (SWIR2, NIR, Blue).

Both study areas are located in the Southeast of the Iberian Peninsula, in the Mediterranean biogeographic region, which is characterized by mean annual precipitations of 600–700 mm with soil hydrological deficit in summer and mean annual temperatures around 15 °C. These areas were dominated by anthropogenic coniferous forests, mainly composed of species of the genus *Pinus* along with certain deciduous species of the genus *Quercus*, and, sclerophyll species such as *Rosmarinus*, *Thymus* or *Juniperus* species in the understory [51]. Due to the differences in post-fire ecological strategies to recover, we selected those patches dominated by *Pinus halepensis* and *Pinus pinaster* according to the Second National Forest Inventory of Spain (SNFI) [52]. Both species are obligate seeders since they have serotinous cones that enable the natural post-fire regeneration [7,53]. The post-fire recovery stages that can be identified in a Mediterranean pine forest 24 years after fire, range from the stand initiation (establishment phase including remnant pines, herbaceous and pine seeding processes) [8] to the stem exclusion (a young regrowth forest composed by shrubs and tree plantlets) [49,54,55]. Since competition between shrubs and trees starts immediately following fire [50,56], vegetation recovery in this study refers to both tree and shrub recovery.

2.2. Data

We downloaded the Landsat TM/ETM+/OLI images from the United States Geological Survey (USGS) Earth Explorer server [57] to build the time series covering the period 1990–2018, including 4 years pre-fire (Path/Row: 200/033, 199/032, 199/033). We selected images from Tier 1 Surface Reflectance products generated from the Landsat Ecosystem Disturbance Adaptive Processing System (LEDAPS) software [58] for TM and ETM + images and Landsat 8 Surface Reflectance Code (LaSRC) for the OLI dataset [59]. We prioritized scenes with less than 10% cloud-cover, within the summer period to minimize the effect of phenological changes. To delimit fire perimeters and to assess fire severity we used two Landsat 5 TM images acquired in 1994 for the Yeste Fire (22 July pre-fire, 23 August post-fire) and for the Requena Fire (29 June pre-fire, 16 August post-fire) (Figure 1). Fire perimeters were constructed by applying the USGS thresholds [39] to the differenced Normalized Burn Ratio (dNBR) [60].

Spatial reference to vegetation types was based upon the Forest Map of Spain, which was made between 1986 and 1997 through aerial photographs and field work [61]. Topographic variables were built from the LiDAR-based Digital Elevation Model (25-m spatial resolution) from the National Geographic Institute (IGN) of Spain [62]. As climatic information we used the Standardized Precipitation-Evapotranspiration Index (SPEI) [63], a multi-scalar drought index that calculates the effect of potential evapotranspiration (PET) on drought severity. Compared to other drought indices, the SPEI has the advantage of combining multi-scalar character with the capacity to include the effects of temperature variability [64]. Data are available for the entire time series (1990–2018) at 1-km resolution. We also downloaded the orthophotos at 0.5-m resolution from the Aerial Orthophotography National Plan, as reference data to assess recovery through time series (Years 2002, 2009, 2010, 2017 and 2018) [65].

2.3. Methods

A flowchart of the methodology is depicted in Figure 2. Firstly, annual composites were created for the available time series. Secondly, spectral indices and TCT was performed to delineate burned areas and derive trajectory metrics of forested pixels based on the LandTrendr segmentation algorithm [32]. Fitted trajectories were categorized according to the change magnitude and duration represented in the sequence of segments of each trajectory using a K-means algorithm. Finally, we analyzed the environmental and contextual drivers of the recovery process. The accuracy assessment of the recovery classes was carried out by visual assessment of vegetation cover in randomly distributed sample plots, in reference to high-resolution orthophotos.

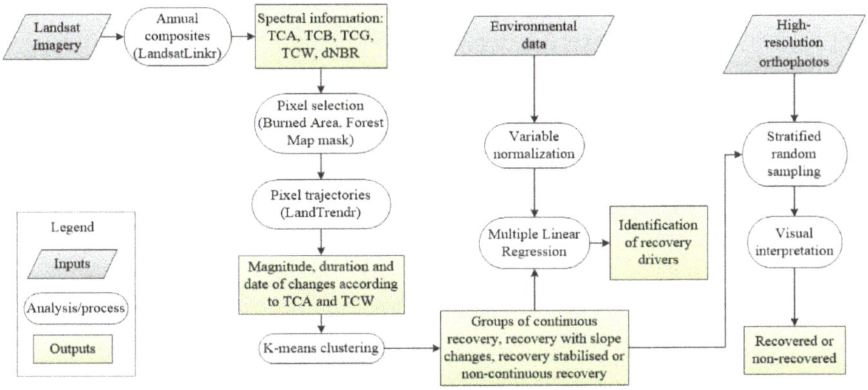

Figure 2. Flowchart of the methodology.

2.3.1. Landsat Time Series

Summer time series was created for the 28-year time period using the closest cloud-free image to the mid of the summer season: Requena median Julian day 209 and Yeste median Julian day 218. The temporal window used to select the images spanned ± 38 days around the reference date (Figure 3) in order to ensure consistency through the time series.

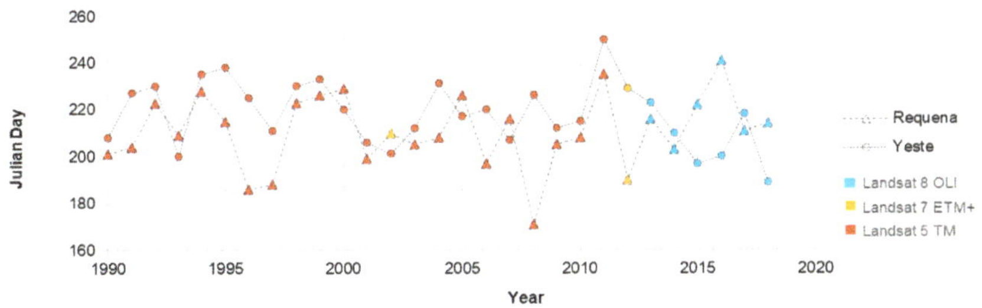

Figure 3. Scene selection dates according to the Julian Day.

We used the R package LandsatLinkr [66,67] to create annual cloud-free image stacks and calculate the TCT components using the coefficients defined for reflectance data [41]: Brightness (TCB), Greenness (TCG) and Wetness (TCW). Subsequently, the angular component of the TCT (TCA) was computed as follows [46]:

$$TCA = \tan^{-1}\left(\frac{TCG}{TCB}\right) \quad (1)$$

2.3.2. Trajectory Segmentation and Clustering

To extract recovery trajectories, we applied the LandTrendr trajectory-based segmentation algorithm [32] to TCA and TCW time series. In this study we opted to use the TCA due to its relation with the percentage of vegetation cover in coniferous and mixed forests [46,47], and the TCW since it is sensitive to canopy moisture and structure [44,45]. LandTrendr goes through the time series and creates a fitted trajectory as a sequence of line segments for each pixel. Firstly, the vertices of each segment were established from an iterative regression process using Ordinary Least Squares (OLS) by estimating the years of change using the TCA and TCW time series. Then, the trajectories are iteratively simplified from a selection process using the angle criterion until a number of segments equal to or less than a user-defined threshold were obtained (segmentation process) [32]. We set this threshold to the maximum available (6) as we attempted to unravel multiple recovery trends. In a second step, the spectral values at candidate vertices were estimated (fitting process), generating a trajectory of connected segments for each pixel. The best model was chosen based on the *p*-value according to the F-statistics ($p < 0.05$). Further details on the segmentation process can be found in Kennedy et al. (2010) [32].

From the derived trajectories, we selected the change magnitude and duration of the segments, at the pixel level, as parameters for the classification. We calculated a recovery ratio (RR) for each segment (Equation (2)), which allows us to describe the recovery rates through time or alternatively for each successional stage defined by the number of segments (Figure 4).

$$\mathrm{RR} = \frac{\text{Magnitude of Change}_x}{\text{Duration of Change}_x} \tag{2}$$

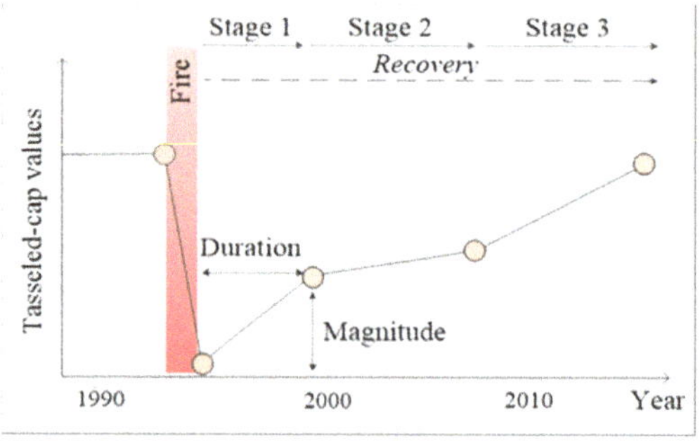

Figure 4. Example of a fitted trajectory.

We performed an unsupervised clustering method, since our first objective was to unravel the different vegetation recovery patterns. K-means clustering [68] is one of the most popular unsupervised machine learning algorithms and has been previously applied to summarize vegetation types and changes [69]. K-means allowed us to group the trajectories (defined by the RR of the segments) in k groups, minimizing the sum of the distances between each trajectory and the centroid of a given class. To define the optimal number of classes we employed the elbow method, which is based on the percentage of variance explained as a function of the number of clusters [70]. This process has been carried out with the scikit-learn library of Python [71]. The clusters obtained were overlapped to the LandTrendr outputs to characterize the categories.

2.3.3. Assessing Driving Factors of Vegetation Recovery

We aimed to explain the influence of fire severity, pre-fire conditions, topographic and climatic variables on recovery ratios through regression analysis. Six explanatory variables obtained from the Landsat imagery as well as auxiliary data sources (see Section 2.2) were derived to model post-fire recovery (Table 1). For the variables that did not meet the assumptions of normality of the residuals and homogeneity of variance, we used log-transformed or rank-transformed data.

Table 1. List of variables used in the regression analysis.

Variable		Units	Description
Dependent	Recovery Ratio$_x$ (RR-TCA$_x$ RR-TCW$_x$)	Z value	Represents the slope of the fitted trajectory at each segment
Explanatory			
Pre-fire conditions	TCA$_{90\text{-}93}$ or TCW$_{90\text{-}93}$	Z value	TCA shows the percent vegetation cover and TCW the moisture and structure before the fire
Fire severity	dNBR	Values between −1 and 1	Represents the short-term post-fire effects on vegetation cover and structure. Severity thresholds proposed by the USGS [39]: Low: $0.1 \leq$ dNBR < 0.27 Moderate-low: $0.27 \leq$ dNBR < 0.44 Moderate-high: $0.44 \leq$ dNBR < 0.66 High: ≥ 0.66
Topography	Elevation	Meters	
	Slope	Percent	
	Aspect	Values between 0 and 1	(TRASP) [72]. Values of 0 correspond to cooler, wetter north-northeastern aspects; values of 1 correspond to hotter, dryer south-southwestern aspects
Climatic Anomalies	Drought index	Z value	(SPEI) [63]. Positive values represent positive water balance and negative values indicate drought conditions

Previous studies have shown strong effects of fire severity on post-fire vegetation recovery [12,16,56]. Fire severity, defined as the degree of ecosystem change caused by a fire with respect to the pre-fire situation [73], was evaluated through the dNBR [60]:

$$dNBR = NBR_{pre-fire} - NBR_{post-fire} \tag{3}$$

$$NBR = \frac{(NIR - SWIR2)}{(SWIR2 + NIR)} \tag{4}$$

Previous work has also addressed the important role of topography in explaining variations in forest establishment following fire [18,20,30]. Other studies have established the relevance of the post-fire climate [16,28,37] since it is related to water availability, and pre-fire vegetation conditions [20] due to its relationship with post-fire seed availability [12]. Regarding the drought index (SPEI), the 3 month-scale (cumulative from June, July and August) was selected since vegetation activity responds predominantly to short drought time-scales [74] and because maximum Pearson correlation coefficients between SPEI aggregated from summer season and TCA–TCW time series were recorded for both study areas.

Firstly, an exploratory regression analysis was carried out to diagnose the suitability of the selected variables. Due to the presence of spatial autocorrelation and heteroscedasticity in our data (the significance of Koenker statistics at the 95% confidence level), we executed a Geographically Weighted Regression (GWR) [75,76], a local regression model that considers spatial heterogeneity in data relationships. Model fitting was conducted using optimized adaptive bi-square kernel bandwidth (according to the corrected Akaike information criterion) [77].

2.3.4. Recovery Assessment

We evaluated the recovery through TCA and TCW trajectories in the short, mid and long-term as Key and Benson (2006) [39] proposed to assess burn severity: 2002 (8 years post-fire), 2009–2010 (15–16 years post-fire) and 2017–2018 (23–24 years post-fire), with 80% of the pre-fire value of TCA and TCW as the recovery threshold [19].

We carried out a stratified validation based on the recovery categories identified with a sample size of 500 plots randomly selected at each phase for each index (a total of 3000 reference plots). Vegetation recovery was evaluated by visual analysis of high resolution orthophotos. To facilitate the visual interpretation, we divided the 30-m pixel of Landsat images equally with a 6 m by 6 m grid (25 cells in one 30-m pixel) and overlaid the grids onto the orthophotos (Figure 5), similarly to Zhao et al. (2016) [78]. As an approximation to the pre-fire fractional cover due to the lack of pre-fire orthophotos, we established as reference the pre-fire fractional cover obtained from the SNFI (mean cover of 42.9% and 47.9% for Requena and Yeste, respectively). A disturbed pixel was considered to have recovered if the tree and shrub cover was at least 40% (i.e., 10/25; Figure 5a,b). Otherwise, that pixel was interpreted as having not recovered (Figure 5c).

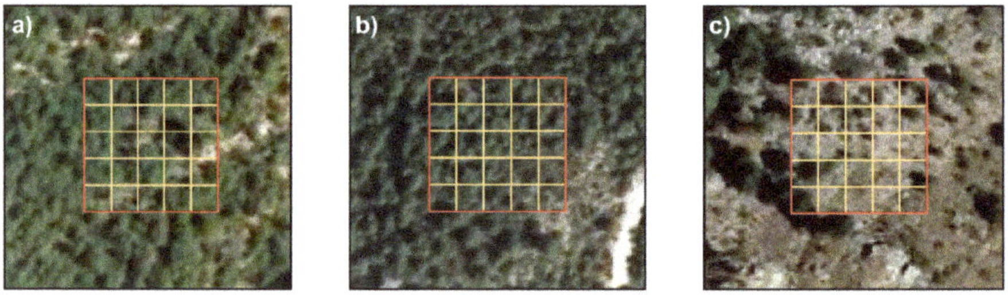

Figure 5. Examples of high resolution orthophotos of recovered (**a**,**b**) and non-recovered pixels (**c**) in 5 by 5 grids.

Four accuracy metrics derived from confusion matrices were computed to validate our vegetation recovery classification (Table 2), the omission error (OE; Equation (5)), the commission error (CE; Equation (6)), the overall accuracy (OA; Equation (7)) and the Dice coefficient (DC; Equation (8)) [79].

Table 2. Confusion matrix example.

	Reference Data		
Estimation	**Recovered**	**Non-Recovered**	**Row Total**
Recovered	P_{11}	P_{12}	P_{1+}
Non-recovered	P_{21}	P_{22}	P_{2+}
Col. total	P_{+1}	P_{+2}	N
OE	P_{21}/P_{+1}		(5)
CE	P_{12}/P_{1+}		(6)
OA	$P_{11} + P_{22}/N$		(7)
DC	$2P_{11}/(P_{1+} + P_{+1})$		(8)

3. Results

3.1. Classification of Post-Fire Trajectories

Four different categories were identified for TCA (CR, CRSC, CRS and NCR) and five for TCW (CR, CR2, CRSC, CRSC2 and CRS). The main characteristics of the categories describing recovery are defined in Table 3.

Table 3. Recovery categories definition.

Category	Acronym	Stages	Description
Continuous Recovery	CR CR2	1 2	Pixels show a continuous increase in the TCA and TCW values since the year of fire (CR) or since the first year post-fire (CR2).
Continuous Recovery with Slope Changes	CRSC CRSC2	4 3	Continuous recovery follows disturbance but slope changes occur through the time series. Changes occur at a different time for TCA and TCW (CRSC and CRSC2).
Continuous Recovery Stabilized	CRS	2	Continuous recovery, which slow down or stop 4–5 years after fire.
Non-continuous Recovery	NCR	3	Recovery process is interrupted in the mid-term followed by a second phase of continuous recovery (only found with TCA).

Recovery dynamics in TCA and TCW tended to be spatially clustered, indicating strong spatial effects (Figure 6). Moreover, there are differences in terms of the magnitude of change, as well as the year in which changes detected between TCA and TCW occurred, and both fires. For the Requena Fire we observed greater homogeneity, with CR as the main category, whereas for the Yeste Fire the CRSC category predominates.

The plots in Figure 7 show the mean fitted trajectories of recovery categories from the TCA and the TCW. In the pre-fire period, the TCA and TCW trajectories did not show significant changes, indicating relative stability in forest cover until the occurrence of fire. Nevertheless, the trend is negative in the case of TCA trajectories, which could be due to a loss of vegetation vigor before the fires occurred. The structure seems to have not changed in the pre-fire period according to stable values in TCW trajectories.

Although burned areas were generally characterized by an increase in spectral values after the fire events, the mean TCA and TCW trajectories showed differences in vegetation recovery between the two fires and across time series. TCA showed faster recovery, with high slopes in trajectories in the short-term, because it is associated with the percent of vegetation cover, whereas TCW showed slower recovery since it is related to the vegetation structure and moisture.

Regarding the TCA categories, the same trends were found in both fires and post-fire mean values tended to overtake pre-fire values in the mid-term. Even so, recovery rates were higher in Requena, as the post-fire mean values reached the pre-fire values before 2005, whereas in Yeste the pre-fire values were reached around 2010. CR pixels correspond to lower disturbance magnitude whereas in CRSC, CRS and NCR the magnitude of change is clearly higher. In the cases of CRSC and NCR the recovery rates were high even though the recovery dynamic was interrupted. CRS showed a particular trajectory as it stabilized in the mid-term and maintained the values across a subsequent phase of slight recovery.

According to the TCW categories, different trends were found between fires with the exception of CR and CRS pixels. In the case of the Yeste Fire the categories CR2 and CRSC continued to decrease until one year after the fire. In the case of Requena Fire, two different categories of CRSC were found since breakpoints in the recovery process occurred in different years. In spite of the trend toward pre-fire values none of the TCW trajectories reached pre-fire conditions after 24 years.

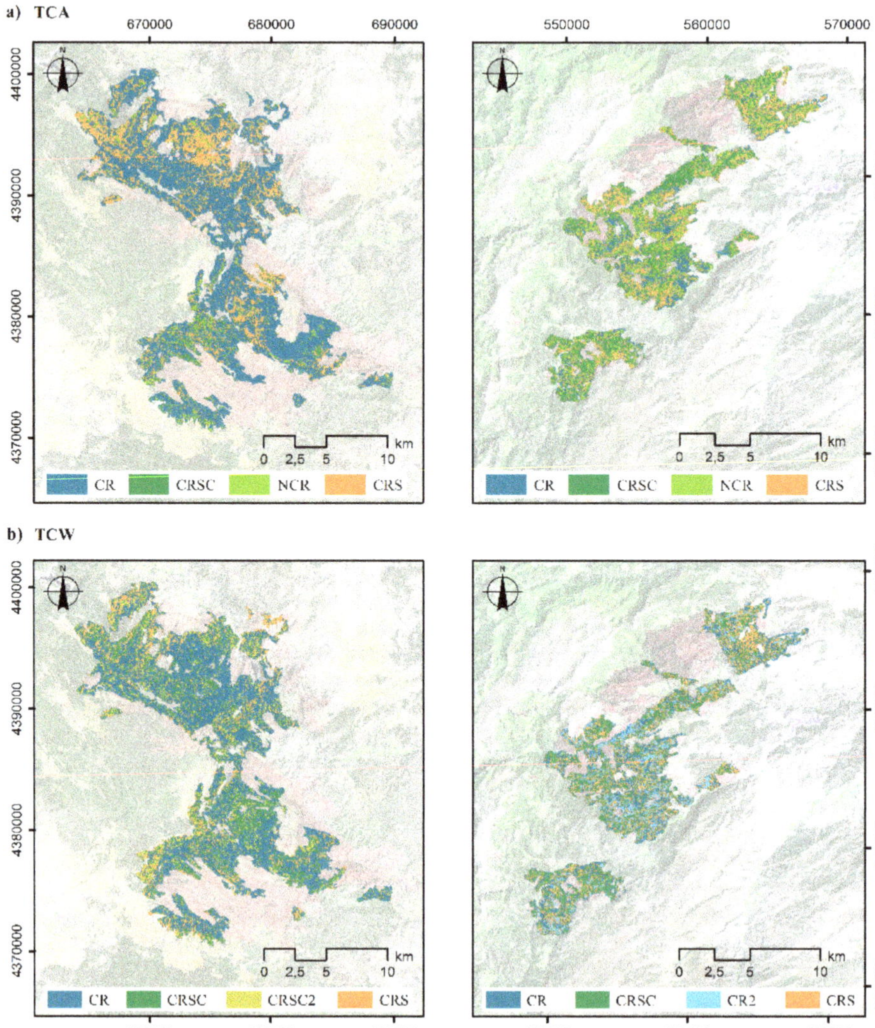

Figure 6. Maps of trajectory categories according to Tasseled Cap Angle (TCA; **a**) and Tasseled Cap Wetness (TCW; **b**). Requena (left), Yeste (right).

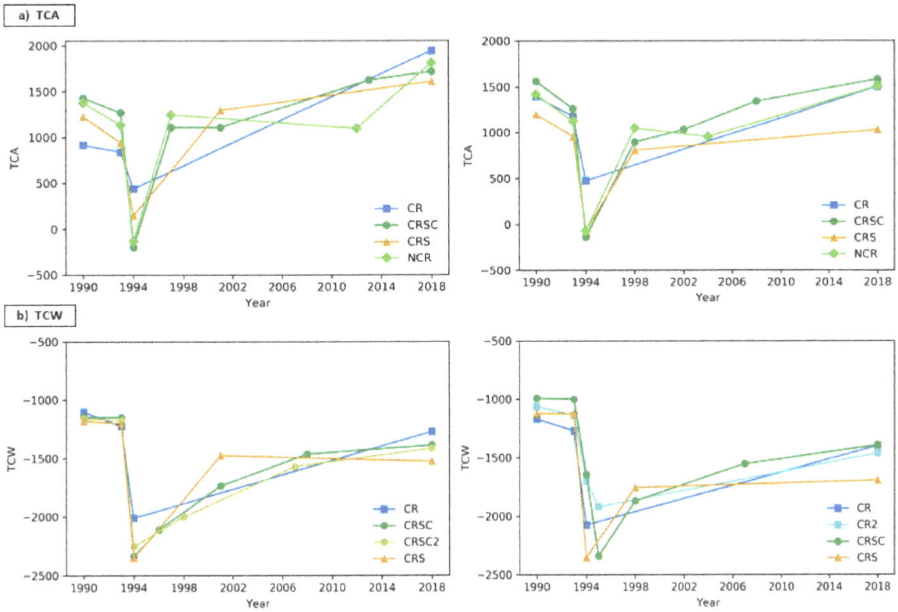

Figure 7. Time series of mean fitted trajectories for each category: (**a**) TCA and (**b**) TCW; Requena (left) and Yeste (right).

3.2. Assessing Drivers of Post-Fire Vegetation Recovery

We summarized the results from regression analysis for TCA and TCW in Tables 4 and 5, respectively. All of the aforementioned variables for predicting the Recovery Ratio ($RR\text{-}TCA_X$ and $RR\text{-}TCW_X$) at each stage were statistically significant at the 95% level. The successional stages correspond to the segments of the fitted trajectories from each category.

With regards to the relationship between the $RR\text{-}TCA_X$ and the predictor variables, different responses were found between the categories but also common trends among recovery classes (Table 4). There was a positive influence of pre-fire conditions since a higher percent of vegetation cover prior to the fire will lead to a higher percent of vegetation cover also after the fire. In addition, climate was positively related to $RR\text{-}TCA_X$ in all stages. Positive water balance resulted in a higher ratio of percent vegetation cover. The coefficients varied for the different stages since post-fire climate increases its explanatory power until stage 3 and drops in the long-term. In relation to severity, high recovery rates were related to high severities whereas low ratios were associated with low-burned pixels. Regarding topographic variables, coefficients varied according to the stage and category. Elevation showed a negative relation as a result of the limiting effects of temperature on vegetation growth, being more important in the short-term and decreasing through subsequent stages. The slope showed a weaker, negative influence as well as the aspect since cooler, wetter north-northeastern aspects (i.e., lower aspect values according to TRASP) were preferred.

Table 4. Modeling results for Tasseled Cap Angle (TCA) recovery categories at each stage.

Category	Variable	Stage 1		Stage 2		Stage 3		Stage 4	
		Coefficient	Standard Error	Coefficient	Standard Error	Coefficient	Standard Error	Coefficient	Standard Error
CR	Intercept	0.662	0.145						
	Pre-fire conditions	0.012	0.005						
	Fire severity	0.061	0.005						
	Elevation	−0.018	0.017						
	Slope	0.003	0.005						
	Aspect	−0,010	0.004						
	Drought Index	0.279	0.492						
		R^2: 0.77; Adjusted R^2: 0.76; AICc: 955.89							
CRSC	Intercept	0.651	0.142	0.384	0.351	−0.847	1.421	−0.565	0.341
	Pre-fire conditions	0.075	0.043	0.055	0.068	0.058	0.277	0.081	0.066
	Fire severity	0.553	0.046	0.265	0.332	0.064	0.362	0.122	0.327
	Elevation	−0.261	0.069	−0.135	0.143	−0.162	0.586	−0.068	0.140
	Slope	−0.046	0.040	0.025	0.058	0.157	0.238	0.043	0.057
	Aspect	0.034	0.034	0.008	0.048	−0.008	0.197	−0.015	0.047
	Drought Index	0.758	0.154	1.143	0.058	2.525	0.130	0.788	0.047
		R^2: 0.77; Adjusted R^2: 0.76; AICc: 1656..55		R^2: 0.81; Adjusted R^2: 0.80; AICc: 4031.20		R^2: 0.75; Adjusted R^2: 0.74; AICc: 11055.78		R^2: 0.73; Adjusted R^2: 0.72; AICc: 4238.93	
CRS	Intercept	−0.495	0.068	0.196	0.053				
	Pre-fire conditions	0.093	0.025	0.024	0.009				
	Fire severity	0.260	0.032	0.040	0.055				
	Elevation	−0.132	0.064	−0.001	0.032				
	Slope	−0.006	0.026	−0.004	0.012				
	Aspect	−0.024	0.022	−0.002	0.008				
	Drought Index	1.849	0.065	0.075	0.086				
		R^2: 0.88; Adjusted R^2: 0.88; AICc: 1746.11		R^2: 0.61; Adjusted R^2: 0.58; AICc: 12831.31					

Table 4. *Cont.*

Category	Variable	Stage 1		Stage 2		Stage 3		Stage 4	
		Coefficient	Standard Error	Coefficient	Standard Error	Coefficient	Standard Error	Coefficient	Standard Error
NCR	Intercept	−0.052	0.101	0.344	0.349	−0.540	0.294		
	Pre-fire conditions	0.079	0.032	0.033	0.064	0.048	0.054		
	Fire severity	0.449	0.039	−0.826	0.353	0.582	0.300		
	Elevation	−0.234	0.063	0.213	0.165	−0.201	0.141		
	Slope	−0.018	0.036	0.051	0.070	−0.001	0.059		
	Aspect	−0.041	0.029	0.021	0.051	−0.032	0.044		
	Drought Index	1.613	0.110	1.065	0.060	0.845	0.231		
		R^2: 0.83; Adjusted R^2: 0.82; AICc: 1527.51		R^2: 0.78; Adjusted R^2: 0.77; AICc: 7082.34		R^2: 0.74; Adjusted R^2: 0.72; AICc: 6085.56			

Table 5. Modeling results for Tasseled Cap Wetness (TCW) recovery categories at each stage.

Category	Variable	Stage 1		Stage 2		Stage 3		Stage 4	
		Coefficient	Standard Error	Coefficient	Standard Error	Coefficient	Standard Error	Coefficient	Standard Error
CR	Intercept	0.416	0.040						
	Pre-fire conditions	−0.008	0.002						
	Fire severity	0.023	0.002						
	Elevation	−0.002	0.005						
	Slope	−0.005	0.002						
	Aspect	0.013	0.002						
	Drought Index	0.121	0.446						
		R^2: 0.71; Adjusted R^2: 0.69; AICc: 955.23							
CR2	Intercept	0.025	0.101	−0.076	0.046				
	Pre-fire conditions	−0.312	0.046	0.049	0.020				
	Fire severity	0.106	0.046	0.006	0.019				
	Elevation	−0.158	0.065	0.027	0.036				
	Slope	0.072	0.036	0.004	0.015				
	Aspect	−0.171	0.039	0.025	0.016				
	Drought Index	0.734	0.156	−0.609	0.170				
		R^2: 0.80; Adjusted R^2: 0.79; AICc: 7870.95		R^2: 0.92; Adjusted R^2: 0.92; AICc: 1227.11					

Table 5. *Cont.*

Category	Variable	Stage 1		Stage 2		Stage 3		Stage 4	
		Coefficient	Standard Error	Coefficient	Standard Error	Coefficient	Standard Error	Coefficient	Standard Error
CRSC	Intercept	−0.322	0.120	0.612	0.214	0.358	0.229	−0.119	0.167
	Pre-fire conditions	−0.116	0.051	−0.091	0.116	−0.008	0.126	0.024	0.092
	Fire severity	0.094	0.044	0.206	0.099	0.140	0.108	0.053	0.079
	Elevation	−0.040	0.072	−0.098	0.248	−0.005	0.270	−0.059	0.198
	Slope	0.041	0.038	0.004	0.091	0.021	0.099	0.014	0.073
	Aspect	−0.148	0.031	0.103	0.070	0.106	0.077	0.050	0.056
	Drought Index	0.122	0.126	0.437	0.088	0.737	0.058	0.640	0.052
		R^2: 0.88; Adjusted R^2: 0.87; AICc: 1959.93		R^2: 0.67; Adjusted R^2: 0.65; AICc: 15831.16		R^2: 0.62; Adjusted R^2: 0.60; AICc: 17032.23		R^2: 0.56; Adjusted R^2: 0.54; AICc: 13009.93	
CRSC2	Intercept	0.765	0.065	0.324	0.198	0.099	0.047		
	Pre-fire conditions	−0.061	0.041	0.085	0.113	−0.008	0.026		
	Fire severity	0.045	0.039	0.139	0.106	0.018	0.025		
	Elevation	−0.127	0.062	−0.166	0.247	−0.017	0.057		
	Slope	0.014	0.030	0.030	0.087	−0.009	0.020		
	Aspect	−0.046	0.027	0.195	0.070	0.025	0.016		
	Drought Index	0.730	0.039	0.735	0.077	−0.150	0.075		
		R^2: 0.81; Adjusted R^2: 0.80; AICc: 1064.20		R^2: 0.69; Adjusted R^2: 0.67; AICc: 11669.50		R^2: 0.70; Adjusted R^2: 0.69; AICc: 1451.47			
CRS	Intercept	0.146	0.061	−0.030	0.074				
	Pre-fire conditions	−0.047	0.028	−0.002	0.031				
	Fire severity	0.014	0.028	0.045	0.031				
	Elevation	−0.030	0.058	−0.005	0.093				
	Slope	0.018	0.022	−0.007	0.026				
	Aspect	−0.021	0.019	0.034	0.021				
	Drought Index	0.244	0.095	0.012	0.121				
		R^2: 0.94; Adjusted R^2: 0.93; AICc: 2685.86		R^2: 0.78; Adjusted R^2: 0.77; AICc: 1967.958					

The relationship between the RR-TCWx and the predictor variables showed different responses reflecting changes in forest structural complexity (Table 5). In this case, a negative relationship with pre-fire values was found because the greater complexity of the forest structure (i.e., higher TCW), the lower the recovery ratio, indicating slower recovery processes of remnant trees and seeding in contrast to the quicker recovery of shrubs. With regard to severity higher values also lead to higher recovery rates. Concerning topographic variables, the slightly positive relationship of the slope might be attributed to pine forest distribution preferably in the foothills. Identically to TCA, the elevation and aspect relationship tended to be negative although in the mid-term stages south-southwestern aspects lead to higher recovery rates. Post-fire climate conditions showed a positive or negative relationship depending on the stage. Enough available moisture post-fire is important for seed germination but abrupt changes in the climate conditions seem to reduce the recovery ratio in the long-term.

The relative importance of the explanatory variables for each trajectory category was assessed through t-statistics (Figures 8 and 9). Post-fire climate in terms of drought had high predictive power in most of the stages and categories according to both TCA and TCW regression analysis. This power increased in the mid-term and long-term while all other variables decreased. Fire severity had the second largest power for explaining percent vegetation cover in the short-term, although its power diminished in the estimation of the recovery in terms of forest structure. Pre-fire conditions also had a higher importance in the short-term, since the seed bank and the seeding processes will depend on the pre-fire forest cover and structure. Topographic variables showed the lowest explanation power. Elevation had higher relative importance in the case of TCA compared to TCW, in which aspect was the most important topographic variable in all stages.

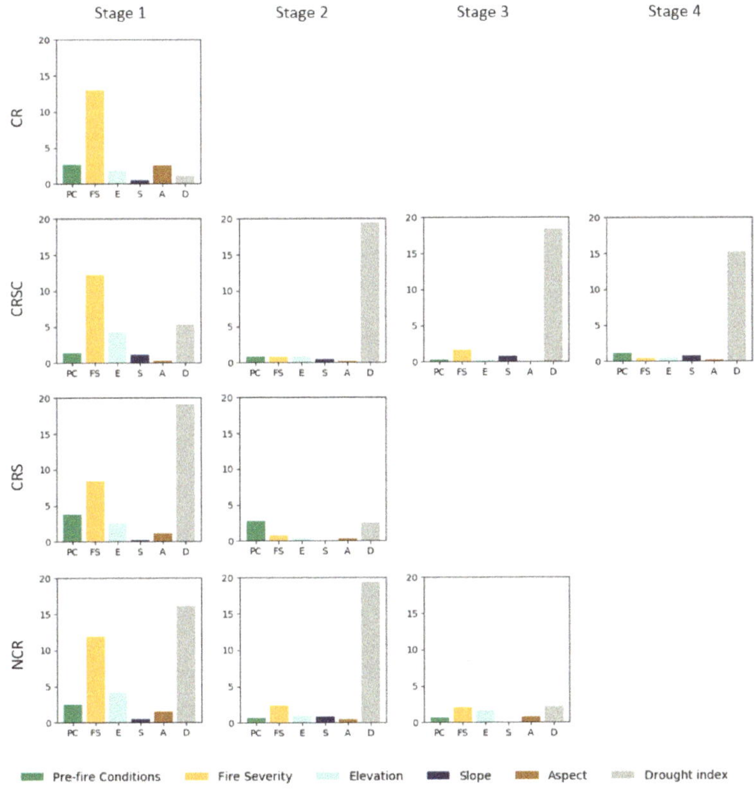

Figure 8. Relative importance of explanatory variables in TCA regression analysis.

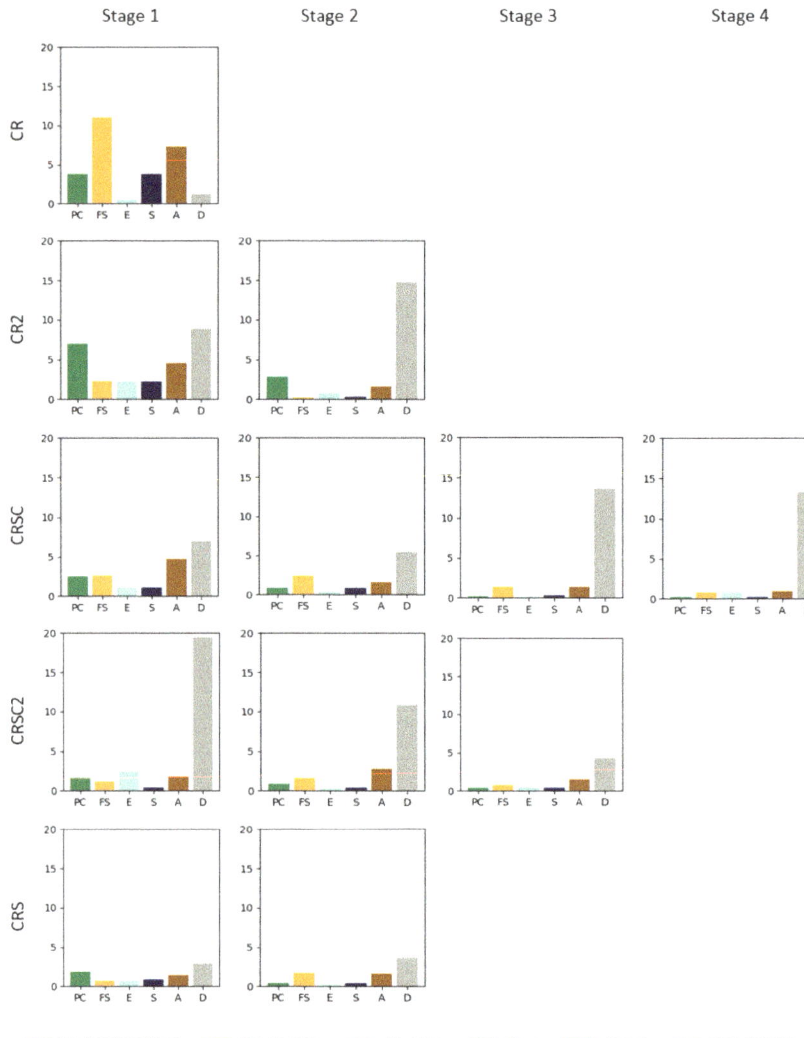

Figure 9. Relative importance of explanatory variables in TCW regression analysis.

3.3. Recovery Estimation Assessment

According to the recovered and non-recovered definitions, all categories showed high accuracy, with OA values ranging from 0.7 to 0.94 (Table 6). The OE and CE of the recovery categories varied, with the highest errors in the short-term (OE of 0.36 for TCW and CE of 0.22 for TCA). Post-fire recovery estimated by TCA shows more balanced errors, although with higher CE due to early successional recovery processes of herbs and shrubs. In contrast, recovery estimated by means of TCW shows higher OE since it is more sensitive to moisture and structure than early soil colonization of herbs. In both cases, more stable recovery classes (CR and CRS) showed higher accuracy than the more disrupted ones (CRSC). DC increased through the time series since forest cover and structure were more clearly defined.

Table 6. Errors metrics for recovery estimation according to TCA and TCW.

Category		2002				2009–2010				2017–2018			
		OA	DC	OE	CE	OA	DC	OE	CE	OA	DC	OE	CE
TCA	CR	0.86	0.91	0.07	0.11	0.90	0.94	0.04	0.07	0.95	0.97	0.00	0.05
	CRSC	0.70	0.84	0.14	0.19	0.82	0.90	0.09	0.11	0.87	0.93	0.07	0.07
	CRS	0.79	0.86	0.05	0.22	0.86	0.92	0.08	0.08	0.94	0.97	0.02	0.05
	NCR	0.74	0.81	0.16	0.20	0.88	0.93	0.10	0.04	0.92	0.96	0.03	0.05
TCW	CR	0.91	0.85	0.21	0.04	0.84	0.88	0.20	0.02	0.93	0.96	0.08	0.00
	CR2	0.82	0.74	0.36	0.13	0.86	0.85	0.21	0.08	0.89	0.93	0.13	0.00
	CRSC	0.82	0.78	0.33	0.06	0.87	0.89	0.15	0.01	0.82	0.91	0.16	0.01
	CRSC2	0.82	0.82	0.29	0.04	0.86	0.91	0.16	0.00	0.88	0.92	0.14	0.00
	CRS	0.86	0.87	0.17	0.10	0.77	0.93	0.12	0.01	0.89	0.93	0.10	0.04

OA—Overall Accuracy; DC—Dice coefficient; OE—omission error and CE—commission error.

4. Discussion

4.1. Post-Fire Recovery Trajectories from LTS

Characterizing post-fire recovery processes is challenging due to the variety of factors driving vegetation recovery, resulting in different recovery dynamics. Post-fire vegetation recovery estimated from spectral data is not a direct measure of actual forest regrowth. However, trends of forest recovery can be quantified by linking spectral change metrics with a reference dataset [50]. Here, we identified different recovery categories according to TCA and TCW recovery ratios at different stages through a 24-year-series in Mediterranean pine forests.

Time series analysis from Landsat data using the LandTrendr segmentation algorithm has been suitable to capture the different post-fire recovery trends. The trajectories extracted revealed continuous and non-continuous recovery processes, allowing us to identify slight changes in the slowly evolving recovery process. This was of great importance in determining slow but more stable recovery processes (CR and CR2) compared to other faster, but also interrupted, recovery processes (CRSC, CRSC2, CRS and NCR), indicating changes in greenery and forest structure throughout the recovering process. Other studies that also employed a Landsat trajectory-based approach with LandTrendr were able to identify different patterns of vegetative regrowth depending on the state, owner category and ecoregion in North America [25] or set recovery levels across sclerophyll forest in Australia [36]. Even though several studies have addressed the analysis of post-fire recovery trends, our findings highlight the importance of defining and grouping recovery patterns to facilitate the understanding of recovery processes.

TCA and TCW have been proved useful to characterize both vegetation cover (TCA) and forest structure (TCW). TCW trajectories were much more gradual, while TCA trajectories tended to saturate around 5 years post-disturbance. Frazier et al. (2015) [44] and Nguyen et al. (2018) [36] also concluded that the TCW contained more detail on the vegetation structure and regrowth in the regeneration processes since it is highly correlated to stand age and structural complexity in mature forest stands. The wetness values rise with an increasing amount of canopy [45], making it more suitable for analyzing mid and long-term recovery. Alternatively, initial increases in vegetation cover were well-characterized with shorter visible and near-infrared wavelengths as used in TCA [43,47,80]. In contrast to the canopy layer, recovery of the understory through both resprouting and seeding is quicker [8,54]. Thus, TCA was more suitable for tracking early stages, suggesting greater sensitivity toward shrub recovery and changes in vegetation condition rather than structure.

The Recovery Ratio (RR) varied across the stages and for each category (Figure 7). Generally, the recovery rates of TCA and TCW were greatest shortly after the fire (Stage 1 in CRSC, CRSC2, CRS and NCR), decreasing afterwards, due to the early post-fire colonization of annual herbs and shrubs [8,56]. In the following stages (2, 3 and 4), the RR according to both TCA and TCW was lower, associated with stem exclusion processes in a young regrowth forest characterized by intense competition among regenerated species [7,49,53]. Our results agree with other studies that also obtained higher recovery rates in the short-term according to spectral vegetation indices in Mediterranean forests [40] and NBR trajectories in pine, mixed conifer and conifer–oak forest [28].

Accordingly, the recovery time for the fitted mean trajectories also varied across categories. TCA categories tended to reach mean pre-fire values quicker, as it tends to saturate earlier (short-mid-term) due to the influence of herbaceous vegetation on TCA. Specifically, in the Requena Fire, the TCA trajectories tended to overtake pre-fire values in the long-term, possibly because these are fire-adapted forests in which fire creates favorable conditions for vegetation germination and regeneration [7]. Nevertheless, TCW trajectories did not reach the previous values 24 years after the fire, suggesting that burned areas did not recover the complexity of the pre-fire forest structure. This agrees with the interval of minimum 15 years to consider a *Pinus halepensis* forest recovered after a fire proposed by Eugenio et al. (2006) [54] since post-fire populations of seeder species does not overpass the reproductive juvenile phase up to 12–20 years after fire [81] and thus, the canopy seed bank is not completely fulfilled [12]. Some studies in Mediterranean ecosystems reported recovery times from remote sensing, which fit with our findings [82]. In this sense, Gouveia et al. (2010) [69] found recovery times of vegetation cover around 3–5 years according to NDVI (highly correlated with TCA), and Fernández-Manso et al. (2016) [17] estimated the time of vegetation cover with VRI between 7 and 20 years depending on fire severity level. However, the estimated recovery times with NBR, which is highly correlated with TCW, were generally longer compared to NDVI [40].

Some of the limitations for the modeling of post-fire vegetation recovery using optical data are related to saturation at high biomass levels. Previous studies found that the saturation of optical indices is reached after 20 years in Mediterranean environments [83]. Likewise, Schroeder et al. (2011) [80], Pickell et al. (2016) [19], Viana-Soto et al. (2017) [37] and Hislop et al. (2018) [38] found post-fire recovery of NDVI or TCA returning back to pre-fire levels rapidly (i.e., around 5–7 years post-fire). Structural information derived from airborne LiDAR data would enable a better characterization of the recovery trajectories and to improve our understanding post-fire vegetation recovery [84,85].

Accuracy Assessment of Post-Fire Recovery

Accuracy assessment of post-fire recovery is often avoided because historical reference datasets are scarce and field data is costly and time-consuming [14]. Here, we used a human interpretation approach to derive reference data of recovered and non-recovered areas, which has been widely employed to derive reference data for disturbance and recovery mapping [23,36,78,86].

All classes showed high accuracy with increasing OA and DC from the short-term to the long-term since forest cover and structure are more clearly defined 24 years post-fire. DC values were slightly lower in the long-term for TCW compared to TCA, which might indicate a recovery process that evolves to a secondary forest with higher shrub domain [12]. This could also be related to the uncertainty associated with signal sensitivity to changes in vegetation cover and biomass [87]. Moreover, the highest accuracy in the more stable categories (CR), in contrast to the more variable (CRSC), also highlights the difficulty in establishing the level of recovery for those more dynamic areas. Nguyen et al. (2018) [36] also pointed out the challenge of determining the post-disturbance recovery level, and identified recovery levels from NBR trajectories after fires according to whether pre-fire conditions were reached or not. Here, we did not distinguish among recovery levels as we were unable to accurately characterize the pre-fire forest structure due to the lack of high-resolution imagery, for which a more extensive reference dataset would also be needed.

The main source of error stemming from OE for both TCA and TCW in the recovered pixels. Our results showed that the TCW had higher accuracy for the non-recovered areas but omitted pixels that had already recovered. Zhao et al. (2016) [78] also found higher OE in the recovered class regarding post-fire and post-harvest forest recovery. Similarly, DeVries et al. (2015) [23] assessed post-harvest regrowth in tropical forests obtaining lower CE in the regrowth class but the highest OE. On the other hand, TCA showed higher CE, as it is more sensitive to fast detection of early recovering processes but also tends to saturate earlier as Schroeder et al. (2011) [80] reported.

4.2. Assessment of Post-Fire Recovery Drivers

Regression modeling of TCA and TCW recovery ratios for the recovery categories identified showed a varied influence of environmental factors, fire severity and pre-fire conditions. The results indicated post-fire climate as one of the most important factors for vegetation recovery in Mediterranean pine forests in Spain. Likewise, Meng et al. (2015) [16], Liu (2016) [88] and Viana-Soto et al. (2017) [37] found that climate conditions in the first post-fire seasons were critical for predicting short-term recovery. Tree regeneration after disturbances in Mediterranean ecosystems could be limited under post-fire drought events since droughts constrain seedling establishment and growth [7,53]. Bright et al. (2019) [28] also reported that post-fire climate explained substantial variation of the degree to which vegetation greenness recovered after a fire. Further analysis between trajectories and post-fire climate revealed that stages of recovery slowdown and even breakpoints coincided with negative SPEI values (i.e., dry or very dry periods). The year of the fires was followed by a slightly humid period, which supported the recovery. However, 5 years after the fires, a new drought event interrupted the recovery, as can be observed very clearly in the stabilization of the recovery in the CRS category from 1999–2000, not only in the TCA trajectories but also in TCW trajectories. The effect of this drought event was also noticeable in the categories of CRSC and CRSC2. Furthermore, the impact of post-fire climate on the recovery process was also shown in the mid-term and long-term. In the case of the Yeste Fire, an extreme drought event in 2005 coincided with the breakpoints in NCR and CRSC, whereas in the Requena fire this event was not as severe as it was in 2012, coinciding with the breakpoints in NCR and CRSC2.

Fire severity was also a key factor in the short-term recovery of pine forests. Some studies also found that fire severity was decisive in recovery dynamics both in mixed-conifer forests [16] and pine forests [18] due to its relation to pine seedling densities after fire, depending on pre-fire vegetation composition, seedling mortality and reestablishment processes [8,53,56]. In this study, the areas were burned at high severity (dNBR > 0.66) as forested areas tended to experience a higher severity compared to herbaceous and shrublands. Moderate-high severity was only found in those pixels of CR, which showed a slower but stable recovery trend. In agreement with the results reported by Shvetsov et al. (2019) [21], a positive relationship between recovery rate and fire severity was found, since recovery rates were higher for the higher severity areas than for high-moderate severity (corresponding to successful recovery). Moderate fire severity sites might result in higher soil organic matter mineralization, and thus in higher post-fire soil fertility that produces faster growth in pine seedlings [89]. Bright et al. (2019) [28] also found that areas burned at higher severities recovered at faster rates, possibly because they are fire-adapted forest in which fire creates favorable conditions for vegetation germination and regeneration. Nevertheless, some studies in Mediterranean pine forest reported that conversion from forest to shrubland occurred in the most xeric sites (south-facing areas) [48] or in those areas with a high severity [12]. In this sense, Baudena et al. (2019) [90] predicted that future potential increases in aridity may drive these fire-prone ecosystems past a tipping point, after which closed forest structure would be replaced by open shrublands.

Topographic variables can also influence post-fire vegetation recovery through its effects on local microclimate, soil and hydrological processes [18,20]. Wittenberg et al. (2007) [91] and Ireland and Petropulous (2015) [18] found that north facing aspects exhibit higher rates of vegetation recovery compared to south facing aspects as we found in the short-term recovery in CRS, NCR and CR categories according to TCA, but also in CRSC according to TCW. The negative influence of elevation was also detected in Mediterranean pine forest [37] and red fir forests [16] that might be attributed to the decreased temperature with elevation. In contrast, Chu et al. (2017) [20] and Shvetsov et al. (2019) [21] reported that topographic variables were the least important factors in explaining the regeneration rate in Siberian forests. In our study, recovery in relation to topographic position did not show any clear pattern. This could be due to the fact that most of the pixels were located at either upper or mid-slope in the foothills and very few were bottom slopes.

Although we found that post-fire climate was the most important variable in explaining post-fire recovery in the mid and long-term, other variables not included could be influencing the recovery process. Further analysis would consider the influence of historical management legacies as well as the distance to seed banks in the post-fire recovery patterns.

5. Conclusions

Time series analysis from a temporal segmentation approach allowed us to unravel and characterize different post-fire recovery trajectories. Although several studies have addressed the estimation of post-fire recovery rates, fewer have been done in defining and characterizing the differences among recovery trends in Mediterranean pine forests. Here, we identified different recovery categories according to TCA trajectories and TCW trajectories, which enabled us to define slow but more stable recovery processes (CR and CR2) compared to other faster but also interrupted recovery processes (CRSC, CRSC2, CRS and NCR). The appraisal of the environmental and contextual drivers of the recovery process showed that fire severity is important to predict the RR in the short-term but post-fire climate in terms of drought better explained the RR in the mid and long-term.

The thermophilous pine forests are the most affected by wildfires in Europe. Increased wildfire activity is expected to continue under warmer and drier conditions, making post-fire vegetation recovery of concern to researchers and forest managers. Since these forests may not be allowed the time to develop into a mature forest that would be able to recover rapidly, the resilience of these ecosystems will therefore be significantly reduced. Hence, a better understanding of fire regimes and forest recovery patterns in different environmental and climatic conditions is needed for developing forest management strategies that enhance forest resilience.

Author Contributions: Conceptualization, A.V.-S., I.A., J.S. and M.G.; methodology, A.V.-S., I.A. and M.G.; formal analysis, A.V.-S.; writing—original draft preparation, A.V.-S.; writing—review and editing, A.V.-S., I.A., J.S. and M.G. All authors have read and agreed to the published version of the manuscript.

Funding: This research was supported by the Spanish Ministry of Science, Innovation and Universities through a FPU doctoral fellowship (FPU17/03260), and by the Department of Economy and Competitiveness under project SERGISAT [CGL2014-57013-C2-1-R].

Conflicts of Interest: The authors declare no conflict of interest.

References

1. Bowman, D.M.J.S.; Balch, J.K.; Artaxo, P.; Bond, W.J.; Carlson, J.M.; Cochrane, M.A.; D'Antonio, C.M.; DeFries, R.S.; Doyle, J.C.; Harrison, S.P.; et al. Fire in the earth system. *Science* **2009**, *324*, 481–484. [CrossRef] [PubMed]
2. Chuvieco, E. Global impacts of fire. In *Earth Observation of Wildland Fires in Mediterranean Ecosystems*; Chuvieco, E., Ed.; Springer: Berlin, Germany, 2009; pp. 1–10. ISBN 9783642017537.
3. Aponte, C.; de Groot, W.J.; Wotton, M. Forest fires and climate change: Causes, consequences and management options. *Int. J. Wildl. Fire* **2016**, *25*, 861–875. [CrossRef]
4. San-Miguel-Ayanz, J.; Durrant, T.; Boca, R.; Liberta', G.; Branco, A.; De Rigo, D.; Ferrari, D.; Maianti, P.; Artes Vivancos, T.; Pfeiffer, H.; et al. *Forest Fires in Europe, Middle East and North Africa 2018*; Publications Office of the European Union: Luxembourg, 2019; ISBN 9789279928321.
5. San-Miguel-Ayanz, J.; Moreno, J.M.; Camia, A. Analysis of large fires in European Mediterranean landscapes: Lessons learned and perspectives. *For. Ecol. Manag.* **2013**, *294*, 11–22. [CrossRef]
6. Keeley, J.E.; Bond, W.J.; Bradstock, R.A.; Pausas, J.G.; Rundel, P.W. *Fire in Mediterranean Ecosystems: Ecology, Evolution and Management*; Cambridge University Press: Cambridge, UK, 2014; Volume 10, ISBN 9780521824910.

7. De las Heras, J.; Moya, D.; Vega, J.A.; Daskalakou, E.; Vallejo, R.; Grigoriadis, N.; Tsitsoni, T.; Baeza, J.; Valdecantos, A.; Fernández, C.; et al. Post-fire management of serotinous pine forests. In *Post-Fire Management and Restoration of Southern European Forests*; Moreira, F., Arianoutsou, M., Corona, P., de las Heras, J., Eds.; Springer: Berlin, Germany, 2012; Volume 24, pp. 121–149. ISBN 978-94-007-2207-1.
8. González-De Vega, S.; de las Heras, J.; Moya, D. Post-fire regeneration and diversity response to burn severity in Pinus halepensis Mill. Forests. *Forests* **2018**, *9*, 299. [CrossRef]
9. Pausas, J.G.; Keeley, J.E. A burning story: The role of fire in the history of life. *Bioscience* **2009**, *59*, 593–601. [CrossRef]
10. Stephens, S.L.; Agee, J.K.; Fulé, P.Z.; North, M.P.; Romme, W.H.; Swetnam, T.W.; Turner, M.G. Managing forests and fire in changing climates. *Science* **2013**, *342*, 41–42. [CrossRef]
11. Lindner, M.; Maroschek, M.; Netherer, S.; Kremer, A.; Barbati, A.; Garcia-Gonzalo, J.; Seidl, R.; Delzon, S.; Corona, P.; Kolström, M.; et al. Climate change impacts, adaptive capacity, and vulnerability of European forest ecosystems. *For. Ecol. Manag.* **2010**, *259*, 698–709. [CrossRef]
12. González-De Vega, S.; De las Heras, J.; Moya, D. Resilience of Mediterranean terrestrial ecosystems and fire severity in semiarid areas: Responses of Aleppo pine forests in the short, mid and long term. *Sci. Total Environ.* **2016**, *573*, 1171–1177. [CrossRef]
13. Gitas, I.; Mitri, G.; Veraverbeke, S.; Polychronaki, A. Advances in remote sensing of post-fire vegetation recovery monitoring—A review. In *Remote Sensing of Biomass—Principles and Applications*; IntechOpen: London, UK, 2012.
14. Banskota, A.; Kayastha, N.; Falkowski, M.J.; Wulder, M.A.; Froese, R.E.; White, J.C. Forest monitoring using Landsat time series data: A review. *Can. J. Remote Sens.* **2014**, *40*, 362–384. [CrossRef]
15. Wulder, M.A.; Loveland, T.R.; Roy, D.P.; Crawford, C.J.; Masek, J.G.; Woodcock, C.E.; Allen, R.G.; Anderson, M.C.; Belward, A.S.; Cohen, W.B.; et al. Current status of Landsat program, science, and applications. *Remote Sens. Environ.* **2019**, *225*, 127–147. [CrossRef]
16. Meng, R.; Dennison, P.E.; Huang, C.; Moritz, M.A.; D'Antonio, C. Effects of fire severity and post-fire climate on short-term vegetation recovery of mixed-conifer and red fir forests in the Sierra Nevada Mountains of California. *Remote Sens. Environ.* **2015**, *171*, 311–325. [CrossRef]
17. Fernandez-Manso, A.; Quintano, C.; Roberts, D.A. Burn severity influence on post-fire vegetation cover resilience from Landsat MESMA fraction images time series in Mediterranean forest ecosystems. *Remote Sens. Environ.* **2016**, *184*, 112–123. [CrossRef]
18. Ireland, G.; Petropoulos, G.P. Exploring the relationships between post-fire vegetation regeneration dynamics, topography and burn severity: A case study from the Montane Cordillera Ecozones of Western Canada. *Appl. Geogr.* **2015**, *56*, 232–248. [CrossRef]
19. Pickell, P.D.; Hermosilla, T.; Frazier, R.J.; Coops, N.C.; Wulder, M.A. Forest recovery trends derived from Landsat time series for North American boreal forests. *Int. J. Remote Sens.* **2016**, *37*, 138–149. [CrossRef]
20. Chu, T.; Guo, X.; Takeda, K. Effects of burn severity and environmental conditions on post-fire regeneration in Siberian Larch forest. *Forests* **2017**, *8*, 76. [CrossRef]
21. Shvetsov, E.G.; Kukavskaya, E.A.; Buryak, L.V.; Barrett, K. Assessment of post-fire vegetation recovery in Southern Siberia using remote sensing observations. *Environ. Res. Lett.* **2019**, *14*, 055001. [CrossRef]
22. Griffiths, P.; Kuemmerle, T.; Baumann, M.; Radeloff, V.C.; Abrudan, I.V.; Lieskovsky, J.; Munteanu, C.; Ostapowicz, K.; Hostert, P. Forest disturbances, forest recovery, and changes in forest types across the carpathian ecoregion from 1985 to 2010 based on landsat image composites. *Remote Sens. Environ.* **2014**, *151*, 72–88. [CrossRef]
23. DeVries, B.; Decuyper, M.; Verbesselt, J.; Zeileis, A.; Herold, M.; Joseph, S. Tracking disturbance-regrowth dynamics in tropical forests using structural change detection and Landsat time series. *Remote Sens. Environ.* **2015**, *169*, 320–334. [CrossRef]
24. Lhermitte, S.; Verbesselt, J.; Verstraeten, W.W.; Veraverbeke, S.; Coppin, P. Assessing intra-annual vegetation regrowth after fire using the pixel based regeneration index. *ISPRS J. Photogramm. Remote Sens.* **2011**, *66*, 17–27. [CrossRef]
25. Kennedy, R.E.; Yang, Z.; Cohen, W.B.; Pfaff, E.; Braaten, J.; Nelson, P. Spatial and temporal patterns of forest disturbance and regrowth within the area of the Northwest Forest Plan. *Remote Sens. Environ.* **2012**, *122*, 117–133. [CrossRef]

26. White, J.C.; Wulder, M.A.; Hermosilla, T.; Coops, N.C.; Hobart, G.W. A nationwide annual characterization of 25 years of forest disturbance and recovery for Canada using Landsat time series. *Remote Sens. Environ.* **2017**, *194*, 303–321. [CrossRef]
27. Frazier, R.J.; .Coops, N.C.; Wulder, M.A.; Hermosilla, T.; White, J.C. Analyzing spatial and temporal variability in short-term rates of post-fire vegetation return from Landsat time series. *Remote Sens. Environ.* **2018**, *205*, 32–45. [CrossRef]
28. Bright, B.C.; Hudak, A.T.; Kennedy, R.E.; Braaten, J.D.; Henareh Khalyani, A. Examining post-fire vegetation recovery with Landsat time series analysis in three western North American forest types. *Fire Ecol.* **2019**, *15*, 8. [CrossRef]
29. Zhu, Z. Change detection using landsat time series: A review of frequencies, preprocessing, algorithms, and applications. *ISPRS J. Photogramm. Remote Sens.* **2017**, *130*, 370–384. [CrossRef]
30. Röder, A.; Hill, J.; Duguy, B.; Alloza, J.A.; Vallejo, R. Using long time series of Landsat data to monitor fire events and post-fire dynamics and identify driving factors. A case study in the Ayora region (eastern Spain). *Remote Sens. Environ.* **2008**, *112*, 259–273. [CrossRef]
31. Hope, A.; Tague, C.; Clark, R. Characterizing post-fire vegetation recovery of California chaparral using TM/ETM+ time-series data. *Int. J. Remote Sens.* **2007**, *28*, 1339–1354. [CrossRef]
32. Kennedy, R.E.; Yang, Z.G.; Cohen, W.B. Detecting trends in forest disturbance and recovery using yearly Landsat time series: 1. LandTrendr—Temporal segmentation algorithms. *Remote Sens. Environ.* **2010**, *114*, 2897–2910. [CrossRef]
33. Huang, C.; Goward, S.N.; Masek, J.G.; Thomas, N.; Zhu, Z.; Vogelmann, J.E. An automated approach for reconstructing recent forest disturbance history using dense Landsat time series stacks. *Remote Sens. Environ.* **2010**, *114*, 183–198. [CrossRef]
34. Verbesselt, J.; Hyndman, R.; Newnham, G.; Culvenor, D. Detecting trend and seasonal changes in satellite image time series. *Remote Sens. Environ.* **2010**, *114*, 106–115. [CrossRef]
35. Zhu, Z.; Woodcock, C.E.; Olofsson, P. Continuous monitoring of forest disturbance using all available Landsat imagery. *Remote Sens. Environ.* **2012**, *122*, 75–91. [CrossRef]
36. Nguyen, T.H.; Jones, S.D.; Soto-Berelov, M.; Haywood, A.; Hislop, S. A spatial and temporal analysis of forest dynamics using Landsat time-series. *Remote Sens. Environ.* **2018**, *217*, 461–475. [CrossRef]
37. Viana-Soto, A.; Aguado, I.; Martínez, S. Assessment of Post-Fire Vegetation Recovery Using Fire Severity and Geographical Data in the Mediterranean Region (Spain). *Environments* **2017**, *4*, 90. [CrossRef]
38. Hislop, S.; Jones, S.; Soto-Berelov, M.; Skidmore, A.; Haywood, A.; Nguyen, T.H. Using Landsat spectral indices in time-series to assess wildfire disturbance and recovery. *Remote Sens.* **2018**, *10*, 460. [CrossRef]
39. Key, C.H.; Benson, N.C. Landscape assessment: Remote sensing of severity, the Normalized Burn Ratio. In *FIREMON: Fire Effects Monitoring and Inventory System*; USDA Forest Service, Rocky Mountain Research Station, Fort Collins: Denver, CO, USA, 2006; pp. 305–325.
40. Morresi, D.; Vitali, A.; Urbinati, C.; Garbarino, M. Forest spectral recovery and regeneration dynamics in stand-replacing wildfires of central Apennines derived from Landsat time series. *Remote Sens.* **2019**, *11*, 308. [CrossRef]
41. Crist, E.P. A TM Tasseled Cap equivalent transformation for reflectance factor data. *Remote Sens. Environ.* **1985**, *17*, 301–306. [CrossRef]
42. Gómez, C.; White, J.C.; Wulder, M.A. Characterizing the state and processes of change in a dynamic forest environment using hierarchical spatio-temporal segmentation. *Remote Sens. Environ.* **2011**, *115*, 1665–1679. [CrossRef]
43. Pflugmacher, D.; Cohen, W.B.; Kennedy, R.E.; Yang, Z. Using Landsat-derived disturbance and recovery history and lidar to map forest biomass dynamics. *Remote Sens. Environ.* **2014**, *151*, 124–137. [CrossRef]
44. Frazier, R.J.; Coops, N.C.; Wulder, M.A. Boreal Shield forest disturbance and recovery trends using Landsat time series. *Remote Sens. Environ.* **2015**, *170*, 317–327. [CrossRef]
45. Hansen, M.J.; Franklin, S.E.; Woudsma, C.; Peterson, M. Forest structure classification in the North Columbia mountains using the Landsat TM Tasseled Cap wetness component. *Can. J. Remote Sens.* **2001**, *27*, 20–32. [CrossRef]
46. Powell, S.L.; Cohen, W.B.; Healey, S.P.; Kennedy, R.E.; Moisen, G.G.; Pierce, K.B.; Ohmann, J.L. Quantification of live aboveground forest biomass dynamics with Landsat time-series and field inventory data: A comparison of empirical modeling approaches. *Remote Sens. Environ.* **2010**, *114*, 1053–1068. [CrossRef]

47. Gómez, C.; Wulder, M.A.; White, J.C.; Montes, F.; Delgado, J.A. Characterizing 25 years of change in the area, distribution, and carbon stock of Mediterranean pines in Central Spain. *Int. J. Remote Sens.* **2012**, *33*, 5546–5573. [CrossRef]
48. Martín-Alcón, S.; Coll, L. Unraveling the relative importance of factors driving post-fire regeneration trajectories in non-serotinous Pinus nigra forests. *For. Ecol. Manag.* **2016**, *361*, 13–22. [CrossRef]
49. Oliver, C.; Larson, B. *Forest Stand Dynamics (Update Edition)*; Wiley: New York, NY, USA, 1996; Volume 42.
50. Bartels, S.F.; Chen, H.Y.H.; Wulder, M.A.; White, J.C. Trends in post-disturbance recovery rates of Canada's forests following wildfire and harvest. *For. Ecol. Manag.* **2016**, *361*, 194–207. [CrossRef]
51. Rivas-Martínez, S. Etages bioclimatiques, secteurs chorologiques et séries de végétation de l'Espagne méditerranéenne. *An. del Jardín Botánico Madr.* **1981**, *37*, 251–268.
52. Ministerio de Agricultura Pesca y Alimentación. Second National Forest Inventory of Spain. Available online: https://www.miteco.gob.es/es/biodiversidad/servicios/banco-datos-naturaleza/informacion-disponible/ifn2.aspx (accessed on 27 January 2020).
53. Fernández-García, V.; Fulé, P.Z.; Marcos, E.; Calvo, L. The role of fire frequency and severity on the regeneration of Mediterranean serotinous pines under different environmental conditions. *For. Ecol. Manag.* **2019**, *444*, 59–68. [CrossRef]
54. Eugenio, M.; Verkaik, I.; Lloret, F.; Espelta, J.M. Recruitment and growth decline in Pinus halepensis populations after recurrent wildfires in Catalonia (NE Iberian Peninsula). *For. Ecol. Manag.* **2006**, *231*, 47–54.
55. Moya, D.; González-De Vega, S.; García-Orenes, F.; Morugán-Coronado, A.; Arcenegui, V.; Mataix-Solera, J.; Lucas-Borja, M.E.; De las Heras, J. Temporal characterisation of soil-plant natural recovery related to fire severity in burned Pinus halepensis Mill. forests. *Sci. Total Environ.* **2018**, *640–641*, 42–51. [CrossRef]
56. Crotteau, J.S.; Morgan Varner, J.; Ritchie, M.W. Post-fire regeneration across a fire severity gradient in the southern Cascades. *For. Ecol. Manag.* **2013**, *287*, 103–112. [CrossRef]
57. United States Geological Survey (USGS). Earth Explorer Server. Available online: https://earthexplorer.usgs.gov/ (accessed on 27 January 2020).
58. Masek, J.G.; Vermote, E.F.; Saleous, N.E.; Wolfe, R.; Hall, F.G.; Huemmrich, K.F.; Gao, F.; Kutler, J.; Lim, T.-K. A Landsat surface reflectance dataset for North America, 1990–2000. *IEEE Geosci. Remote Sens. Lett.* **2006**, *3*, 68–72. [CrossRef]
59. Vermote, E.; Justice, C.; Claverie, M.; Franch, B. Preliminary analysis of the performance of the Landsat 8/OLI land surface reflectance product. *Remote Sens. Environ.* **2016**, *185*, 46–56. [CrossRef]
60. Miller, J.D.; Thode, A.E. Quantifying burn severity in a heterogeneous landscape with a relative version of the delta Normalized Burn Ratio (dNBR). *Remote Sens. Environ.* **2007**, *109*, 66–80. [CrossRef]
61. Ministerio de Agricultura Pesca y Alimentación. Forest Map of Spain. Available online: https://www.miteco.gob.es/es/biodiversidad/servicios/banco-datos-naturaleza/informacion-disponible/mfe200.aspx (accessed on 27 January 2020).
62. Spain National Geographic Institute (IGN). Digital Elevation Model 25-m. Available online: http://centrodedescargas.cnig.es/CentroDescargas/index.jsp (accessed on 7 February 2020).
63. Vicente-Serrano, S.M.; Tomas-Burguera, M.; Beguería, S.; Reig, F.; Latorre, B.; Peña-Gallardo, M.; Luna, M.Y.; Morata, A.; González-Hidalgo, J.C. A high resolution dataset of drought indices for Spain. *Data* **2017**, *2*, 22. [CrossRef]
64. Vicente-Serrano, S.M.; Beguería, S.; López-Moreno, J.I. A multiscalar drought index sensitive to global warming: The standardized precipitation evapotranspiration index. *J. Clim.* **2010**, *23*, 1696–1718. [CrossRef]
65. Spain National Geographic Institute (IGN). Aerial Orthophotography National Plan. Available online: http://centrodedescargas.cnig.es/CentroDescargas/index.jsp (accessed on 7 February 2020).
66. Braaten, J.D.; Cohen, W.B.; Yang, Z. LandsatLinkr. Zenodo. Available online: https://zenodo.org/record/1231029#.XrPS5MARWUk (accessed on 20 January 2020).
67. Vogeler, J.C.; Braaten, J.D.; Slesak, R.A.; Falkowski, M.J. Extracting the full value of the Landsat archive: Inter-sensor harmonization for the mapping of Minnesota forest canopy cover (1973–2015). *Remote Sens. Environ.* **2018**, *209*, 363–374. [CrossRef]
68. Hartigan, J.A.; Wong, M.A. Algorithm AS 136: A K-Means Clustering Algorithm. *J. R. Stat. Soc. Ser. C Appl. Stat.* **1979**, *28*, 100–108. [CrossRef]
69. Gouveia, C.; DaCamara, C.C.; Trigo, R.M. Post-fire vegetation recovery in Portugal based on spot/vegetation data. *Nat. Hazards Earth Syst. Sci.* **2010**, *10*, 673–684. [CrossRef]

70. Purnima, B.; Arvind, K. EBK-Means: A Clustering Technique based on Elbow Method and K-Means in WSN. *Int. J. Comput. Appl.* **2014**, *105*, 17–24.
71. Varoquaux, G.; Buitinck, L.; Louppe, G.; Grisel, O.; Pedregosa, F.; Mueller, A. Scikit-learn. *GetMobile Mob. Comput. Commun.* **2015**, *19*, 29–33. [CrossRef]
72. Roberts, D.W.; Cooper, S.V. Concepts and techniques of vegetation mapping. Pages 90–96 BT—Compilers. Land classifications based on vegetation: Applications for resource management. In *USDA Forest Service General Technical Report INT-257*; Ferguson, D.E., Morgan, P., Johnson, F.D., Eds.; Intermountain Research Station: Ogden, UT, USA, 1989.
73. Lentile, L.B.; Holden, Z.A.; Smith, A.M.S.; Falkowski, M.J.; Hudak, A.T.; Morgan, P.; Lewis, S.A.; Gessler, P.E.; Benson, N.C. Remote sensing techniques to assess active fire characteristics and post-fire effects. *Int. J. Wildl. Fire* **2006**, *15*, 319–345. [CrossRef]
74. Vicente-Serrano, S.M.; Gouveia, C.; Camarero, J.J.; Beguería, S.; Trigo, R.; López-Moreno, J.I.; Azorín-Molina, C.; Pasho, E.; Lorenzo-Lacruz, J.; Revuelto, J.; et al. Response of vegetation to drought time-scales across global land biomes. *Proc. Natl. Acad. Sci. USA* **2013**, *110*, 52–57. [CrossRef]
75. Fotheringham, A.S.; Brunsdon, C.; Charlton, M. *Geographically weighted regression: The analysis of spatially varying relationships*; John Wiley & Sons: Hoboken, NJ, USA, 2003; ISBN 0-471-49616-2.
76. Brunsdon, C.; Fotheringham, A.S.; Charlton, M.E. Geographically Weighted Regression: A Method for Exploring Spatial Nonstationarity. *Geogr. Anal.* **2010**, *28*, 281–298. [CrossRef]
77. Hurvich, C.M.; Simonoff, J.S.; Tsai, C.-L. Smoothing parameter selection in nonparametric regression using an improved Akaike information criterion. *J. R. Stat. Soc. Ser. B Stat. Methodol.* **1998**, *60*, 271–293. [CrossRef]
78. Zhao, F.R.; Meng, R.; Huang, C.; Zhao, M.; Zhao, F.A.; Gong, P.; Yu, L.; Zhu, Z. Long-term post-disturbance forest recovery in the greater yellowstone ecosystem analyzed using Landsat time series stack. *Remote Sens.* **2016**, *8*, 898. [CrossRef]
79. Congalton, R.G.; Green, K. *Assessing the Accuracy of Remotely Sensed Data: Principles and Practices*; Lewis Publishers: Boca Raton, FL, USA, 1999; ISBN 0873719867.
80. Schroeder, T.A.; Wulder, M.A.; Healey, S.P.; Moisen, G.G. Mapping wildfire and clearcut harvest disturbances in boreal forests with Landsat time series data. *Remote Sens. Environ.* **2011**, *115*, 1421–1433. [CrossRef]
81. Thanos, C.A.; Daskalakou, E.N. Reproduction in Pinus halepensis and P. brutia. In *Ecology, Biogeography and Management of Pinus Halepensis and P. Brutia Forest Ecosystems in the Mediterranean Basin*; Backhuys Publishers: Leiden, The Netherlands, 2000; pp. 79–90.
82. Viana Soto, A.; Aguado, I.; Salas, J.; García, M. Classification of post-fire recovery trajectories using Landsat time series in the Mediterranean region: Spain. In *Earth Resources and Environmental Remote Sensing/GIS Applications*; International Society for Optics and Photonics: Strasbourg, France, 2019; Volume 1115607, p. 6.
83. Tanase, M.; de la Riva, J.; Santoro, M.; Pérez-Cabello, F.; Kasischke, E. Sensitivity of SAR data to post-fire forest regrowth in Mediterranean and boreal forests. *Remote Sens. Environ.* **2011**, *115*, 2075–2085. [CrossRef]
84. Martín-Alcón, S.; Coll, L.; De Cáceres, M.; Guitart, L.; Cabré, M.; Just, A.; González-Olabarría, J.R. Combining aerial LiDAR and multispectral imagery to assess postfire regeneration types in a Mediterranean forest. *Can. J. For. Res.* **2015**, *45*, 856–866. [CrossRef]
85. Bolton, D.K.; Coops, N.C.; Wulder, M.A. Characterizing residual structure and forest recovery following high-severity fire in the western boreal of Canada using Landsat time-series and airborne lidar data. *Remote Sens. Environ.* **2015**, *163*, 48–60. [CrossRef]
86. Cohen, W.B.; Yang, Z.; Kennedy, R. Detecting trends in forest disturbance and recovery using yearly Landsat time series: 2. TimeSync—Tools for calibration and validation. *Remote Sens. Environ.* **2010**, *114*, 2911–2924. [CrossRef]
87. Storey, E.A.; Stow, D.A.; Roberts, D.A. Evaluating uncertainty in Landsat-derived postfire recovery metrics due to terrain, soil, and shrub type variations in southern California. *GIScience Remote Sens.* **2020**, *57*, 352–368. [CrossRef]
88. Liu, Z. Effects of climate and fire on short-term vegetation recovery in the boreal larch forests of northeastern China. *Sci. Rep.* **2016**, *6*, 37572. [CrossRef]
89. Pausas, J.G.; Ouadah, N.; Ferran, A.; Gimeno, T.; Vallejo, R. Fire severity and seedling establishment in Pinus halepensis woodlands, eastern Iberian Peninsula. *Plant Ecol.* **2002**, *169*, 205–213. [CrossRef]

90. Baudena, M.; Santana, V.M.; Baeza, M.J.; Bautista, S.; Eppinga, M.B.; Hemerik, L.; Garcia Mayor, A.; Rodriguez, F.; Valdecantos, A.; Vallejo, V.R.; et al. Increased aridity drives post-fire recovery of Mediterranean forests towards open shrublands. *New Phytol.* **2020**, *225*, 1500–1515. [CrossRef] [PubMed]
91. Wittenberg, L.; Malkinson, D.; Beeri, O.; Halutzy, A.; Tesler, N. Spatial and temporal patterns of vegetation recovery following sequences of forest fires in a Mediterranean landscape, Mt. Carmel Israel. *Catena* **2007**, *71*, 76–83. [CrossRef]

remote sensing

MDPI

Article

Can Landsat-Derived Variables Related to Energy Balance Improve Understanding of Burn Severity From Current Operational Techniques?

Alfonso Fernández-Manso [1,2], Carmen Quintano [2,3,4,*] and Dar A. Roberts [2]

1 Agrarian Science and Engineering Department, University of León, Av. Astorga s/n., 24400 Ponferrada, Spain; alfonso.manso@unileon.es
2 Department of Geography, University of California, Santa Barbara, CA 93106, USA; dar@geog.ucsb.edu
3 Electronic Technology Department, University of Valladolid, Paseo del Cauce, 59, 47011 Valladolid, Spain
4 Sustainable Forest Management Research Institute, University of Valladolid-Spanish National Institute for Agricultural and Food Research and Technology, 34004 Palencia, Spain
* Correspondence: carmen.quintano@uva.es

Received: 20 February 2020; Accepted: 9 March 2020; Published: 10 March 2020

Abstract: Forest managers rely on accurate burn severity estimates to evaluate post-fire damage and to establish revegetation policies. Burn severity estimates based on reflective data acquired from sensors onboard satellites are increasingly complementing field-based ones. However, fire not only induces changes in reflected and emitted radiation measured by the sensor, but also on energy balance. Evapotranspiration (ET), land surface temperature (LST) and land surface albedo (LSA) are greatly affected by wildfires. In this study, we examine the usefulness of these elements of energy balance as indicators of burn severity and compare the accuracy of burn severity estimates based on them to the accuracy of widely used approaches based on spectral indexes. We studied a mega-fire (more than 450 km^2 burned) in Central Portugal, which occurred from 17 to 24 June 2017. The official burn severity map acted as a ground reference. Variations induced by fire during the first year following the fire event were evaluated through changes in ET, LST and LSA derived from Landsat data and related to burn severity. Fisher's least significant difference test (ANOVA) revealed that ET and LST images could discriminate three burn severity levels with statistical significance (uni-temporal and multi-temporal approaches). Burn severity was estimated from ET, LST and LSA using thresholding. Accuracy of ET and LST based on burn severity estimates was adequate ($\kappa = 0.63$ and 0.57, respectively), similar to the accuracy of the estimate based on dNBR ($\kappa = 0.66$). We conclude that Landsat-derived surface energy balance variables, in particular ET and LST, in addition to acting as useful indicators of burn severity for mega-fires in Mediterranean ecosystems, may provide critical information about how energy balance changes due to fire.

Keywords: burn severity; Landsat; Mediterranean; energy balance; evapotranspiration; land surface temperature; land surface albedo; dNBR

1. Introduction

Fire is one of the major disturbance processes in many ecosystems and is particularly prevalent in Mediterranean ecosystems, where it impacts biodiversity [1,2] and other important ecosystem properties. The frequency, severity and size of fires have been increasing in recent decades in Mediterranean ecosystems, due mainly to climate change and modifications in land use [3]. To evaluate fire damage, one of the most commonly used metrics is fire/burn severity [4]. Burn severity groups short- and long-term fire effects on vegetation and soil [5]. Key and Benson [6] distinguished between initial assessment of burn severity, which aims to map burn severity immediately after fire, from an extended assessment, whose goal is to map burn severity when survivorship and mortality are detectable. Increases in burn severity due to changes in climate and land cover may modify vegetation characteristics (composition, resilience, structure) and soil attributes, even in Mediterranean ecosystems that are well adapted to fire [7,8].

Fire reduces evapotranspiration (ET) and increases surface runoff following rainfall [9]. Usually, the increase in surface runoff is lower than the decrease in ET, causing an increase in soil moisture documented by different studies (e.g., [10,11]). However, forested areas affected by high burn severity show a decrease in soil moisture. In this case, the increase in surface runoff exceeds the reduction in ET and leads to drier post-fire soils [9,12]. Post-fire changes in vegetation, soil and water balance corroborate wildfires impact on energy balance [13–15]. Fire-induced modifications in vegetation structure and species composition alter latent heat flux and other variables of the energy balance equation [16,17]. Changes in Land Surface Temperature (LST), an important parameter of ET, have been recently identified as a potential indicator of burn severity [18–21]. Similarly, previous research studies showed the relationship between fire damage and land surface albedo (LSA) [20,22–24]. LSA determines the total amount of solar radiation that an ecosystem absorbs [25] and it is subject to seasonality [20]. To our knowledge, there is a lack of studies relating to all three elements of energy balance (ET, LST, and LSA) to burn severity, in particular in Mediterranean forest ecosystems. However, understanding their interactions is critical to evaluating the role of fire on Earth. The loss of vegetation modifies rainfall and water availability (more than 40% of rainfall over Earth's land is due to ET on average) and has an impact on warming at local scales as well [26]. Forest fires may have an influence on regional climate by modifying energy balance [27,28], and some studies suggest that this may affect climate at global scales [29,30].

There is a variety of approaches to compute LST from thermal satellite data with a satisfactory accuracy (1K or better) and spatial resolution (30 × 30m) [31,32]. Similarly, LSA may be computed from reflective satellite data with a low residual error (less than 0.02) at the same fine spatial resolution [33,34]. However, many previous studies relating to fire effects and ET [13,15,35,36] were based on surface fluxes measured in the field, as it is difficult to estimate ET from satellite data with an adequate accuracy and spatial resolution. Thus, these results are only valid for vegetation similar to the vegetation analyzed by the instrument on a local scale [37]. ET requires a relatively high number of parameters to be computed with reasonable accuracy. Thus, a high temporal resolution (daily frequency or higher) is needed, whereas a fine spatial resolution is advisable. For this reason, different satellite sensors combined with meteorological ground stations are usually needed to estimate ET from satellite data [38]. Among the remote sensing-based ET models, surface energy balance techniques based on LST from thermal infrared data are widely used. Mapping EvapoTranspiration at high Resolution with Internalized Calibration (METRIC) [39] is a well-established surface energy balance-based model [40]. One of the main limitations of METRIC is, however, the need for a well-trained expert to calibrate and run the model [41]. The calibration of METRIC requires the identification of maximum and minimum ET within a satellite image. Some automatic calibration algorithms for the METRIC model have, however, been developed to avoid this limitation (e.g., [42,43]). Among them, the Earth Engine Evapotranspiration Flux (EEFlux) application was designed and implemented in the Google Earth Engine (GEE) platform [44]. The ECOsystem Spaceborne Thermal Radiometer Experiment on Space Station (ECOSTRESS) mission has ET as one of the primary output variables [45,46], although at the present it only collects data over the conterminous United States (CONUS) as well as key biomes and agricultural zones around the world and selected FLUXNET [47] validation sites.

Similarly, remote sensing techniques provide a widely used alternative to field-measured burn damage [14,48–53]. Operational examples include the European Forest Fire Information System (EFFIS) and the Monitoring Trends in Burn Severity (MTBS) project in the USA. Among the most widely used techniques to estimate burn severity from satellite data, spectral indices stand out due to their simplicity [52–58]. In particular, using thresholds to classify differenced Normalized Burn Ratio (dNBR) [6] has become a standard to estimate burn severity from optical remotely sensed data [59], specifically from Landsat data [57,60,61]. Additionally, differenced Normalized Difference Vegetation Index (dNDVI) may be used to map burn severity when it is not possible to calculate dNBR due to sensor characteristics [62]. As most of the sensors onboard Unmanned Aerial Vehicles (UAVs) do not record information in shortwave infrared wavelengths, NDVI, and in particular dNDVI, is emerging as an accurate index to estimate burn severity [63]. However, some studies have pointed out some limitations of using spectral indices as a base to estimate burn severity: their interpretation in terms of fire damage is not easy due to the absence of standardized units [64,65] and due to a nonlinear relationship to burn severity [66].

In this context, this research work aims to analyze the relationship between burn severity and variations in ET, LST and LSA after large fires in Mediterranean forests at fine spatial resolution and to verify whether they may act as indicators of burn severity (initial and extended) in these ecosystems. Spectral indexes widely used to estimate burn severity (dNDVI, dNBR) were used as additional variables. In particular, our specific objectives are: (1) to evaluate the changes due to fire in the five variables (ET, LST, LSA, NDVI and NBR) over time: immediately after fire, a month after fire, two months after fire, and one year after fire; (2) to determine whether the five variables can be regarded as indicators of burn severity (initial and extended assessments) from both uni-temporal and multi-temporal perspectives and how many burn severity levels may be distinguished by each variable; (3) to identify pre-fire factors that may also cause changes in the variables: in particular, pre-fire species composition (highly related to fuel model), climate and topography; and 4) to determine the accuracy of the burn severity estimates from these variables. We structured Results and Discussion sections by three main questions that are closely related to these specific objectives: (1) How does fire modify ET, LST, LSA, NDVI and NBR?; (2) How are ET, LST, LSA, NDVI and NBR post-fire images influenced by burn severity and pre-fire factors?; and (3) Is it possible to accurately estimate burn severity from the ET, LST and LSA images?

2. Materials

2.1. Study Area

Our study area is the mega-fire of Pedrogão Grande and environs that burned in 2017 in central Portugal (Figure 1, left). The Joint Research Centre of the European Commission included this fire in its annual report as the fire that caused a total of 66 deaths and important losses in term of forest areas and buildings [67]. In particular, 263 structures were inventoried as being damaged by Ribeiro et al. [68]. The Pedrogão Grande fire burned 458.93 km^2 between June 17th 2017 and June 24th 2017 [69]. After months with low rainfall and high temperatures, 80% of the region was in extreme drought on June 17th according to the Instituto Português do Mar e da Atmosfera (IPMA). With these conditions, a multi-cell thunderstorm emerged, and lightning and strong winds caused the fire to burn out of control [70,71].

Climate in the area can be described as a mix of Csa (typical Mediterranean) and Csb (humid Mediterranean) Köppen-Geiger classes [72]. Rainy winters and hot and dry summers characterize the typical Mediterranean climate (annual precipitation is between 400–750mm). The humid Mediterranean climate, however, shows a higher annual precipitation (750–2000mm), with summers that are mild and dry.

The highest elevations are to the north and northwest of the affected area. In progression to the south, lower elevations predominate (approximately 400m around Pedrógão Grande). Orography is very heterogeneous. Very steep slopes with values between 45% and 60% or higher predominate in the north and northeast, whereas the area is flatter, with gentle slopes between 0% and 15%, in the south and southwest (basin of Zêzere river). In the extreme Southeast, there are slopes between 30% and 40%.

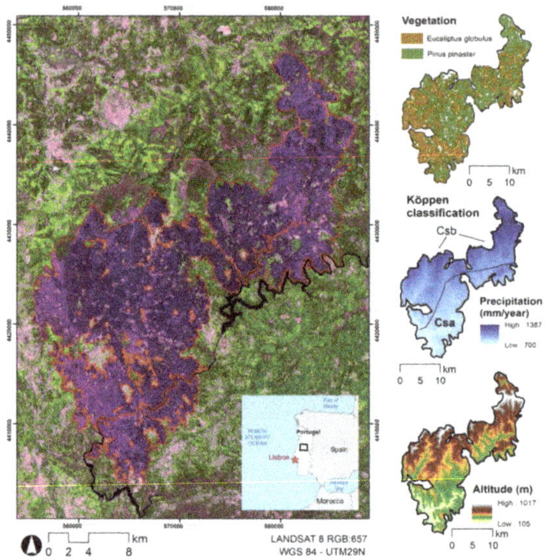

Figure 1. Left: location of study area; **right**: vegetation species map (**upper**), climatic map (**center**); altitude map (**lower**).

The Pedrógão Grande area is integrated into the Interior North Pinhal Zone, the largest forest spot in Europe before the fire. It is contained between the basins of the Zêzere and Unhais rivers and the banks of the Pêra and Mega rivers. In this area, plantations of eucalyptus (*Eucalyptus globulus* Labill.), and maritime pine forest (*Pinus pinaster* Ait.) are dominant (Figure 1, upper right). However, it is still possible to find cork oak forests (*Quercus suber* L.), chestnut trees (*Castanea sativa* Mill.), strawberry trees (*Arbutus unedo* L.) and shrubs such as gorse (*Ulex europaeus* L.), red heather (*Erica australis* L.), white heather (*Erica arborea* L.), and broom (*Cytisus scoparius* L.). Acacias and olive trees have great importance in the local economy. Agricultural areas were located close to villages, although both agriculture and grazing activities were low (1–3 livestock units per ha) [71].

2.2. Materials

We downloaded USGS Landsat 8 Operational Land Imager (OLI)/ Thermal InfraRed Sensor (TIRS) surface reflectance images (path/row 204/32, higher-level data product on demand) acquired on 15 June 2017; 1 July 2017; 17 July 2017; 2 August 2017; 18 August 2017; 19 September 2017; 5 August 2018; and 8 October 2018. From them, we computed NDVI, NBR, LST and LSA. METRIC daily ET images were obtained by using the software developed by University of Nebraska-Lincoln [44].

The official fire perimeter and burn severity map (three severity levels) were used as ground reference (estimated geometric accuracy 5m CE90 or better). They were based on Copernicus—Emergency Management Service maps (built using SPOT 6 and 7 imagery) verified in the field and from photointerpretation of Sentinel 2 MSI data by the Portuguese Study Center of Forest Fires [69].

As potential pre-fire factors that may also influence the variables related to energy balance, we included in the study: pre-fire species composition, climate and topography. The official Portuguese Land Cover Map (Carta de Ocupação do Solo, COS15) enabled us to define pre-fire vegetation species composition. COS15 has 48 classes in vector format and has a minimum cartographic unit of one hectare and a minimum distance between lines of 20 meters [73]. Köppen-Geiger climate classes [72] of the study area were extracted from the updated world map of the Köppen-Geiger climate classification [74]. Finally, a 30m Advanced Spaceborne Thermal Emission and Reflection Radiometer (ASTER) Global

Digital Elevation Model Version 2 (GDEM V2) provided by USGS (https://search.earthdata.nasa.gov) was employed to derive three topographic factors (elevation, aspect and slope).

3. Methods

3.1. LST Calculation

As in previous studies [75,76], LST was calculated based on the single-channel method proposed by Yu et al. [77]. This method is based on the radiative transfer equation (Equation (1)). See the review work of Li et al. [32] for detailed information on satellite-based methods to obtain LST. Calculations were implemented in the R statistical program (release 3.4.3) [78]

$$B(T_s) = \left[\frac{(B(T) - I\uparrow)}{\tau(\theta)} - I\downarrow(1-\varepsilon)\right]\frac{1}{\varepsilon} \tag{1}$$

where $B(T_s)$ represents ground radiance; B(T), radiance of thermal band 10 with brightness temperature T; $\tau(\theta)$, atmospheric transmittance for a view zenith angle θ; $I\downarrow$, and $I\uparrow$ respectively, denote downwelling and upwelling path radiance; and ε, land surface emissivity calculated from Sobrino et al. [79]. LST is computed from ground radiance according to Planck's law using Equation (2)

$$LST = \frac{K_2}{\ln(1 + K_1/B(T_s))} \tag{2}$$

where K1 and K2 are thermal constants defined in metadata of Landsat 8 OLI/TIRS images.

3.2. LSA Calculation

Similar to previous works [23,24], LSA was estimated from surface reflectance based on Liang [80] and Liang et al. [34]. We only considered the total shortwave broadband LSA (LSA) (Equation (3))

$$LSA = 0.356\alpha_1 + 0.130\alpha_3 + 0.373\alpha_4 + 0.085\alpha_5 + 0.072\alpha_7 - 0.0018 \tag{3}$$

where α_i indicates the albedo in the 'i' Landsat 8 OLI band that is approximated by reflectance.

3.3. ET (METRIC Model)

We include here a summary of the METRIC model. More detailed information about the METRIC-model can be found in Allen et al. [39] and Irmak et al. [81]. In the METRIC model, the energy consumed by the ET process is determined from satellite imagery using the surface energy balance equation (Equation (4)) on a pixel by pixel basis

$$LE = R_n - G - H \tag{4}$$

where LE represents latent energy consumed by actual ET (i.e., heat energy used by water in its phase change from liquid to gas during the ET process); R_n, net radiation flux density (W/m^2); G, ground heat flux density or sensible heat conducted into the ground) (W/m^2); and H, sensible heat flux density convected into the air (W/m^2). In METRIC, these three key variables (R_n, G and H) are estimated from satellite data. In particular: R_n from narrow-band reflectance and surface temperature; G, from R_n, surface temperature, and vegetation indices; and H, from surface temperature ranges, surface roughness, and wind speed (see [39]). Next, instantaneous ET (ET_i, expressed in mm/h) is calculated for each pixel using Equation (5)

$$ET_i = 3600\frac{LE}{\lambda\rho_w} \tag{5}$$

where 3600 is a constant that converts seconds to hours; ρ_w is the density of water (Kg/m^3); and λ, is the latent heat of vaporization, expressed in J/Kg, and estimated from surface temperature, expressed in K.

From ET_i, reference ET fraction (ET_rF) is calculated as the ratio of ET_i to the instantaneous tall crop reference ET (ET_r) computed from weather data (Equation (6)). Alfalfa is the tall crop used for METRIC as the tall alfalfa reference resembles maximum energy-limited ET from a well-watered, extensive surface of vegetation.

$$ET_rF = \frac{ET_i}{ET_r} \tag{6}$$

ET_rF is used for extrapolating instantaneous ET to daily ET (ET) expressed in mm/day (Equation (7)). In the METRIC process, it is assumed that the instantaneous ET_rF calculated at image time is constant for the 24-hour period. This assumption is generally valid for agricultural crops (developed to maximize photosynthesis and thus stomatal conductance). For some native vegetation under water stress, however, ET_rF may reduce its value during the afternoon. In this case, ET_rF is modeled from instantaneous ET_rF by developing the appropriate functions.

$$\mathrm{ET} = \mathrm{C}_{rad} \times ET_rF \times ET_{r-24} \tag{7}$$

where C_{rad} represent the correction term used in sloping terrain, in horizontal areas C_{rad} equals 1.0; and ET_{r-24}, cumulative 24 h ET_r for the day of the image.

3.4. Database Construction

As a preliminary step to database construction, NBR [82] and NDVI [83] were computed from Landsat surface reflectance images. Additionally, we calculated the difference between pre- and post-fire situations to perform a multi-temporal analysis. In particular, we subtracted post-fire variables from pre-fire ones for ET, LSA, NBR and NDVI, and pre-fire variables from post-fire ones for LST. Variables ending with "_i" represent the initial assessment (difference between pre-fire -06/15/17- and immediately after fire images -07/01/17-), and "_e" an extended assessment (difference between pre-fire -06/15/17- and 1-year after fire images -08/05/18-).

We applied a mean 3 × 3 filter to satellite-derived images before extracting the digital values for the sampling points to minimize positional errors [6,19]. As recommended by Congalton and Green [84], a stratified random sampling was adopted. The number of training samples was related to the surface that each burn severity level occupies on the ground reference map: 2532 samples for high burn severity, 1638 for moderate burn severity and 721 for low burn severity. In addition, the zone outside the fire perimeter was sampled to define the unburned class; we defined 1213 sampling points for the unburned class. A kurtosis normality test enabled us to check its normality.

3.5. Statistical Analysis

After evaluating the statistical characteristics of the variables of interest (ET, LST, LSA NDVI and NBR) over time as a preliminary analysis, we aim to determine how spatial variation in the five variables is related to burn severity. One-way Analysis of Variance (ANOVA) was applied by grouping values into each burn severity level. A Fisher's least significant difference (LSD) test determined which sample means were significantly different from the others. These analyses were performed at different moments to analyze the temporal evolution of variables: immediately after fire, a month after fire, two months after fire and 1-year after fire (summer and fall). ANOVA analysis enabled us to identify the number of burn severity levels that may be distinguished with statistical significance from the images and to determine whether those variables could be used as indicators of burn severity. In addition, to the uni-temporal post-fire analysis, a similar multi-temporal analysis was performed by subtracting the pre-fire image from the immediately after fire and 1-year after fire images (initial and extended assessment respectively).

Pre-fire conditions also have an influence on the variables of interest. To evaluate the contribution of those determining factors on the variables, a multifactor analysis of variance was performed. It searched for significant interactions amongst the factors. The p-value tests the statistical significance of the contribution of each measured factor, having removed the effects of all other factors. We included as pre-fire factors: dominant vegetation species; Köppen climate category and topographic variables (in particular, elevation, slope and aspect). Vegetation species is usually related to fire damage, although it is not a true measure of fuel state [85]. Specifically, three categories were taken into account in the vegetation species factor—eucalyptus, pine and shrubland—and two Köppen classes in the climate factor—Csa (Mediterranean climate) and Csb (humid Mediterranean climate). Regarding topographic factors, we defined five classes in elevation: 0- 300m, 301-600m, 601-900m and 901m-highest value; five classes in slope: 0°–5°, 5°–10°, 10°–20°, 20°–30°, 30°–highest value; and eight classes in aspect: north (0°–22.5° and 337.5°–360°), northeast (22.5°–67.5°), east (67.5°–112.5°), southeast (112.5°–157.5°), south (157.5°–202.5°), southwest (202.5°–247.5°), west (247.5°–292.5°), and northwest (292.5°–337.5°).

The relationship amongst the five variables was evaluated as well. As in previous studies (e.g., [19,86,87] Pearson's correlation analysis was utilized to analyze the interactions among the five variables taken into account.

3.6. Burn Severity Mapping

Burn severity (initial assessment) was mapped from both uni-temporal (immediately after fire images) and multi-temporal (pre- and post-fire difference images) perspectives. Classification of the continuous images into four classes—high, moderate and low burn severity levels and unburned—was based on thresholds. The halfway values between the mean values of consecutive classes (computed from the training samples) acted as initial thresholds. These initial thresholds were refined manually to minimize the number of errors of classification [88].

Our study used error matrices and κ statistics [84] to measure the accuracy of each burn severity estimate. In addition, Overall Accuracy (OA), Producer's Accuracy (PA) (omission error) and User's Accuracy (UA) (commission error) were computed as well. Prior to error matrix computation, a new stratified random disproportional sampling was applied to the ground reference and to the classified images. In this case we defined 2357 sampling points for high burn severity, 1694 for moderate burn severity, 661 for low burn severity level, and 1140 for the unburned class.

4. Results

4.1. How Does Fire Modify ET, LST, LSA, NDVI and NBR?

Notable differences between burned and unburned areas in every post-fire variable are visually apparent (Figure 2). The images immediately after the fire (second column of Figure 2) showed such differences more clearly. The images one month and two months after the fire (third and fourth column) also enabled us to clearly distinguish burned from unburned areas. Moreover, the two months after the fire image (19 September 2017) enabled us to visually identify a new forest fire in the lower right corner. Even 1-year after fire (right column), burned areas were readily identified in every image except LSA. Moreover, a large forest fire (332 km^2) which happened in Ponte das Portelinhas (Sertã) in 2 November 2017 is still evident in the 1-year after fire images. In addition, the capability of ET images to clearly delineate riparian vegetation along the Zêzere River should be noted. Spatial differences inside of the burn perimeter could be noticed as well in the five variables (more clearly in the images closer to the fire date), which suggests sensitivity to burn severity levels.

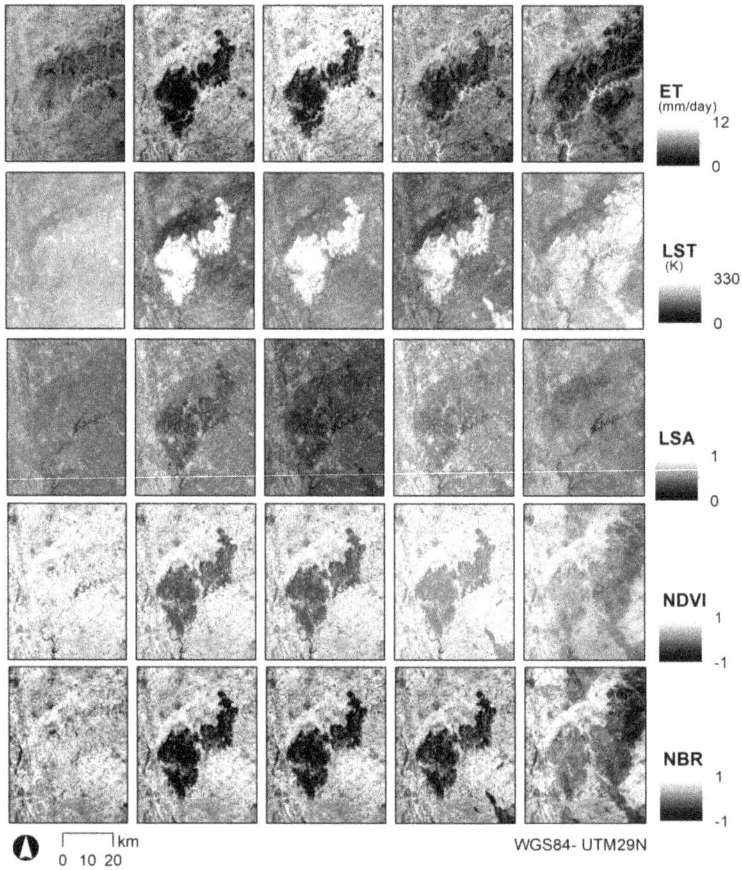

Figure 2. Temporal evolution of evapotranspiration (ET) (**upper row**), land surface temperature (LST) (**second row**), land surface albedo (LSA) (**third row**); normalized difference vegetation index (NDVI) (**forth row**) and normalized burn ratio (NBR) (**lower row**) images. Left column: 06/15/17 second column: 07/01/17, third column: 08/02/17, fourth column 09/19/17, right column 08/05/18.

Table 1 displays the statistical characteristics (mean values and standard deviation) of ET, LST, LSA, NDVI and NBR for burned and unburned samples at different times: before the fire (only unburned samples), immediately after fire, one month, two months and 1-year after fire. Knowing the mean values of the burned and unburned samples for each image enabled us to minimize the influence of external and meteorological parameters in the fire-induced changes in the five variables. Fire caused an important initial ET decrease, which eventually reduced its value. ET declined seasonally in unburned samples as well. Patterns of LST variation after fire were similar to the ET trend, but showed a reverse pattern. LST values increased in burned samples. The increase was higher immediately after fire, although it could be observed even a year after the fire. Regarding LSA, burned areas decreased immediately after they burned relative to unburned areas, with the difference decreasing over time. A year after fire, LSA in burned samples increased relative to unburned samples, mainly due to vegetation regrowth. Finally, both NDVI and NBR showed the same tendency: a strong decrease in the burned samples immediately after the fire. These differences declined in the year after fire image. The standard deviation of burned samples immediately after fire was the highest of all images and implies a high spatial heterogeneity that could be related to burn severity.

Table 1. Temporal evolution of statistical characteristics of ET. LST. LSA. NDVI and NBR for burned and unburned samples.

		06/15/17		07/01/17		08/02/17		09/19/17		08/05/18	
		μ	σ	μ	σ	μ	σ	μ	σ	μ	σ
ET (mm/day)	Burned	-	-	1.62	1.62	1.62	0.95	1.52	0.98	1.76	1.01
	Unburned	5.58	1.34	6.65	1.16	3.47	0.94	3.26	0.79	3.13	1.21
LST (K)	Burned	-	-	315.79	4.90	319.51	4.72	308.96	3.99	315.25	2.47
	Unburned	305.13	3.80	299.36	3.24	304.10	3.91	298.38	4.05	309.57	4.19
LSA	Burned	-	-	0.08	0.02	0.09	0.03	0.09	0.02	0.16	0.01
	Unburned	0.12	0.03	0.12	0.03	0.11	0.03	0.10	0.03	0.15	0.01
NDVI	Burned	-	-	0.29	0.12	0.30	0.12	0.37	0.11	0.41	0.07
	Unburned	0.72	0.12	0.73	0.13	0.72	0.13	0.68	0.14	0.51	0.11
NBR	Burned	-	-	-0.12	0.24	-0.06	0.22	0.07	0.21	0.29	0.10
	Unburned	0.57	0.16	0.55	0.17	0.55	0.18	0.55	0.22	0.45	0.15

μ: mean value in sampling plots; σ: standard deviation in sampling plots; ET: evapotranspiration; LST: land surface temperature; LSA: land surface albedo; NDVI: normalized difference vegetation index; NBR: normalized burn ratio

4.2. How Are ET, LST, LSA, NDVI and NBR Post-Fire Images Influenced by Burn Severity and Pre-Fire Factors?

The results of one-way ANOVA applied to each post-fire image (Table 2) indicated that the five variables tested showed significant differences ($p < 0.01$) among the mean values of each burn severity level when the 2017 images were taken into account, with the exception of LSA, whose low burn severity and unburned mean values could not be distinguished in the last two 2017 images. Therefore, three burn severity levels could be distinguished in all immediately post-fire and a month after fire images.

Table 2. Fisher's least significant difference test for post-fire ET, LST, LSA, NDVI and NBR images and burn severity levels.

Evapotranspiration (ET, mm/day)														
S	07/01/17		07/17/17		08/02/17		08/18/17		09/19/17		08/05/18		10/08/18	
	μ	HG	μ	HG	μ	HG	μ	HG	μ	HG	μ	HG	μ	HG
H	0.94	a	1.32	a	1.45	a	0.98	a	1.16	a	1.50	a	1.03	a
M	1.89	b	1.52	b	1.70	b	1.19	b	1.74	b	1.90	b	1.17	b
L	3.36	c	1.73	c	2.04	c	1.51	c	2.26	c	2.31	c	1.45	c
U	6.65	d	2.80	d	3.47	d	2.76	d	3.26	d	3.13	d	1.84	d
Land Surface Temperature (LST, K)														
S	07/01/17		07/17/17		08/02/17		08/18/17		09/19/17		08/05/18		10/08/18	
	μ	HG	μ	HG	μ	HG	μ	HG	μ	HG	μ	HG	μ	HG
H	318.7	a	325.9	a	322.1	a	321.6	a	310.9	a	315.8	a	302.9	a
M	314.5	b	321.8	b	318.4	b	318.2	b	308.1	b	315.1	b	302.5	a
L	309.4	c	317.5	c	314.3	c	314.6	c	305.0	c	313.5	c	300.8	b
U	299.1	d	307.6	d	303.9	d	305.6	d	298.4	d	309.5	d	297.7	c

Table 2. *Cont.*

Land Surface Reflectance (LSA)														
S	**07/01/17**		**07/17/17**		**08/02/17**		**08/18/17**		**09/19/17**		**08/05/18**		**10/08/18**	
	μ	**HG**	μ	**HG**	μ	**HG**	μ	**HG**	μ	**HG**	μ	**HG**	μ	**HG**
H	0.07	a	0.08	a	0.08	a	0.08	a	0.08	a	0.16	a	0.10	a
M	0.08	b	0.10	b	0.10	b	0.10	b	0.09	b	0.16	a	0.10	a
L	0.10	c	0.11	c	0.11	c	0.11	c	0.10	c	0.16	a	0.10	a
U	0.12	d	0.12	d	0.11	d	0.11	c	0.10	c	0.15	b	0.08	b
Normalized Difference Vegetation Index (NDVI)														
S	**07/01/17**		**07/17/17**		**08/02/17**		**08/18/17**		**09/19/17**		**08/05/18**		**10/08/18**	
	μ	**HG**	μ	**HG**	μ	**HG**	μ	**HG**	μ	**HG**	μ	**HG**	μ	**HG**
H	0.21	a	0.21	a	0.25	a	0.28	a	0.33	a	0.40	a	0.54	a
M	0.30	b	0.29	b	0.32	b	0.34	b	0.38	b	0.41	a	0.55	a
L	0.49	c	0.44	c	0.48	c	0.49	c	0.51	c	0.45	b	0.62	b
U	0.74	d	0.71	d	0.73	d	0.72	d	0.69	d	0.51	c	0.71	c
Normalized Burn Ratio (NBR)														
S	**07/01/17**		**07/17/17**		**08/02/17**		**08/18/17**		**09/19/17**		**08/05/18**		**10/08/18**	
	μ	**HG**	μ	**HG**	μ	**HG**	μ	**HG**	μ	**HG**	μ	**HG**	μ	**HG**
H	-0.28	a	-0.21	a	-0.21	a	-0.15	a	1.10	a	0.28	a	0.29	a
M	-0.05	b	0.00	b	0.01	b	0.06	b	1.69	b	0.29	a	0.30	a
L	0.23	c	0.22	c	0.26	c	0.29	c	2.26	c	0.35	b	0.41	b
U	0.57	d	0.55	d	0.56	d	0.56	d	3.25	d	0.46	c	0.53	c

S: burn severity level; H: high burn severity level; M: moderate burn severity level; L: low burn severity level; U: unburned; μ: mean value; HG: homogeneous groups, different letters in each HG column indicate a group of means within which there are no significant differences, showing significant differences among group means; white background: four levels were discriminated; light grey background: three levels were discriminated; dark grey background: two levels were discriminated

Regarding the 2018 images, only ET and LST could discriminate three burn severity levels with statistical significance. ET displayed significant differences among all burn severity levels in the two 2018 images (summer and fall). LST, however, was not distinct between high and moderate burn severity levels in the image acquired in fall 2018 (8th October 2018). Similarly, NDVI and NBR only discriminated two burn severity levels (high-moderate and low) in the 2018 images. LSA showed the worst performance in the ANOVA analysis as it could only discriminate between the burned and unburned class in both 2018 images. Consequently, ET and LST were the only variables that enabled us to distinguish the three burn severity levels in the 1-year after fire summer image, and ET could also discriminate the three burn severity levels in the 1-year after fire fall image.

Table 3 displays the results of the one-way ANOVA from a multi-temporal perspective. Regarding the initial assessment (difference between pre-fire and immediately post-fire image), the five variables (ET, LST, LSA, NBR and NDVI) performed similarly. All of them could discriminate the three burn severity levels with statistical significance. However, regarding the extended assessment (difference between pre-fire and 1-year after fire image), only dET_e and dLST_e could discriminate them. dLSA_e, dNBR_e, and dNDVI_e enabled only the discrimination of two burn severity levels (high-moderate and low). These results are in accordance with the results of the one-way ANOVA from uni-temporal perspective (Table 2).

Table 3. Fisher's least significant difference test for the difference between pre-fire and post-fire ET, LST, LSA, NBR and NDVI images and burn severity levels.

Initial Assessment (06/15/17–07/01/17)										
S	dET_i		dLST_i		dLSA_i		dNBR_i		dNDVI_i	
	μ	HG	μ	HG	μ	HG	μ	HG	μ	HG
H	3.76	a	9.52	a	0.05	a	0.78	a	0.46	a
M	3.02	b	5.57	b	0.04	b	0.56	b	0.38	b
L	1.52	c	1.04	c	0.03	c	0.32	c	0.22	c
U	-1.07	d	-5.77	d	0.01	d	0.02	d	-0.01	d
Extended assessment (06/15/17 - 08/05/18)										
S	dET_e		dLST_e		dLSA_e		dNBR_e		dNDVI_e	
	μ	HG	μ	HG	μ	HG	μ	HG	μ	HG
H	3.20	a	6.84	a	-0.03	a	0.22	a	0.28	a
M	3.00	b	6.27	b	-0.04	a	0.23	a	0.28	a
L	2.57	c	4.98	c	-0.03	b	0.20	b	0.26	b
U	2.46	d	4.44	d	-0.03	b	0.12	c	0.21	c

S: burn severity level; H: high burn severity level; M: moderate burn severity level; L: low burn severity level; U: unburned; μ: mean value; HG: homogeneous groups, different letters in each HG column indicate a group of means within which there are no significant differences, showing significant differences among group means; white background: four levels were discriminated; light grey background: three levels were discriminated; dark grey background: two levels were discriminated

Table 4 displays the results of the multifactor analysis of variance for ET, LST, LSA, NDVI and NBR that used post-fire burn severity, pre-fire vegetation species, climate, elevation, slope and aspect as factors. We observed that burn severity contributes to more than 80% of ET, LST, NDVI and NBR images (see Table 4, initial assessment), with pre-fire vegetation species and aspect being the next two factors with a high contribution. The total variance explained for the factors in those four variables was approximately 90%. With regards to the extended assessment (images acquired 1-year after fire), the total explained variance showed a decrease when compared to the initial assessment. Burn severity contribution also decreased. The topographic contribution increased for all of the variables. The vegetation species factor also increased its contribution to all variables except ET. Climate only had a relevant contribution to ET variance. LSA displayed the lowest values for the total explained variance. In addition, the contribution of burn severity to LSA was relatively low compared to its contribution to the rest of the variables, and the contribution of pre-fire vegetation was relatively high.

Table 4. Multifactor analysis of variance for ET, LST, LSA, NDVI and NBR. Initial and extend assessment.

Initial Assessment (07/01/17)										
	ET		LST		LSA		NDVI		NBR	
Factors	p-value	%	p-value	%	p-value	%	p-value	%	p-value	%
Burn severity	0.000	83.90	0.000	83.87	0.000	40.53	0.000	85.60	0.000	85.66
Climate	0.107	0.00	0.000	0.43	0.197	0.00	0.415	0.00	0.170	0.00
Vegetation	0.000	1.49	0.000	0.77	0.000	17.47	0.000	1.60	0.000	1.22
Elevation	0.085	0.44	0.000	1.65	0.000	1.27	0.000	0.79	0.000	0.94
Slope	0.000	0.58	0.000	0.07	0.000	0.93	0.007	0.63	0.009	1.07
Aspect	0.000	6.20	0.000	4.44	0.000	8.75	0.000	2.04	0.000	0.24

Table 4. *Cont.*

Total explained		92.61		91.23		68.95		90.66		89.13
Total error		7.39		8.76		31.05		9.34		10.87
Extended assessment (08/05/17)										
	ET		**LST**		**LSA**		**NDVI**		**NBR**	
Factors	**p-value**	**%**	**p-value**	**%**	**p-value**	**%**	**p-value**	**%**	**p-value**	**%**
Burn severity	0.000	37.32	0.000	47.24	0.000	2.52	0.000	27.13	0.000	32.47
Climate	0.000	8.01	0.564	0.00	0.567	0.00	0.074	0.00	0.323	0.00
Vegetation	0.000	0.65	0.000	6.02	0.000	20.13	0.000	10.70	0.000	11.08
Elevation	0.000	10.49	0.000	4.38	0.000	6.17	0.002	5.78	0.006	4.17
Slope	0.000	4.23	0.000	1.75	0.000	5.32	0.000	0.94	0.001	1.06
Aspect	0.000	4.98	0.000	8.85	0.000	6.43	0.000	3.05	0.000	4.03
Total explained		65.68		68.24		40.57		47.60		52.81
Total error		34.32		31.75		59.43		52.40		47.19

p-value: p-values test the statistical significance of each of the factors, p-value < 0.01 indicates a significant contribution at the 99% confidence level; %: percentage of variance contributed by each of the factors

Table 5 summarizes the mean values of ET, LST, LSA, NDVI and NBR before the fire for every class, defined in climate, vegetation species and topographic factors. As expected, higher mean values of ET, NDVI and NBR (and lower of LST) were found in the humid Mediterranean climate (Csb class), at lower elevation (0–300m, 300–600m classes) and at lower slope (0°–5°, 5°–10°) ranges because of a high water availability due to the close proximity to the Zêzere River, and on North aspect slopes (more humid). Regarding vegetation species, differences in mean values of the variables were smaller.

Table 5. Mean values of pre-fire (06/15/17) ET, LST, LSA, NDVI and NBR images in different classes taken into account for climate, vegetation species, elevation, slope and aspect.

		ET (mm/day)	**LST (K)**	**LSA**	**NDVI**	**NBR**
Climate	**Csa**	4.85	307.44	0.13	0.62	0.44
	Csb	5.31	306.64	0.13	0.63	0.44
Vegetation	**Eucalyptus**	4.92	307.47	0.12	0.66	0.49
	Pine	4.82	308.03	0.11	0.67	0.49
	Shrub	4.74	307.98	0.12	0.65	0.43
Elevation (m)	**0-300**	6.34	305.82	0.14	0.64	0.46
	301-600	5.35	307.51	0.13	0.66	0.50
	601-900	4.94	306.61	0.13	0.64	0.47
	>900	3.69	308.21	0.13	0.56	0.33
Slope (°)	**0-5**	5.63	305.57	0.12	0.65	0.46
	5-10	5.20	306.72	0.13	0.63	0.44
	10-20	4.87	307.59	0.13	0.62	0.43
	20-30	4.88	307.57	0.14	0.62	0.44
	>30	4.82	307.74	0.14	0.61	0.42

Table 5. *Cont.*

		ET (mm/day)	LST (K)	LSA	NDVI	NBR
Aspect	North	5.42	305.83	0.13	0.66	0.48
	Northeast	4.97	307.33	0.14	0.64	0.45
	East	4.73	308.16	0.14	0.63	0.43
	Southeast	4.70	308.44	0.14	0.60	0.40
	South	4.92	307.68	0.13	0.61	0.42
	Southwest	5.03	307.20	0.13	0.60	0.42
	West	5.38	306.12	0.13	0.62	0.44
	Northwest	5.49	305.54	0.13	0.64	0.46

4.3. Is it Possible to Estimate Accurately Burn Severity From the ET, LST and LSA Images?

A summary of the accuracy measures of the burn severity estimates from each variable taken into account (uni-temporal and multi-temporal perspective) is displayed in Table 6. From a uni-temporal perspective, all burn severity estimates except the LSA-based one showed adequate accuracy. The post-fire ET-based estimate achieves the highest κ statistic (0.63), followed by the post-fire the NBR-based estimate ($\kappa = 0.61$). When a multi-temporal perspective was adopted, both burn severity estimates based on dET and dLST decreased their accuracy slightly, whereas burn severity estimates based on dNDVI and dNBR increased their accuracy slightly. Consequently, some differences in accuracy were observed in this case: κ statistic was 0.55 in dET-based estimate, and 0.51 in dLST-based, compared to 0.65 and 0.66 in dNDVI- and dNBR-based ones, respectively. The dLSA-based burn severity estimate showed approximately the same accuracy as the LSA-based one in the uni-temporal perspective (too low to be acceptable).

Table 6. Accuracy summary of the burn severity estimates from ET, LST, LSAT, NDVI and NBR (uni-temporal and multi-temporal perspective).

	Uni-Temporal Perspective (07/01/2017)					Multi-Temporal Perspective (06/15/17–07/01/2017)				
	ET	LST	LSA	NDVI	NBR	dET_i	dLST_i	dLSA_i	dNDVI_i	dNBR_i
κ	0.63	0.57	0.44	0.59	0.61	0.55	0.52	0.45	0.65	0.66
PA	0.69	0.65	0.49	0.67	0.70	0.66	0.63	0.58	0.70	0.71
UA	0.68	0.66	0.51	0.70	0.69	0.62	0.60	0.55	0.72	0.70
OA	0.73	0.69	0.61	0.72	0.72	0.67	0.64	0.60	0.75	0.76

Figure 3 provides a visual comparison of burn severity (initial assessment) maps based on immediately post-fire ET and dNBR_i (used as a reference). From it, we can verify a high level of similarity between the two burn severity maps. A small variation in burn severity level actually associated to aspect changes may be observed in the upper right corner of the ET-based map. We should remember that this aspect contributed 6.20% to the variance in immediately after fire ET image (Table 4).

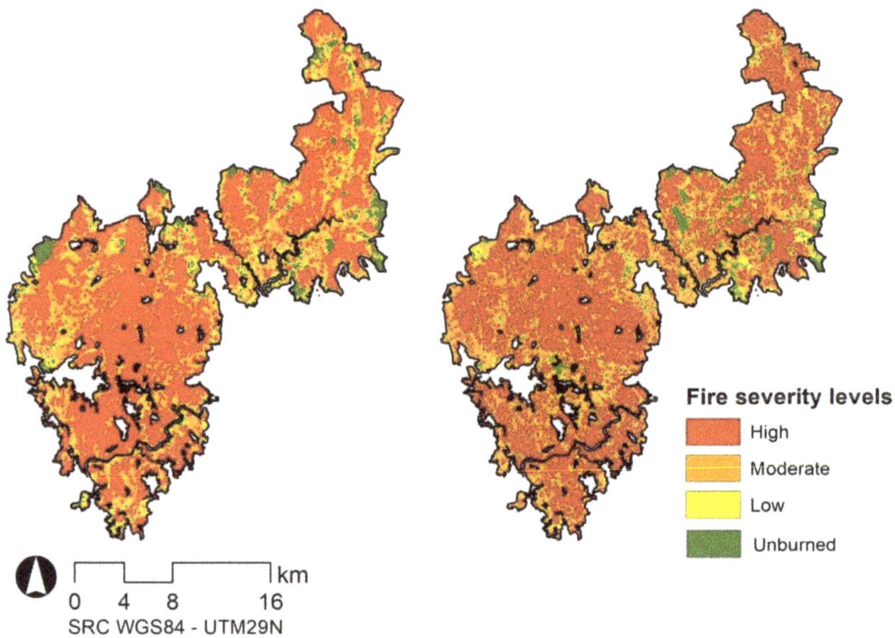

Figure 3. Burn severity maps. Initial assessment. **Left**: based on ET (uni-temporal perspective); **right**: based on dNBR_i (multi-temporal perspective).

5. Discussion

5.1. How Does Fire Modify ET, LST, LSA, NDVI and NBR?

Our results agree with previous studies showing that ET decreases after a fire event (e.g. [9,89] in semi-arid; [15,90] in boreal forests; and [13,35,91] in temperate conifer forests). In Mediterranean ecosystem, Sánchez et al. [17] studied the temporal evolution of changes insurface energy balance over an eleven-year period following fire in a maritime pine-shrub mixed forest (Spain). They observed that, after 6 to 7 years, energy fluxes recovered to their pre-fire values in shrubs, while maritime pine forest needed 9 years. Häusler et al. [37], in a forested area near our study area, observed that, similar to maritime pine forest, ET decreased after fire in the Mediterranean Eucalyptus dominated forests, in particular at moderate and high burn severity levels. In eucalypt forests, but in Australia, Nolan et al. [92] found that ET decreased after fire and was higher in forests burned at high severity compared to moderate levels.

Patterns of variation due to fire in LST were similar to ET but inversed. The increase in LST was higher at high burn severity levels than in moderate and low ones. This increase was also observed one year after the fire. These results are consistent with previous studies concluding that LST shows an increase immediately after wildfire that may persist even years after the event [19,21,93–97]. Liu et al. [25] observed that burned forested areas showed a higher LST mean and variability than unburned areas in boreal forest. In our study, variability was higher in burned areas than in unburned ones in only the first year after the fire. In contrast, variance in unburned areas in 2018 exceeded the variability of burned areas. The 2018 decrease in LST variance in burned areas suggests a reduction in the number of burn severity levels that may be differentiated between, as burn severity determines the magnitude of post-fire changes in LST [90].

Immediately after a fire, ash and charcoal accumulation over soil and charred surfaces causes a decrease in LSA [98,99] whose amplitude is influenced by burn severity [22]. However, both 2018

LSA images displayed higher values for burned areas than for unburned ones. The principal reason for this increase is strong regeneration of Mediterranean vegetation following a fire event [100]. Shrubs begin to regrow in a few weeks after the fire event [101]. E. globulus, one of the dominant species in our study area, has a high resprouting capacity, mainly due to a high epicormic bud initiation potential and to the presence of a lignotuber [102]. By contrast, P. pinaster, the other dominant species, is a seeder, with a rapid seed dispersal from the beginning of fire until some months after its extinction [103]. LSA values in the summer 2018 summer image were also observed to be higher than those of the fall image, suggesting that LSA is greatly influenced by seasonality [20]. Ecosystem type and fire regimes, however, also have an important influence on LSA changes in duration, sign and magnitude [25,99].

Mean NDVI and NBR values in burned and unburned surfaces displayed a similar trend: a fall immediately after the fire, whose amplitude followed a decreasing pattern over time. This pattern responds to the changes in radiation recorded by satellite sensors due to both the removal of healthy vegetation and the presence of ash and charcoal [104]. Healthy vegetation absorbs red radiation and reflects NIR radiation. Conversely, burned vegetation reflects more radiation in red and SWIR wavelengths and absorbs NIR radiation, making their spectral signature flatter than the signature of healthy vegetation [6]. However, NDVI is not the most appropriate index to map burned areas in Mediterranean-type regions [105], as it has not been designed to discriminate ash and char, predominant materials in areas recently affected by fire [106]. The main advantage of NDVI is that it can be computed from almost all existing sensors [107]. Although thresholding the differenced form of NBR (dNBR) [6] has become a standard method to estimate burn severity due to its easy computation [60], some limitations of this method have been pointed out (see [66]) and new alternatives have been proposed. Among them, the ones based on char fraction from unmixing reflective bands stand out [66,75,108–110].

5.2. How Are ET, LST, LSA, NDVI and NBR Post-Fire Images Influenced by Burn Severity and Pre-Fire Factors?

We clearly observed the dependence of ET, LST, NDVI and NBR on burn severity in our study (Table 4). Fisher's least significant difference showed that, from an uni-temporal point of view, the three energy balance related variables and the two spectral indices discriminated three burn severity levels with statistical significance the year of the fire event. Conversely, 1-year after fire, only ET (summer and fall images) and LST (summer image) enabled such discrimination (Table 2). From a multi-temporal approach, results similar to the uni-temporal procedure were obtained. Although energy balance variables showed a strong dependence on burn severity, pre-fire vegetation species, and especially topographic factors, also had a relatively important contribution. Climate (Köppen-Geiger classes, [72]) did not contribute to explain the variance in any of the variables used in our study, except ET in the extended assessment. Our study area is a transition zone between Mediterranean (Csa) and humid Mediterranean (Csb), and there is no clear border between these two climatic classes. We believe that the climatic fuzziness resulted in the climate having almost no influence on the five studied variables.

Vegetation species type has an important influence on energy and water exchange between atmosphere and land [25,89,111]. However, in our study vegetation species had a low to moderate contribution to ET, LST, NDVI and NBR in the initial assessment, and a higher contribution to the extended one. We found very similar ET and LSA values in the main pre-fire vegetation species, although broadleaf trees usually show higher values in ET and LSA than coniferous trees [25]. However, Eucalyptus cannot be =considered a broadleaf tree, and thus its ET values are close to the ET values found in pines (Table 5). With regards to topographic factors, they had a reasonable contribution to ET and LSA, and a lower one to the spectral indices. Aspect was the most influential topographic factor in the initial assessment, with elevation together with aspect forming the most relevant topographic factors in the extended one. Previous studies [92,112] have already demonstrated the important effect of topography on ET, through its influence on evaporative demand and forest structure. Furthermore, topography also influences the recovery pattern ET following the fire event, as vegetation regrow relies partly on elevation [91].

5.3. Is it Possible to Accurately Estimate Burn Severity From the ET, LST and LSA Images?

From a uni-temporal perspective, burn severity estimates based on ET achieved the highest accuracy, with (κ = 0.63) being the acceptable accuracy of LST-based estimates (κ = 0.57). From a multi-temporal perspective, both dET_i- and dLST_i-based estimates showed a lower accuracy than dNDVI_i- and dNBR_i-based ones (multi-temporal approach) (Table 6). The obtained accuracy values were in accordance with accuracy values found in previous studies (e.g., [18,60,75,113–117]).

In our study, burn severity estimates based on variables related to energy balance reached similar accuracies to estimates based on commonly used spectral indices. However, according to Parks et al. [118], these variables should be included to model burn severity across broad regions, due to the difficulty in interpreting models based on spectral indices [64,118,119]. Moreover, these variables provide critical information about the alterations in energy balance caused by fire, and in particular, information on water balance. Fire has a clear impact on stand water cycle [120] and hydrology at watershed scale [121]. Forest wildfires, importantly, alter the hydrological processes, which may impact on the growth of the forest and climate at regional scale [122,123].

5.4. Final Considerations and Future Work

Most previous studies on the impact of fire on energy balance have been based on field-measured surface fluxes, and are only valid at very local scales [13,15,91,92,124]. Others rely on variables derived from coarse spatial resolution satellites (e.g., MODIS), and are only valid at global scales [90,125,126]. There are very few studies (e.g., [17,37]) that use satellite-data-derived variables at a fine spatial resolution. The availability of these data makes it possible not only to work at a local-regional scale, but also to study the potential of these energy balance variables as indicators of burn severity. In our study, ANOVA analysis showed that ET, LST and LSA could discriminate three burn severity levels with statistical significance at initial assessment (both from uni- and multi-temporal perspectives). Burn severity estimates from post-fire ET not only reached almost the same accuracy as the widely used method proposed by Key and Benson [6], but also highlighted the important impact of fire on water balance. Since forests absorb and keep rainfall to make it available during dry season [127], wildfires alter the volume stored by forests by removing the vegetation cover [128]. The consequences of that change on water balance may include: modifications of ET from burned soil and vegetation versus unburned ones [37]; alterations in soil moisture [129]; an increase in the volume of precipitation [130]; a post-fire increase in soil erosion (due to higher surface runoff) [131]; lower water quality caused by an increase in suspended sediments [132] and modifications to surface hydrology [128]. Furthermore, unlike spectral indices, ET has a physical meaning, which makes it easier to interpret than spectral indices. Moreover, procedures based on variables related to energy balance such as ET or LST, estimated from fine-scale remotely sensed data, might reduce over-dependence on spectral indices to infer burn severity [65].

Thus, future research studies about the relationship between burn severity and satellite-derived variables related to water balance should be focused on three areas. First, the classification procedure of the variables related to energy balance needs to be revisited, and other classification algorithms should be tested. Second, future studies should evaluate the integration of post-fire energy balance measures with widely used spectral information to increase the accuracy of burn severity estimates. Third, it is necessary to verify the potential use of these variables as indicators of burn severity in other ecosystems and fire regimes.

6. Conclusions

Wildfires not only modify surface reflectance but also induce changes in energy balance. Our study proved the usefulness of Landsat-derived variables related to energy balance to estimate burn severity with a similar accuracy to other well-established satellite-based methods, with the added advantage of providing a better understanding of fire-induced changes in energy balance. We analyzed the post-fire temporal changes in energy fluxes by means of three variables—ET, LST and LSA—within the area affected by a mega-fire in Mediterranean forests, contributing to the evaluation of their important ecological consequences. Although the relationship between these variables and burn severity has been previously recognized, this is the first study that showed their potential as indicators of burn severity and that used them to estimate accurately burned severity.

Our results revealed that fire has a strong impact on energy balance variables at a regional-local scale during the first year after fire: ET decreased and LST and LSA increased. We showed that the fire-induced changes in these variables were related to burn severity. Three burn severity levels could be distinguished with statistical significance from ET and LST images using uni-temporal and multi-temporal approaches. Thus, the proven relationship between ET and fire damage may provide a physical meaning to burn severity in terms of changes in water balance.

We mapped burn severity from each potential indicator using thresholds based on the mean values for each class. Accuracy of burn severity estimated using ET was the highest from an uni-temporal perspective, although the dNBR-based burn severity map had the highest accuracy from a multitemporal one. We conclude that the three Landsat-derived energy balance variables analyzed (especially ET and LST) can be seen as effective indicators of burn severity in mega-fires in Mediterranean ecosystems. In addition, these variables in themselves provide important measures of the impact of fire on energy balance. The proposed method for estimating burn severity may help advance burn severity assessment and our understanding of burn severity patterns. Finally, it takes advantage of new satellites with a similar spatial resolution to Landsat, equipped to provide an accurate estimate of daily ET, such as ECOSTRESS.

Author Contributions: Conceptualization, A.F.-M. and C.Q.; Formal analysis, A.F.M. and C.Q.; Methodology, A.F.M. and C.Q.; Supervision, D.A.R.; Writing—original draft, A.F.M. and C.Q.; Writing—review and editing, A.F.M., C.Q. and D.A.R. All authors have read and agree to the published version of the manuscript.

Funding: This research was funded by the Spanish Ministry of Economy and Competitiveness (FIRESEVES project, AGL2017-86075-C2-1-R), and the Regional Government of Castile and León (SEFIRECYL project, LE001P17).

Acknowledgments: The two first authors were supported as research visitors at VIPER Lab. (University of California, Santa Barbara) by Spanish Education Ministry grants (Salvador de Madariaga program, codes PRX19/00021 and PRX19/00052).

Conflicts of Interest: The authors declare no conflict of interest.

References

1. Bowman, D.M.J.S.; Balch, J.K.; Artaxo, P.; Bond, W.J.; Carlson, J.M.; Cochrane, M.A.; D'Antonio, C.M.; Defries, R.S.; Doyle, J.C.; Harrison, S.P.; et al. Fire in the earth system. *Science* **2009**, *324*, 481–548. [CrossRef] [PubMed]
2. Scott, A.C.; Bowman, D.M.; Bond, W.J.; Pyne, S.J.; Alexander, M.E. *Fire on Earth: An Introduction*; John Wiley & Sons: Chichester, UK, 2013.
3. Leblon, B.; San-Miguel-Ayanz, J.; Bourgeau-Chavez, L.; Kong, M. Remote sensing of wildfires. In *Land Surface Remote Sensing: Environment and Risks*; Elsevier: Amsterdam, The Netherlands, 2016; pp. 55–95. ISBN 978-1-78548-105-5. [CrossRef]
4. Meng, R.; Zhao, F. A review for recent advances in burned area and burn severity mapping. In *Remote Sensing of Hydrometeorological Hazards*; Petropoulos, G.P., Islam, T., Eds.; Taylor & Francis: Abingdon, UK, 2017.
5. Jain, T.B.; Pilliod, D.; Graham, R.T. Tongue-tied. Confused meanings for common fire terminology can lead to fuels mismanagement. A new framework is needed to clarify and communicate the concepts. *Wildfire* **2004**, *4*, 22–26.
6. Key, C.H.; Benson, N.C. *Landscape Assessment (LA) Sampling and Analysis Methods*; USDA Forest Service General Technical Reports; Rocky Mountain Research Station: Fort Collins, CO, USA, 2006; RMRS-GTR-164-CD.
7. Marcos, E.; Villalón, C.; Calvo, L.; Luis-Calabuig, E. Short-term effects of experimental burning on the soil nutrient in the Cantabrian heathlands. *Ecol. Eng.* **2009**, *35*, 820–828. [CrossRef]

8. González-de-Vega, S.; de las Heras, J.; Moya, D. Resilience of Mediterranean terrestrial ecosystems and fire severity in semiarid areas: Responses of Aleppo pine forests in the short, mid and long term. *Sci. Total Environ.* **2016**, *573*, 1171–1177. [CrossRef]
9. Atchley, A.L.; Kinoshita, A.M.; Lopez, S.R.; Trader, L.; Middleton, R. Simulating surface and subsurface water balance changes due to burn severity. *Vadose Zone J.* **2018**, *17*, 180099. [CrossRef]
10. Cardenas, M.B.; Kanarek, M.R. Soil moisture variation and dynamics across a wildfire burn boundary in a loblolly pine (Pinus taeda) forest. *J. Hydrol.* **2014**, *519*, 490–502. [CrossRef]
11. Moody, J.A.; Ebel, B.A. Infiltration and runoff generation processes in fire-affected soils. *Hydrol. Process.* **2014**, *28*, 3432–3453. [CrossRef]
12. Moody, J.A.; Martin, D.A.; Haire, S.L.; Kinner, D.A. Linking runoff response to burn severity after a wildfire. *Hydrol. Process.* **2008**, *22*, 2063–2074. [CrossRef]
13. Clark, K.L.; Skowronski, N.; Gallagher, M.; Renninger, H.; Schäfer, K. Effects of invasive insects and fire on forest energy exchange and evapotranspiration in the New Jersey pinelands. *For. Meteorol.* **2012**, *166*, 50–61. [CrossRef]
14. Lentile, L.; Holden, Z.; Smith, A.; Falkowski, M.; Hudak, A.; Morgan, P.; Lewis, S.; Gessler, P.; Benson, N. Remote sensing techniques to assess active fire characteristics and post-fire effects. *Int. J. Wildland Fire* **2006**, *15*, 319–345. [CrossRef]
15. Rocha, A.V.; Shaver, G.R. Postfire energy exchange in arctic tundra: The importance and climatic implications of burn severity. *Glob. Chang. Biol.* **2011**, *17*, 2831–2841. [CrossRef]
16. Randerson, J.T.; Liu, H.; Flanner, M.G.; Chambers, S.D.; Jin, Y.; Hess, P.G.; Pfister, G.; Mack, M.C.; Treseder, K.K.; Welp, L.R.; et al. The impact of boreal forest fire on climate warming. *Science* **2006**, *314*, 1130–1132. [CrossRef] [PubMed]
17. Sánchez, J.M.; Bisquert, M.; Rubio, E.; Caselles, V. Impact of land cover change induced by a fire event on the surface energy fluxes derived from remote sensing. *Remote Sens.* **2015**, *7*, 14899–14915. [CrossRef]
18. Quintano, C.; Fernández-Manso, A.; Roberts, D.A. Burn severity mapping from Landsat MESMA fraction images and Land Surface Temperature. *Remote Sens. Environ.* **2017**, *190*, 83–95. [CrossRef]
19. Quintano, C.; Fernández-Manso, A.; Calvo, L.; Marcos, E.; Valbuena, L. Land surface temperature as potential indicator of burn severity in forest Mediterranean ecosystems. *Int. J. Appl. Earth Obs.* **2015**, *36*, 1–12. [CrossRef]
20. Veraverbeke, S.; Verstraeten, W.W.; Lhermitte, S.; Van de Kerchove, R.; Goossens, R. Assessment of post-fire changes in land surface temperature and surface albedo, and their relation with fire-burn severity using multi-temporal MODIS imagery. *Int. J. Wildland Fire* **2012**, *21*, 243–256. [CrossRef]
21. Vlassova, L.; Pérez-Cabello, F.; Rodrigues, M.; Montorio, R.; García-Martín, A. Analysis of the relationship between land surface temperature and wildfire severity in a series of Landsat images. *Remote Sens.* **2014**, *6*, 6136–6162. [CrossRef]
22. Beringer, J.; Hutley, L.; Tapper, N.; Coutts, A.; Kerley, A.; O'Grady, A. Fire impacts on surface heat, moisture and carbon fluxes from a tropical savanna in northern Australia. *Int. J. Wildland Fire* **2003**, *12*, 333–340. [CrossRef]
23. Quintano, C.; Fernández-Manso, A.; Fernández-García, V.; Marcos, E.; Calvo, L. Changes on albedo after a large forest fire in Mediterranean ecosystems. In Proceedings of the Remote Sensing and Modeling of Ecosystems for Sustainability XII (SPIE 9610), San Diego, CA, USA, 11–12 August 2015. [CrossRef]
24. Quintano, C.; Fernandez-Manso, A.; Marcos, E.; Calvo, L. Burn severity and post-fire land surface albedo relationship in Mediterranean forest ecosystems. *Remote Sens.* **2019**, *11*, 2309. [CrossRef]
25. Liu, Z.; Ballantyne, A.P.; Cooper, L.A. Increases in land surface temperature in response to fire in Siberian boreal forests and their attribution to biophysical processes. *Geophys. Res. Lett.* **2018**, *45*, 6485–6494. [CrossRef]
26. Ellison, D.; Morris, C.E.; Locatelli, B.; Sheil, D.; Cohen, J.; Murdiyarso, D.; Gutierrez, V.; van Noordwijkl, M.; Creed, I.F.; Pokorny, J.; et al. Trees, forests and water: Cool insights for a hot world. *Sci. Tech.* **2017**, *43*, 51–61. [CrossRef]
27. Amiro, B.D.; MacPherson, J.I.; Desjardins, R.L. BOREAS flight measurements of forest-fire effects on carbon dioxide and energy fluxes. *Agric. For. Meteorol.* **1999**, *96*, 199–208. [CrossRef]

28. Dintwe, K.; Okin, G.S.; Xue, Y. Fire-induced albedo change and surface radiative forcing in sub-Saharan Africa savanna ecosystems: Implications for the energy balance. *J. Geophys. Res. Atmos.* **2017**, *122*, 6186–6201. [CrossRef]
29. Alkama, R.; Cescatti, A. Biophysical climate impacts of recent changes in global forest cover. *Science* **2016**, *351*, 600–604. [CrossRef] [PubMed]
30. Liu, Z.; Ballantyne, A.P.; Cooper, L.A. Biophysical feedback of global forest fires on surface temperature. *Nat. Commun.* **2019**, *10*, 214. [CrossRef]
31. Benmechet, A.; Abdellaoui, A.; Hamou, A. A comparative study of land surface temperature retrieval methods from remote sensing data. *Can. J. Remote Sens.* **2013**, *39*, 59–73. [CrossRef]
32. Li, Z.-L.; Tang, B.-H.; Wu, H.; Ren, H.; Yan, G.; Wan, Z.; Trigo, I.F.; Sobrino, J.A. Satellite-derived land surface temperature: Current status and perspectives. *Remote Sens. Environ.* **2013**, *131*, 14–37. [CrossRef]
33. Coakley, J.A., Jr. Reflectance and albedo, surface. In *Encyclopedia of the Atmosphere*; Holton, J.R., Curry, J.A., Eds.; Academic Press: Cambridge, CA, USA, 2002; pp. 1914–1923.
34. Liang, S.; Shuey, C.J.; Russ, A.L.; Fang, H.; Chen, M.; Walthall, C.L.; Daughtry, C.S.T.; Hunt, R., Jr. Narrowband to broadband conversions of land surface albedo: II. Validation. *Remote Sens. Environ.* **2002**, *84*, 25–41. [CrossRef]
35. Dore, S.; Kolb, T.E.; Montes-Helu, M.; Eckert, S.E.; Sullivan, B.W.; Hungate, B.A.; Kaye, J.P.; Hart, S.C.; Koch, G.W.; Finkral, A. Carbon and water fluxes from ponderosa pine forests disturbed by wildfire and thinning. *Ecol. Appl.* **2010**, *20*, 663–683. [CrossRef]
36. Montes-Helu, M.C.; Kolb, T.; Dore, S.; Sullivan, B.; Hart, S.C.; Koch, G.; Hungate, B.A. Persistent effects of fire-induced vegetation change on energy partitioning and evapotranspiration in ponderosa pine forests. *Agric. For. Meteorol.* **2009**, *149*, 491–500. [CrossRef]
37. Häusler, M.; Nunes, J.P.; Soares, P.; Sánchez, J.M.; Silva, J.M.N.; Warneke, T.; Keizer, J.J.; Pereira, J.M.C. Assessment of the indirect impact of wildfire (severity) on actual evapotranspiration in eucalyptus forest based on the surface energy balance estimated from remote-sensing techniques. *Int. J. Remote Sens.* **2018**, *39*, 6499–6524. [CrossRef]
38. Van der Tol, C.; Norberto-Parodi, G. Guidelines for remote sensing of evapotranspiration. In *Evapotranspiration—Remote Sensing and Modeling*; Irmak, A., Ed.; InTech: Croatia, Balkans, 2011; pp. 227–250.
39. Allen, R.G.; Tasumi, M.; Trezza, R. Satellite-based energy balance for mapping evapotranspiration with internalized calibration (METRIC)-Model. *J. Irrig. Drain. Eng.* **2007**, *133*, 380–394. [CrossRef]
40. De la Fuente-Sáiz, D.; Ortega-Farí, S.; Fonseca, D.; Ortega-Salazar, S.; Kilic, A.; Allen, R. Calibration of METRIC model to estimate energy balance over a drip-irrigated apple orchard. *Remote Sens.* **2017**, *9*, 670. [CrossRef]
41. Foolad, F.; Blankenau, P.; Kilic, A.; Allen, R.G.; Huntington, J.L.; Erickson, T.A.; Ozturk, D.; Morton, C.G.; Ortega, S.; Ratcliffe, I.; et al. Comparison of the automatically calibrated Google Evapotranspiration Application—EEFlux and the manually calibrated METRIC application. *Preprints* **2018**. [CrossRef]
42. Allen, R.G.; Burnett, B.; Kramber, W.; Huntington, J.; Kjaersgaard, J.; Kilic, A.; Kelly, C.; Trezza, R. Automated calibration of the METRIC-landsat evapotranspiration process. *J. Am. Water Resour. Assoc.* **2013**, *49*, 563–576. [CrossRef]
43. Morton, C.G.; Huntington, J.L.; Pohll, G.M.; Allen, R.G.; McGwire, K.C.; Bassett, S.D. Assessing calibration uncertainty and automation for estimating evapotranspiration from agricultural areas using METRIC. *J. Am. Water Resour. Assoc.* **2013**, *49*, 549–562. [CrossRef]
44. Allen, R.; Morton, C.; Kamble, B.; Kilic, A.; Huntington, J.; Thau, D.; Gorelick, N.; Erickson, T.; Moore, R.; Trezza, R.; et al. EEFlux: A Landsat-based Evapotranspiration mapping tool on the Google Earth Engine. In Proceedings of the 2015 Irrigation Symposium: Emerging Technologies for Sustainable Irrigation, Long Beach, CA, USA, 10–12 November 2015; ASABE: St. Joseph, MI, USA Publication No. 701P0415; Paper Number 43511.
45. Fisher, J.B.; Hook, R.; Allen, R.G.; Anderson, M.C.; French, A.N.; Hain, C.R.; Hulley, G.; Wood, E.F. The ECOsystem Spaceborne Thermal Radiometer Experiment on Space Station (ECOSTRESS): Science motivation. In Proceedings of the American Geophysical Union Fall Meeting, San Francisco, CA, USA, 15–19 December 2014. Abstract id. H31J-07.

46. Fisher, J.B.; Melton, F.; Middleton, E.; Hain, C.; Anderson, M.; Allen, R.; McCabe, M.F.; Hook, S.; Baldocchi, D.; Townsend, P.A.; et al. The future of evapotranspiration: Global requirements for ecosystem functioning, carbon and climate feedbacks, agricultural management, and water resources. *Water Resour. Res.* **2017**, *53*, 2618–2626. [CrossRef]
47. Baldocchi, D.; Falge, E.; Gu, L.; Olson, R.; Hollinger, D.; Running, S.; Anthoni, P.; Bernhofer, C.; Davis, K.; Evans, R.; et al. FLUXNET: A new tool to study the temporal and spatial variability of ecosystem-scale carbon dioxide, water vapor, and energy flux densities. *Bull. Am. Meteorol. Soc.* **2001**, *82*, 2415–2434. [CrossRef]
48. Cabral, A.I.R.; Silva, S.; Silva, P.C.; Vanneschi, L.; Vasconcelos, M.J. Burned area estimations derived from Landsat ETM+ and OLI data: Comparing genetic programming with maximum likelihood and classification and regression Trees. *ISPRS J. Photogramm. Remote Sens.* **2018**, *142*, 94–105. [CrossRef]
49. Chen, G.; Metz, M.R.; Rizzo, D.M.; Dillon, W.W.; Meentemeyer, R.K. Object-based assessment of burn severity in diseased forests using high-spatial and high-spectral resolution MASTER airborne imagery. *ISPRS J. Photogramm. Remote Sens.* **2015**, *102*, 38–47. [CrossRef]
50. Fernández-Manso, A.; Quintano, C.; Roberts, D.A. Burn severity analysis in Mediterranean forests using maximum entropy model trained with EO-1 Hyperion and LiDAR data. *ISPRS J. Photogramm. Remote Sens.* **2019**, *155*, 102–118. [CrossRef]
51. Veraverbeke, S.; Dennison, P.; Gitas, I.; Hulley, G.; Kalashnikova, O.; Katagis, T.; Kuai, L.; Meng, R.; Roberts, D.; Stavros, N. Hyperspectral remote sensing of fire: State-of-the-art and future perspectives. *Remote Sens. Environ.* **2018**, *216*, 105–121. [CrossRef]
52. Franco, M.G.; Mundo, I.A.; Veblen, T.T. Field-validated burn-severity mapping in north Patagonian forests. *Remote Sens.* **2020**, *12*, 214. [CrossRef]
53. Fornacca, D.; Ren, G.; Xiao, W. Evaluating the best spectral indices for the detection of burn scars at several post-fire dates in a mountainous region of Northwest Yunnan, China. *Remote Sens.* **2018**, *10*, 1196. [CrossRef]
54. Arnett, J.T.T.R.; Coops, N.C.; Daniels, L.D.; Falls, R.W. Detecting forest damage after a low-severity fire using remote sensing at multiple scales. *Int. J. Appl. Earth Obs.* **2015**, *35*, 239–246. [CrossRef]
55. Fernández-Manso, A.; Fernandez-Manso, O.; Quintano, C. SENTINEL-2A red-edge spectral indices suitability for discriminating burn severity. *Int. J. Appl. Earth Obs.* **2016**, *50*, 170–175. [CrossRef]
56. Lu, B.; He, Y.; Tong, A. Evaluation of spectral indices for estimating burn severity in semiarid grasslands. *Int. J. Wildland Fire* **2015**, *25*, 147–157. [CrossRef]
57. Quintano, C.; Fernández-Manso, A.; Fernández-Manso, O. Combination of Landsat and Sentinel-2 MSI data for initial assessing of burn severity. *Int. J. Appl. Earth Obs.* **2018**, *64*, 221–225. [CrossRef]
58. Tanase, M.; de la Riva, J.; Pérez-Cabello, F. Estimating burn severity in Aragón pine forest using optical based indices. *Can. J. Forest Res.* **2011**, *41*, 863–872. [CrossRef]
59. Szpakowski, D.M.; Jensen, J.L.R. A review of the applications of remote sensing in fire ecology. *Remote Sens.* **2019**, *11*, 2638. [CrossRef]
60. Soverel, N.O.; Perrakis, D.D.B.; Coops, N.C. Estimating burn severity from Landsat dNBR and RdNBR indices across western Canada. *Remote Sens. Environ.* **2010**, *114*, 1896–1909. [CrossRef]
61. Stambaugh, M.C.; Hammer, L.D.; Godfrey, R. Performance of burn-severity metrics and classification in oak woodlands and grasslands. *Remote Sens.* **2015**, *7*, 10501–10522. [CrossRef]
62. Zhu, Z.; Key, C.; Ohlen, D.; Benson, N. Evaluate Sensitivities of Burn-Severity Mapping Algorithms for Different Ecosystems and Fire Histories in the United States. In *Evaluate Sensitivities of Burn-Severity Mapping Algorithms for Different Ecosystems and Fire Histories in the United States*; Final Report to the Joint Fire Science Program; USGS (United States Geological Survey): Reston, VA, USA; National Park Service Fire Management Program Center: Boise, ID, USA, 2006; Project: JFSP 01-1-4-12.
63. Carvajal-Ramírez, F.; da Silva, J.R.M.; Agüera-Vega, F.; Martínez-Carricondo, P.; Serrano, J.; Moral, F.J. Evaluation of fire severity indices based on pre- and post-fire multispectral imagery sensed from UAV Remote Sens. *Remote Sens.* **2019**, *11*, 993. [CrossRef]
64. Harvey, B.J.; Donato, D.C.; Turner, M.G. Drivers and trends in landscape patterns of stand-replacing fire in forests of the US Northern Rocky Mountains (1984–2010). *Landsc. Ecol.* **2016**, *31*, 2367–2383. [CrossRef]
65. Morgan, P.; Keane, R.E.; Dillon, G.K.; Jain, T.B.; Hudak, A.T.; Karau, E.C.; Sikkink, P.G.; Holden, Z.A.; Strand, E.K. Challenges of assessing fire and burn severity using field measures, remote sensing and modelling. *Int. J. Wildland Fire* **2014**, *23*, 1045–1060. [CrossRef]

66. Lentille, L.B.; Smith, A.M.S.; Hudak, A.T.; Morgan, P.; Bobbitt, M.J.; Lewis, S.A.; Robichaud, P.R. Remote sensing for prediction of 1-year post-fire ecosystem condition. *Int. J. Wildland Fire* **2009**, *18*, 594–608. [CrossRef]
67. San-Miguel-Ayanz, J.; Durrant, T.; Boca, R.; Libertà, G.; Branco, A.; de Rigo, D.; Ferrari, D.; Maianti, P.; Vivancos, T.A.; Costa, H.; et al. *Forest Fires in Europe, Middle East and North Africa 2017*; European Comission, Joint Research Centre: Ispra, Italy, 2018; ISBN 978-92-79-92831-4. [CrossRef]
68. Ribeiro, L.M.; Rodrigues, A.; Lucas, D.; Viegas, D.X. The large fire of Pedrógão Grande (Portugal) and its impact on structures. *Adv. For. Fire Res.* **2018**, 852–858. [CrossRef]
69. ADAI/LAETA. *O Complexo de Incêndios de Pedrógão Grande e Concelhos Limítrofes, Iniciado a 17 de Junho de 2017*; Universidade de Coimbra Official Report; Universidade de Coimbra: Coimbra, Portugal, 2017. (In Portuguese)
70. Pinto, P.; Silva, A. Advances in Forest Fire Research; Viegas, D.X., Ed. ADAI/CEIF University of Coimbra, Portugal; 2018; pp. In *Atmospheric Flow and a Large Fire Interaction: The Unusual Case of Pedrogão Grande, Portugal (17 June 2017)*; Viegas, D.X., Ed.; ADAI/CEIF; University of Coimbra: Coimbra, Portugal, 2018; pp. 922–932. [CrossRef]
71. *CTI, Análise e Apuramento Dos Factos Relativos Aos Incêndios Que Ocorreram Em Pedrogão Grande, Castanheira de Pêra, Ansião, Alvaiázere, Figueiró dos Vinhos, Arganil, Góis, Penela, Pampilhosa da Serra, Oleiros e Sertã, Entre 17 e 24 de Junho de 2017*; Assembleia da República: Lisboa, Portugal, 2017.
72. Köppen, W. *Das Geographische System der Klimate, 1–44*; Gebrüder Borntraeger: Berlin, Germany, 1936.
73. DGT. *Especificações Técnicas da Carta de Uso e Ocupação do Solo de Portugal Continental Para 1995, 2007, 2010 e 2015*; Relatório Técnico; Direção-Geral do Território: Lisboa, Portugal, 2018.
74. Peel, M.C.; Finlayson, B.L.; McMahon, T.A. Updated world map of the Köppen-Geiger climate classification. *Hydrol. Earth Syst. Sci.* **2007**, *11*, 1633–1644. [CrossRef]
75. Quintano, C.; Fernández-Manso, A.; Calvo, L.; Roberts, D.A. Vegetation and soil fire damage analysis based on species distribution modeling trained with multispectral satellite data. *Remote Sens.* **2019**, *11*, 1832. [CrossRef]
76. Fernández-García, V.; Quintano, C.; Taboada, A.; Marcos, E.; Calvo, L.; Fernández-Manso, A. Remote sensing applied to the study of fire regime attributes and their influence on post-fire greenness recovery in pine ecosystems. *Remote Sens.* **2018**, *10*, 733. [CrossRef]
77. Yu, X.; Guo, X.; Wu, Z. Land surface temperature retrieval from Landsat 8 TIRS—Comparison between radiative transfer equation-based method, split window algorithm and single channel method. *Remote Sens.* **2014**, *6*, 9829–9852. [CrossRef]
78. R Core Team. R: A Language and Environment for Statistical Computing, URL. Available online: http://www.R-project.org (accessed on 23 October 2019).
79. Sobrino, J.A.; Jiménez-Muñoz, J.C.; Sòria, G.; Romaguera, M.; Guanter, L.; Moreno, J.; Plaza, A.; Martínez, P. Land surface emissivity retrieval from different VNIR and TIR sensors. *IEEE T. Geosci. Remote* **2008**, *48*, 316–327. [CrossRef]
80. Liang, S. Narrowband to broadband conversions of land surface albedo. I Algorithms. *Remote Sens. Environ.* **2000**, *76*, 213–238. [CrossRef]
81. Irmak, A.; Allen, R.G.; Kjaersgaard, J.; Huntington, J.; Kamble, B.; Trezza, R.; Ratcliffe, I. Operational remote sensing of ET and challenges. In *Evapotranspiration—Remote Sensing and Modeling*; Irmak, A., Ed.; InTech: Rijeka, Croatia, 2011; p. 526.
82. López-García, M.J.; Caselles, V. Mapping burns and natural reforestation using Thematic Mapper data. *Geocarto Int.* **1991**, *1*, 31–37. [CrossRef]
83. Tucker, C.J. Red and photographic infrared linear combinations formonitoring vegetation. *Remote Sens. Environ.* **1979**, *8*, 127–150. [CrossRef]
84. Congalton, R.G.; Green, K. *Assessing the Accuracy of Remotely Sensed Data. Principles and Practices*, 2nd ed.; CRC Press: Boca Ratón, FL, USA, 2009.
85. Weatherspoon, C.P.; Skinner, C.N. An assessment of factors associated with damage to tree crowns from the 1987 wildfires in northern California. *For. Sci.* **1995**, *41*, 430–451.
86. Godwin, D.R.; Kobziar, L.N. Comparison of burn severities of consecutive large-scale fires in Florida sand pine scrub using satellite imagery analysis. *Fire Ecol.* **2011**, *7*, 99–113. [CrossRef]

87. Miler, J.D.; Thode, A.E. Quantifying burn severity in a heterogeneous landscape with a relative version of the delta Normalized Burn Ratio (dNBR). *Remote Sens. Environ.* **2007**, *109*, 66–80. [CrossRef]
88. Botella-Martínez, M.A.; Fernández-Manso, A. Study of post-fire severity in the Valencia region comparing the NBR, RdNBR and RBR indexes derived from Landsat 8 images. *Span. J. Remote Sens. Rev. Asoc. Española Teledetección* **2017**, *49*, 33–47.
89. Poon, P.K.; Kinoshita, A.M. Spatial and temporal evapotranspiration trends after wildfire in semi-arid landscapes. *J. Hydrol.* **2018**, *559*, 71–83. [CrossRef]
90. Li, X.; Zhang, H.; Yang, G.; Ding, Y.; Zhao, J. Post-Fire vegetation succession and surface energy fluxes derived from remote sensing. *Remote Sens.* **2018**, *10*, 1000. [CrossRef]
91. Roche, J.W.; Goulden, M.L.; Bales, R.C. Estimating evapotranspiration change due to forest treatment and fire at the basin scale in the Sierra Nevada, California. *Ecohydrology* **2018**, *11*, e1978. [CrossRef]
92. Nolan, R.H.; Lane, P.N.J.; Benyon, R.G.; Bradstock, R.A.; Mitchell, P.J. Changes in evapotranspiration following wildfire in resprouting eucalypt forests. *Ecohydrology* **2014**, *7*, 1363–1377. [CrossRef]
93. Huang, S.; Jin, S.; Dahal, D.; Chen, X.; Young, C.; Liu, H.; Liu, S. Reconstructing satellite images to quantify spatially explicit land surface change caused by fires and succession: A demonstration in the Yukon River Basin of interior Alaska. *ISPRS J. Photogramm. Remote Sens.* **2013**, *79*, 94–105. [CrossRef]
94. Lambin, E.; Goyvaerts, K.; Petit, C. Remotely-sensed indicators of burn-ing efficiency of savannah and forest fires. *Int. J. Remote Sens.* **2003**, *24*, 3105–3118. [CrossRef]
95. Mölders, N.; Kramm, G. Influence of wildfire induced land-cover changes on clouds and precipitation in interior Alaska—A case of study. *Atmos. Res.* **2007**, *82*, 142–168. [CrossRef]
96. Veraverbeke, S.; Van de Kerchove, R.; Verstraeten, W.; Lhermitte, S.; Goossens, R. Fire-induced changes in vegetation, albedo and land surface temperature assessed with MODIS. In Proceedings of the EARSeL Symposium 2010 Remote Sensing for Science, Education, and Natural and Cultural Heritage, Paris, France, 31 May–1 June 2010; University of Oldenburg: Oldenburg, Germany, 2010; pp. 431–438.
97. Wendt, C.; Beringer, J.; Tapper, N.; Hutley, L. Local boundary-layer development over burnt and unburnt tropical savanna: An observational study. *Bound. Layer Meteorol.* **2007**, *124*, 291–304. [CrossRef]
98. Amiro, B.D.; Orchansky, A.L.; Barr, A.G.; Black, T.A.; Chambers, S.D.; Chapin, F.S., III; Goulden, M.L.; Litvak, M.; Liu, H.P.; McCaughey, J.H.; et al. The effect of post-fire stand age on the boreal forest energy balance. *Agric. For. Meteorol.* **2006**, *140*, 41–50. [CrossRef]
99. Gatebe, C.K.; Ichoku, C.M.; Poudyal, R.; Román, M.O.; Wilcox, E. Surface albedo darkening from wildfires in northern sub-Saharan Africa. Environ. *Res. Lett.* **2014**, *9*, 065003. [CrossRef]
100. Tessler, N.; Wittenberg, L.; Greenbaum, N. Vegetation cover and species richness after recurrent forest fires in the Eastern Mediterranean Ecosystem of Mount Carmel, Israel. *Sci. Total Environ.* **2016**, *572*, 1395–1402. [CrossRef]
101. Bastos, A.; Gouveia, C.M.; DaCamara, C.C.; Trigo, R.M. Modelling post-fire vegetation recovery in 899 Portugal. *Biogeo Sci.* **2011**, *8*, 3593–3607. [CrossRef]
102. Whittock, S.P.; Apiolaza, L.A.; Kelly, C.M.; Potts, B.M. Genetic control of coppice and lignotuber development in Eucalyptus globulus. *Aust. J. Bot.* **2003**, *51*, 57–68. [CrossRef]
103. Herranz, J.M.; Martínez-Sánchez, J.J.; Marín, A.; Ferrandis, P. Post-fire regeneration of Pinus halepensis Miller in a semi-arid area in Albacete province (south-eastern Spain). *Ecoscience* **1997**, *4*, 86–90. [CrossRef]
104. Kontoes, C.C.; Poilve, H.; Florsch, G.; Keramitsoglou, I.; Paralikidis, S. A comparative analysis of a fixed thresholding vs. A classification tree approach for operational burn scar detection and mapping. *Int. J. Appl. Earth Obs.* **2009**, *11*, 299–316. [CrossRef]
105. Pereira, J.M.C.; Sa, A.C.L.; Sousa, A.M.O.; Martín, M.P.; Chuvieco, E. Regional-scale burnt area mapping in Southern Europe using NOAA-AVHRR 1 km data. In *Remote Sensing of Large Wildfires in the European Mediterranean Basin*; Chuvieco, E., Ed.; Springer: Berlin, Germany, 1999; pp. 139–155.
106. Martín, M.P.; Gómez, I.; Chuvieco, E. Performance of a burned-area index (BAIM) for mapping Mediterranean burned scars from MODIS data. In Proceedings of the 5th International Workshop on Remote Sensing and GIS Applications to Forest Fire Management: Fire Effects Assessment, Zaragoza, Spain, 16–18 June 2005; Universidad de Zaragoza: Zaragoza, Spain, 2005.
107. Salvador, R.; Valeriano, J.; Pons, X.; Díaz-Delgado, R. A semiautomatic methodology to detect fire scars in shrubs and evergreen forest with Landsat MSS time series. *Int. J. Remote Sens.* **2000**, *21*, 655–671. [CrossRef]

108. Quintano, C.; Fernández-Manso, A.; Roberts, D.A. Multiple endmember spectral mixture analysis (MESMA) to map burn severity levels from Landsat images in Mediterranean countries. *Remote Sens. Environ.* **2013**, *136*, 76–88. [CrossRef]
109. Tane, Z.; Roberts, D.; Veraverbeke, S.; Casas, A.; Ramirez, C.; Ustin, S. Evaluating endmember and band selection techniques for multiple endmember spectral mixture analysis using post-fire imaging spectroscopy. *Remote Sens.* **2018**, *10*, 389. [CrossRef]
110. Veraverbeke, S.; Hook, S.J. Evaluating spectral indices and spectral mixture analysis for assessing fire severity, combustion completeness and carbon emissions. *Int. J. Wildland Fire* **2013**, *22*, 707–720. [CrossRef]
111. Rogers, B.M.; Soja, A.J.; Goulden, M.L.; Randerson, J.T. Influence of tree species on continental differences in boreal fires and climate feedbacks. *Nat. Geosci.* **2015**, *8*, 228–234. [CrossRef]
112. Goulden, M.L.; Bales, R.C. Mountain runoff vulnerability to increased evapotranspiration with vegetation expansion. *Proc. Natl. Acad. Sci. USA* **2014**, *111*, 14071–14075. [CrossRef]
113. Venâncio, P. Open Source, Open Data e Citizen Science contributos para a Avaliação da Catástrofe de Pedrógão Grande e Góis. In SASIG 2017. In Proceedings of the Encontro Nacional de Software Aberto para Sistemas de Informação Geográfica, Porto, Portugal, 20–22 November 2017.
114. Cocke, A.E.; Fulé, P.Z.; Crouse, J.E. Comparison of burn severity assessments using differenced normalized burn ratio and ground data. *Int. J. Wildland Fire* **2005**, *14*, 189–198. [CrossRef]
115. Cansler, C.A.; McKenzie, D. How robust are burn severity indices when applied in a new region? Evaluation of alternate field-based and remote-sensing methods. *Remote Sens.* **2012**, *4*, 456–483. [CrossRef]
116. Amos, C.; Petropoulos, G.P.; Ferentinos, K.P. Determining the use of Sentinel-2A MSI for wildfire burning & severity detection. *Int. J. Remote Sens.* **2019**, *40*, 905–930. [CrossRef]
117. Tran, B.N.; Tanase, M.A.; Bennett, L.T.; Aponte, C. Evaluation of spectral indices for assessing fire severity in australian temperate forests. *Remote Sens.* **2018**, *10*, 1680. [CrossRef]
118. Parks, S.A.; Holsinger, L.M.; Panunto, M.H.; Jolly, W.M.; Dobrowski, S.Z.; Dillon, G.K. High-severity fire: Evaluating its key drivers and mapping its probability across western US forests. *Environ. Res. Lett* **2018**, *13*, 044037. [CrossRef]
119. Harvey, B.J.; Andrus, R.A.; Anderson, S.C. Incorporating biophysical gradients and uncertainty into burn severity maps in a temperate fire-prone forested region. *Ecosphere* **2019**, *10*. [CrossRef]
120. Ewers, B.E.; Gower, S.T.; Bond-Lamberty, B.; Wang, C. Effects of stand age and tree species composition on transpiration and canopy conductance of boreal forest stands. *Plant Cell Environ.* **2005**, *28*, 660–678. [CrossRef]
121. Valeo, C.; Beaty, K.; Hesslein, R. Influence of forest fires on climate change studies in the central boreal forest of Canada. *J. Hydrol.* **2003**, *280*, 91–104. [CrossRef]
122. Bond-Lamberty, B.; Peckham, S.D.; Gower, S.T.; Ewers, B.E. Effects of fire on regional evapotranspiration in the central Canadian boreal forest. *Glob. Chang. Biol.* **2009**, *15*, 1242–1254. [CrossRef]
123. Cai, X.; Riley, W.J.; Zhu, Q.; Tang, J.; Zeng, Z.; Bisht, G.; Randerson, J.T. Improving representation of deforestation effects on evapotranspiration in the E3SM land model. *J. Adv. Model. Earth Syst.* **2019**. [CrossRef]
124. Nolan, R.H.; Lane, P.N.J.; Benyon, R.G.; Bradstock, R.A.; Mitchell, P.J. Trends in evapotranspiration and streamflow following wildfire in resprouting eucalypt forests. *J. Hydrol.* **2015**, *524*, 614–624. [CrossRef]
125. García-Llamas, P.; Suárez-Seoane, S.; Taboada, A.; Fernández-García, V.; Fernández-Guisuraga, J.M.; Fernández-Manso, A.; Quintano, C.; Marcos, E.; Calvo, L. Assessment of the influence of biophysical properties related to fuel conditions on fire severity using remote sensing techniques: A case study on a large fire in NW Spain. *Int. J. Wildland Fire* **2019**, *28*, 512–520. [CrossRef]
126. Ha, W.; Kolb, T.E.; Springer, A.E.; Dore, S.; O'Donnell, F.C.; Martínez, R.; López, S.; Koch, G.W. Evapotranspiration comparisons between eddy covariance measurements and meteorological and remote-sensing-based models in disturbed ponderosa pine forests. *Ecohydrology* **2014**, *8*, 1335. [CrossRef]
127. Qazi, N.Q.; Bruijnzeel, L.A.; Rai, S.P.; Ghimire, C.P. Impact of forest degradation on streamflow regime and runoff response to rainfall in the Garhwal Himalaya, Northwest India. *Hydrolog. Sci. J.* **2017**, *62*, 1114–1130. [CrossRef]
128. Pimentel, R.; Arheimer, B. Wildfire impact on Boreal hydrology: Empirical study of the Västmanland fire 2014 (Sweden). *Hydrol. Earth Syst. Sci. Discuss.* **2018**, 1–26. [CrossRef]

129. Van der Ent, R.J.; Wang-Erlandsson, L.; Keys, P.W.; Savenije, H.H.G. Contrasting roles of interception and transpiration in the hydrological cycle—Part 2: Moisture recycling. *Earth Syst. Dynam.* **2014**, *5*, 471–489. [CrossRef]
130. Jeffery, M.L.; Yanites, B.J.; Poulsen, C.J.; Ehlers, T.A. Vegetation-precipitation controls on Central Andean topography. *J. Geophys. Res. Earth* **2014**, *119*, 1354–1375. [CrossRef]
131. Cerda, A.; Robichaud, P.R. *Fire Effects on Soils and Restoration Strategies*; CRC Press: Boca Ratón, FL, USA, 2009.
132. Smith, H.G.; Sheridan, G.J.; Lane, P.N.J.; Nyman, P.; Haydon, S. Wildfire effects on water quality in forest catchments: A review with implications for water supply. *J. Hydrol.* **2011**, *396*, 170–192. [CrossRef]

remote sensing

MDPI

Article

A New Model for Transfer Learning-Based Mapping of Burn Severity

Zhong Zheng [1,2,3,4], Jinfei Wang [2,*], Bo Shan [2], Yongjun He [2], Chunhua Liao [2], Yanghua Gao [3] and Shiqi Yang [3]

1 College of Resources and Environment, Chengdu University of Information Technology, Chengdu 610225, China; zhengzhong@cuit.edu.cn
2 Department of Geography, The University of Western Ontario, London, ON N6A 5C2, Canada; bshan3@uwo.ca (B.S.); yhe563@uwo.ca (Y.H.); cliao33@uwo.ca (C.L.)
3 Chongqing Institute of Meteorological Sciences, Chongqing 401147, China; gaoyanghua@sina.com (Y.G.); yangshiqi1980@sina.com (S.Y.)
4 School of Geographical Sciences, Southwest University, Chongqing 400715, China
* Correspondence: jfwang@uwo.ca

Received: 29 January 2020; Accepted: 19 February 2020; Published: 21 February 2020

Abstract: In recent years, global forest fires have occurred more frequently, seriously destroying the structural functions of forest ecosystem. Mapping the burn severity after forest fires is of great significance for quantifying fire's effects on landscapes and establishing restoration measures. Generally, intensive field surveys across burned areas are required for the effective application of traditional methods. Unfortunately, this requirement could not be satisfied in most cases, since the field work demands a lot of personnel and funding. For mapping severity levels across burned areas without field survey data, a semi-supervised transfer component analysis-based support vector regression model (SSTCA-SVR) was proposed in this study to transfer knowledge trained from other burned areas with field survey data. Its performance was further evaluated in various eco-type regions of southwestern United States. Results show that SSTCA-SVR which was trained on source domain areas could effectively be transferred to a target domain area. Meanwhile, the SSTCA-SVR could maintain as much spectral information as possible to map burn severity. Its mapped results are more accurate (RMSE values were between 0.4833 and 0.6659) and finer, compared to those mapped by ΔNDVI-, ΔLST-, ΔNBR- (RMSE values ranged from 0.7362 to 1.1187) and SVR-based models (RMSE values varied from 1.7658 to 2.0055). This study has introduced a potentially efficient mechanism to map burn severity, which will speed up the response of post-fire management.

Keywords: transfer learning model; SSTCA; burn severity; forest fire; SVR

1. Introduction

Forests contain up to 80% of the planet's above ground biomass carbon, which plays a very important role in maintaining the security of global ecosystem [1,2]. Annually, millions of hectares of global forest areas are severely damaged by fires [3]. Mapping the spatial distribution of damage levels not only facilitates the quantification of fire's impact on landscapes, but also effectively guides the implementation of post-fire restoration efforts [4,5].

Burn severity is often considered a measure for describing fire's impacts on forest landscape, which is commonly measured by a quantitative metric: composite burn index (CBI) [6,7]. For mapping the spatial distribution of this variable, the field survey method is supposed to be direct and accurate, but requires a great deal of time, money, and human resources [8]. With large-scale and low-cost characteristics, remote sensing-based methods can greatly improve the efficiency of mapping burn severity [9,10].

Generally, remote sensing-based methods for mapping burn severity include: predictor variables-based regression models [5,7,9], spectral mixture analysis [4,11], and radiative transfer models [6,12,13]. Regarding predictor variables, spectral indices are widely employed in most reported models [14–22]. One of spectral indices is the normalized burn ratio (NBR) which can be calculated from near- and shortwave-infrared spectral bands of remotely sensed data [17]. The differenced NBR (ΔNBR) was further derived from pre- and post-fire images and it has been recognized as a reference for mapping burn severity [4,23]. Following that, a relative modification of ΔNBR (RdNBR) was developed to remove the burn severity misclassification of low vegetated pixels [24]. Based on the Moderate Resolution Imaging Spectroradiometer Fire Radiative Power data (i.e., MODIS FRP), Heward, et al. [25] further tested the potential of fire intensity measures to predict burn severity. For avoiding several difficulties associated with the denominator of RdNBR, Parks, et al. [26] proposed and evaluated a new relative version of ΔNBR, i.e., relativized burn ratio (RBR).

As another new and attractive predictor variable, the land surface temperature (LST) has recently been applied to map burn severity [27–33]. To be specific, Lentile et al. [27] argued that thermal infrared bands might be a predictor variable of burn severity. After that, the relationship between LST differences and burn severity levels was assessed, separately by using MODIS [31] and Landsat images [34]. In Mediterranean forest ecosystems, Quintano et al. [28] first confirmed that summer post-fire LST might be a valuable predictor variable for mapping burn severity. Meanwhile, Chen et al. [35] stated that the mid to thermal infrared spectral bands from MASTER data (i.e., MODIS/ASTER airborne simulator data) could augment the visible to shortwave infrared spectral bands from Landsat in burn severity assessment. Following that, Zheng et al. [30] proposed a new burn severity index combining LST with enhanced vegetation index (EVI) together in the Western United States.

Spectral mixture analysis (SMA) was also utilized to generate fraction images for estimating burn severity. For example, Fernández-Manso et al. [36] employed SMA in northwestern Spain to obtain fraction images as input data for object-based image analysis, which achieved a high overall accuracy and improved accuracy of individual classes. Then, Quintano et al. [11] used multiple endmember spectral mixture analysis (MESMA) to overcome the linear-SMA limitation of fixed endmembers. More recently, Quintano et al. [4] further concluded that combining fraction images of MESMA and LST can map burn severity accurately in Mediterranean countries.

Moreover, to propose a physically based approach despite of biomes characteristics, Chuvieco, et al. [6] applied the radiative transfer models (RTM) to simulate spectral signatures for assessing burn severity. Subsequently, De Santis and Chuvieco [13] applied the inversion of a simulation model to estimate burn severity levels in central Spain. By combining two commonly used RTM models (i.e., PROSPECT at leaf level and GeoSail at canopy level), De Santis, et al. [12] further improved simulation models to estimate burn severity from satellite data in three Mediterranean forest fires.

If dense field survey data are presented as training samples, methods mentioned above will perform well in different challenging scenarios. But, because of the limitation of some resources, the amount of field survey data usually is limited. In a previous study, Zheng et al. [8] proposed a SVR-based model with multi-temporal satellite data to maximize the generalization ability of trained models. Results indicated that, in small sample size scenarios, the SVR-based model could achieve better generalization ability due to reduced complexity. Unfortunately, the more common reality is that field survey data are not available in most cases, since the field work demands a lot of personnel and funding. Thus, there is an urgent need to develop a new model for transferring knowledge trained from other burned areas with field survey data [37].

The main purpose of this study was to propose a transfer learning model for mapping severity levels across burned areas where field survey data are not available. This model is based on a domain adaptation approach (i.e., semi-supervised Transfer Component Analysis, SSTCA [38]) which can transfer trained knowledge on source domain areas (i.e., burned areas with field survey data) to a target domain area (i.e., a burned area where no field survey data are available). The main objectives of this study are to: (1) Develop a transfer learning-based model (i.e., SSTCA-SVR) for mapping burn severity, (2) then test its performance and compare it with some predictor variables- and SVR-based models, (3) and consequently evaluate its reliability in different eco-type regions using a cross-validation procedure.

2. Methodology

The SSTCA-SVR consists of the following three steps (see in Figure 1): (a) the SSTCA was applied at first to project a set of shared and invariant features from source and target domains [39]. (b) Based on the projected features of field survey data from source domain areas, support vector regression (SVR) model was trained and the burn severity level of each pixel from the target domain area was estimated. (c) The performance of SSTCA-SVR was validated and compared with that of other models (i.e., predictor variables-based regression models and SVR model with original spectral features). Its main operations are described in the following sections.

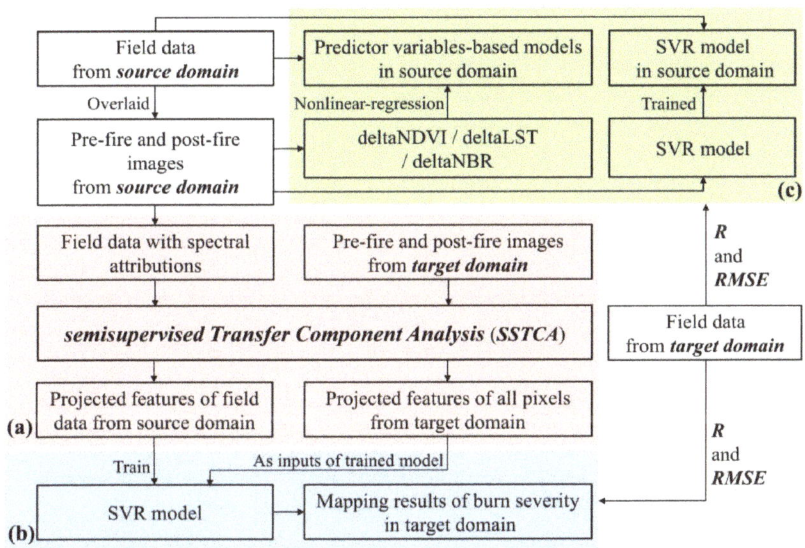

Figure 1. The flowchart of methodology: (**a**) domain adaptation; (**b**) burn severity mapping; and (**c**) performance validation and comparisons.

2.1. Domain Adaptation

Generally, a model for burn severity mapping is trained and applied in the same burned area, which means that training data and targeting data all come from the same domain area. Thus, the feature of these data should have the same distribution:

$$P(Y_{Tr}|X_{Tr}) = P(Y_{Ta}|X_{Ta}) \tag{1}$$

where X_{Tr} represents the feature space $\{x_{Tr1}, x_{Tr2}, \ldots\ldots, x_{Trn}\}$ of ***training data***, Y_{Tr} is their corresponding output which could be obtained using the field survey, and $P(X)$ means a marginal probability of feature distributions. While, X_{Ta} is the feature space $\{x_{Ta1}, x_{Ta2}, \ldots\ldots, x_{Tan}\}$ of ***target data*** and Y_{Ta} is their corresponding output which needs to be estimated.

The corresponding outputs of training data are typically difficult to obtain because of the resources required [40]. Thus, there is a strong desire to make the use of any related data (i.e., ***source data***) [38].

However, the distributions of source data and target data vary, since these data are collected under different background conditions:

$$P(Y_S|X_S) \neq P(Y_{Tr}|X_{Tr}) \tag{2}$$

where X_S is the feature space $\{x_{S1}, x_{S2}, \ldots \ldots, x_{Sn}\}$ of source data and Y_S are their corresponding outputs which have already been obtained in previous field survey.

Here, a transformation Φ is needed to project original spectral features of source and target data to an invariant subspace which can be shared:

$$P(\Phi(X_S)) \approx P(\Phi(X_{Ta})) \tag{3}$$

This means that the distributions of source and target data are similar:

$$P(Y_S|\Phi(X_S)) \approx P(Y_{Ta}|\Phi(X_{Ta})) \tag{4}$$

Hence, the model which was trained on projected features of source data could be applied to estimate corresponding outputs of target data with projected subspace feature as inputs. Next, the key issue is to find an appropriate transformation Φ using the SSTCA, which could be defined using a supervision process [38]. The resulting function of SSTCA should fulfill the following three objectives:

***Objective 1*:** To minimize the distance between the distributions of $\Phi(X_S)$ and $\Phi(X_{Ta})$ in subspace after using the transformation Φ, as well as to preserve the variances of initial data.

The similarity of projected features in subspace should be great, which could be measured using a distance metric: *Maximum Mean Discrepancy* (i.e., Formula (A1) in Appendix A). By virtue of well-known *kernel matrix* (i.e., Formula (A2) in Appendix A) and coefficient L (i.e., Formula (A3)) in Appendix A), this distance can be written as a matrix trace (i.e., Formula (A4) in Appendix A). Further, considering the use of original kernel matrix $\boldsymbol{K}$ and transformation matrix $\boldsymbol{W}$, kernel matrix between projected data (K^*) could be calculated as Formula (A5) in Appendix A. Consequently, we could rewrite the *Maximum Mean Discrepancy* of projected source ($X_S{}^*$) and projected target ($X_{Ta}{}^*$) data as Formula (A6) in Appendix A.

Meanwhile, the statistical properties of original data should be preserved, since it might be useful for the target supervised learning job. The variance matrix of projected samples (Σ^*) could be quantitatively calculated as Formula (A7) in Appendix A.

Based on ***Objective 1***, the unsupervised version of SSTCA (i.e., TCA) could be expressed as:

$$\begin{array}{ll} \min_W & tr(W^\top KLKW) + \mu tr(W^\top W) \\ s.t. & W^\top KHKW = I_m \end{array} \tag{5}$$

where $tr(W^\top W)$ is a regularization term which is introduced for controlling the complexity of W by adjusting μ.

***Objective 2*:** To maximize the dependence between projected features of source data and their corresponding outputs. This is to make full use of outputs of available source data for defining an appropriate transformation Φ. Specifically, this objective could be expressed as:

$$tr(H(KWW^\top K)HK^*_{YY}) = tr(W^\top KHK^*_{YY}HKW) \tag{6}$$

where K^*_{YY} can be calculated as Formula (A9) in Appendix A.

***Objective 3*:** To preserve the local geometrical structure. This is to ensure that close data in the original spectral feature space still are close in the projected subspace. This objective could be expressed as:

$$\frac{1}{(n_S + n_{Ta})^2} \sum_{i,j} M_{i,j} \|X^*_i - X^*_j\|^2 = \frac{1}{(n_S + n_{Ta})^2} tr(W^\top K \mathcal{L} KW) \tag{7}$$

where $\mathcal{L}$ is the graph Laplacian matrix which can be computed as Formula (A10) in Appendix A.

Integrated with ***Objective 2*** and ***Objective 3***, the ***Objective 1***-based TCA can be updated as a semi-supervised version, i.e., the SSTCA:

$$\begin{array}{ll} \min_W & tr(W^\top KLKW) + \mu tr(W^\top W) + \frac{\lambda}{(n_S+n_{Ta})^2} tr(W^\top K \mathcal{L} KW) \\ s.t. & W^\top KHK^*_{YY}HKW = I_m \end{array} \quad (8)$$

The Formula (8) could be further reformulated as trace maximization which could be solved by introducing Lagrange multipliers and a supervision process. The SSTCA was implemented in this study using "a domain adaptation toolbox" of Matlab 2011b [41] and the Eigendecompose matrix is Formula (A12) in Appendix A. In this matrix, m leading eigenvectors is selected to build the matrix W for transformation Φ.

2.2. Burn Severity Mapping

After applying the transformation Φ, original spectral features of source data are projected to a shared subspace. Then, parameters of traditional SVR (i.e., w^*) are trained on these projected features (i.e., $X^*_{S,i}$) [42–44]. The basic principle of SVR model could be seen in Formula (A13) in Appendix A, which was executed in this study using the LIBSVM Toolbox of Matlab 2011b.

Based on the trained SVR model (i.e., $Y_{S,i} = f(X^*_{S,i}, w^*, b)$), $X^*_{S,i}$ could be replaced by the projected features of target data (i.e., $X^*_{Ta,j}$), since they distribute similarly in the projected subspace:

$$Y_{Ta,j} = f(X^*_{Ta,j}, w^*, b) \quad (9)$$

2.3. Performance Validation and Comparisons

Subsequently, a validation procedure was conducted to test the performance of SSTCA-SVR. More specifically, for field survey data from the target domain area, their estimated values were calculated using SSTCA-SVR and two accuracy measurements of R and $RMSE$ were then computed. The performance of SSTCA-SVR was then evaluated against some predictor variables-based regression models (i.e., ΔNDVI, ΔLST, and ΔNBR) as well as that of SVR model with original spectral features.

3. Data and Processing

3.1. Study Area

The Bear Fire, located in the Southwest of United States, was selected as the target domain area in this study (see in Figure 2). This fire occurred at the Dinosaur National Monument on 27 June 2002. After 8 days of burning, the final burned area was about 18.62 km^2. This area was situated in the Highland (alpine) region with a low annual precipitation (<300 mm), which was mainly covered by trees, shrubs, and grasses.

Jasper Fire, Mule Fire, and PW03-Wolf Fire were chosen as the source domain areas where training data were from. The Jasper Fire, alarmed on 24 August 2000, was located in the Black Hills National Forest. Its burned area was about 336 km^2. The local region was Semiarid Steppe climate and the primary tree species was ponderosa pine. The Mule Fire was caused by lightning strike on 11 July 2002 and occurred in Northern Rockies region. Its climate also belongs to Highland (alpine) with annual precipitation between 750 and 1150 mm. Predominant tree species in this area were subalpine fir, Engelmann spruce, and Douglas-fir. The Pw03-Wolf Fire includes two nearby fires and both of them were located in the Yosemite National Park of California region. These burned area featured Mediterranean climate with a high annual precipitation being between 804 and 1722 mm. The Chaparral and Sierra Mixed Conifer were primary vegetation species in this area. More descriptions about abovementioned fires could be found in reported studies [8,30,45].

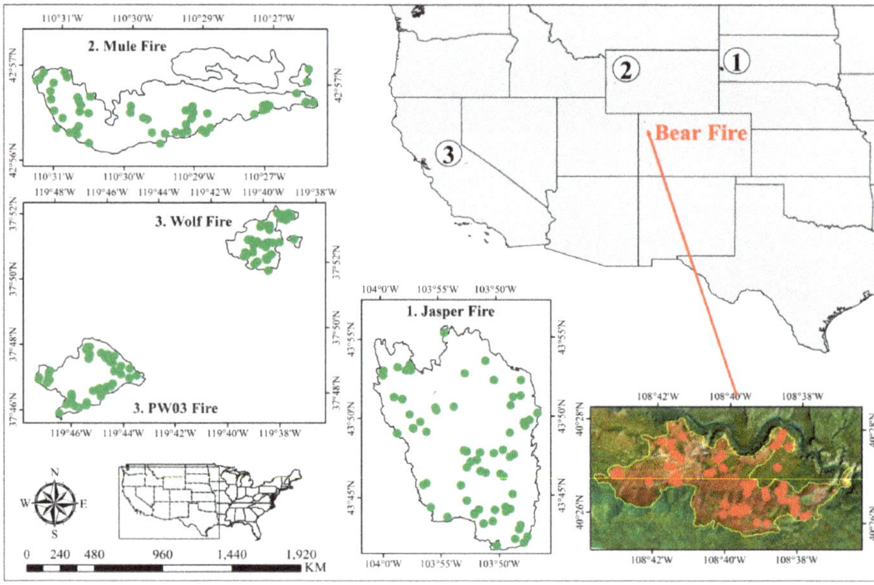

Figure 2. The location of study area in the southwest of United States. Red plots are testing data from target domain area and Green plots are training data from source domain areas.

3.2. Remotely Sensed Data

Three sets of remotely sensed data (i.e., Landsat 5 TM, Landsat 7 ETM+, and Terra MODIS) were obtained in this study (see in Table 1). For each fire, TM or ETM+ satellite images, close to the collection time of field survey data, were selected as post-fire remotely sensed data. The pre-fire remotely sensed data were selected according to the principle of "minimal moisture contents and phenology difference." All TM and ETM+ remotely sensed data were downloaded from the U.S. Geological Survey EarthExplorer, which have already undergone Level L1G processing [46]. Subsequently, reflectance transformation and atmospheric correction were done with the ENVI 5.0 software.

Table 1. Summary of study area and data.

Fire Name	Bear	Jasper	Mule	Pw03-Wolf
Regional Eco-types	Southwest	Central	Northern Rockies	California
Alarmed Date	27 June 2002	24 August 2000	11 July 2002	3 October 2002 and 11 July 2002
Regional Climate	Highland (alpine)	Semiarid Steppe	Highland (alpine)	Mediterranean
Annual Precipitation	<300 mm	440 to 673 mm	750 to 1150 mm	804 to 1722 mm
Elevations Range	1700 to 2740 m	1050 to 2207 m	1981 to 3353 m	657 to 3997 m
Sensors	TM and MODIS	TM/ETM+ and MODIS	TM and MODIS	TM/ETM+ and MODIS
Pre-fire Image	13 June 2002	1 May 2000	21 September 2001	18 October 2001
Post-fire Image	31 May 2003	31 May 2002	27 September 2003	16 October 2003
CBI Collection Date	28–30 May 2003	14–26 May 2002	8–11 September 2003	15 August–30 October 2003
Burned Area Size	18.62 km^2	336.37 km^2	11.86 km^2	13.61 km^2

To calculate LST for each fire, one MODIS product (i.e., total precipitable water, MOD05_L2. v051) was employed as atmospheric water vapor content for application of generalized single-channel algorithm. At first, this product was georeferenced using the MODIS Conversion Toolkit tool of ENVI 5.0 software. It was then resampled for matching the spatial resolution of thermal infrared band. A second MODIS product (i.e., Daily Land Surface Temperature/Emissivity, MOD11A1.v005) was applied to check the rightness of Landsat-based LST calculations [30,47]. All these MODIS data were obtained from the NASA's Level 1 and Atmosphere Archive and Distribution System (LAADS Web) [48].

3.3. Field Survey Data

The CBI field survey data were collected during the first or second following growing season for extended assessment for each fire [7]. All field survey data were obtained from the Joint Fire Science Program (JFSP) project which was undertaken by National Park Service and the US Geological Survey [45]. The specific number of field survey data for each fire is: 55 for Bear Fire, 66 for Jasper Fire, 55 for Mule Fire, and 78 for Pw03-Wolf Fire. Based on the recorded locations during data collections, each CBI field survey data was matched with corresponding pixel on remote sensing images. In a 3 × 3 matrix centered on matched pixel, the spectral value of each band was extracted and averaged to reduce the influence of GPS positioning error [30].

3.4. Cross-Validation Setting

First, the Bear Fire was selected as the target domain area and all CBI field survey data were employed as testing data to validate the accuracy of mapping results of burn severity. Meanwhile, Jasper, Mule, and PW03-Wolf Fire were chosen as the source domain areas. This means all CBI field survey data in these fires were applied to train predictor variables-based regression, SVR, and SSTCA-SVR models. Then, to further evaluate the performance of SSTCA-SVR in different eco-type regions, a cross-validation procedure was repeated by sequentially changing the target domain area and the source domain areas. Four scenarios were finally formed as follows (see in Table 2).

Table 2. Evaluation scenarios of cross-validation.

Scenarios	Target Domain Area	Source Domain Areas
Scenario 1	Bear Fire	Jasper, Mule, and PW03-Wolf Fires
Scenario 2	Jasper Fire	Bear, Mule, and PW03-Wolf Fires
Scenario 3	Mule Fire	Bear, Jasper, and PW03-Wolf Fires
Scenario 4	Pw03-Wolf Fire	Bear, Jasper, and Mule Fires

4. Results

4.1. Parameters Analysis of SSTCA

The SSTCA was applied in this study as domain adaptation approach to project original spectral features of target and source data. As described in Section 2, it mainly involved six parameters, i.e., μ, γ, kernel function, k, number of projected features, and λ. Specifically, μ was used for adjusting the complexity of transformation matrix in Formula (5) and γ was employed to balance the outputs dependence and the data variance in Formula (A9) in Appendix A. They were respectively fixed to 1 and 0.5 based on the preliminary tests and some reported researches [38,39]. Meanwhile, several preliminary trials suggested that the kernel function in Formula (A2) in Appendix A should be set to "Linear" and the optimum number of k neighbors in Formula (A11) in Appendix A was 100 which was also consistent with the previous research [39]. Therefore, only two parameters (i.e., number of projected features and λ) need to be further fitted for SSTCA in this study.

Figure 3a shows that, when the number of projected features varied from 1 to 14, mean RMSE value decreased and then increased. The minimum RMSE value was 0.6229, when the number of projected features was 10. Thus, the optimum number of projected features on Bear Fire should be 10. After this setting, Figure 3b further displays that optimal value of parameter λ should be 0.1, when it was chosen between 0 and 1. At last, optimal parameters (i.e., $\mu = 1$, $\gamma = 0.5$, kernel = linear, $k = 100$, number of projected features = 10, and $\lambda = 0.1$) were fixed in SSTCA for projecting the original spectral features of source domain areas (i.e., Jasper, Mule, and PW-03 Fire) and the target domain area (i.e., Bear Fire) to an invariant and shared subspace.

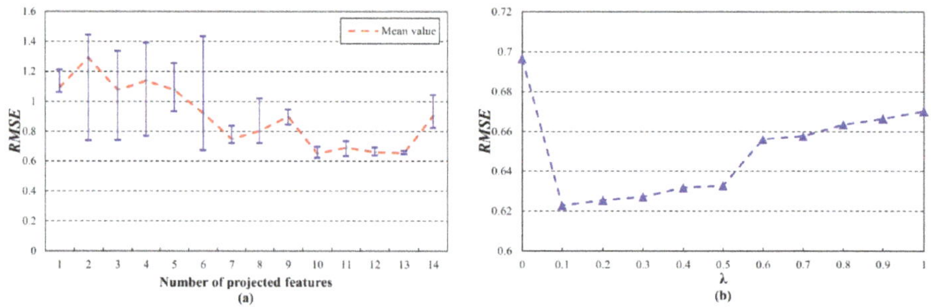

Figure 3. Performance of SSTCA-SVR on Bear Fire with respect to two parameters: (**a**) the number of projected-features and (**b**) λ. Note that the SSTCA-SVR has the lowest RMSE value, with number of projected features set to 10. After this setting, the optimal value for λ was 0.1.

4.2. Visual Analysis of Projected Features

Across the target domain area (i.e., blue dots) and source domain areas (i.e., red dots), original spectral feature space of remote sensing data (i.e., spectral bands of pre- and post-fire images) was plotted. Figure 4 illustrates that, in most scatter plots, blue and red dots have different distributions. This means that a difference of each band in original spectral feature space was evident.

Based on the optimal parameters selected in Section 4.1, ten projected features were transformed using SSTCA. Figure 5 shows that the distributions of blue dots were similar with that of red dots in most scatter plots. This means that, in these projected feature spaces, the model trained from data of source domain areas could be transferred to data of the target domain area. Figure 5 also displays that, after domain adaptation of SSTCA, data from source and target domain areas have the maximum variance, which means that transformation matrix of the fixed SSTCA have fulfilled the ***objective 1*** in Section 2 (i.e., preserving the variances of initial data in Formula (7)). This maximized variance of data would be very useful for following supervised learning job.

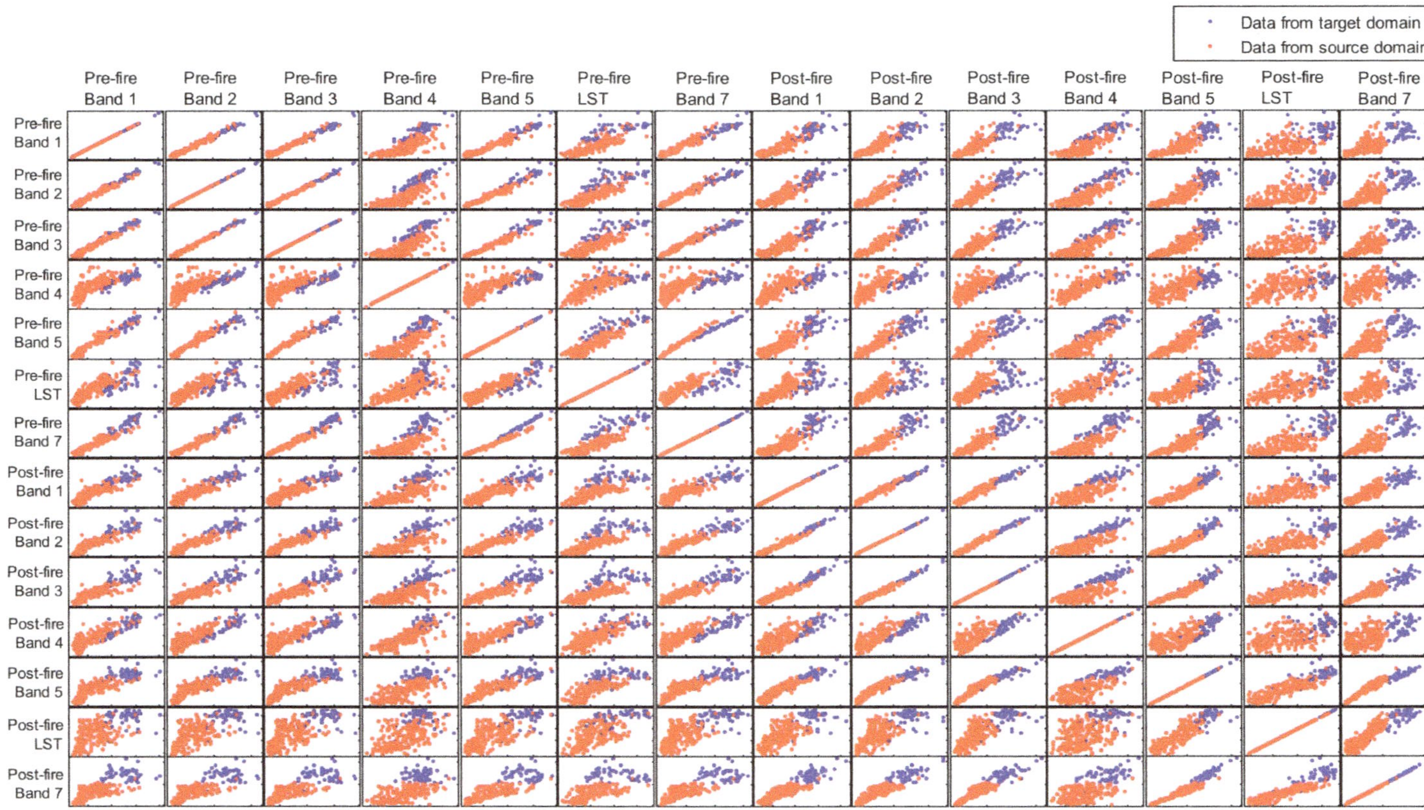

Figure 4. The scatter plot matrix of original spectral feature space over Bear Fire. Each panel is one scatter plot of a pair of spectral bands. Blue dots are field survey data from the target domain area: Bear Fire (n = 55). Red dots are field survey data from source domain areas: Jasper, Mule, and PW03-Wolf Fires (n = 199). Note that the distributions of data from target domain area and source domain areas are different in original spectral feature spaces.

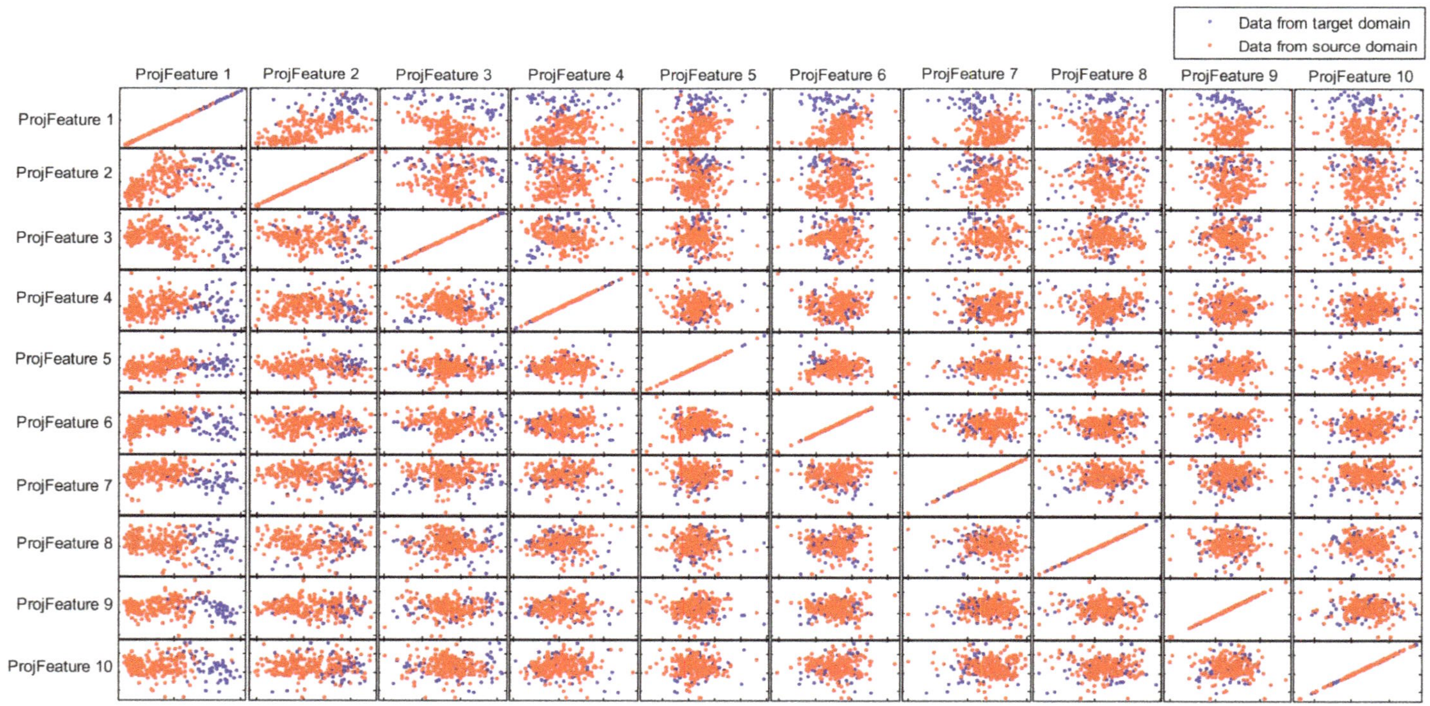

Figure 5. The scatter plot matrix of projected feature spaces over Bear Fire. Each panel is one scatter plot of a pair of projected features. Blue dots are field survey data from the target domain area: Bear Fire (n = 55). Red dots are field survey data from source domain areas: Jasper, Mule, and PW03-Wolf Fires (n = 199). Note that the distributions of data from target domain area and source domain areas are similar in most projected feature spaces.

4.3. Mapping Results of Burn Severity

As a specific example, mapped results of burn severity across Bear Fire were chosen for analysis. From a visual perspective, Figure 6 exhibits that the spatial distributions of burn severity mapped by using different methods were similar as a whole. However, the mapped spatial distribution of CBI value using ΔLST-based model was coarser as shown in Figure 6b. This was because LST value was calculated from thermal infrared data of Landsat TM/ETM+ which has a lower spatial resolution (i.e., 60 m/120 m). In contrast, it can be observed from Figure 6 that spatial distributions of burn severity mapped by SSTCA-SVR were finer than that of other predictor variables-based empirical fitting models. Although SVR model also has a fine burn severity results, its values were generally higher than that of SSTCA-SVR.

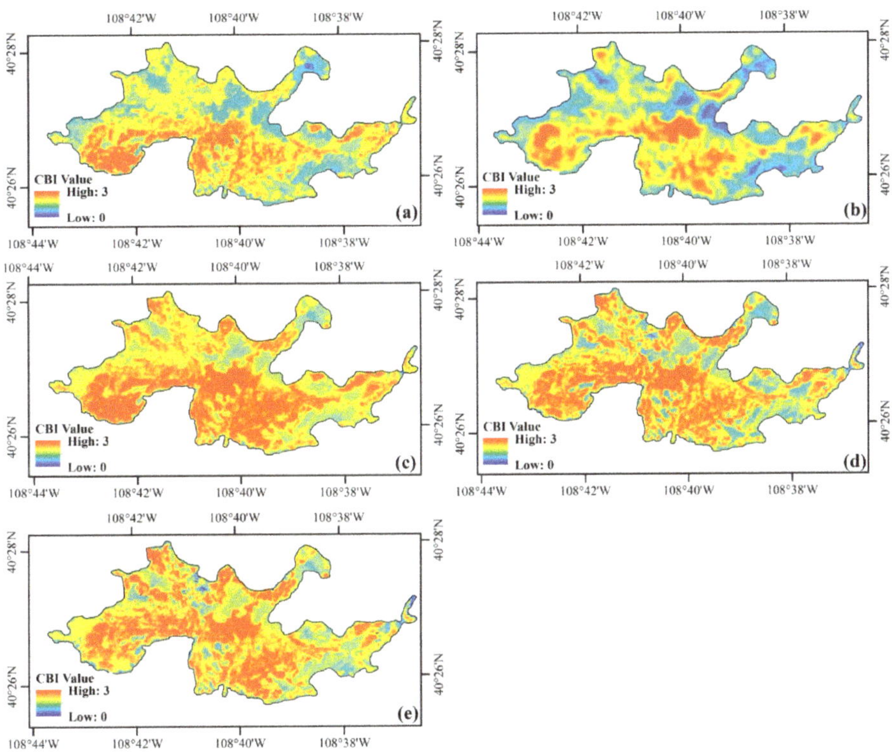

Figure 6. Mapped results of burn severity levels across Bear Fire using different models: (**a**) ΔNDVI-based regression model; (**b**) ΔLST-based regression model; (**c**) ΔNBR-based regression model; (**d**) SVR model; and (**e**) SSTCA-SVR model.

From a quantitative perspective, the field survey data of Bear Fire were employed as testing data to validate the accuracy of mapped results in Figure 6. Statistical results of accuracy in Table 3 revealed that SSTCA-SVR has the best performance at mapping burn severity with the lowest error value (i.e., RMSE is 0.6229). In contrast, we found that mapping result of SVR has the lowest accuracy with the highest RMSE value of 1.7658. This was consistent with the visual analysis result shown in Figure 6. The RMSE values of other predictor variables-based regression models were between 0.7362 and 1.1187.

Table 3. Accuracies of mapped results across Bear Fire using different models.

Predictor Variable	Empirical Fitting's Parameters			Accuracy		Model	Parameters		Accuracy	
	a	b	c	*R*	*RMSE*		*C*	*ε*	*R*	*RMSE*
ΔNDVI	0.5963	4.1368	1.0724	0.63	**1.1187**	SVR	97.0059	0.2176	0.76	**1.7658**
ΔLST	2.3073	2.5626	1.4793	0.69	**0.7362**	SSTCA-SVR	55.7152 *N.P.F.* = 10	0.0412 *λ* = 0.1	0.80	**0.6229**
ΔNBR	−1.3456	3.5138	0.9056	0.73	**0.9732**					

Note: *N.P.F.* means the number of projected features and the nonlinear regression model for ΔNDVI, ΔLST, and ΔNBR are CBI = a × $(x)^2$ + b × x + c (all *p* values < 0.05).

4.4. Cross-Validation in Different Scenarios

In order to further verify the reliability of SSTCA-SVR model in different eco-type regions, a cross-validation has been applied to sequentially exchange target and source domain data. Similar procedures in Sections 4.1–4.3 were operated in each scenario. Table 4 illustrates that SSTCA-SVR has the best performance at mapping burn severity in all scenarios (RMSE values are between 0.4833 and 0.6659). On the contrary, accuracies of mapped results by using SVR were the worst (RMSE value varies from 1.7658 to 2.0055), which was consistent with the above results in Figure 6 and Table 3. Moreover, ΔNBR-based regression model has a similar performance (i.e., RMSE value is 0.5917) with SSTCA-SVR (i.e., RMSE value is 0.5824) in Scenario 4.

Table 4. Accuracies of mapped results in different scenarios.

Models		Scenario 1	Scenario 2	Scenario 3	Scenario 4
ΔNDVI	a	0.5963	−4.1698	−2.825	−1.5803
	b	4.1368	5.789	4.445	4.6692
	c	1.0724	1.0724	1.4011	1.4085
	R	0.63	0.62	0.62	0.75
	RMSE	**1.1187**	**0.6316**	**0.8683**	**0.7422**
ΔLST	a	2.3073	−3.8144	1.7842	4.2049
	b	2.5626	5.6159	3.356	2.5752
	c	1.4793	1.2141	1.5525	1.3911
	R	0.69	0.58	0.58	0.38
	RMSE	**0.7362**	**0.8814**	**0.9763**	**0.7733**
ΔNBR	a	−1.3456	−3.4368	−3.2525	−2.5958
	b	3.5138	5.0849	4.5846	4.1274
	c	0.9056	0.8436	1.1096	1.1155
	R	0.73	0.69	0.71	0.74
	RMSE	**0.9732**	**0.5123**	**0.8282**	**0.5917**
SVR	*C*	97.0059	97.0059	97.0059	97.0059
	ε	0.2176	0.125	0.0412	0.2176
	R	0.76	0.76	0.61	0.60
	RMSE	**1.7658**	**1.9093**	**1.1464**	**2.0055**
SSTCA-SVR	*N.P.F.*	10	5	7	8
	λ	0.1	0.9	0.5	0.1
	C	55.7152	97.0059	55.7152	10.5561
	ε	0.0412	0.0136	0.0412	0.2176
	R	0.80	0.76	0.77	0.72
	RMSE	**0.6229**	**0.4833**	**0.6659**	**0.5824**

Note: *N.P.F.* means the Number of projected features and the nonlinear regression model for ΔNDVI, ΔLST, and ΔNBR are CBI = a × $(x)^2$ + b × x + c (all *p* values < 0.05).

5. Discussion

In this study, a transfer learning-based model was developed to map the burn severity of forest fire. Our findings suggest that, based on the domain adaptation of SSTCA approach, the proposed model could transfer knowledge from other burned areas. This model will be very helpful for mapping severity levels across burned areas where field survey data are not available. This research has advanced our capability to use remotely sensed data to quantitatively measure the damage levels after forest fire.

In original spectral features space, the data distributions of source and target domain areas indeed are different, since the remotely sensed data with temporal and spatial variations have different spectral signature. Thus, models which were trained on data from source domain areas should not be directly applied to map burn severity of a target domain area. This might result in a poor performance, which has been confirmed in experimental results.

Experimental results further indicated that, using SSTCA approach, original spectral features of target and source domains can be projected to a new subspace. In this subspace, data distributions of source and target domains actually are similar, which means that trained knowledge from source domain areas could be transferred to the target domain area. This fact also has been verified by subsequent experimental results. Therefore, to build a model for mapping burn severity of burned areas without field survey data, the domain adaptation is needed at first.

Moreover, results in this study have revealed that some predictor variables-based regression models (i.e., ΔNDVI, ΔLST, and ΔNBR) also have a certain transferring ability across target and source domain areas. This result agrees with that of a reported study [37], in which the overall ΔNBR-derived model has the transferability across multiple western Canadian landscapes. For spectral indices (e.g., NDVI and NBR), their calculations mainly are based on the differencing and ratio operations of spectral bands. This can reduce spectral changing of each original band caused by environmental variations to a certain extent. As for LST, it was retrieved in this study using the generalized single-channel algorithm from the radiative transfer equation [49,50]. During its retrieval process, variations of local environmental factors (e.g., atmospheric water vapor content, vegetation cover, and land surface emissivity) have also been considered as inputs of retrieval algorithm. Therefore, LST variations caused by environmental variations have been reduced and the transferring ability of LST-based model can also be observed across different domain areas.

In essence, the domain adaptation in the proposed model performs a similar function as the calculation of predictor variables (i.e., ΔNDVI, ΔLST, and ΔNBR), since they all have extracted several new features from original spectral features space. However, the calculation of predictor variables was based on limited spectral bands, which would lead to a lower transferring ability. In contrast, SSTCA method has projected all original spectral bands to exclude the difference of spectral signature. This means that more spectral information as possible have been utilized to map burn severity. Therefore, SSTCA-SVR model could map the burn severity finer and its obtained results have higher accuracy [8].

Compared to the LST-based model, SSTCA-SVR model not only has higher accuracy, but also higher efficiency. The reason for this is that SSTCA-SVR could be considered one type of machine learning algorithms, in which all model parameters could be learned by using supervision process and optimization procedure. Therefore, SSTCA-SVR model is relatively simple and efficient without setting of complex physical parameters and supporting of external environmental data.

Lastly, this research has introduced a potentially efficient mechanism to map burn severity, which could speed up the response of post-fire management. Unlike traditional mapping process which requires field survey data to build the regression model, only remote sensing data in the target domain area and field survey data in the source domain areas are needed as inputs of SSTCA-SVR. This means that burn severity levels can be mapped quickly with a certain degree of reliability, as long as appropriate remote sensing data are obtained across target burned areas. There is no need to wait for the collection of field survey data and external support data.

We should acknowledge that the transferring ability of SSTCA-SVR is limited, when climatic variations between source domain area (e.g., Pw03-Wolf Fire is located in the Mediterranean climate region) and target domain areas (e.g., Bear, Jasper, and Mule Fires are located in the Highland and Semiarid Steppe climate regions) are very high. It thus can be found in Scenario 4 that the accuracy of

SSTCA-SVR is similar with that of ΔNBR-based regression model. But, for a specific target domain area (e.g., in African savanna), this limitation could be overcome to a certain extent if some historical field survey data in the similar climate region are available.

Moreover, considering the availability of CBI-field survey data, the performance of SSTCA-SVR was only tested in the Midwestern United States. Its transferring ability needs to be further validated in different types of forest ecosystems around the world. Subsequent studies also are required to explore the spatial scale adaptability of SSTCA-SVR across different geographic regions in the future.

6. Conclusions

In this study, a transfer learning-based model was proposed to map the severity levels across the burned areas without field survey data. We have evaluated the performance of proposed model in different eco-type regions of Southwestern United States. Results indicate that original spectral features with different characteristic could be projected to an invariant and shared subspace by using SSTCA approach. Based on these projected features, the SVR model which was trained on source domain areas could be transferred to a target domain area. Compared to that of ΔNDVI-, ΔLST-, ΔNBR-, and SVR-based models, mapped results of CBI values using by SSTCA-SVR were more fine and accurate. This research has further developed our capability to use remotely sensed data to map the burn severity of forest fire. Moreover, this study has introduced a potentially efficient mechanism to map the burn severity to speed up the response of post-fire management.

Author Contributions: Z.Z. performed the experiment designing, the data analysis, as well as the manuscript writing; J.W. supervised the research work; B.S. helped with revisions. All authors contributed to writing the paper. All authors have read and agreed to the published version of the manuscript.

Funding: This work has been funded by the National Natural Science Foundation of China under Grant number 41801315, 41771535, the program of China Scholarship Council under Grant number 201908510029, the Open Research Fund of Chongqing Meteorological Bureau under Grant number KFJJ-201705, the Project Supported by the Scientific Research Foundation of CUIT under Grant number KYTZ201742, and the Scientific Research Foundation of the Education Department of Sichuan Province under Grant number 18ZB0128.

Acknowledgments: The authors would like to acknowledge the data made available by the U.S. Geological Survey, the U.S. Geological Survey's National Center for Earth Resources Observation and Science, the USDA Forest Service Remote Sensing Applications Center for making remote sensing data available, and the Joint Fire Science Program (JFSP) project undertaken by the National Park Service and the US Geological Survey. We wish to thank the editor and anonymous reviewers for their thoughtful and helpful comments.

Conflicts of Interest: The authors declare no conflict of interest.

Appendix A

$$MMD(X_S, X_{Ta}) = \left\| \frac{1}{n_S}\sum_{i=1}^{n_s} \Phi(X_{S_i}) - \frac{1}{n_{Ta}}\sum_{j=1}^{n_{Ta}} \Phi(X_{Ta_j}) \right\|_{\mathcal{H}}^{2} \tag{A1}$$

where $\|\cdot\|_{\mathcal{H}}^2$ is the L-2 norm.

$$K = \begin{bmatrix} K_{S,S} & K_{S,Ta} \\ K_{Ta,S} & K_{Ta,Ta} \end{bmatrix} \in \mathbb{R}^{(n_s+n_{Ta})\times(n_s+n_{Ta})} \tag{A2}$$

where its elements can be calculated as $K_{i,j} = \Phi(X_i)^\top \Phi(X_j)$.

$$\begin{cases} L_{i,j} = 1/n_S^2 & \text{if} : x_i, x_j \in X_S \\ L_{i,j} = 1/n_{Ta}^2 & \text{if} : x_i, x_j \in X_{Ta} \\ L_{i,j} = -1/n_S n_{Ta} & \text{otherwise} \end{cases} \tag{A3}$$

$$MMD(X_S, X_{Ta}) = tr(KL) \tag{A4}$$

$$K^* = KWW^\top K \tag{A5}$$

$$MMD(X_S^*, X_{Ta}^*) = tr((KWW^\top K)L) = tr(W^\top KLKW) \tag{A6}$$

$$\begin{aligned} \Sigma^* &= \frac{1}{n_S+n_{Ta}} \sum_{i,j=1}^{n_S+n_{Ta}} (X_i^* - \overline{X}^*)(X_j^* - \overline{X}^*)^\top \\ &= W^\top KHKW \end{aligned} \tag{A7}$$

where H is the centering matrix:

$$H = I_{n_S+n_{Ta}} - 11^\top(n_S + n_{Ta}) \in \mathbb{R}^{(n_S+n_{Ta})\times(n_S+n_{Ta})} \tag{A8}$$

While, $I_{n_S+n_{Ta}}$ is the identify matrix ($I_{n_S+n_{Ta}} \in \mathbb{R}^{(n_S+n_{Ta})\times(n_S+n_{Ta})}$) and 1 is the column vector with all 1's ($1 \in \mathbb{R}^{n_S+n_{Ta}}$).

$$K_{YY}^* = \gamma K_{YY} + (1-\gamma)I \tag{A9}$$

where K_{YY} is to maximize the outputs dependence and maximize the data variance. The $\gamma \geq 0$ is used as the tradeoff parameter to balance the two terms.

$$\mathcal{L} = D - M \tag{A10}$$

where D is a diagonal matrix with elements $D_{i,i} = \sum_j^{n_S+n_{Ta}} M_{i,j}$ and the $M_{i,j}$ can be calculated as:

$$M_{i,j} = \begin{cases} \exp(-d_{i,j}^2/2\sigma^2) & \text{if} : x_i \text{ is one of the } k\text{−nearest neighbors of } x_j \\ 0 & \text{otherwise} \end{cases} \tag{A11}$$

$$(K(L+\lambda)\mathcal{L}K + \mu I)^{-1} KHK_{YY}^* HK \tag{A12}$$

$$\begin{aligned} \underset{w^*,b}{\text{minimize}} \quad & \tfrac{1}{2}\|w^*\|^2 + C\sum_{i=1}^{n}\left(\xi_i + \xi_i^*\right) \\ \text{s.t.} \quad & \left|Y_{S,i} - f(X_{S,i}^*, w^*, b)\right| \leq \varepsilon \end{aligned} \tag{A13}$$

where $\|w^*\|^2$ represents the model complexity, slack variables (ξ_i, ξ_i^*) are deviation measurements, and tradeoff constant C determines tradeoff between these two terms.

References

1. Dixon, R.K.; Solomon, A.M.; Brown, S.; Houghton, R.A.; Trexier, M.C.; Wisniewski, J. Carbon pools and flux of global forest ecosystems. *Science* **1994**, *263*, 185–190. [CrossRef] [PubMed]
2. Park, H.-J.; Lim, S.-S.; Yang, H.I.; Lee, K.-S.; Park, S.-I.; Kwak, J.-H.; Kim, H.-Y.; Oh, S.-W.; Choi, W.-J. Co-elevated CO_2 and temperature and changed water availability do not change litter quantity and quality of pine and oak. *Agric. For. Meteorol.* **2020**, *280*, 107795. [CrossRef]
3. Zheng, Z.; Huang, W.; Li, S.; Zeng, Y. Forest fire spread simulating model using cellular automaton with extreme learning machine. *Ecol. Model.* **2017**, *348*, 33–43. [CrossRef]
4. Quintano, C.; Fernandez-Manso, A.; Roberts, D.A. Burn severity mapping from landsat mesma fraction images and land surface temperature. *Remote Sens. Environ.* **2017**, *190*, 83–95. [CrossRef]
5. Chen, G.; Metz, M.R.; Rizzo, D.M.; Dillon, W.W.; Meentemeyer, R.K. Object-based assessment of burn severity in diseased forests using high-spatial and high-spectral resolution master airborne imagery. *ISPRS J. Photogramm. Remote Sens.* **2015**, *102*, 38–47. [CrossRef]
6. Chuvieco, E.; Riaño, D.; Danson, F.M.; Martin, P. Use of a radiative transfer model to simulate the postfire spectral response to burn severity. *J. Geophys. Res. Biogeosci.* **2006**, *111*. [CrossRef]
7. Key, C.H.; Benson, N.C. Landscape assessment (la): Sampling and analysis methods. In *Rocky Mountain Research Station, USDA*; Forest Service: Fort Collins, CO, USA, 2006; pp. LA-1–LA-51.
8. Zheng, Z.; Zeng, Y.; Li, S.; Huang, W. Mapping burn severity of forest fires in small sample size scenarios. *Forests* **2018**, *9*, 608. [CrossRef]

9. Schepers, L.; Haest, B.; Veraverbeke, S.; Spanhove, T.; Vanden Borre, J.; Goossens, R. Burned area detection and burn severity assessment of a heathland fire in belgium using airborne imaging spectroscopy (apex). *Remote Sens.* **2014**, *6*, 1803–1826. [CrossRef]
10. He, Y.; Chen, G.; De Santis, A.; Roberts, D.A.; Zhou, Y.; Meentemeyer, R.K. A disturbance weighting analysis model (dwam) for mapping wildfire burn severity in the presence of forest disease. *Remote Sens. Environ.* **2019**, *221*, 108–121. [CrossRef]
11. Quintano, C.; Fernández-Manso, A.; Roberts, D.A. Multiple endmember spectral mixture analysis (mesma) to map burn severity levels from landsat images in mediterranean countries. *Remote Sens. Environ.* **2013**, *136*, 76–88. [CrossRef]
12. De Santis, A.; Chuvieco, E.; Vaughan, P.J. Short-term assessment of burn severity using the inversion of prospect and geosail models. *Remote Sens. Environ.* **2009**, *113*, 126–136. [CrossRef]
13. De Santis, A.; Chuvieco, E. Burn severity estimation from remotely sensed data: Performance of simulation versus empirical models. *Remote Sens. Environ.* **2007**, *108*, 422–435. [CrossRef]
14. Epting, J.; Verbyla, D.; Sorbel, B. Evaluation of remotely sensed indices for assessing burn severity in interior alaska using landsat tm and etm+. *Remote Sens. Environ.* **2005**, *96*, 328–339. [CrossRef]
15. Veraverbeke, S.; Hook, S.; Hulley, G. An alternative spectral index for rapid fire severity assessments. *Remote Sens. Environ.* **2012**, *123*, 72–80. [CrossRef]
16. van Wagtendonk, J.W.; Root, R.R.; Key, C.H. Comparison of aviris and landsat etm+ detection capabilities for burn severity. *Remote Sens. Environ.* **2004**, *92*, 397–408. [CrossRef]
17. Key, C.H.; Benson, N.C. The Normalized Burn Ratio (nbr): A Landsat tm Radiometric Measure of Burn Severity. Available online: http://www.nrmsc.usgs.gov/research/ndbr.htm (accessed on 22 May 2015).
18. Garcia, M.L.; Caselles, V. Mapping burns and natural reforestation using thematic mapper data. *Geocarto Int.* **1991**, *6*, 31–37. [CrossRef]
19. Cocke, A.E.; Fulé, P.Z.; Crouse, J.E. Comparison of burn severity assessments using differenced normalized burn ratio and ground data. *Int. J. Wildland Fire* **2005**, *14*, 189–198. [CrossRef]
20. Roy, D.P.; Boschetti, L.; Trigg, S.N. Remote sensing of fire severity: Assessing the performance of the normalized burn ratio. *IEEE Geosci. Remote Sens. Lett.* **2006**, *3*, 112–116. [CrossRef]
21. Escuin, S.; Navarro, R.; Fernandez, P. Fire severity assessment by using nbr (normalized burn ratio) and ndvi (normalized difference vegetation index) derived from landsat tm/etm images. *Int. J. Remote Sens.* **2008**, *29*, 1053–1073. [CrossRef]
22. Zheng, Z. Study on the risk, spread and assessment of forest fire based on the model and remote sensing. *Acta Geod. Cartogr. Sin.* **2019**, *48*, 133.
23. Soverel, N.O.; Perrakis, D.D.B.; Coops, N.C. Estimating burn severity from landsat dnbr and rdnbr indices across western canada. *Remote Sens. Environ.* **2010**, *114*, 1896–1909. [CrossRef]
24. Miller, J.D.; Thode, A.E. Quantifying burn severity in a heterogeneous landscape with a relative version of the delta normalized burn ratio (dnbr). *Remote Sens. Environ.* **2007**, *109*, 66–80. [CrossRef]
25. Heward, H.; Smith, A.M.S.; Roy, D.P.; Tinkham, W.T.; Hoffman, C.M.; Morgan, P.; Lannom, K.O. Is burn severity related to fire intensity? Observations from landscape scale remote sensing. *Int. J. Wildland Fire* **2013**, *22*, 910–918. [CrossRef]
26. Parks, S.; Dillon, G.; Miller, C. A new metric for quantifying burn severity: The relativized burn ratio. *Remote Sens.* **2014**, *6*, 1827–1844. [CrossRef]
27. Lentile, L.B.; Holden, Z.A.; Smith, A.M.S.; Falkowski, M.J.; Hudak, A.T.; Morgan, P.; Lewis, S.A.; Gessler, P.E.; Benson, N.C. Remote sensing techniques to assess active fire characteristics and post-fire effects. *Int. J. Wildland Fire* **2006**, *15*, 319–345. [CrossRef]
28. Quintano, C.; Fernández-Manso, A.; Calvo, L.; Marcos, E.; Valbuena, L. Land surface temperature as potential indicator of burn severity in forest mediterranean ecosystems. *Int. J. Appl. Earth Obs. Geoinf.* **2015**, *36*, 1–12. [CrossRef]
29. Fernández-García, V.; Santamarta, M.; Fernández-Manso, A.; Quintano, C.; Marcos, E.; Calvo, L. Burn severity metrics in fire-prone pine ecosystems along a climatic gradient using landsat imagery. *Remote Sens. Environ.* **2018**, *206*, 205–217. [CrossRef]
30. Zheng, Z.; Zeng, Y.; Li, S.; Huang, W. A new burn severity index based on land surface temperature and enhanced vegetation index. *Int. J. Appl. Earth Obs. Geoinf.* **2016**, *45*, 84–94. [CrossRef]

31. Veraverbeke, S.; Verstraeten, W.W.; Lhermitte, S.; Van De Kerchove, R.; Goossens, R. Assessment of post-fire changes in land surface temperature and surface albedo, and their relation with fire–burn severity using multitemporal modis imagery. *Int. J. Wildland Fire* **2012**, *21*, 243–256. [CrossRef]
32. Marcos, E.; Fernández-García, V.; Fernández-Manso, A.; Quintano, C.; Valbuena, L.; Tárrega, R.; Luis-Calabuig, E.; Calvo, L. Evaluation of composite burn index and land surface temperature for assessing soil burn severity in mediterranean fire-prone pine ecosystems. *Forests* **2018**, *9*, 494. [CrossRef]
33. Harris, S.; Veraverbeke, S.; Hook, S. Evaluating spectral indices for assessing fire severity in chaparral ecosystems (southern california) using modis/aster (master) airborne simulator data. *Remote Sens.* **2011**, *3*, 2403–2419. [CrossRef]
34. Vlassova, L.; Pérez-Cabello, F.; Mimbrero, M.; Llovería, R.; García-Martín, A. Analysis of the relationship between land surface temperature and wildfire severity in a series of landsat images. *Remote Sens.* **2014**, *6*, 6136–6162. [CrossRef]
35. Chen, G.; Metz, M.R.; Rizzo, D.M.; Meentemeyer, R.K. Mapping burn severity in a disease-impacted forest landscape using landsat and master imagery. *Int. J. Appl. Earth Obs. Geoinf.* **2015**, *40*, 91–99. [CrossRef]
36. Fernández-Manso, O.; Quintano, C.; Fernández-Manso, A. Combining spectral mixture analysis and object-based classification for fire severity mapping. *For. Syst.* **2009**, *18*, 296–313. [CrossRef]
37. Soverel, N.O.; Coops, N.C.; Perrakis, D.D.B.; Daniels, L.D.; Gergel, S.E. The transferability of a dnbr-derived model to predict burn severity across 10 wildland fires in western canada. *Int. J. Wildland Fire* **2011**, *20*, 518–531. [CrossRef]
38. Pan, S.J.; Tsang, I.W.; Kwok, J.T.; Yang, Q. Domain adaptation via transfer component analysis. *IEEE Trans. Neural Netw.* **2011**, *22*, 199–210. [CrossRef]
39. Matasci, G.; Volpi, M.; Kanevski, M.; Bruzzone, L.; Tuia, D. Semisupervised transfer component analysis for domain adaptation in remote sensing image classification. *IEEE Trans. Geosci. Remote Sens.* **2015**, *53*, 3550–3564. [CrossRef]
40. Yan, K.; Kou, L.; Zhang, D. Learning domain-invariant subspace using domain features and independence maximization. *IEEE Trans. Cybern.* **2018**, *48*, 288–299. [CrossRef]
41. Yan, K. A Domain Adaptation Toolbox. Available online: https://www.github.com/viggin/domain-adaptation-toolbox (accessed on 9 December 2019).
42. Cherkassky, V.; Ma, Y. Practical selection of svm parameters and noise estimation for svm regression. *Neural Netw.* **2004**, *17*, 113–126. [CrossRef]
43. Vapnik, V.N. An overview of statistical learning theory. *IEEE Trans. Neural Netw.* **1999**, *10*, 988–999. [CrossRef]
44. Vapnik, V.N. *Statistical Learning Theory*; Wiley: New York, NY, USA, 1998; Volume 2.
45. Zhu, Z.; Key, C.; Ohlen, D.; Benson, N.C. Evaluate sensitivities of burn-severity mapping algorithms for different ecosystems and fire histories in the united states. In *Final Report to the Joint Fire Science Program*; JFSP 01-1-4-12; Joint Fire Science Program: Boise, IA, USA, 2006.
46. USGS. Earthexplorer. Available online: http://earthexplorer.usgs.gov/ (accessed on 20 December 2013).
47. Srivastava, P.K.; Majumdar, T.J.; Bhattacharya, A.K. Surface temperature estimation in singhbhum shear zone of india using landsat-7 etm+ thermal infrared data. *Adv. Space Res.* **2009**, *43*, 1563–1574. [CrossRef]
48. NASA. Level 1 and Atmosphere Archive and Distribution System (LAADS). Available online: http://earthexplorer.usgs.gov/ (accessed on 20 October 2014).

49. Zhou, J.; Li, J.; Zhang, L.; Hu, D.; Zhan, W. Intercomparison of methods for estimating land surface temperature from a landsat-5 tm image in an arid region with low water vapour in the atmosphere. *Int. J. Remote Sens.* **2012**, *33*, 2582–2602. [CrossRef]
50. Zhang, Z.; He, G.; Wang, M.; Long, T.; Wang, G.; Zhang, X. Validation of the generalized single-channel algorithm using landsat 8 imagery and surfrad ground measurements. *Remote Sens. Lett.* **2016**, *7*, 810–816. [CrossRef]

remote sensing

MDPI

Article

Field-Validated Burn-Severity Mapping in North Patagonian Forests

María Guadalupe Franco [1,*], Ignacio A. Mundo [1,2] and Thomas T. Veblen [3]

1 Laboratorio de Dendrocronología e Historia Ambiental, IANIGLA–CONICET, Mendoza CC330-M5502IRA, Argentina; iamundo@mendoza-conicet.gob.ar

2 Facultad de Ciencias Exactas y Naturales, Universidad Nacional de Cuyo, Mendoza M5502JMA, Argentina

3 Department of Geography, University of Colorado, Boulder, CO 80309, USA; thomas.veblen@colorado.edu

* Correspondence: mgfranco@mendoza-conicet.gob.ar

Received: 10 December 2019; Accepted: 6 January 2020; Published: 8 January 2020

Abstract: Burn severity, which can be reliably estimated by validated spectral indices, is a key element for understanding ecosystem dynamics and informing management strategies. However, in North Patagonian forests, where wildfires are a major disturbance agent, studies aimed at the field validation of spectral indices of burn severity are scarce. The aim of this work was to develop a field validated methodology for burn-severity mapping by studying two large fires that burned in the summer of 2013–2014 in forests of *Araucaria araucana* and other tree species. We explored the relation between widely used spectral indices and a field burn-severity index, and we evaluated index performance by examining index sensitivity in discriminating burn-severity classes in different vegetation types. For those indices that proved to be suitable, we adjusted the class thresholds and constructed confusion matrices to assess their accuracy. Burn severity maps of the studied fires were generated using the two most accurate methods and were compared to evaluate their level of agreement. Our results confirm that reliable burn severity estimates can be derived from spectral indices for these forests. Two severity indices, the delta normalized burn ratio (dNBR) and delta normalized difference vegetation index (dNDVI), were highly related to the fire-induced changes observed in the field, but the strength of these associations varied across the five different vegetation types defined by tree heights and tree and tall shrub species regeneration strategies. The thresholds proposed in this study for these indices generated classifications with global accuracies of 82% and Kappa indices of 70%. Both the dNBR and dNDVI classification approaches were more accurate in detecting high severity, but to a lesser degree for detecting low severity burns. Moderate severity was poorly classified, with producer and user errors reaching 50%. These constraints, along with detected differences in separability, need to be considered when interpreting burn severity maps generated using these methods.

Keywords: wildfire; *Araucaria araucana*; Landsat 8 OLI; normalized burn ratio; normalized difference vegetation index; char soil index; mid-infrared burned index; classification thresholds

1. Introduction

Burn severity, defined as the magnitude of ecological change associated with a wildfire [1], is a key element for understanding ecosystem dynamics as it influences processes that shape community structure and composition, nutrient availability, and wildlife habitat [2–4]. At broader scales, the spatial distribution of fire effects can determine species replacement and alter landscape dynamics [5,6]. Consequently, the assessment of fire effects is essential for understanding ecosystem resilience and designing management strategies.

Remote sensing allows for the spatial description of burn severity while overcoming cost and logistic limitations of extensive field surveys [7], and numerous mapping methodologies have been proposed and tested (see review by Meng and Zhao [8]). These are based on changes in the spectral signature of fire-affected surfaces, mainly associated with moisture and chlorophyll content of vegetation: red and near-infrared (NIR) reflectance decreases and short-wave-infrared (SWIR) reflectance increases [9]. Several indices combining these bands have been tested for burn severity mapping including the normalized difference vegetation index (NDVI) [10], the normalized burn ratio (NBR) [11], the mid-infrared burned index (MIRBI) [12], and the char soil index (CSI) [13]. As fire effects are dependent on pre-existing vegetation, bi-temporal approaches use the difference between pre- and post-fire indices; however, this requires precautions related to imagery selection to minimize the differences in phenology, solar angle, geometric correction, and sensor calibration [7]. Uni-temporal analyses, based only on post-fire observations, are less affected by these issues but have problems in discriminating burned areas from other spectrally similar areas [14]. Several studies have demonstrated high correlations between field-based reference observations of burn severity and post-fire and differenced versions of spectral indices [15–17]. Other protocols use a multitemporal approach to improve the reference data (e.g., the multitemporal $dNBR_{MT}$ [18] that relies on time series to select reference pixels) or avoid limited image availability due to quality issues (e.g., the relativized burn ratio (RBR) [19], which uses compositing techniques to obtain pre- and post-fire data).

The sensitivity of these methods varies across different vegetation types [20,21] so they need to be field validated to ensure that reliable information is generated. The validation should compare the spectral index to the equivalent surface process or property [1]. Remotely sensed burn severity has proven to be reliable in forests, savannahs, shrublands, and grasslands of different regions [13,22,23]. Some studies have also explored the variation in the sensitivity of indices for detecting changes in different vegetation groups within the same region and have proposed group-specific parameters for burn severity estimation [24–26]; grouping criteria have often considered structure, composition, and/or regeneration processes (e.g., contrasting resprouting species and species regenerating mainly from seed). However, field validation still needs to be conducted in many regions with complex vegetation mosaics including North Patagonia.

Fire is a major factor shaping landscapes in North Patagonia and fire-induced vegetation dynamics have been described for the most widespread community types ranging from closed-canopy tall forests of *Nothofagus* species (either evergreen or deciduous species), dense tall shrublands dominated by resprouting woody species <12 m tall, to grassy open woodlands of the conifer *Austrocedrus chilensis* [27,28]. Regeneration strategies (e.g., obligate seeding tall trees versus resprouting shrubs) and community flammability have been associated with positive fire feedbacks that maintain a mosaic of pyrophobic forests and pyrophytic shrublands over much of the region [29–31]. While forests dominated by fire-sensitive obligate seeders are vulnerable to fire-induced conversion to shrublands, some forests in North Patagonia are dominated by *Araucaria araucana*, a thick-barked fire-resistant species also capable of resprouting following fire [32–34]. Thus, the general model of tall forest conversion to more flammable shrub-dominated vegetation types in northern Patagonia needs to be examined for vegetation types characterized by *A. araucana*. In addition, fire activity in *Araucaria*-dominated forests has been shown to be sensitive to both interannual climatic variation and to variability in the frequency of human-set ignitions [35]. Projected shifts in fire regimes caused by climate and land cover changes raise concerns regarding forest resilience in relation to probable increases in the frequency, extent, and severity of fires in North Patagonia [36–38] for which improved methods of assessing fire effects related to burn severity are essential.

Araucaria araucana-dominated forests are characterized by a mixed severity fire regime including both high and low severity burns [34,39]. Burn severity varies spatially over relatively short distances, leading to a complex mosaic of differently structured vegetation patches across the landscape. Mixed forests of *A. araucana* and *Nothofagus pumilio* or *N. antarctica* are typically controlled by moderate- to high-severity fires that originate from stands with scattered surviving *A. araucana* surrounded by small multi-aged groups of *A. araucana* and a single post-fire cohort of *Nothofagus spp.* [34]. Post-fire establishment of the obligate seeder *N. pumilio* decreases under conditions of higher severity [40]; if the accompanying species is *N. antarctica*, a post-fire cohort is rapidly created by vigorous resprouts that reach 1 m in height within three years [39]. In terms of the understory, higher severity induces decreases in richness and abundance and changes in composition, with exotic species being more abundant in situations of low severity [41]. Regeneration in burned areas is mainly controlled by burn severity and basal area, with influences of topography and biotic interactions, but fire refugia also play a role as seed sources for post-fire dynamics in the longer term [42]. As an important driver of the dynamics of these forests, burn severity is generally assessed in the ecological studies of fire events [43–45]. However, field-validation of burn severity mapping methodologies for Patagonian forests is scarce [46]. In the case of *Araucaria*-dominated forests, the relative version of dNBR (RdNBR) was tested using reference field data gathered 10 years after a fire in Chile [42], but no analogous work is available for the Argentinian forests that are under drier climatic conditions.

The aim of this work was to develop field-validated burn severity maps of two recent (2013–2014) wildfires that affected a complex mosaic of *A. araucana*-dominated vegetation types in North Patagonia in order to propose a replicable mapping methodology for the region. We address the following questions:

1. How do burn-severity estimators derived from satellite images relate to field-based measurements of fire impact in *A. araucana* vegetation types defined by stand structure and species composition?
2. What are the classification thresholds that best discriminate severity classes across the range of *A. araucana* vegetation types?
3. How well do fire-severity maps that are based on the different best performing spectral indices agree?

First, we tested the association between field measurements of burn severity (four years following the fire events) and widely used indices based on remotely-sensed data and conducted a sensitivity analysis to select the best estimators of burn severity across a range of vegetation types. Then, we adjusted the class thresholds for each selected index and assessed their accuracy, contrasting it against other thresholds proposed in previous publications. Finally, we generated burn severity maps for the two studied wildfire events that affected *A. araucana* forests by applying the most accurate methodologies and evaluated the agreement between them.

2. Materials and Methods

2.1. Study Area

North Patagonian forests grow on the hills and valleys of the Andes in Chile and Argentina, within a steep moisture gradient that decreases from 4000 mm annual precipitation in the west to 500 mm on the eastern side [32,47]. The precipitation is mostly concentrated between May and September and occurs both as rain and snowfall; summer is the dry season. Mean monthly temperatures range from 4 °C to 15 °C, and maximum values reach up to 35 °C in summer. These conditions favor the occurrence of wildfires in summer, which can be more extreme during drought periods. Electric storms are the main natural cause of ignition, but currently, most fires are human-ignited [35,48].

These forests present numerous endemics and are mainly dominated by species of the genus *Nothofagus* [32,49]. In Argentina, large portions of Patagonian forests are protected under the National Parks Administration. Most of the tree dominant species are relatively sensitive to fire because of their thin to moderately thick bark. Among them, the more fire-sensitive tree species are obligate seeders, whereas the fire-resistant trees and shrubs resprout after fire [50]. All of these species are shade-intolerant and present a regeneration mode that is linked to disturbances. The structure of these forests varies from tall pure or mixed-species stands to <12 m tall stands of multi-stemmed shrubs and small trees including two-tiered stands. Wildfires are a widespread disturbance agent, due partially to the fire-promoting features of some species of the tree canopy and the understory. Consequently, this landscape has been historically shaped by fires, and the mosaic of vegetation types largely reflects the extent and severity of the past events [32,50].

This study focused on two large wildfires that burned between December 2013 and January 2014, near Lakes Ñorquinco and Rucachoroy, in the northernmost area of the Argentinian Patagonian forests (from 39°06′S to 39°13′S and from 71°05′W to 71°27′W; Figure 1). The Ñorquinco fire burned an area of 1626 ha and the Rucachoroy fire extended over 1336 ha, mainly on the valley bottoms and moderate to very steep hillsides with a northern- and southern-aspect, respectively. Both events occurred in summer on days with strong westerly winds within a period of extreme drought, which resulted in widespread intense fires.

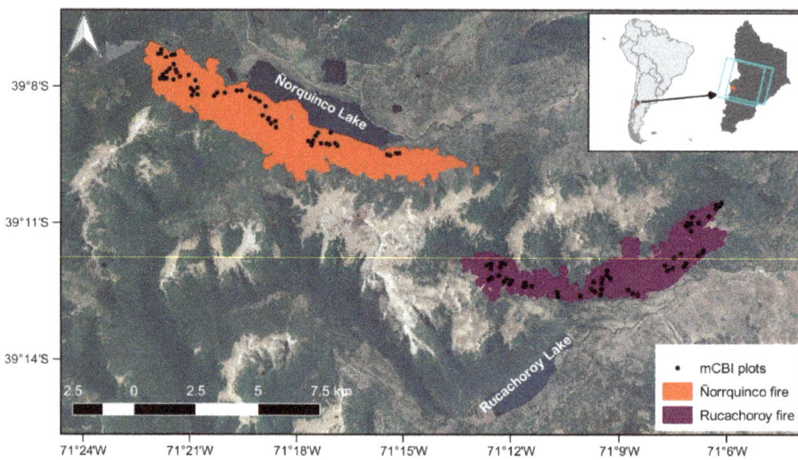

Figure 1. Location of study area, footprint of Landsat 8 OLI images (cyan squares), perimeters of burned areas in Ñorquinco and Rucachoroy (red and purple polygons), and 111 burn-severity validation plots (black dots).

2.2. Remote Sensing Data

We used Landsat 8 OLI images downloaded from the USGS repository. As both fires occurred in nearby areas, the same set of images was used in the assessment of both events (Path 232, Row 87; Figure 1). Images with low cloud cover were selected from the Level 1 Landsat Collection, which includes terrain correction, corresponding to the end of the summers that preceded (7 April) and followed the fires up until the year of field survey (23 March 2014; 26 March 2015; 28 March 2016; 15 March 2017; 2 March 2018; Table 1). In this way, the first post-fire image corresponds to the end of the growing season when the fires occurred, minimizing the effect of early recovery or invasion, and images of the following years might reflect either of these processes and/or delayed mortality. The selection of close-to-anniversary images allows the comparison of pre- and post-fire images with no concern about the phenology and solar angle effects. All image processing was conducted in QGIS 2.18.20 [51].

Images were converted to surface reflectance with the semi-automatic classification plugin, using dark-object subtraction. Red, near-infrared (NIR), and short-wave infrared (SWIR1 and SWIR2) bands were used to generate indices that are commonly used for burn severity estimation: normalized burn ratio (NBR) [11], normalized difference vegetation index (NDVI) [10], mid-infrared burned index [12] and the char soil index (CSI) [13]. The differenced version of the indices (dNDVI, dNBR, dMIRBI, dCSI) were also included in this work as they measure a magnitude of change, which is closer to the definition of burn severity. These indices have been widely applied in burn severity mapping, with different accuracies in contrasting vegetation types [26,52,53]. To compensate for the effect of the misalignment between the plot and grid centers, index values associated with each mCBI plot were sampled through bilinear interpolation.

Table 1. Field and remote sensing data: identification and date of wildfires, date of field survey, number of plots, and capture dates of the Landsat 8 OLI images.

Wildfires		Field Survey		Landsat 8 OLI images		
ID	Date	No. of Plots	Date	Path–Row	Pre-Fire	Post-Fire
Ñorquinco Rucachoroy	29/12/13 to 20/01/14	55 56	20/02/18 to 02/03/18	232–87	07/04/13	23/03/14; 26/03/15; 28/03/16; 15/03/17; 02/03/18

2.3. Field-Based Reference Burn Severity Data

Field metrics of burn severity were collected between February and March 2018 in accessible areas of the Ñorquinco and Rucachoroy fires. Prior to the fieldwork, preliminary severity maps were developed using reference dNBR thresholds [11], and were used to identify homogeneous patches. Uniformity is important due to the offset between the circular plot and grid cell centers. As these points are rarely aligned, surveyed information usually combines data registered in more than one pixel. Therefore, this precaution in plot location improves the relationship between the field and remotely sensed data. One hundred and eleven 15 m radius plots were distributed in areas of low, moderate, and high severity across all vegetation types. The number of plots was determined following Chuvieco [54] for the error estimation of classified images with a 95% probability and maximum allowed error of 5%.

Extended assessment, that is sampling the year following the fire, is expected to enable accounting for delayed mortality or survival of vegetation. Due to the timing of the fieldwork (four years after the fires), this analysis constitutes a long-term assessment. One of the dominant species of these forests (*A. araucana*) is documented to resprout from epicormic and protected terminal buds following fire damage, but many sprouts dry in the following years and only some individuals continue regenerating the crown [39,55]. Consequently, gathering field data a few years after fire events (in this case, in the fourth growing season after the events) should allow for a better characterization of fire impact on these forests. However, the long time lapse from burning to field sampling impedes the evaluation of fire effects on the components of vegetation that recover more quickly (i.e., herbs and low shrubs), limiting the direct application of indices designed for more immediate surveys.

We based our field protocol on the composite burn index (CBI) [11]. In the summer of 2016–2017, we conducted a preliminary assessment where we surveyed 20 plots distributed along the severity gradient and the different vegetation types affected in both fires. Based on this information, we developed a modified CBI index (mCBI) that better fit the specific conditions of this study (Table 2). The modifications included the exclusion of the fast recovering components (herbs and low shrubs) and the elimination of variables that required knowledge of some attributes of the pre-burn vegetation in the plots (e.g., pre-burn presence of small herbs and shrubs and standing dead trees), as a consequence of the lack of pre-fire data and the high structural variability of these forests. Consequently, low severity was characterized by high proportions of green foliage, low char height on trunks, and unburned litter and woody debris; high severity was defined by the opposite conditions; and moderate severity implied an intermediate situation (Figure 2). The proportion of dead fire-killed trees was recorded in every plot.

Figure 2. Photographs of burned *Araucaria araucana* forests showing a gradient of fire damage, four years after fire. (**a**) Low severity, with intact crowns and low char height on trunks, and with both dead and resprouting shoots of the bamboo *Chusquea culeou*; (**b**) moderate severity, with an intermediate proportion of dead trees, partially burned crowns and trunks and no remains of living pre-fire tall shrubs; and (**c**) high severity, with almost no surviving trees, completely charred trunks and no remains of living pre-fire shrubs.

Table 2. Assessment table of the modified composite burn index (mCBI): ranking scale for each factor of the four strata (substrates, tall shrubs, intermediate trees, big trees) corresponding to unburned, low, moderate, and high severity.

Strata	Severity Scale						
	Unburned	Low		Moderate		High	
	0	0.5	1	1.5	2	2.5	3
			Substrates				
Litter	intact	charred	absent	absent	absent	absent	absent
Duff	intact	Intact	partially charred	charred	scarce	very scarce	absent
Light Fuels	intact	partially charred	charred/absent	absent	absent	absent	absent
Medium Fuels	intact	Intact	partially charred	partially charred	charred/absent	absent	absent
Heavy Fuels	intact	Intact	intact	intact	partially charred	charred	absent
			Tall shrubs (1 to 6 m)				
% altered foliage	0%	5%	10%	30%	50%	75%	100%
% resprout	100%	95%	90%	60%	30%	<10%	<1%
			Intermediate trees (subcanopy)				
% green foliage	100%	90%	80%	60%	40%	<10%	0%
% black foliage	0%	5%	20%	30%	60%	>85%	100%
% charred trunk height	0%	10%	30%	40%	60%	70%	>90%
			Big trees (upper canopy)				
% green foliage	100%	90%	80%	60%	40%	<10%	0%
% black foliage	0%	5%	20%	30%	60%	>85%	100%
% charred trunk height	0%	10%	30%	40%	60%	70%	>90%

2.4. Data Analyses

2.4.1. Sensitivity Analysis

We explored the relationship between the field and remotely sensed estimates of burn severity by using scatter plots and correlation analysis. Correlation between mCBI and percentage of dead trees was analyzed to assess the performance of the index. Pearson correlation tests have been used in similar studies [15,56]; we used Spearman tests because the data did not meet the assumption of normality, tested using the 'fitdistrplus' function in R. Due to the four-year gap between the fires and the field survey, and considering that both early and delayed resprouting occurs in these forests, we evaluated the most useful acquisition date for severity mapping based on the strength of the association between mCBI and image-derived indices. Indices showing a correlation higher than 0.70 (absolute values) were then subjected to a sensitivity analysis. This was assessed through a separability index [57], calculated as:

$$S_{i-ii} = \frac{\mu_i - \mu_{ii}}{sd_i - sd_{ii}} \quad (1)$$

where $S_{i\text{-}ii}$ is the separability of two consecutive severity classes (low-moderate or moderate-high); μ_i and μ_{ii} are the mean values of the index for each severity class; and sd_i and sd_{ii} are the respective standard deviations. Overall separability was obtained as a weighted mean, following Nguyen Tran et al. [26], to assign higher importance to the detection of high severity (0.35 $S_{\text{low-moderate}}$; 0.75 $S_{\text{moderate-high}}$). To determine constraints on sensitivity related to species-specific reproductive and other traits, this analysis was also conducted separately for each vegetation type (described below) and then summarized as a mean value for each severity index. The higher the value of S, the better the discrimination between two classes; a value of 1 or higher is desired for indices to be used for mapping [58].

Vegetation types were discriminated based on canopy height and regeneration strategy of the dominant species (Figure 3). Tall forests were dominated by trees, with a single closed canopy layer and heights between 30–40 m; multi-stemmed structures of <15 m in height occurred in low forests or two-tiered stands when combined with tall trees. Dominant species were either obligate seeders (S) or resprouters (R). Five vegetation types were identified: Tall/S, dominated by *N. pumilio*; Tall/R, including stands of *N. obliqua* or *A. araucana*; Tall/RS, characterized by mixed stands of *N. pumilio-A. araucana*; Two-tiered/R, codominated by *N. antarctica* and *A. araucana*; and Low/R, dominated *by N. antarctica*. *A. araucana* forests varied from open- to closed-canopy stands and, although it can regenerate both by seedlings and resprouts, the crown recovery is the fastest recovery mechanism. The 3 to 6 m tall *Chusquea culeou* native bamboo occurs in varying quantities in all vegetation types and acts as a highly flammable ladder fuel because it is mostly very fine (grass leaves) or fine (<24 mm diameter) material, and when the culms die following mast events, it leaves massive amounts of fine dead fuels [50].

Figure 3. Photographs of the five vegetation types: (**a**) low/R, dominated by *Nothofagus antarctica*; (**b**) tall/R, dominated by *Araucaria araucana*; (**c**) tall/S, dominated by *N. pumilio*; (**d**) tall/SR, codominated by *N. pumilio* and *A. araucana* with the bamboo *Chusquea culeou*; (**e**) two-tiered/R, codominated by *A. araucana* and *N. antarctica*.

2.4.2. Thresholding and Accuracy Assessment

To define the class thresholds, we followed two different approaches using mCBI as a reference with class limits of 1.25 (low/moderate) and 2.25 (moderate/high) [59]. First, we fitted linear and polynomial models. Linear and non-linear models have been widely applied for describing relationships between field severity data and spectral indices [17,60,61]. Polynomial models appear to better describe the non-linear relation between these variables [62], but the improvement may be only marginal [17]. Having graphically checked assumptions of normality, homoscedasticity, and independence, best-fit models were selected following the criterion of lowest Akaike score and class thresholds were derived. As an alternative strategy, we determined thresholds by an iterative procedure: increasing values for each threshold (low/moderate and moderate/high) were combined until the highest proportion of samples was correctly classified [42,59].

To select the best performing method for mapping, we compared the accuracies of the classifications derived from the adjusted thresholds and others recommended by Key and Benson [11] and by Assal et al. [42]. The latter study was conducted in *Araucaria*-dominated forests in Chile, similar to the vegetation in our study area. We applied a bootstrapping method, sampling with reposition, to generate 111 confusion tables and the associated accuracy indices (total accuracy, Kappa, and the producer and user errors for each severity class). Based on this information, we conducted Kruskal–Wallis non-parametric variance analyses to evaluate significance of apparent differences in accuracy.

2.5. Burn Severity Mapping

The two most accurate methods were used to generate burn severity maps for each event. Perimeters of the burned areas were obtained through supervised classification on dNBR and dNDVI layers and manual editing using technical reports (M. Mermoz, pers. com.) and high resolution images as references. Finally, we assessed the agreement between the maps by contrasting the information generated on the proportion of area affected under each severity class.

3. Results

3.1. Index Sensitivity

Field measurements of burn severity were highly and significantly correlated (0.87, $p < 0.001$; Table 3), indicating that mCBI is a good estimator of post-fire tree mortality. The correlation between mCBI and the tested indices was strongest for data acquired at the end of the summer of 2013–2014 when the fires occurred (>0.70) and declined in the subsequent years for all burn severity indices, except for MIRBI, dMIRBI, and dCSI, which consistently presented low correlations (<0.60; Figure 4). These indices were discarded from further analyses. In the year of the end of the growing season when the fire occurred, the strongest Spearman correlation with mCBI was that of the NDVI (−0.85), followed by NBR, CSI, dNBR, and dNDVI (Table 3). NDVI also had the strongest correlation with %DT (−0.83), whereas the correlations of the other four indices were weaker but similar (0.77 to 0.78).

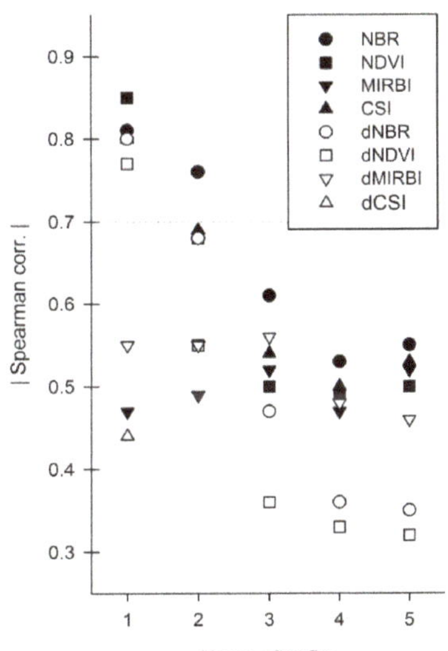

Figure 4. Correlations between the mCBI and remote burn severity indices, for post-fire images acquired 0 to 4 years after fire. The Spearman coefficient is presented in absolute values. All correlations are statistically significant ($p < 0.001$).

Table 3. Spearman correlation between the field and remote burn severity estimates for the end of the season when the fires occurred. * Asterisks denote significant correlations ($p < 0.001$).

	mCBI	%DT	NBR	NDVI	CSI	dNBR	dNDVI
mCBI	1	0.87 *	−0.81 *	−0.85 *	−0.80 *	0.80 *	0.77 *
%DT	0.87 *	1	−0.77 *	−0.83 *	−0.78 *	0.77 *	0.77 *

%DT: Percentage of dead trees.

The highest sensitivity was shown by NDVI and dNBR, but the indices performed differently when vegetation type was considered. In general, separability was around 1 for tall or low stature vegetation types, regardless of the regeneration strategy, but was always under 0.6 for the two-tiered forest (Table 4). In resprouter-dominated forests, NDVI showed higher separability, whereas NBR was more sensitive for obligate seeders. Severity in stands with both types of species was well detected by all indices (separability > 1). When these partial values were summarized, the most sensitive index was dNBR and the separability of all the others was below 1. dNDVI presented a more balanced sensitivity across vegetation type than NDVI and was particularly more sensitive for the two-tiered stands.

Table 4. Mean separability of indices calculated for the complete dataset (All), discriminated for each vegetation type: low canopy, dominated by resprouting species (low/R); tall canopy, dominated by resprouting species (tall/R); tall canopy, dominated by seeding species (tall/S); tall canopy, codominated by resprouting and seeding species (tall/R-S); two tiered canopy, dominated by resprouting species (two-tiered/R) and summarized as a general mean (Overall).

		Plots	NBRpost	NDVIpost	CSIpost	dNBR	dNDVI
All		*111*	*0.95*	*1.06*	*0.90*	*1.10*	*0.93*
Canopy	**Regen.**						
Low	R	30	0.92	1.04	0.94	0.94	1.03
Tall	R	25	1.00	1.21	0.92	0.98	1.03
Tall	S	18	1.00	0.73	0.80	1.12	0.89
Tall	RS	27	1.58	1.64	1.40	1.56	1.05
Two tiered	R	11	0.43	0.23	0.29	0.58	0.40
Overall			*0.99*	*0.97*	*0.87*	*1.03*	*0.88*

R: resprouter-dominated stand; S: obligate seeder-dominated stand; RS: mixed stand of resprouters and obligate seeders.

3.2. Thresholds and Accuracy of Selected Indices

Three severity indices were selected for thresholding: dNBR, dNDVI, and CSI. For all indices, the best model was a second order polynomial, with an adjusted R^2 of 0.77 for dNBR, 0.71 for dNDVI, and 0.70 for CSI (all $p < 0.001$). The model-derived thresholds were moderately close to those obtained through the iteration process, and for dNBR, the latter approach produced class limits that were the most similar to the field-based reference classes (Table 5).

Table 5. Accuracy assessment of burn severity classifications. Values in brackets indicate the classification thresholds for low-moderate/moderate-high severity. H and p values correspond to the Kruskal–Wallis variance test for each accuracy index and lowercase letters indicate significantly different groups.

			Low		Moderate		High	
Classification	**Total Acc.**	**Kappa**	**UE**	**PE**	**UE**	**PE**	**UE**	**PE**
dNBR Ref. 1	77 ± 1	64 ± 2	15 ± 2	11 ± 2	55 ± 2	21 ± 2	1 ± 1	27 ± 2
(270/600)	B	c	B	c	bc	f	d	a
dNBR Ref. 2	82 ± 1	68 ± 2	15 ±2	11 ±2	46 ± 2	49 ± 2	11 ±1	11 ± 1
(275/500)	De	d	B	c	d	c	c	e
dNBR It.	83 ± 1	70 ± 2	17 ± 2	7 ± 2	41 ± 2	49 ± 2	11 ± 1	11 ± 1
(300/500)	E	d	Ab	d	e	c	c	e
dNDVI It.	83 ± 1	69 ± 2	20±2	3 ±2	33 ±2	59 ±2	13 ±1	9 ± 1
(200/300)	E	d	A	e	f	b	b	ef
CSI It.	73 ± 1	54 ± 2	12 ± 2	39 ± 2	64 ± 2	44 ± 2	11 ± 1	16 ± 1
(1070/920)	A	a	C	a	a	d	c	c
dNBR Mod.	77 ± 1	61 ± 2	6 ± 2	26 ± 2	58 ±2	33 ± 2	11 ± 1	19 ± 2
(205/550)	B	b	D	b	b	e	c	b
dNDVI Mod.	80 ± 1	65 ± 2	6 ± 2	28 ± 2	52 ± 2	32 ± 2	11 ± 1	14 ± 2
(116/323)	C	c	D	b	c	e	bc	d
CSI Mod.	81 ± 1	66 ± 2	17 ± 2	7 ± 2	43 ± 2	67 ± 2	15 ± 2	8 ± 1
(138/75)	D	c	B	d	de	a	a	f
H	372	310	310	267	353	249	575	539
p	*<0.0001*	*<0.0001*	*<0.0001*	*<0.0001*	*<0.0001*	*<0.0001*	*<0.0001*	*<0.0001*

dNBR Ref. 1: thresholds as in Key and Benson (2006); dNBR Ref. 2: thresholds as in Assal et al. (2018). It.: Iterative thresholding; Mod.: Model-derived thresholding. Total Acc.: total accuracy; UE: user's error; PE: producer's error.

The most accurate classifications were those generated using the iteratively obtained thresholds for the indices dNBR and dNDVI, with total accuracies over 80% and Kappa indices around 70%. Reference class limits taken from the study by Assal et al. [42] in an area in Chile close to our study site produced a similarly accurate categorization (Table 5). These classifications were significantly more accurate than the other tested ones. In terms of class accuracy, the general pattern was shared among all classifications: the lowest errors were found in the high severity class, followed by low severity, while the moderate severity class was the least accurate one (Table 5). For the most accurate protocols, the producer's error for the high severity class ranged from 9% to 11% and the user's error varied between 11% and 13%. These errors took values of 3–11% and 15–20% in the case of the low severity class and reached levels of 49–59% and 33–46% for moderate severity, respectively.

3.3. Burn Severity Maps

Burn severity maps were generated using dNDVI and dNBR indices with thresholds obtained by the iteration process (Figure 5). In the case of the Ñorquinco fire, most of the area was burned with high severity (60–70%) while the rest of the area was evenly distributed between moderate and low severity. The agreement between the two maps for this event was 81%, with most of the disagreement occurring in the moderate burn class (42% coincidence). In the case of the Rucachoroy fire, the area was burned in proportions of 50%, 30%, and 20% with high, low and moderate severity, respectively. Again, the two maps coincided in categorizing 81% of the area and the moderate class was the one with the lowest agreement (52% coincidence). Considering dNBR maps as a reference, misclassification was more often due to an underestimation of burn severity (15% of the burned area) rather than an overrating (6%) and mainly occurred in the moderate class (Table 6).

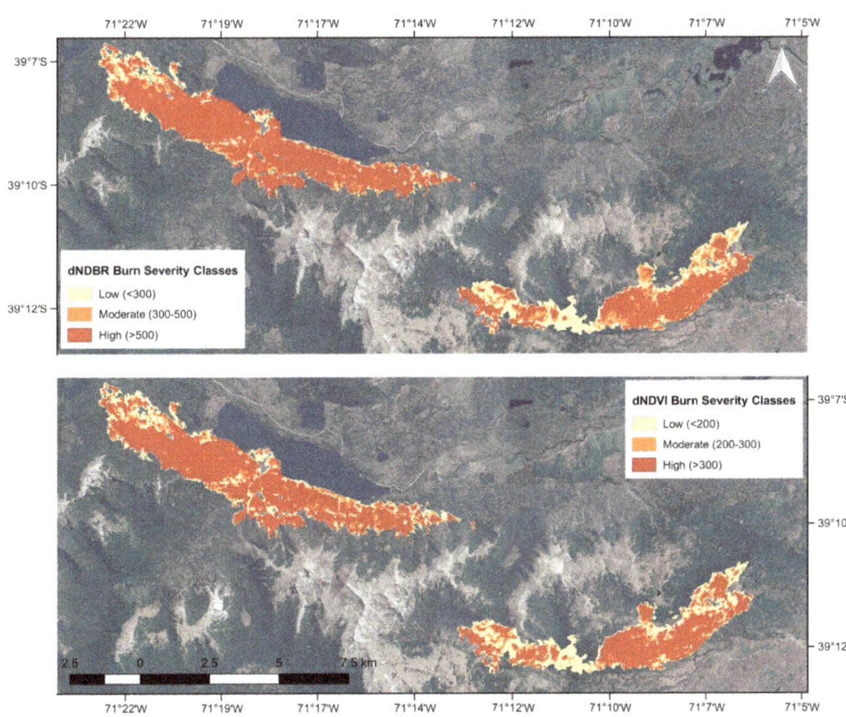

Figure 5. Burn severity maps based on delta normalized burn ratio (dNBR) and delta normalized difference vegetation index (dNDVI). Classification thresholds are indicated in the legend, in brackets.

Table 6. Confusion matrix comparing the severity maps based on dNBR and dNDVI with iteratively defined thresholds including the area burned in the Ñorquinco and Rucachoroy fires. Area expressed in hectares.

		dNBR It.				
		Low	**Moderate**	**High**	**Total (ha)**	**Agreement**
dNDVI It.	Low	658	183	18	859	77%
	Moderate	47	288	267	603	48%
	High	0	136	1703	1839	93%
	Total (ha)	706	608	1988	3302	
	Agreement	93%	47%	86%		

4. Discussion

Our results, based on the collection of field-based reference data, confirm that reliable burn severity estimates can be derived from spectral indices for *Araucaria araucana*-dominated vegetation types in North Patagonia. We demonstrate that the most commonly employed indices (dNBR and dNDVI) corresponded highly with fire-induced changes observed in the field, although this association was not equally strong across different vegetation types. Time since fire influences this correspondence, with the correlations being higher when spectral indices are obtained from the first post-fire growing season. Based on these indices, we were able to determine class thresholds that produced accurate classifications and identify their constraints as well as finally propose a validated burn severity mapping methodology for the region.

4.1. From Index Suitability to Index Selection

The strong relations of NBR, NDVI, dNBR, dNDVI, and CSI with mCBI suggest the potential use of these spectral indices for burn severity estimation in North Patagonian forests. Previous research in Mediterranean pure and mixed-species forest stands showed that NBR performed better than NDVI for severity estimation and that the association was always weaker in the case of bi-temporal indices [16,52]. A stronger relation to the post-fire uni-temporal indices was also detected in boreal forests and wetlands, where correlation to field severity metrics was higher for NBR than for the other twelve spectral indices including NDVI and CSI [15]. In Australian sclerophyllous forests and woodlands, dNBR outperformed dNDVI in discriminating low severities in a 'cool' fire but not in a 'hot' fire where classifications where similar [63]. dNDVI was also weaker than dNBR in estimating severity across different vegetation types in the USA, ranging from grasslands and shrublands to conifer and deciduous forests [64,65]. In contrast, CSI was better than NBR for mapping severity in African savannas [13] and MRIBI, CSI, or NDVI were more suitable in wet heath, dry heath, or coniferous stands of European heathlands, respectively [25]. In our study, MIRBI was the only index that showed a weak relation to field severity metrics and was discarded from further analyses.

In all cases, the strength of these relations decreased over time, indicating that burn severity in these forests is better expressed by reflectance in the first growing season following fire events. In *Pinus ponderosa* dominated forests, the dNDVI correlation with CBI decreased after the second post-fire year, but the dNBR correlation was strong five years later [66]. In North Patagonian forests, the heterogeneity at plot and landscape levels resulted in multiple post-fire scenarios combining various processes. Patches dominated by vigorous resprouters would be expected to mask severity faster than stands of slow resprouters or obligate seeders; a similar situation would occur in open stands where the growth of grasses can be strongly enhanced. In contrast, delayed mortality and slow recovery, as exhibited by *A. araucana*, would be presumed to be limited or not easily detected in the first period following the fire, leading to an under- or overestimation of severity. Our results suggest that the heterogeneity in species composition and stand structures across this landscape generates a mixture of diverging post-fire reflectance trends, weakening the potential of spectral indices to assess burn severity as time since fire increases. This heterogeneity also generates variability within the growing season, so we compared images at the end of the summer to ensure all species were actively growing. Future research should assess the performance of these indices using a multitemporal approach that could summarize the variability or integrate the information of several post-fire seasons. In the U.S., composite-derived dNBR, RBR, and RdNBR indices outperformed the bi-temporal versions, although improvement in validation statistics was variable in some cases [67].

In our study in North Patagonia, all indices showed an overall suitable sensitivity, but with variable sensitivities across different vegetation types as has been found in studies conducted in other regions. Severity detection by dNBR and dNDVI has been found to perform better in forests than in grasslands and shrublands of temperate and boreal regions in the USA [64], with differences between coniferous- and deciduous-dominated stands [68]. Vegetation class also determined the potential of spectral indices for severity mapping in a heterogeneous landscape in Australia, which was higher for forest covered areas than for grasslands and shrublands [24]. In this same region, index suitability changed along with the regeneration strategy of the dominant species: dNBR was better for resprouter-dominated forests and dNDVI was superior for those composed by obligate seeders [26]. In contrast, our results showed that dNBR sensitivity to burn severity was higher in obligate seeder-dominated stands whereas dNDVI was more sensitive in stands dominated by resprouting species. Differences in tree life forms might explain these opposite results: while the studied Australian forests were all dominated by evergreen angiosperms (*Eucalyptus* spp.), in our study areas, the dominant trees included deciduous angiosperms and one evergreen conifer, so differences in spectral behavior could be expected. Coinciding with Nguyen Tran et al. [26], CSI sensitivity was lower in all vegetation classes in our study. Regardless of the variation, the separability values were around the expected for indices with good discriminatory power (>1) [58] for nearly all vegetation types.

The effect of vegetation type on indices suitability in our study and in previous similar studies suggests that burn severity estimation could be improved if calibration is conducted separately for each vegetation type. Allen and Sorbel [68] suggest using different scales for coniferous versus deciduous boreal forests. Schepers et al. [25] and Nguyen Tran et al. [26] followed this approach and adjusted predictive models or class thresholds using the best-performing index per vegetation group. This was not feasible in our research because the number of mCBI plots, when stratified by vegetation type, was not sufficient to conduct robust accuracy analyses. However, the investigation of sensitivity differences is relevant for gauging the constraints of the severity estimates.

4.2. From Thresholding to Validated Classifications

In our study, second order polynomials were the best fitting models for all indices, as was previously described for this vegetation type in Chile [42] and other forested regions [16,66,69]. Different models have performed better in other forest types: the sigmoidal model in temperate forests in the USA [70], saturated growth model in Canadian boreal forests [71], and third order polynomial in Russian forests [61], etc. It has also been suggested that using non-linear- instead of linear-models might not refine the severity prediction in spite of the improvement of the fitting parameters [72]. In the cases of dNDVI and dNBR, we found that the model-derived classifications were less accurate than those generated using the iteratively defined thresholds.

All severity estimations in our study performed within the accuracy range reviewed by French et al. [73] with a mean accuracy of 73%, ranging from 59% to 95%. Fire impact has been estimated with similar accuracy in Australian forests, where dNBR performed better than the dNDVI (Kappa index of 77% and 57%, respectively) [63]. In temperate North American forests, dNBR detection of severity was accomplished with accuracy ranging from 62 to 66% [21,70]. The best dNBR and dNDVI classifications presented global accuracies higher than 80% and Kappa indices around 70%, indicating that severity assessment through these methods is reliable. Nevertheless, global indicators should never represent the only evaluating criteria [59]. Unbalanced performance of severity classifications is common, especially on the low and moderate classes [74]. Moderate severity was the least accurately detected class by Soverel et al. [22] in Canadian forests as well as by Kurbanov et al. [61] in Russia. Coinciding with Cocke et al. [21], these studies were best at classifying areas corresponding to high severity. Our results show that, in all cases, high severity was the best discriminated class, with balanced user's and producer's errors, and moderate severity was poorly classified (errors around 50%). Low severity class was satisfactorily distinguished, but alternately, one of the two errors was always higher. This is probably due to the high variability of spectral indices in the moderate class, associated with the heterogeneity of fire effects and vegetation types that the field severity indices summarize. This class includes different fire effects on surface and understory modifications that might not be captured by passive sensors [7]. These constraints should be considered when extracting conclusions from the spectral estimation of burn severity.

When subsequent validations are conducted for the same region, the confidence in remote sensing assessments is expected to increase and, ideally, reach a point where the need for field data is minimized [11]. For forests dominated by *A. araucana*, similar to our study area, the study conducted nearby in Chile by Assal et al. [42] is the only precedent of field validated severity mapping and the RdNBR thresholds they proposed are similar to the dNBR limits we determined. Moreover, the accuracy of the classifications derived from either set of values in our study is higher than the one reported in their work (68% global accuracy and 52% Kappa index), most likely because our study was based on field data that was acquired closer to the date of the fires. Overall, these results indicate that comparable severity assessments can be executed in these forests. The higher accuracy and the different value of our proposed thresholds in comparison to those fixed by Key and Benson [11] sustain the idea of a need to calibrate the methodology to each unique region.

4.3. Application: Burn Severity Mapping

The two most accurate classification protocols (dNBR and dNDVI) were used to generate burn severity maps of the two fire events in the *Araucaria araucana*-dominated vegetation of North Patagonia. The agreement between these maps reflects the strengths and weaknesses described in the preceding sections: coincidence was high for the high and low severity classes, which were more accurately detected, but declined to 50% for moderately burned areas. Regardless of the spatial distribution, both maps produced the same estimation of proportion of area burned. Similarity in the spatial patterns of NDVI- and NBR-derived severity estimates was detected by Lhermitte et al. [75] in African savannas, who found differences when comparing severity maps generated using the bi-temporal approach and the control pixel method. Comparing maps allows for a demonstration of the impact of methodology in the thematic presentation of burn severity and, thus, can identify the limitations of the generated information. Our results indicate that either approach is similarly reliable for severity mapping in North Patagonia. However, the dissimilar association of NDVI and NBR to different vegetation types could determine the selection of one over the other. We found that greater sensitivity to burn severity was shown by NDVI in areas dominated by resprouting tall shrub species and NBR in obligate seeder stands of tall trees. Further studies specifically comparing the performance of these indices in areas dominated by species of different regeneration strategies would be needed to confirm if one index is preferable, depending on the predominant vegetation type.

5. Conclusions

This work explored the performance of the most frequently used spectral indices for burn severity mapping in North Patagonian forests. Our findings suggest that reliable estimates can be obtained from the dNBR and dNDVI using the adjusted thresholds. As in other forested regions, field-validated maps increase the confidence in these analyses and contribute to a replicable methodology and comparable results. This is particularly important in Patagonian forests where burn severity is usually assessed on a per-fire event basis, with no standardized methods or specific comparison with previously applied thresholds. However, interpretation of the results of our study should consider limitations related to the low-accuracy identification of moderate severity as well as the differential sensitivities of indices for vegetation types of different structure, species composition, and contrasting regeneration strategies of resprouters versus seeders. Similar research including a greater range of vegetation types throughout the Andean-Patagonian region is necessary for generating a burn-severity mapping methodology that performs well across these vegetation types.

Author Contributions: All authors contributed to the conception of the work. Methodology was defined by M.G.F. and I.A.M. Acquisition and curation of data and formal analyses were conducted by M.G.F., the latter under the supervision of I.A.M. All authors participated in the discussion of results. M.G.F. prepared the first draft of this manuscript and I.A.M. and T.T.V. substantially revised it. I.A.M. and T.T.V. were responsible for funding acquisition. All authors have read and agreed to the published version of the manuscript.

Funding: This research was funded by the SECTYP-Universidad Nacional de Cuyo (grant number M043) and the National Science Foundation (grant number 1832483). The article processing charge was funded by the National Science Foundation (grant number 1832483).

Acknowledgments: The authors thank the National Parks Administration for sampling permission and assistance during fieldwork. We also thank Mónica Mermoz for sharing the fire signature shapes and dNDVI maps of the Ñorquinco and Ruchachoroy fires. We thank three anonymous reviewers for helping to improve the final version of this manuscript.

Conflicts of Interest: The authors declare no conflict of interest.

References

1. Lentile, L.B.; Holden, Z.A.; Smith, A.M.S.; Falkowski, M.J.; Hudak, A.T. Remote sensing techniques to assess active fire characteristics and post fire effects. *Int. J. Wildland Fire* **2006**, *194*, 319–345. [CrossRef]
2. Smucker, K.M.; Hutto, R.L.; Steele, B.M. Changes in bird abundance after wildfire: Importance of fire severity and time since fire. *Ecol. Appl.* **2005**, *15*, 1535–1549. [CrossRef]
3. Shenoy, A.; Johnstone, J.F.; Kasischke, E.S.; Kielland, K. Persistent effects of fire severity on early successional forests in interior Alaska. *For. Ecol. Manag.* **2011**, *261*, 381–390. [CrossRef]
4. Adkins, J.; Sanderman, J.; Miesel, J. Soil carbon pools and fluxes vary across a burn severity gradient three years after wildfire in Sierra Nevada mixed-conifer forest. *Geoderma* **2019**, *333*, 10–22. [CrossRef]
5. Turner, M.G.; Hargrove, W.; Gardner, R.H.; Romme, W.H. Effects of fire on landscape heterogeneity in Yellowstone National Park, Wyoming. *J. Veg. Sci.* **1994**, *5*, 731–742. [CrossRef]
6. Morgan, P.; Hardy, C.C.; Swetnam, T.W.; Rollins, M.G.; Long, D.G. Mapping fire regimes across time and space: Understanding coarse and fine-scale fire patterns. *Int. J. Wildland Fire* **2001**, *10*, 329–342. [CrossRef]
7. De Santis, A.; Chuvieco, E. Burn severity estimation from remotely sensed data: Performance of simulation versus empirical models. *Remote Sens. Environ.* **2007**, *108*, 422–435. [CrossRef]
8. Meng, R.; Zhao, F. Remote sensing of fire effects: A review for recent advances in burned area and burn severity mapping. In *Remote Sensing of Hydrometeorological Hazards*; Petropoulos, G.P., Islam, T., Eds.; CRC Press: Boca Raton, FL, USA, 2017; pp. 261–283.
9. White, J.D.; Ryan, K.C.; Key, C.H.; Running, S.W. Remote sensing of forest fire severity and vegetation recovery. *Int. J. Wildland Fire* **1996**, *6*, 125. [CrossRef]
10. Rouse, W.J.; Haas, R.H.; Schell, J.A.; Deering, D.W. Monitoring vegetation systems in the Great Plains with ERTS. In *NASA SP-251, Third ERTS-1 Symposium*; NASA: Washington, DC, USA, 1974; pp. 309–317.
11. Key, C.H.; Benson, N.C. *Landscape Assessment: Sampling and Analysis Methods*; General Technical Report RMRS-GTR-164-CD; USDA Forest Service, Rocky Mountain Research Station: Fort Collins, CO, USA, 2006.
12. Trigg, S.N.; Flasse, S. An evaluation of different bi-spectral spaces for discriminating burned shrub-savannah. *Int. J. Remote Sens.* **2001**, *22*, 2641–2647. [CrossRef]
13. Smith, A.M.S.; Wooster, M.J.; Drake, N.A.; Dipotso, F.M.; Falkowski, M.J.; Hudak, A.T. Testing the potential of multi-spectral remote sensing for retrospectively estimating fire severity in African Savannahs. *Remote Sens. Environ.* **2005**, *97*, 92–115. [CrossRef]
14. Pereira, J.M.C. A comparative evaluation of NOAA/AVHRR vegetation indexes for burned surface detection and mapping. *IEEE Trans. Geosci. Remote Sens.* **1999**, *37*, 217–226. [CrossRef]
15. Epting, J.F.; Verbyla, D.L.; Sorbel, B. Evaluation of remotely sensed indices for assessing burn severity in interior Alaska using Landsat TM and ETM+. *Remote Sens. Environ.* **2005**, *96*, 328–339. [CrossRef]

16. Veraverbeke, S.; Lhermitte, S.; Verstraeten, W.W.; Goossens, R. Evaluation of pre/post-fire differenced spectral indices for assessing burn severity in a Mediterranean environment with Landsat Thematic Mapper. *Int. J. Remote Sens.* **2011**, *32*, 3521–3537. [CrossRef]
17. Fernández-García, V.; Santamarta, M.; Fernández-Manso, A.; Quintano, C.; Marcos, E.; Calvo, L. Burn severity metrics in fire-prone pine ecosystems along a climatic gradient using Landsat imagery. *Remote Sens. Environ.* **2018**, *206*, 205–217. [CrossRef]
18. Veraverbeke, S.; Lhermitte, S.; Verstraeten, W.W.; Goossens, R. A time-integrated MODIS burn severity assessment using the multi-temporal differenced normalized burn ratio (dNBRMT). *Int. J. Appl. Earth Obs. Geoinf.* **2011**, *13*, 52–58. [CrossRef]
19. Parks, S.A.; Dillon, G.K.; Miller, C. A new metric for quantifying burn severity: The Relativized Burn Ratio. *Remote Sens.* **2014**, *6*, 1827–1844. [CrossRef]
20. Epting, J.F. Remote Sensing of Burn Severity and the Interactions between Burn Severity, Topography and Vegetation in Interior Alaska. Master's Thesis, University of Alaska Fairbanks, Fairbanks, Alaska, 2004.
21. Cocke, A.E.; Fulé, P.Z.; Crouse, J.E. Comparison of burn severity assessments using Differenced Normalized Burn Ratio and ground data. *Int. J. Wildland Fire* **2005**, *14*, 189. [CrossRef]
22. Soverel, N.O.; Perrakis, D.D.B.; Coops, N.C. Estimating burn severity from Landsat dNBR and RdNBR indices across western Canada. *Remote Sens. Environ.* **2010**, *114*, 1896–1909. [CrossRef]
23. Lu, B.; He, Y.; Tong, A. Evaluation of spectral indices for estimating burn severity in semiarid grasslands. *Int. J. Wildland Fire* **2016**, *25*, 147–157. [CrossRef]
24. Hammill, K.A.; Bradstock, R.A. Remote sensing of fire severity in the Blue Mountains: Influence of vegetation type and inferring fire intensity. *Int. J. Wildland Fire* **2006**, *15*, 213–226. [CrossRef]
25. Schepers, L.; Haest, B.; Veraverbeke, S.; Spanhove, T.; Borre, J.V.; Goossens, R. Burned area detection and burn severity assessment of a heathland fire in Belgium Using Airborne Imaging Spectroscopy (APEX). *Remote Sens.* **2014**, *6*, 1803–1826. [CrossRef]
26. Nguyen Tran, B.; Tanase, M.A.; Bennett, L.T.; Aponte, C. Evaluation of spectral indices for assessing fire severity in Australian temperate forests. *Remote Sens.* **2018**, *10*, 1680. [CrossRef]
27. Kitzberger, T.; Veblen, T.T. Fire-induced changes in northern Patagonian landscapes. *Landsc. Ecol.* **1999**, *14*, 1–15. [CrossRef]
28. Mermoz, M.; Kitzberger, T.; Veblen, T.T. Landscape influences on occurrence and spread of wildfires in Patagonian forests and shrublands. *Ecology* **2005**, *86*, 2705–2715. [CrossRef]
29. Kitzberger, T.; Aráoz, E.; Gowda, J.H.; Mermoz, M.; Morales, J.M. Decreases in fire spread probability with forest age promotes alternative community states, reduced resilience to climate variability and large fire regime shifts. *Ecosystems* **2012**, *15*, 97–112. [CrossRef]
30. Paritsis, J.; Holz, A.; Veblen, T.T.; Kitzberger, T. Habitat distribution modeling reveals vegetation flammability and land use as drivers of wildfire in SW Patagonia. *Ecosphere* **2013**, *4*, 1–20. [CrossRef]
31. Tiribelli, F.; Kitzberger, T.; Morales, J.M. Changes in vegetation structure and fuel charasteristics along post-fire succession promote alternative stable states and positive fire-vegetation feedbacks. *J. Veg. Sci.* **2018**, *29*, 147–156. [CrossRef]
32. Veblen, T.T.; Kitzberger, T.; Raffaele, E.; Lorenz, D.C. Fire history and vegetation changes in northern Patagonia, Argentina. In *Fire and Climatic Changes in Temperate Ecosystems of the Western Americas*; Veblen, T.T., Baker, W., Montenegro, G., Swetnam, T.W., Eds.; Springer: New York, NY, USA, 2003; pp. 265–295.
33. Fajardo, A.; González, M.E. Replacement patterns and species coexistence in an Andean *Araucaria-Nothofagus* forest. *J. Veg. Sci.* **2009**, *20*, 1176–1190. [CrossRef]
34. González, M.E.; Veblen, T.T.; Sibold, J.S. Influence of fire severity on stand development of *Araucaria araucana–Nothofagus pumilio* stands in the Andean cordillera of south-central Chile. *Austral Ecol.* **2010**, *35*, 597–615. [CrossRef]
35. Mundo, I.A.; Kitzberger, T.; Roig, F.A.; Villalba, R.; Barrera, M.D. Fire history in the *Araucaria araucana* forests of Argentina: Human and climate influences. *Int. J. Wildland Fire* **2013**, *22*, 194–206. [CrossRef]
36. Holz, A.; Veblen, T.T. Variability in the Southern Annular Mode determines wildfire activity in Patagonia. *Geophys. Res. Lett.* **2011**, *38*, 1–6. [CrossRef]
37. Veblen, T.T.; Holz, A.; Paritsis, J.; Raffaele, E.; Kitzberger, T.; Blackhall, M. Adapting to global environmental change in Patagonia: What role for disturbance ecology? *Austral Ecol.* **2011**, *36*, 891–903. [CrossRef]

38. González, M.E.; Lara, A. Large fires in the Andean *Araucaria* forests: When a natural ecological process becomes a threat. *Oryx* **2015**, *49*, 394. [CrossRef]
39. Burns, B.R. Fire-induced dynamics of *Araucaria araucana-Nothofagus antarctica* forest in the Southern Andes. *J. Biogeogr.* **1993**, *20*, 669. [CrossRef]
40. González, M.E.; Szejner, M.; Muñoz, A.A.; Silva, J. Incendios catastróficos en bosques andinos de Araucaria-Nothofagus: Efecto de la severidad y respuesta de la vegetación. *Bosque Nativ.* **2009**, *46*, 12–17.
41. Urrutia-Estrada, J.; Fuentes-Ramírez, A.; Hauenstein, E. Diferencias en la composición florística en bosques de *Araucaria-Nothofagus* afectados por distintas severidades de fuego. *Gayana Botánica* **2018**, *75*, 625–638. [CrossRef]
42. Assal, T.J.; González, M.E.; Sibold, J. Burn severity controls on post-fire *Araucaria-Nothofagus* regeneration in the Andean Cordillera. *J. Biogeogr.* **2018**, *45*, 2483–2494. [CrossRef]
43. Peñaloza Rubio, A.R. Zonificación de la Severidad de un Incendio Natural y su Descripción Topográfica Cuantitativa en el Parque Nacional Tolhuaca, IX Región, Chile. Bachelor's Thesis, Facultad de Ciencias Forestales, Universidad Austral de Chile, Valdivia, Chile, 2006.
44. Mermoz, M.; Müller, M.; Nuñez, C.; Pastore, H.; Ramilo, E. *Evaluación ecológica del incendio "El Cristo" Área Futalaufquen-Parque Nacional Los Alerces Marzo-Abril de 2015*; Administración de Parques Nacionales, Intendencia del Parque Nacional Los Alerces: Chubut, Argentina, 2016.
45. De la Barrera, F.; Barraza, F.; Favier, P.; Ruiz, V.; Quense, J. Megafires in Chile 2017: Monitoring multiscale environmental impacts of burned ecosystems. *Sci. Total Environ.* **2018**, *637–638*, 1526–1536. [CrossRef]
46. Navarro Cerrillo, R.M.; Olave Ortiz, F.; Hayas, A.; Castillo, M. Metodología para la elaboración de un plan de restauración postincendio en Chile: La experiencia del Parque Nacional de Torres del Paine The experience of the Torres del Paine National Park. *An. Inst. Patagon.* **2015**, *43*, 53–73.
47. De Fina, A.L. El clima de la región de los bosques andino-patagónicos argentinos. In *La Región de Los Bosques Andino-Patagónicos. Sinopsis General. Colección Científica Del INTA N° 10*; Dimitri, M.J., Ed.; INTA: Buenos Aires, Argentina, 1972; pp. 35–58.
48. Veblen, T.T.; Kitzberger, T.; Raffaele, E.; Mermoz, M.; González, M.E.; Sibold, J.S.; Holz, A. The historical range of variability of fires in the Andean Patagonian *Nothofagus* forest region. *Int. J. Wildland Fire* **2008**, *17*, 724–741. [CrossRef]
49. Tortorelli, L.A. *Maderas y Bosques Argentinos*; Editorial ACME: Buenos Aires, Argentina, 1956.
50. Kitzberger, T.; Perry, G.L.W.; Paritsis, J.; Gowda, J.H.; Tepley, A.J.; Holz, A.; Veblen, T.T. Fire-vegetation feedbacks and alternative states: Common mechanisms of temperate forest vulnerability to fire in southern South America and New Zealand. *N. Z. J. Bot.* **2016**, *54*, 247–272. [CrossRef]
51. QGIS Development Team. QGIS Geographic Information System. Open Source Geospatial Foundation Project. 2018. Available online: http://qgis.osgeo.org (accessed on 10 December 2019).
52. Escuin, S.; Navarro, R.; Fernández, P. Fire severity assessment by using NBR (Normalized Burn Ratio) and NDVI (Normalized Difference Vegetation Index) derived from LANDSAT TM/ETM images. *Int. J. Remote Sens.* **2008**, *29*, 1053–1073. [CrossRef]
53. Stambaugh, M.C.; Hammer, L.D.; Godfrey, R. Performance of burn-severity metrics and classifications in oak woodlands and grasslands. *Remote Sens.* **2015**, *7*, 10501–10522. [CrossRef]
54. Chuvieco, E. *Fundamentals of Satellite Remote Sensing: An Environmental Approach*, 2nd ed.; CRC Press: Boca Raton, FL, USA, 2016; ISBN 9781498728072.
55. Fuentes-Ramírez, A.; Arroyo-Vargas, P.; Del Fierro, A.; Pérez, F. Post-fire response of *Araucaria araucana* (Molina) K. Koch: Assessment of vegetative resprouting, seed production and germination. *Gayana Botánica* **2019**, *76*, 119–122. [CrossRef]
56. Diaz-Delgado, R. Efecto de la recurrencia de los incendios sobre la resiliencia post- incendio de las comunidades vegetales de Cataluña a partir de imágenes de satélite. *Ecosistemas* **2003**, *12*, 10.
57. Kaufman, Y.J.; Remer, L.A. Detection of forests using Mid-IR reflectance: An application for aerosol studies. *IEEE Trans. Geosci. Remote Sens.* **1994**, *32*, 672–683. [CrossRef]
58. Veraverbeke, S.; Harris, S.; Hook, S. Evaluating spectral indices for burned area discrimination using MODIS/ASTER (MASTER) airborne simulator data. *Remote Sens. Environ.* **2011**, *115*, 2702–2709. [CrossRef]
59. Miller, J.D.; Thode, A.E. Quantifying burn severity in a heterogeneous landscape with a relative version of the delta Normalized Burn Ratio (dNBR). *Remote Sens. Environ.* **2007**, *109*, 66–80. [CrossRef]

60. Boucher, J.; Beaudoin, A.; Hébert, C.; Guindon, L.; Bauce, E. Assessing the potential of the differenced Normalized Burn Ratio (dNBR) for estimating burn severity in eastern Canadian boreal forests. *Int. J. Wildland Fire* **2017**, *26*, 32–45. [CrossRef]
61. Kurbanov, E.; Vorobyev, O.; Leznin, S.; Polevshiova, Y.; Demisheva, E. Assessment of burn severity in Middle Povozhje with Landsat multitemporal data. *Int. J. Wildland Fire* **2017**, *26*, 772–782. [CrossRef]
62. Cansler, C.A.; McKenzie, D. How robust are burn severity indices when applied in a new region? Evaluation of alternate field-based and remote-sensing methods. *Remote Sens.* **2012**, *4*, 456–483. [CrossRef]
63. Chafer, C.J. A comparison of fire severity measures: An Australian example and implications for predicting major areas of soil erosion. *Catena* **2008**, *74*, 235–245. [CrossRef]
64. Zhu, Z.; Key, C.H.; Ohlen, D.; Benson, N.C. *Evaluate Sensitivities of Burn-Severity Mapping Algorithms for Different Ecosystems and Fire Histories in the United States*; Final report to the Joint Fire Science Program; USGS, National Center of Earth Resources Observation and Science: Sioux Falls, SD, USA, 2006.
65. Holden, Z.A.; Morgan, P.; Smith, A.M.S.; Vierling, L. Beyond Landsat: A comparison of four satellite sensors for detecting burn severity in ponderosa pine forests of the Gila Wilderness, NM, USA. *Int. J. Wildland Fire* **2010**, *19*, 449–458. [CrossRef]
66. Chen, X.; Vogelmann, J.E.; Rollins, M.G.; Key, C.H.; Yang, L.; Huang, C.; Shi, H. Detecting post-fire burn severity and vegetation recovery using multitemporal remote sensing spectral indices and field-collected composite burn index data in a ponderosa pine forest. *Int. J. Remote Sens.* **2011**, *32*, 7905–7927. [CrossRef]
67. Parks, S.A.; Holsinger, L.M.; Voss, M.A.; Loehman, R.A.; Robinson, N.P. Mean Composite Fire Severity Metrics Computed with Google Earth Engine Offer Improved Accuracy and Expanded Mapping Potential. *Remote Sens.* **2018**, *10*, 879. [CrossRef]
68. Allen, J.L.; Sorbel, B. Assessing the differenced Normalized Burn Ratio's ability to map burn severity in the boreal forest and tundra ecosystems of Alaska's national parks. *Int. J. Wildland Fire* **2008**, *17*, 463–475. [CrossRef]
69. Soverel, N.O.; Coops, N.C.; Perrakis, D.D.B.; Daniels, L.D.; Gergel, S.E. The transferability of a dNBR-derived model to predict burn severity across 10 wildland fires in western Canada. *Int. J. Wildland Fire* **2011**, *20*, 518–531. [CrossRef]
70. Picotte, J.J.; Robertson, K.M. Validation of remote sensing of burn severity in south-eastern US ecosystems. *Int. J. Wildland Fire* **2011**, *20*, 453–464. [CrossRef]
71. Hall, R.J.; Freeburn, J.T.; de Groot, W.J.; Pritchard, J.M.; Lynham, T.J.; Landry, R. Remote sensing of burn severity: Experience from western Canada boreal fires. *Int. J. Wildland Fire* **2008**, *17*, 476–489. [CrossRef]
72. Murphy, K.A.; Reynolds, J.H.; Koltun, J.M. Evaluating the ability of the differenced Normalized Burn Ratio (dNBR) to predict ecologically significant burn severity in Alaskan boreal forests. *Int. J. Wildland Fire* **2008**, *17*, 490–499. [CrossRef]
73. French, N.H.F.; Kasischke, E.S.; Hall, R.J.; Murphy, K.A.; Verbyla, D.L.; Hoy, E.E.; Allen, J.L. Using Landsat data to assess fire and burn severity in the North American boreal forest region: An overview and summary of results. *Int. J. Wildland Fire* **2008**, *17*, 443–462. [CrossRef]
74. Miller, J.D.; Knapp, E.E.; Key, C.H.; Skinner, C.N.; Isbell, C.J.; Creasy, R.M.; Sherlock, J.W. Calibration and validation of the relative differenced Normalized Burn Ratio (RdNBR) to three measures of fire severity in the Sierra Nevada and Klamath Mountains, California, USA. *Remote Sens. Environ.* **2009**, *113*, 645–656. [CrossRef]
75. Lhermitte, S.; Verbesselt, J.; Verstraeten, W.W.; Veraverbeke, S.; Coppin, P. Assessing intra-annual vegetation regrowth after fire using the pixel based regeneration index. *ISPRS J. Photogramm. Remote Sens.* **2011**, *66*, 17–27. [CrossRef]

remote sensing

MDPI

Article

Using Long-Term SAR Backscatter Data to Monitor Post-Fire Vegetation Recovery in Tundra Environment

Zhiwei Zhou [1,2,*], Lin Liu [2], Liming Jiang [1], Wanpeng Feng [3,4] and Sergey V. Samsonov [5]

1 State Key Laboratory of Geodesy and Earth's Dynamics, Institute of Geodesy and Geophysics, Chinese Academy of Sciences, Wuhan 430077, China; jlm@whigg.ac.cn
2 Earth Science System Programme, The Chinese University of Hong Kong, Hong Kong 999077, China; liulin@cuhk.edu.hk
3 Guangdong Provincial Key Laboratory of Geodynamics and Geohazards, School of Earth Sciences and Engineering, Sun Yat-Sen University, Guangzhou 510275, China; fengwp5@mail.sysu.edu.cn
4 Southern Marine Science and Engineering *Guangdong* Laboratory (Zhuhai), Zhuhai 519000, China
5 Canada Centre for Mapping and Earth Observation, Natural Resources Canada, Ottawa, ON K1A 0E4, Canada; sergey.samsonov@canada.ca
* Correspondence: Zhiwei.Zhou@whigg.ac.cn; Tel.: +86-27-68881355

Received: 24 August 2019; Accepted: 21 September 2019; Published: 25 September 2019

Abstract: Wildfires could have a strong impact on tundra environment by combusting surface vegetation and soil organic matter. For surface vegetation, many years are required to recover to pre-fire level. In this paper, by using C-band (VV/HV polarization) and L-band (HH polarization) synthetic aperture radar (SAR) images acquired before and after fire from 2002 to 2016, we investigated vegetation change affected by the Anaktuvuk River Fire in Arctic tundra environment. Compared to the unburned areas, C- and L-band SAR backscatter coefficients increased by up to 5.5 and 4.4 dB in the severely burned areas after the fire. Then past 5 years following the fire, the C-band SAR backscatter differences decreased to pre-fire level between the burned and unburned areas, suggesting that vegetation coverage in burned sites had recovered to the unburned level. This duration is longer than the 3-year recovery suggested by optical-based Normalized Difference Vegetation Index (NDVI) observations. While for the L-band SAR backscatter after 10-year recovery, about 2 dB higher was still found in the severely burned area, compared to the unburned area. The increased roughness of the surface is probably the reason for such sustained differences. Our analysis implies that long records of space-borne SAR backscatter can monitor post-fire vegetation recovery in Arctic tundra environment and complement optical observations.

Keywords: arctic tundra fire; vegetation recovery; C- and L-band SAR; SAR backscatter

1. Introduction

Climate change is occurring rapidly at high latitudes, where surface air temperature has increased at twice the rate of the rest of the globe in the past several decades [1–3]. Such warming has resulted in pronounced environmental changes in the Arctic [4–6]. During that period, it has caused a dramatic increase in the frequency of large wildfires in boreal and tundra regions [7–9]. Wildfires strongly affect ecosystem in tundra environments, such as carbon balance of tundra [10,11], land surface albedo [12], active layer thickness change [13], vegetation shifts [14], permafrost thaw and thermokarst development [13,15], and even wildlife habits [16].

By burning surface vegetation and soil organic matter, fire can rapidly transfer large stocks of aboveground biomass and below-ground soil carbon to the atmosphere and change the ecosystem carbon balance in the short-term. Additionally, the surface albedo balance is also changed by burning some, or all, of the insulating layer of moss and soil organic matter, which shields the underlying permafrost from warm summer temperatures [12]. Then the active layer of burned zone is destabilized in summer, which may cause decomposition of deeper soil carbon stored below ground and results in additional carbon transfer to the atmosphere in the long-term after the initial combustion [17–19]. Thermokarst development induced by fire may accelerate the above process and results in carbon transfer to the atmosphere and to the tundra gully as dissolved organic carbon [20,21]. Since soil carbon underlain by permafrost in the Arctic is nearly twice the total amount of carbon as is in the atmosphere [22,23], increased fire frequency and severity in Arctic tundra could act as positive feedback to global climate warming [24].

Wildfire regimes could exert landscape-scale controls on vegetation structure and composition [25]. Wildfire also directly affects seedling recruitment success and migration [26]. Fire may alter vegetation successional trajectory and prepare seedbeds for shrub or tree invasion. Therefore, increased fire frequency and severity in Arctic tundra is expected to facilitate both shrub expansion and tree migration [25,27]. Wildfires also consume caribou forage lichens and exert strong effects on post-fire lichen recovery for decades [28]. Such a change in successional trajectory could alter the structure of the vegetation and ecosystem functions, the ecosystem may undergo a regime shift into an alternate state, and ecosystem services may be degraded [29].

Thus, information on post-fire vegetation regrowth in tundra is of great interest for scientists and land managers dealing with environmental issues (e.g., carbon budget, climate impact, ecosystem services). For monitoring post-fire vegetation recovery in expansive and remote tundra area, satellite remote sensors could be suitable tools to conduct the temporal analysis, since they are suitable to continuously collect data for long-term monitoring and temporal analysis in a large area. Tanase et al. [30] suggest that the suitability of remote sensing data is determined by two factors: (a) sensitivity to the changes in radiometric response due to the development of vegetation and (b) the spatial resolution of satellite imagery, which provides the ability to identify the features of the response patterns.

Normalized Difference Vegetation Index (NDVI) images derived from optical satellite observations have been most frequently used measurements for monitoring, analyzing, and mapping vegetation changes in time and space. However, NDVI is more sensitive to changes in leaf area than to changes in structure and overall biomass. The relationship between NDVI and leaf area index (LAI) varies both seasonally and inter-annually. Furthermore, NDVI saturates at high values of LAI [31]. This implies that NDVI is only sensitive to vegetation development in the first few years after fire disturbance. In a study on arctic tundra fires recovery by NDVI, Barrett et al., [14] showed that NDVI quickly recovered to the range of pre-fire level in 3 years.

Imaging radars may have the potential to complement traditional optical data in mapping vegetation recovery in the terms of vegetation structure [32–39], because the radar backscattered signal is highly sensitive to structural properties of the vegetation. In addition, radar sensors have all-weather, and day and night observing capability (independent of solar radiation). Furthermore, as long as more than 25 years' archives of spaceborne synthetic aperture radar (SAR) data and planned future SAR missions, as well as global coverage, provide potential to monitor post-fire vegetation recovery in Arctic tundra.

The radar backscatter from vegetation is the result of complex interactions between microwave electromagnetic energy and the ground and vegetation scattering components in a resolution cell. The intensity of backscatter from vegetation depends on several factors: (1) the amount and geometric properties of vegetation on site, (2) surface dielectric and roughness properties, and (3) radar frequency, polarization, and look direction [40–42]. Generally, due to the penetration capability, short wavelengths of 2–6 cm, such as X-band mostly interact with leaves/needles, twigs, and small branches, and C-band mostly interacts with leaves and small secondary branches [43]. In contrast, long wavelengths of approximately 24 cm, such as L-band, radar waves interact mostly with primary branches, tree trunks, and even underground surface [41]. There is an approximately linear response of backscatter to increasing forest biomass reaching saturation as a function of forest type and structure [42]. Therefore, radar backscatter is expected to change before and after a fire, as the fire removes most of the vegetation components and may leave the ground or tree trunks exposed. Specifically, the backscatter increases with burn severity for L-band at HH polarized (SAR data acquired under wet condition) [44], and for X- and C-bands at HH and VV polarizations, while the backscatter decreases with burn severity for all frequency at cross-polarization [45]. The backscatter increases with incidence angle at co-polarization and decreases with incidence angle at cross-polarization in burned sites [45]. For example, using C-band SAR (HV polarization), previous studies show 3–6 dB increase in backscatter coefficients between fire scars and adjacent undisturbed forest and seasonal changes in backscatter coefficients in a burned forest [30,43].

Although SAR data usage for forest recovery after wildfires is well studied, it is limited used for tundra vegetation recovery. Using L-band HV data, it is estimated that approximately 40–50 years are required for the recovery of forest structure after disturbance [30]. The effects of wildfire on tundra vegetation are poorly studied by using SAR data. As post-fire vegetation regrowth in tundra environment was mostly investigated by fieldwork periodically [8,16,25,46], or projection based on post-fire soil biota [26], remotely sensed data are rarely used [14]. Using nearly two-decades ERS-1/2 C-band dataset (VV polarization), Jenkins et al. [47] demonstrated the capability of SAR backscatter in detection and monitoring of fire disturbance in the Alaska tundra. The study suggested that there was a 3-dB backscatter coefficient increase between burned and unburned area, and also four to five years of landscape recovery was observed by SAR backscatters.

The objective of this study is to monitor post-fire vegetation recovery in Arctic tundra environment by using long-term SAR backscatter data. In this study, we not only use C-band dataset but also L-band dataset. Compared to NDVI information, radar backscatter information provides new insights on vegetation recovery regarding the structure and density [30,48].

2. Materials and Methods

2.1. Study Area

The Anaktuvuk River Fire (ARF) burned 1039 km^2 of Arctic tundra on the North Slope of the Brooks Range in the late summer and early fall of 2007. Figure 1a shows the fire perimeters obtained from the Alaska Interagency Coordination Center Large Fire Database [49]. This fire has an order of magnitude larger than the average fire size in the historic record for the North Slope [16]. The ARF resulted in the release of approximately 2.1×10^3 kg of carbon to the atmosphere, an amount similar in magnitude to the annual net carbon sinks for the entire Arctic tundra biome averaged over the last quarter of the twentieth century [10]. The fire occurred primarily in the Arctic Foothills physiographic province of the North Slope, where the ecosystem is described as upland, shrubby tussock tundra. The climate in the central North Slope is cold in winter (−25 °C mean high in January) and relatively cool in summer (20 °C mean high in July, vegetation growing season). The mean annual temperature and precipitation of this study area are −10 °C and 30 cm, respectively [10,16].

Figure 1b shows differenced Normalized Burn Ratio (dNBR) calculated from a pair of pre-fire Landsat Enhanced Thematic Mapper-plus (ETM+) image acquired on 30 June 1999, and post-fire Landsat Thematic Mapper (TM) image acquired on 14 June 2008 [50]. The dNBR values indicate burn severity within the fire boundary: 11% of the area did not burn (dNBR < 300), while within the burned area, approximately 47% of the ARF burned at high severity (dNBR > 600) and 35% and 18% at moderate to low severity (300 < dNBR < 600) [16]. Three sites located in severely and moderately burned, and unburned area were chosen to examine the SAR backscatter change before and after fire (Figure 1b). These three sites were placed less than 7 km away to each other and had similar pre-fire vegetation features [51].

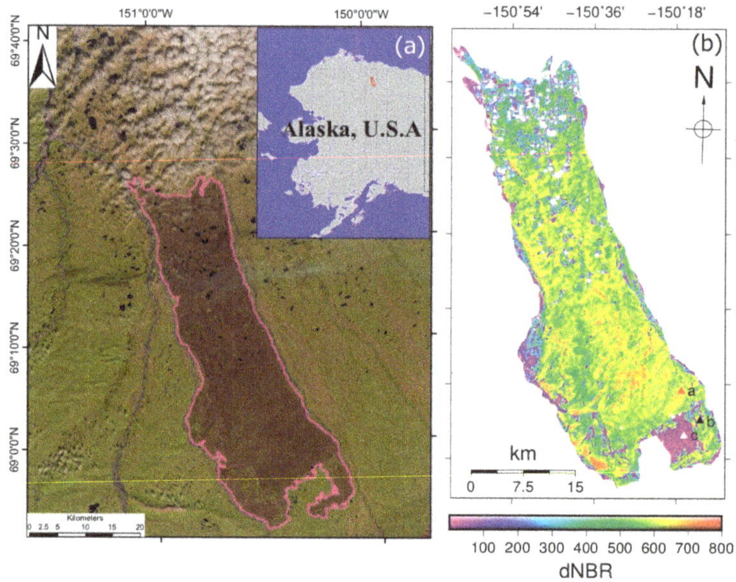

Figure 1. The location and burn severity of the Anaktuvuk River Fire (ARF). (**a**) the False-color Landsat-5 image acquired on 17 July 2008, the pink polygon is the fire boundary and the red polygon of upper left map shows the location of the fire area in Alaska. (**b**) differenced Normalized Burn Ratio (dNBR) within the burn scar [50]. The red, black, and white triangles (labeled as 'a', 'b', and 'c') mark the locations of severely and moderately burned, and unburned sites to be examined in our analysis, respectively.

Prior to the fire, 54% of the burned area was classified as upland moist acidic tundra (soil pH < 5.5), 15% as moist nonacidic tundra (soil pH > 5.5), and 30% as shrubland [16]. The vegetation in the burned area is dominated by the tussock-forming sedges [50]. The dominant shrub types include willow, dwarf birch, Labrador tea, blueberry, and prostrate shrubs lowbush cranberry, bearberry, and crowberry [16].

2.2. Remote Sensing Data

We only used SAR amplitude images acquired by C-band from the European Remote Sensing-2 (ERS-2) and RADARSAT-2 (RS-2) satellites, L-band from Advanced Land Observing Satellite-1/2 (ALOS-1/2) Phased Array-type L-band SAR (PALSAR) sensors. To compare the difference of SAR amplitude before and after the fire, these images were acquired from 2002 to 2017, with different incidence angles and polarization modes (Table 1). The ERS-2 and ALOS-1 PALSAR datasets contain images acquired before and after the fire, and the RS-2 and ALOS-2 PALSAR datasets spanned nearly 10 years after the fire. ERS-2, RS-2, and ALOS-1 PALSAR Level-1.5 Geo-referenced amplitude SAR images used in this study were provided by the Alaska Satellite Facility (ASF) [52].

All SAR images were firstly radiometrically calibrated and then radiometric variability associated with topography corrected using a Digital Elevation Model (DEM). Next, speckle noise was reduced by using a 3 × 3 window Lee filter. Finally, all SAR images were geocoded to WGS-84 projection. The ERS-2 and ALOS-1 images were processed by the MapReady package (v3.1.24) [53]. The RS-2 and ALOS-2 were processed by the Sentinel Application Platform (SNAP, v5.0). The resulting images were calibrated to sigma nought (σ°, backscatter coefficient) format. A 30 m ASTER DEM was used for radiometric terrain correction caused by local terrain variability in both MapReady and SNAP. Figure 2 shows a series of SAR backscatter coefficient images in both C- and L-bands that spanned the period of before and after the fire. Before the fire, there was no difference in backscatter between the burned and unburned areas (visually represented by brightness in Figure 2a,b). An obviously brighter (increased backscatter) fire scars can be observed in the burned area in 2008 in both the C- and L-band SAR images (Figure 2c,d). Moreover, fire scars become lighter five years after the area burned in C-band (Figure 2e) and after 10 years in L-band SAR (Figure 2f).

Table 1. List of synthetic aperture radar (SAR) images used in this study. Note: the resolution is equal in range and azimuth direction.

Satellite	ERS-2	RADARSAT-2	ALOS-1	ALOS-2
Band/Wavelength	C/5.6 cm	C/5.6 cm	L/23.6 cm	L/23.6 cm
Number of images	52	34	30	14
Acquisition period (DD/MM/YYYY)	06/01/2002–31/10/2010	17/01/2011–08/02/2016	03/06/2006–17/03/2010	17/11/2014–19/07/2017
Incidence angle (degree)	23.0	27.0	38.0	36.8
Polarization	VV	HV	HH	HH
Resolution (m)	12.5	50.0	12.5	6.25

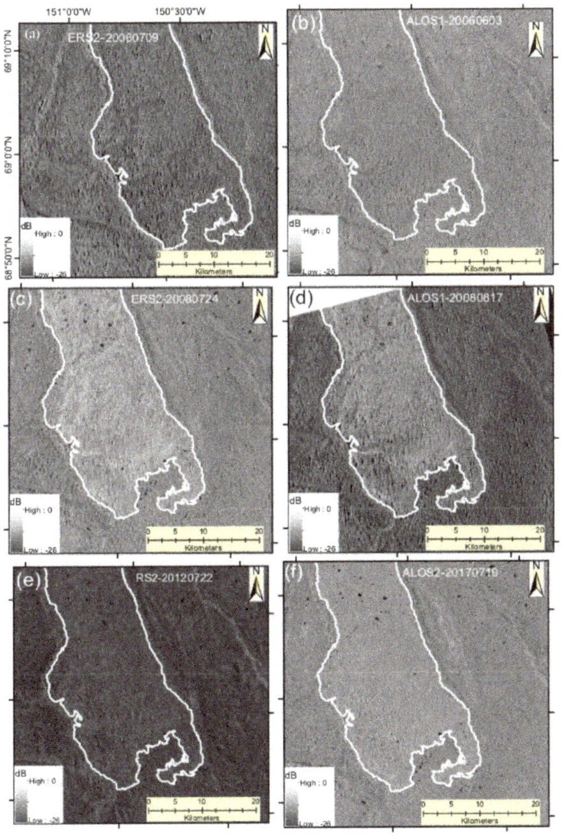

Figure 2. Images of backscatter coefficient that span before and after the fire. The white polygon marks the fire boundary. (**a**,**b**) present the images acquired from ERS2 and ALOS-1 before the fire. (**c**–**f**) present the post-fire images acquired from ERS2, ALOS-1, RS-2, and ALOS-2, respectively.

Due to the inherent speckle noise caused by the nature of the SAR systems, a single pixel of SAR data should not be used to directly relate to field variable. Instead, a group of pixels in azimuth and range direction (called multi-looking process) are usually averaged to mitigate speckle noise. In this study, for all C- and L-band data, a mean of approximately 150 m × 150 m (e.g., 3 × 3 and 12 × 12 looks in both azimuth and range direction for RS-2 and ALOS-1 based on the spatial resolution) area centered on the unburned, moderately burned, and severely burned sites are chosen as the control, moderate, and severe burn sites, respectively (see their locations in Figure 1b). Five-year (2002–2007) pre-fire data were used to compare with 6-year (2008–2010, 2014–2017) post-fire data.

To aid the interpretation of changes of SAR backscatters, NDVI of the whole fire zone and soil moisture collected from the severely burned, moderately burned, and unburned zones are used as ancillary information. We used the NDVI products generated from the Moderate Resolution Imaging Spectroradiometer (MODIS) and distributed by the Land Processes Distributed Active Archive Center (LP DACC) of NASA/USGS [54]. More specifically, the MOD13 Q1 products obtained during growing seasons (June, July, and August) from 2002 to 2016 were used. The accuracy of this product is now within ±0.025, and the derived NDVI accuracy is within ±0.025 [55]. We chose these days because mid-July is close to the end of the growing season. The MOD13 Q1 product is a global product with a spatial resolution of 250 m and a temporal interval of 16 days.

2.3. SAR Backscattering Mechanisms before and after a Fire

In order to evaluate the sensitivity of radar backscatter to the post-fire vegetation regrowth, it is important to identify the dominant scattering mechanisms before and after a fire. Figure 3 illustrates the dominant backscatter scattering mechanisms in tundra environment, including pre-fire volume scattering (Figure 3a) and post-fire surface scattering (Figure 3b) [43]. Before a fire, the C-/L-band radar waves interact with the canopy of the tussock-forming sedges, in this case, most of the C-/L-band radar backscatter originates from the volume scattering with the canopy layer (Figure 3a). Thus, the backscatter is more sensitive to canopy changes than the changes of ground moisture and roughness. After a fire, the vegetation and/or part of organic soil are burned, leaving the ground exposed. In this case, radar waves primarily interact with the ground surface (Figure 3b). In such case of surface scattering, the backscatter strongly depends on the roughness and soil moisture content of the ground. In a few years after a fire, the radar waves interact more and more with newly regenerated vegetation. Then the effects of soil moisture and roughness on backscatter gradually diminish because of the attenuation due to the growing vegetation. Once the vegetation density and structure reach the pre-fire level, the volume scattering becomes the dominant mechanism again (Figure 3a).

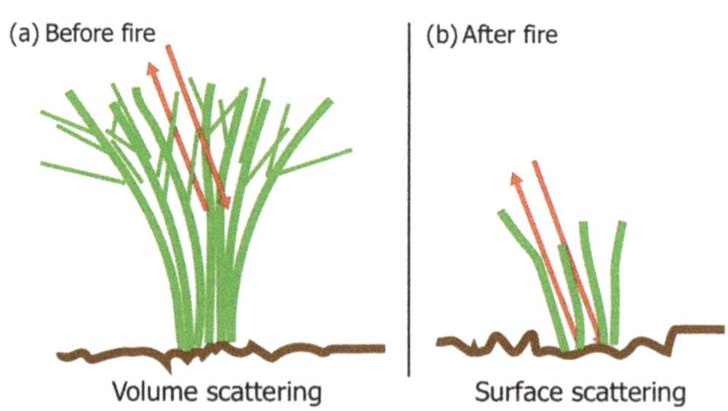

Figure 3. Dominant radar backscattering mechanisms in tundra environment. (**a**) volume scattering within tussock-forming sedges before a fire, (**b**) surface scattering from rougher ground after a fire.

3. Results

3.1. NDVI Changes

Figure 4 shows the NDVI on 12 or 13 July between 2002 and 2016. Generally, sparse vegetation may result in moderate NDVI values (approximately 0.2–0.5). High NDVI values (approximately 0.6–0.9) correspond to dense vegetation. A clear decrease of surface greenness can be observed following the ARF in 2007 (Figure 4g), and the greenness quickly recovered by 3 years after the fire (Figure 4i).

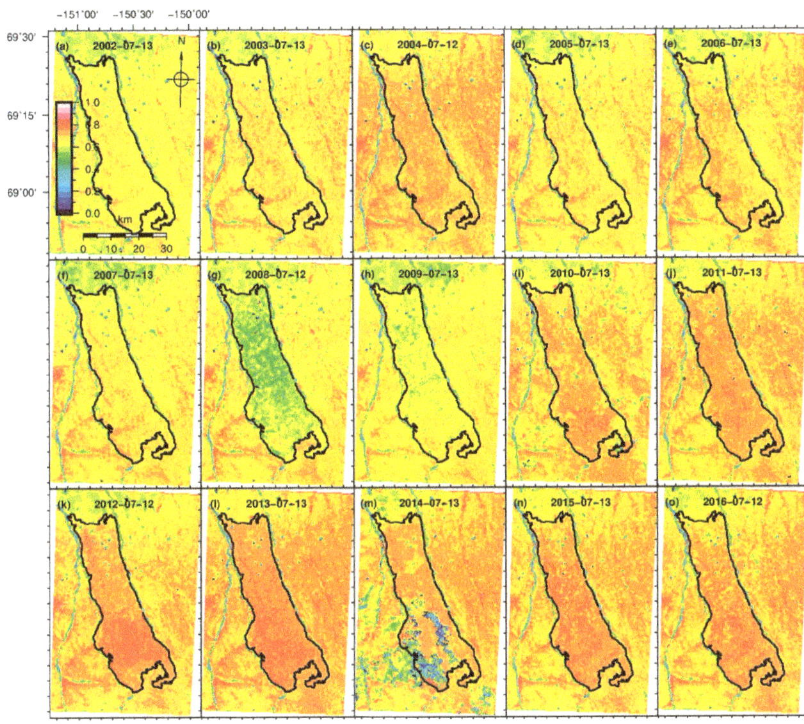

Figure 4. Maps of Normalized Difference Vegetation Index (NDVI) from July 2002 to July 2016. The black polygon in each figure marks the fire boundary. Note that the NDVI image on 13 July 2014 is contaminated by cloud cover.

Figure 5 presents a quantitative analysis of NDVI changes at the severely burned and unburned sites. Each site presents an area of approximately 1.25 km × 1.25 km. At both sites, NDVI showed a clear growing pattern (between 0.5 and 0.7) and reached the highest value in mid-July or early August (Figure 5a). The pre-fire NDVI values were similar at both sites. After the fire, an obvious reduction of NDVI can be observed at the burned site in 2008 (Figure 5a), and the difference between the severely burned and unburned sites was up to 0.2 (Figure 5b). Then a quick recovery occurred in the subsequent 3 years. The post-fire NDVI difference reached zero again from 2010 onward.

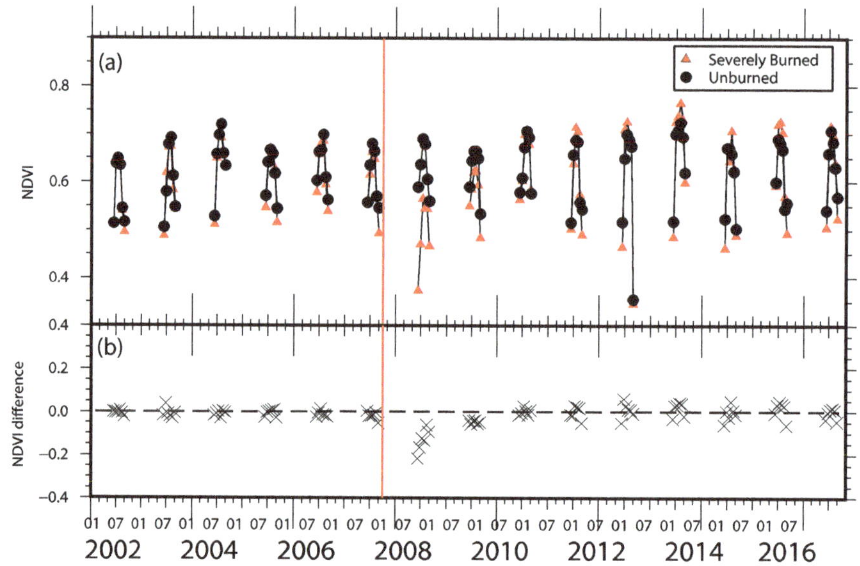

Figure 5. Time series of NDVI at the severely burned and unburned sites (**a**) and their differences (**b**) severely burned minus unburned. The vertical red line marks the end of the ARF.

3.2. SAR Backscatter Changes

In Figure 6, we present plots of mean SAR backscatter pre- and post-fire trend. Figure 7 shows C- and L-band backscatter difference (only in growing season) between burned and unburned, and its corresponding statistics analysis is plotted in Figure 8. Similar annual and inter-annual temporal trend can be observed for all sites and bands. The backscatter coefficient changes seasonally from −18 dB to −9 dB for L-band (Figure 7a) and from −25 dB to −12 dB for C-band (Figure 7b).

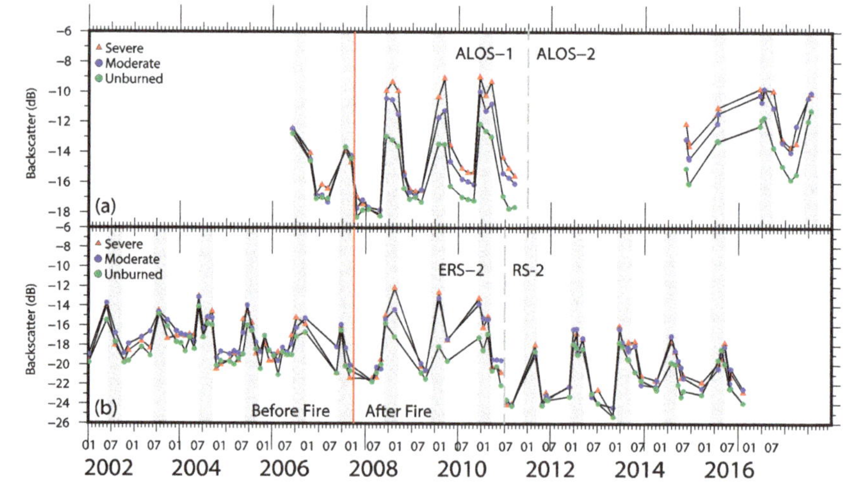

Figure 6. Time series of backscatter coefficients at the severely and moderately burned and unburned sites. (**a**) for ALOS-1 and ALOS-2, (**b**) for ERS-2 and RS-2. The red vertical line marks the end of the fire. The gray bars denote the growing seasons (June, July, August). The gray dashed lines separate ALOS-1 and ALOS-2 in (**a**), ERS-2 and RS-2 in (**b**).

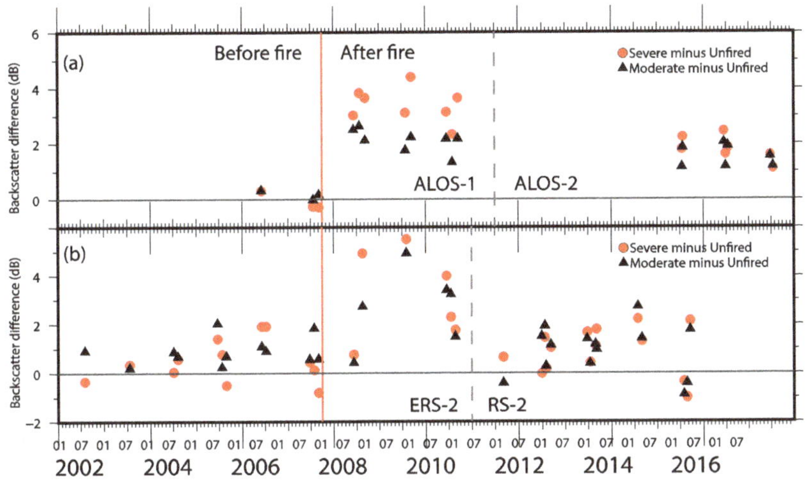

Figure 7. Time series of backscatter difference between burned and unburned sites. (**a**) for ALOS-1 and ALOS-2, (**b**) for ERS-2 and RS-2.

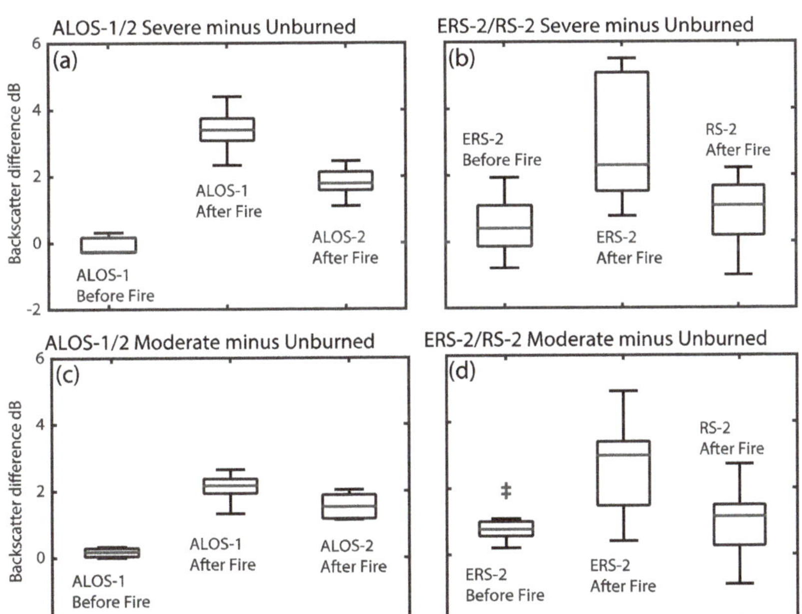

Figure 8. Statistical summary of the backscatter differences. For each box, the central line is the median, the upper and lower edges of the box are the 25th and 75th percentiles, the whiskers represent the minimum and maximum values, and the plus symbols mark the outliers. (**a**,**b**) present ALOS-1/2 and ERS-2/RS-2 backscatter difference between severe and unburned area, (**c**,**d**) present the ALOS-1/2 and ERS-2/RS-2 backscatter difference between moderate and unburned area.

We observe the following temporal characteristics of backscatter changes before the fire. Firstly, the temporal trend characteristics of severely and moderately burned sites are very similar to those of the unburned site for both C- and L-bands (Figure 6). Secondly, in the growing seasons before the fire, most of the C- and L-band backscatter coefficients at the severely and moderately burned sites showed little dynamics: close to 0 dB variation for L-band and <1 dB for C-band (Figures 7 and 8). Thirdly, the backscatter coefficients from both C- and L-band showed strong seasonal changes, increased in summer and decreased in winter. This is mainly caused by seasonal changes in vegetation, soil moisture, and soil freeze/thaw status. In the fall, the soil and vegetation freeze and the leaves fall from deciduous shrubs, reducing the soil moisture and overall backscattering, which causes backscatter to decrease. During the summer, the trend of soil and vegetation changes reverses, resulting in increasing the soil moisture and over backscattering, causing backscatter increase in summer. Combining the above three points, these suggest that the backscatter is mainly from the similar vegetation canopy property in the summer.

We observe the following temporal characteristics of backscatter coefficient changes after fire. Firstly, comparing to backscatter coefficient value in growing season before fire, a clear increase in backscatter coefficient value occurred in growing season after fire for both C- and L-band between 2008 and 2010 (Figures 7 and 8). During the growing season, the mean backscatter coefficient differences between severely burned and unburned site is nearly 3 dB (with maximum of 5.5 dB) and 1 dB (with maximum of 2 dB) for ERS-2 and RS-2 C-band, respectively; nearly 3 dB (with maximum of 4.4 dB) and 2 dB (with maximum of 2.5 dB) for ALOS-1 and ALOS-2 L band, respectively. The differences between moderately burned and unburned site were relatively lower for both C- and L-bands. Secondly, different change patterns for C- and L-bands can be observed between 2008 and 2016. C-band backscatter coefficient in growing season decreased from 2008 to 2010 (Figure 7b). Furthermore, a clear drop of backscatter values can be found from ERS-2 records to RS-2 records. This decline is contributed by polarization mode (HH to HV) and incidence angle (23 degree to 27 degree) change after the fire. However, there was no decreasing trend after 2012 (Figure 7b), and three sites present almost similar temporal change pattern after 2012 (Figure 6b). This suggests that after 5-year regrowth, radar waves from the severely and moderately burned sites interacted with similar vegetation canopy level as unburned site again.

For L-band, there was no decreasing trend of backscatter coefficient in severely and moderately burned sites between 2008 and 2010 during the growing season (Figures 6a and 7a). This suggests the vegetation in the severely and moderately burned sites did not recover to the similar state as in unburned site in 2010. The radar signal is primarily contributed by the roughness and soil moisture content of the ground in this stage. However, a slow decreasing trend can be observed between 2008 and 2016, and about 2 dB difference between unburned and moderately/severely sites still can be found in 2016 (Figures 7a and 8a,c). This is mainly because the radar signal interacts with the regenerated vegetation canopy, and the effects of roughness and soil moisture on backscatter gradually diminished. We should note that there is 1.2 degrees (at mid swath) difference in the incidence angle between ALOS-1 and ALOS-2 SAR acquisitions. Since they are in the same HH polarization modes, the different observation angle could have limited influence on SAR backscatter values after the fire.

4. Discussion

4.1. Inconsistency between NDVI Trend and Field Observations

In the ARF burned area, the NDVI changes suggest approximately 3 years of vegetation recovery. Barrett et al. [14] reported a similar 3-year NDVI recovery after another tundra fire on the western North Slope of Alaska. However, field investigations over three years after the ARF found the vegetation cover in the burned areas was still sparser than in the unburned areas (35–50% cover for graminoid and 33–42% cover for shrubs). Moreover, field investigations found that the most significant change in vegetation species in the burned areas was graminoid, whose cover increased from 11% in 2008 to 45% in 2010 [16]. Shrub cover also increased in the burned areas, but slowly, from 4% to 11% for deciduous shrubs, and from 4% to 13% for ericaceous shrubs [16]. These field data suggest that the vegetation did not return to pre-fire level after three years of recovery, which contradicts the NDVI observations.

4.2. SAR Backscatter before and after the Fire

The change of surface components due to fire causes SAR backscatter coefficient change. Our results show that the backscatter value of severe burn area increased up to 5.5 dB and 4.4 dB after the fire for C- and L-band, respectively. Similar backscatter coefficient differences are also found in different Alaska tundra fire sites using ERS-1 C-band dataset [47]. Over the three years of recovery, the L-band backscatter coefficient showed less change than the C-band (Figures 7 and 8). One possible reason is that regenerated vegetation is not dense as pre-fire. L-band radar waves can penetrate these newly regrown land cover, backscatter mainly still depends on roughness and soil moisture of the ground. While C-band backscatter from severely and moderately burned sites indicate a slow decrease trend with the regrowth of vegetation, but it does not return to the same level as the pre-fire level. C-band radar waves interact with new regenerated vegetation, resulting in reduction of the effect of roughness and soil moisture on the backscatter. Combining C- and L-band backscatter different changes over the three years of recovery, we could infer that: (1) the vegetation recovery did not return to the pre-fire status in vegetation structure and density; (2) C-band is more sensitive than L-band to the vegetation recovery in tundra environment.

After nearly 10 years of vegetation regrowth, the backscatter from RS-2 C-band and ALOS-2 L-band showed different change pattern. The C-band backscatter coefficient of all the sites showed almost the same changing pattern after 2011. This implies that the C-band radar waves returned from three sites interacts with a similar surface property after 2011. In other words, the vegetation property in severe burn and moderate burn had recovered to a similar level as in unburned site in 2012. The L-band backscatter from the severe burn and moderate burn sites was still about 2 dB higher than that from the unburned site. Using multitemporal airborne Light Detection and Ranging (LiDAR) dataset acquired in 2009 and 2014, Jones et al. [15] found surface roughness increased in burned area caused by thermokarst development. Therefore, the soil moisture and increased roughness probably contribute to the higher backscatter value in severely and moderately burned sites for ALOS-2 L-band. As L-band has stronger penetration capability, ground roughness and moisture have stronger effects on the returned radar signal than C-band. Combining these observations, we are able to infer that it takes 5 years for vegetation in the burned area to recover to a similar level as unburned site. We note that Jenkins et al. (2014) also suggested that 4–5 years are required for landscape recovery in tundra environment.

4.3. Limitations of SAR Data

However, compositions or new species vegetation recovered after the fire are unable to differentiate by SAR data. Fieldwork is necessary to investigate the vegetation compositions or shifts. A severe burn is able to facilitate colonization by new species when there is strong combustion of the pre-fire seedbed and high plant propagates. A 17-year study of post-fire vegetation recovery in Arctic tundra shows that there is abundance of graminoid species and an absence of Betulanana [14]. With an increase in fire activity and the conditions that contribute to increased fire severity, fire-related shifts in vegetation type will become more common [14]. Therefore, combining SAR imagery and ground observations could be an effective way to investigate the post-fire vegetation recovery in Arctic tundra environment.

5. Conclusions

Post-fire vegetation recovery in Arctic tundra environment is important to the balance of tundra ecosystem, such as carbon storage, active layer thickness, and vegetation composition. Effective measurements are required to monitoring the post-fire vegetation regrowth. In this paper, we investigate the potential of C- and L-band SAR backscatter for monitoring post-fire vegetation recovery after the 2007 Anaktuvuk River fire. We found that vegetation in burned area has probably returned to the pre-fire level after 5-year recovery, and C-band is more suitable to monitor such recovery process. The ground soil moisture is a major component of the total backscatter power for both C- and L-bands after fire. Yet, SAR data alone cannot quantify the vegetation composition and shifts, and additional field observations are required toward a thorough understanding of the composition.

Author Contributions: Conceptualization, Z.Z. and L.L.; data curation, Z.Z.; formal analysis, Z.Z. and L.L.; funding acquisition, L.L. and L.J.; investigation, Z.Z. and L.L.; methodology, Z.Z.; project administration, L.L. and L.J.; resources, L.L., W.F. and S.V.S.; supervision, L.L.; validation, Z.Z.; visualization, Z.Z.; writing—original draft, Z.Z.; writing—review and editing, L.L., L.J., and W.F.

Funding: This work is supported by the Strategic Priority Research Program of the Chinese Academy of Sciences (Grant No. XDA19070104), the National Natural Science Foundation of China (Grant No. 41804017), the Key Research Program of Frontier Sciences, CAS (QYZDB-SSW-DQC027 and QYZDJ-SSW-DQC042), the National Key R & D Program of China (2017YFA0603103), the Hong Kong Research Grants Council Grants CUHK14300815, and CUHK Direct Grant for Research 4053206.

Acknowledgments: We thank Benjamin Jones for providing the dNBR data and the Canadian Space Agency for providing the RS-2 ScanSAR Wide images.

Conflicts of Interest: The authors declare no conflict of interest.

References

1. Hinzman, L.D.; Bettez, N.D.; Bolton, W.R.; Chapin, F.S.; Dyurgerov, M.B.; Fastie, C.L.; Griffith, B.; Hollister, R.D.; Hope, A.; Huntington, H.P.; et al. Evidence and implications of recent climate change in northern alaska and other arctic regions. *Clim. Chang.* **2005**, *72*, 251–298.
2. Vavrus, S.J.; Holland, M.M.; Jahn, A.; Bailey, D.A.; Blazey, B.A. Twenty-first-century arctic climate change in ccsm4. *J. Clim.* **2012**, *25*, 2696–2710.
3. ACIA. Impacts of a warming arctic-arctic climate impact assessment. In *Impacts of a Warming Arctic-Arctic Climate Impact Assessment, by Arctic Climate Impact Assessment*; Cambridge University Press: Cambridge, UK, 2004; Volume 1, p. 144. ISBN 0521617782.
4. Schuur, E.A.G.; Bockheim, J.; Canadell, J.G.; Euskirchen, E.; Field, C.B.; Goryachkin, S.V.; Hagemann, S.; Kuhry, P.; Lafleur, P.M.; Lee, H.; et al. Vulnerability of permafrost carbon to climate change: Implications for the global carbon cycle. *BioScience* **2008**, *58*, 701–714.
5. Phoenix, G.K.; Bjerke, J.W. Arctic browning: Extreme events and trends reversing arctic greening. *Glob. Chang. Biol.* **2016**, *22*, 2960–2962. [PubMed]
6. Hu, F.S.; Higuera, P.E.; Walsh, J.E.; Chapman, W.L.; Duffy, P.A.; Brubaker, L.B.; Chipman, M.L. Tundra burning in alaska: Linkages to climatic change and sea ice retreat. *J. Geophys. Res. Biogeosci.* **2010**, *115*, G04002.
7. Kasischke, E.S.; Turetsky, M.R. Recent changes in the fire regime across the north american boreal region—Spatial and temporal patterns of burning across canada and alaska. *Geophys. Res. Lett.* **2006**, *33*, L09703.
8. Bret-Harte, M.S.; Mack, M.C.; Shaver, G.R.; Huebner, D.C.; Johnston, M.; Mojica, C.A.; Pizano, C.; Reiskind, J.A. The response of arctic vegetation and soils following an unusually severe tundra fire. *Philos. Trans. R. Soc. B Biol. Sci.* **2013**, *368*, 20120490.
9. Jones, B.M.; Breen, A.L.; Gaglioti, B.V.; Mann, D.H.; Rocha, A.V.; Grosse, G.; Arp, C.D.; Kunz, M.L.; Walker, D.A. Identification of unrecognized tundra fire events on the north slope of alaska. *J. Geophys. Res. Biogeosci.* **2013**, *118*, 1334–1344.
10. Mack, M.C.; Bret-Harte, M.S.; Hollingsworth, T.N.; Jandt, R.R.; Schuur, E.A.G.; Shaver, G.R.; Verbyla, D.L. Carbon loss from an unprecedented arctic tundra wildfire. *Nature* **2011**, *475*, 489–492.
11. Jiang, Y.; Rastetter, E.B.; Shaver, G.R.; Rocha, A.V.; Zhuang, Q.; Kwiatkowski, B.L. Modeling long-term changes in tundra carbon balance following wildfire, climate change, and potential nutrient addition. *Ecol. Appl.* **2017**, *27*, 105–117.
12. French, N.H.F.; Whitley, M.A.; Jenkins, L.K. Fire disturbance effects on land surface albedo in alaskan tundra. *J. Geophys. Res. Biogeosci.* **2016**, *121*, 841–854.
13. Liu, L.; Jafarov, E.E.; Schaefer, K.M.; Jones, B.M.; Zebker, H.A.; Williams, C.A.; Rogan, J.; Zhang, T. Insar detects increase in surface subsidence caused by an arctic tundra fire. *Geophys. Res. Lett.* **2014**, *41*, 3906–3913.
14. Barrett, K.; Rocha, A.V.; van de Weg, M.J.; Shaver, G. Vegetation shifts observed in arctic tundra 17 years after fire. *Remote Sens. Lett.* **2012**, *3*, 729–736.

15. Jones, B.M.; Grosse, G.; Arp, C.D.; Miller, E.; Liu, L.; Hayes, D.J.; Larsen, C.F. Recent arctic tundra fire initiates widespread thermokarst development. *Sci. Rep.* **2015**, *5*, 15865. [PubMed]
16. Jandt, R.R.; Miller, E.A.; Yokel, D.A.; Bret-Harte, M.S. *Findings of Anaktuvuk River Fire Recovery Study*; US Bureau of Land Management: Fairbanks, AK, USA, 2012.
17. Rocha, A.V.; Loranty, M.M.; Higuera, P.E.; Mack, M.C.; Hu, F.S.; Jones, B.M.; Breen, A.L.; Rastetter, E.B.; Goetz, S.J.; Shaver, G.R. The footprint of alaskan tundra fires during the past half-century: Implications for surface properties and radiative forcing. *Environ. Res. Lett.* **2012**, *7*, 044039.
18. De Baets, S.; van de Weg, M.J.; Lewis, R.; Steinberg, N.; Meersmans, J.; Quine, T.A.; Shaver, G.R.; Hartley, I.P. Investigating the controls on soil organic matter decomposition in tussock tundra soil and permafrost after fire. *Soil Biol. Biochem.* **2016**, *99*, 108–116.
19. Hu, F.S.; Higuera, P.E.; Duffy, P.; Chipman, M.L.; Rocha, A.V.; Young, A.M.; Kelly, R.; Dietze, M.C. Arctic tundra fires: Natural variability and responses to climate change. *Front. Ecol. Environ.* **2015**, *13*, 369–377.
20. Jolivel, M.; Allard, M. Thermokarst and export of sediment and organic carbon in the sheldrake river watershed, nunavik, canada. *J. Geophys. Res. Earth Surf.* **2013**, *118*, 1729–1745.
21. Natali, S.; Schuur, T.; Mauritz, M.; Schade, J.; Celis, G.; Crummer, G.; Johnston, C.; Krapek, J.; Pegoraro, E.; Salmon, V.; et al. Permafrost thaw and soil moisture drive co2 and ch4 release from upland tundra. *J. Geophys. Res. Biogeosci.* **2015**, *120*, 525–537.
22. Hugelius, G.; Strauss, J.; Zubrzycki, S.; Harden, J.W.; Schuur, E.A.G.; Ping, C.L.; Schirrmeister, L.; Grosse, G.; Michaelson, G.J.; Koven, C.D.; et al. Estimated stocks of circumpolar permafrost carbon with quantified uncertainty ranges and identified data gaps. *Biogeosciences* **2014**, *11*, 6573–6593.
23. Tarnocai, C.; Canadell, J.G.; Schuur, E.A.G.; Kuhry, P.; Mazhitova, G.; Zimov, S. Soil organic carbon pools in the northern circumpolar permafrost region. *Glob. Biogeochem. Cycles* **2009**, *23*, GB2023.
24. Zimov, S.A.; Davydov, S.P.; Zimova, G.M.; Davydova, A.I.; Schuur, E.A.G.; Dutta, K.; Chapin, F.S. Permafrost carbon: Stock and decomposability of a globally significant carbon pool. *Geophys. Res. Lett.* **2006**, *33*, L20502.
25. Racine, C.; Jandt, R.; Meyers, C.; Dennis, J. Tundra fire and vegetation change along a hillslope on the seward peninsula, alaska, U.S.A. *Arct. Antarct. Alp. Res.* **2004**, *36*, 1–10.
26. Hewitt, R.E.; Hollingsworth, T.N.; Stuart Chapin, F., III; Lee Taylor, D. Fire-severity effects on plant–fungal interactions after a novel tundra wildfire disturbance: Implications for arctic shrub and tree migration. *BMC Ecol.* **2016**, *16*, 25.
27. Pearson, R.G.; Phillips, S.J.; Loranty, M.M.; Beck, P.S.A.; Damoulas, T.; Knight, S.J.; Goetz, S.J. Shifts in arctic vegetation and associated feedbacks under climate change. *Nat. Clim. Chang.* **2013**, *3*, 673–677.
28. Joly, K.; Jandt, R.R.; Klein, D.R. Decrease of lichens in arctic ecosystems: The role of wildfire, caribou, reindeer, competition and climate in north-western alaska. *Polar Res.* **2009**, *28*, 433–442.
29. Folke, C.; Carpenter, S.; Walker, B.; Scheffer, M.; Elmqvist, T.; Gunderson, L.; Holling, C.S. Regime shifts, resilience, and biodiversity in ecosystem management. *Annu. Rev. Ecol. Evol. Syst.* **2004**, *35*, 557–581.
30. Tanase, M.; de la Riva, J.; Santoro, M.; Pérez-Cabello, F.; Kasischke, E. Sensitivity of sar data to post-fire forest regrowth in mediterranean and boreal forests. *Remote Sens. Environ.* **2011**, *115*, 2075–2085.
31. Wang, Q.; Adiku, S.; Tenhunen, J.; Granier, A. On the relationship of ndvi with leaf area index in a deciduous forest site. *Remote Sens. Environ.* **2005**, *94*, 244–255.
32. Bourgeau-Chavez, L.L.; Harrell, P.A.; Kasischke, E.S.; French, N.H.F. The detection and mapping of alaskan wildfires using a spaceborne imaging radar system. *Int. J. Remote Sens.* **1997**, *18*, 355–373.
33. Bourgeau-Chavez, L.L.; Garwood, G.; Riordan, K.; Cella, B.; Alden, S.; Kwart, M.; Murphy, K. Improving the prediction of wildfire potential in boreal alaska with satellite imaging radar. *Polar Rec.* **2007**, *43*, 321–330.
34. French, N.H.F.; Bourgeau-Chavez, L.L.; Wang, Y.; Kasischke, E.S. Initial observations of radarsat imagery at fire-disturbed sites in interior alaska. *Remote Sens. Environ.* **1999**, *68*, 89–94.
35. Harrell, P.A.; Bourgeau-Chavez, L.L.; Kasischke, E.S.; French, N.H.F.; Christensen, N.L., Jr. Sensitivity of ers-1 and jers-1 radar data to biomass and stand structure in alaskan boreal forest. *Remote Sens. Environ.* **1995**, *54*, 247–260.
36. Kasischke, E.S.; Bourgeau-Chavez, L.L.; Christensen, N.L., Jr.; Haney, E. Observations on the sensitivity of ers-1 sar image intensity to changes in aboveground biomass in young loblolly pine forests. *Remote Sens.* **1994**, *15*, 3–16.

37. Way, J.; Paris, J.; Kasischke, E.; Slaughter, C.; Viereck, L.; Christensen, N.; Dobson, M.C.; Ulaby, F.; Richards, J.; Milne, A. The effect of changing environmental conditions on microwave signatures of forest ecosystems: Preliminary results of the march 1988 alaskan aircraft sar experiment. *Int. J. Remote Sens.* **1990**, *11*, 1119–1144.
38. Kasischke, E.S.; Bourgeau-Chavez, L.L.; Johnstone, J.F. Assessing spatial and temporal variations in surface soil moisture in fire-disturbed black spruce forests in interior alaska using spaceborne synthetic aperture radar imagery—Implications for post-fire tree recruitment. *Remote Sens. Environ.* **2007**, *108*, 42–58.
39. Minchella, A.; Del Frate, F.; Capogna, F.; Anselmi, S.; Manes, F. Use of multitemporal sar data for monitoring vegetation recovery of mediterranean burned areas. *Remote Sens. Environ.* **2009**, *113*, 588–597.
40. Zebker, H.A.; Villasenor, J. Decorrelation in interferometric radar echoes. *Geosci. Remote Sens. IEEE Trans.* **1992**, *30*, 950–959.
41. Imhoff, M.L. Radar Backscatter/Biomass Saturation: Observations and Implications for Global Biomass Assessment. In Proceedings of the IGARSS'93—IEEE International Geoscience and Remote Sensing Symposium, Tokyo, Japan, 18–21 August 1993; IEEE: Piscataway, NJ, USA, 1993; pp. 43–45.
42. Dobson, M.C.; Ulaby, F.T.; LeToan, T.; Beaudoin, A.; Kasischke, E.S.; Christensen, N. Dependence of radar backscatter on coniferous forest biomass. *Geosci. Remote Sens. IEEE Trans.* **1992**, *30*, 412–415.
43. Smith, G.; Dammert, P.B.; Askne, J. Decorrelation mechanisms in c-band sar interferometry over boreal forest. In Proceedings of the Satellite Remote Sensing Iii, Taormina, Italy, 17 December 1996; pp. 300–310.
44. Tanase, M.; Santoro, M.; Riva, J.D.L.; Kasischke, E.; Korets, M.A. L-band sar backscatter prospects for burn severity estimation in boreal forests. In Proceedings of the Esa Living Planet Symposium, Bergen, Norway, 28 June–2 July 2010.
45. Tanase, M.A.; Santoro, M.; Riva, J.D.L.; rez-Cabello, F.P.; Toan, T.L. Sensitivity of x-, c-, and l-band sar backscatter to burn severity in mediterranean pine forests. *IEEE Trans. Geosci. Remote Sens.* **2010**, *48*, 3663–3675.
46. Racine, C.H.; Johnson, L.A.; Viereck, L.A. Patterns of vegetation recovery after tundra fires in northwestern alaska, U.S.A. *Arct. Alp. Res.* **1987**, *19*, 461–469.
47. Jenkins, K.L.; Bourgeau-Chavez, L.L.; French, H.N.; Loboda, V.T.; Thelen, J.B. Development of methods for detection and monitoring of fire disturbance in the alaskan tundra using a two-decade long record of synthetic aperture radar satellite images. *Remote Sens.* **2014**, *6*, 6347–6364.
48. Luckman, A.; Baker, J.; Honzák, M.; Lucas, R. Tropical forest biomass density estimation using jers-1 sar: Seasonal variation, confidence limits, and application to image mosaics. *Remote Sens. Environ.* **1998**, *63*, 126–139.
49. AICC. Fire History in Alaska. Available online: https://fire.ak.blm.gov/predsvcs/maps.php (accessed on 7 December 2016).
50. Jones, B.M.; Kolden, C.A.; Jandt, R.; Abatzoglou, J.T.; Urban, F.; Arp, C.D. Fire behavior, weather, and burn severity of the 2007 anaktuvuk river tundra fire, north slope, alaska. *Arct. Antarct. Alp. Res.* **2009**, *41*, 309–316.
51. Rocha, A.V.; Shaver, G.R. Burn severity influences postfire co2 exchange in arctic tundra. *Ecol. Appl.* **2011**, *21*, 477–489. [PubMed]
52. ASF. Vertx: The Alaska Satellite Facility's Data Portal for Remotely Sensed Imagery of the Earth. Available online: https://vertex.daac.asf.alaska.edu (accessed on 1 June 2016).
53. ASF. Mapready. Available online: https://www.asf.alaska.edu/data-tools/mapready/ (accessed on 1 June 2016).
54. NASA. Earth Data. Available online: https://search.earthdata.nasa.gov (accessed on 20 November 2016).
55. NASA. Mod13q1v006. Available online: https://lpdaac.usgs.gov/products/mod13q1v006/ (accessed on 1 August 2017).

remote sensing

MDPI

Technical Note

Quantitative Analysis of Forest Fires in Southeastern Australia Using SAR Data

Aqil Tariq [1], Hong Shu [1], Qingting Li [2], Orhan Altan [3], Mobushir Riaz Khan [4], Muhammad Fahad Baqa [5] and Linlin Lu [5,*]

1 State Key Laboratory of Information Engineering in Surveying, Mapping and Remote Sensing (LIESMARS), Wuhan University, Wuhan 430079, China; aqiltariq@whu.edu.cn (A.T.); shu_hong@whu.edu.cn (H.S.)
2 Airborne Remote Sensing Center, Aerospace Information Research Institute, Chinese Academy of Sciences, Beijing 100094, China; liqt@radi.ac.cn
3 Department of Geomatics, Istanbul Technical University, Istanbul 36626, Turkey; oaltan@itu.edu.tr
4 School of Environmental Science, Charles Sturt University, Albury-Wodonga 2640, Australia; mobkhan@csu.edu.au
5 Key Laboratory of Digital Earth Science, Aerospace Information Research Institute, Chinese Academy of Sciences, Beijing 100094, China; 2252293808@mails.ucas.edu.cn
* Correspondence: lull@radi.ac.cn

Abstract: Prescribed burning is a common strategy for minimizing forest fire risk. Fire is introduced under specific environmental conditions, with explicit duration, intensity, and rate of spread. Such conditions deviate from those encountered during the fire season. Prescribed burns mostly affect surface fuels and understory vegetation, an outcome markedly different when compared to wildfires. Data on prescribed burning are crucial for evaluating whether land management targets have been reached. This research developed a methodology to quantify the effects of prescribed burns using multi-temporal Sentinel-1 Synthetic Aperture Radar (SAR) imagery in the forests of southeastern Australia. C-band SAR datasets were specifically used to statistically explore changes in radar backscatter coefficients with the intensity of prescribed burns. Two modeling approaches based on pre- and post-fire ratios were applied for evaluating prescribed burn impacts. The effects of prescribed burns were documented with an overall accuracy of 82.3% using cross-polarized backscatter (VH) SAR data under dry conditions. The VV polarization indicated some potential to detect burned areas under wet conditions. The findings in this study indicate that the C-band SAR backscatter coefficient has the potential to evaluate the effectiveness of prescribed burns due to its sensitivity to changes in vegetation structure.

Keywords: prescribed burns; SAR; fire impact; radar burn ratio; post-fire restoration; change detection

Citation: Tariq, A.; Shu, H.; Li, Q.; Altan, O.; Khan, M.R.; Baqa, M.F.; Lu, L. Quantitative Analysis of Forest Fires in Southeastern Australia Using SAR Data. *Remote Sens.* **2021**, *13*, 2386. https://doi.org/10.3390/rs13122386

Academic Editors: Alfonso Fernández-Manso and Carmen Quintano

Received: 10 May 2021
Accepted: 17 June 2021
Published: 18 June 2021

1. Introduction

Wildfires are a global agent for environmental change [1,2], and prescribed burning is an important strategy to minimize the harmful impact of wildfires globally [3]. Prescribed burning has proven effective in various regions, including Southern Europe [4], North America [5,6], and Australia [7–9]. In order to meet particular environmental protection targets, controlled fires are required for prescribed burning. A fire is set under specific environmental conditions, over an explicit region, and with a specific length, intensity, and spread rate [3]. Prescribed burning was first implemented in forests with the aim of minimizing wildfire risks in Australia in the late 1950s [9]. The prescribed fire is used for: (i) fuel reduction to moderate the adverse effects of wildfires on biodiversity, water, and soil [10–12], and (ii) land management [13], including logging restoration, forest maintenance, and conservation of biodiversity [14].

The intensity of prescribed burns varies between low to moderate in southeastern Australia [15]. The rate of fire spread depends on the seasons. For example, there are fewer fire ignitions in spring, which is from September to December, whereas there are more ignitions in autumn, from March to May, when weather conditions are mild. Broad-scale adoption of prescribed burning as a conservation measure has altered the structure and pattern of the fuel period, substantially decreasing the frequency and magnitude of unplanned fires, along with the severity of impacts on land, environmental factors, and vegetation in this area [9,11]. However, the effects of prescribed burning should be monitored to confirm whether defined consumption, area coverage, and fire intensity levels were achieved. In various habitats, coarse woody debris is a crucial structural feature for forest biodiversity and functioning [16], and can be impacted by the burning season, severity, and frequency of prescribed fire treatments [17,18]. Historical data on prescribed burning effects expressed as intensity or severity of fire are important for determining the degree to which the goals of land/wildfire management have been accomplished and also for informing the reporting process aimed at quantifying atmospheric emissions from fire.

The degree or magnitude of environmental changes caused by fire can be defined as the severity of fire and can be calculated by bio-loss [19]. These environmental trends are generally measured by severity indices, including the Composite Burn Index (CBI), to assess the rate of burning for vegetation areas in various sections [20]. Nevertheless, field investigations are expensive and their application over large areas is impractical. Remote sensing methods have been widely used to collect information on post-fire landscape conditions over broad spatial scales [21–23], and spectral indices from multi-spectral remote sensing data are commonly used for assessing burn severity at the spatial grain of the sensor [24,25]. However, reflectance-based indices have difficulty in yielding accurate results due to the influence of multiple factors, such as forest sensitivity, plant phenology, or solar elevation, rather than structure [26,27].

Considering their sensitivity to scattering elements and capacity to provide direct measurements of vegetation structure, active sensors such as radar and LiDAR may overcome the above limitations [28,29]. Various SAR-based techniques are useful to examine the effects of fire on the three-dimensional structures of canopy and the strata of lower plants, as radar waves have the ability to penetrate through canopy. The possibility to estimate the effects of forest fires via radar-based measurements, such as polarimetric decomposition coefficients, interferometric coherence, and backscatter intensities, has been demonstrated in previous studies [30,31]. These studies obtained satisfactory estimation accuracies in wildfire-affected forests and emphasized the need for model calibration with reference datasets. Burned area algorithms based on temporal differences between pre- and post-fire radar backscatter values may also suffer from the temporal decorrelation effect that must also be accounted for. A recent study also revealed that undesirable effects caused by topography and varying environmental circumstances could be minimized with bi-temporal approaches [30].

Since prescribed burns primarily affect surface fuels and understory vegetation, their effect on radar backscattering is largely unknown and may vary from that of wildfires. The feasibility of analysis methods developed for wildfires has barely been evaluated when applied to controlled burns. Due to the introduction of thresholds for distinguishing different groups of fires categorized by their severity, their applications are further complicated. In addition, it is difficult to distinguish between high-intensity wildfire events and controlled burns, where the impact of fire ranges from low to moderate. Since SAR data provide a wealth of information on fire effects even for fires at low severity levels, C-band SAR data hold great potential for assessing the impacts of prescribed burns [22]. The objectives of this study include developing and applying a framework for forest fire analysis in areas impacted by controlled burns, and evaluating the effect of SAR sensor polarization and environmental conditions on analysis in comparison with wildfires.

2. Study Area

This study was conducted in Croajingolong National Park in southeastern Australia (36°13′30″ S to 39°24′42″ S, 149°47′53″ E to 150°18′12″) (Figure 1) [32]. One of only twelve world biosphere reserves in Australia, Croajingolong National Park in the state of Victoria is adjacent to Nadgee Nature Reserve in New South Wales. The study area has an oceanic Mediterranean climate with mean annual precipitation of about 1200 mm, which can be divided into a warm and dry season (October to April) and a cool and wet season (May to September) with >70% of precipitation [33]. The monthly average temperature can be as high as 27.2 °C, usually in January [34]. The above-ground biomass (AGB) ranges between 250 and 350 t/ha for mature Croajingolong National Park, with a further tall woody understory of 10–20 t/ha and woody debris of 50–70 t/ha. Southeastern Australia's forests in Croajingolong are distinguished by breeding plants (70% of the species, specifically the main species of tree) from post-fire stems [35]. With high-intensity fires, the trees may withstand defoliation, destruction of stem, and total top scorch, and some particular trees regenerate apically dominant crowns within five years. Within a month of burning, epicormics shoots emerge when fire-destroyed branches and upper stem parts weakened by fire eventually split and fall within the first two years. Fire-defoliated crowns of trees are typically destroyed, with shoots of epicormics restricted to the lower stem. With 35% to 45% of the total trees impacted by complete crown scorch, crown mortality also occurs. Depending on the species, crown replacement is between 30% and 95% within five years, and it is impossible to differentiate plants with substituted tops from those that have not experienced top death. In the lesser plant layers, up to 85% of classes can achieve flowering in the first two years after prescribed burns. Therefore, in the first year after the fire, certain improvements in forest composition can be predicted, as leaves emerge from the stem and main branches of both eucalyptus and shrub plants. Between 11 November and 30 December 2019, a scheduled burn was carried out, and approximately 88,500 hectares of tall open karri forest were burned in Croajingolong (Figure 1).

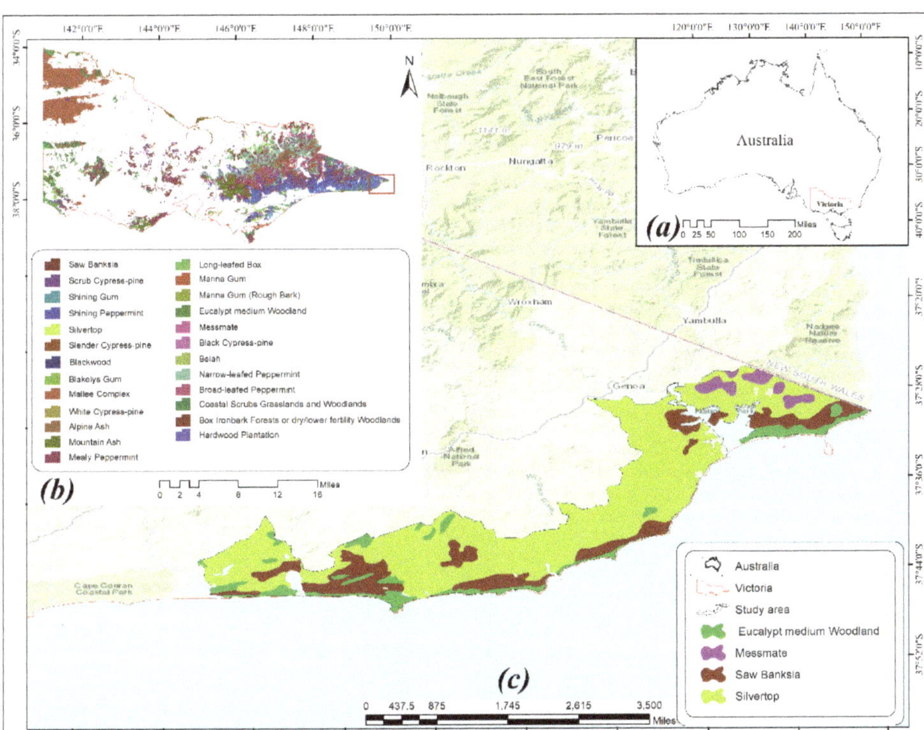

Figure 1. Geographic location of the study area: (**a**) the study area in Australia, (**b**) main forest distribution in Victoria, Australia, and (**c**) geographical location and main forest distribution in the study area.

The goal of the planned burn was to moderate the danger of wildfires by reducing fuel such as dead wood, leaf litter, bark, and shrubs, preventing extreme scorching or defoliation of tree crowns in the forest with young regrowth. The flow diagram in Figure 2 summarizes the approaches developed in this study.

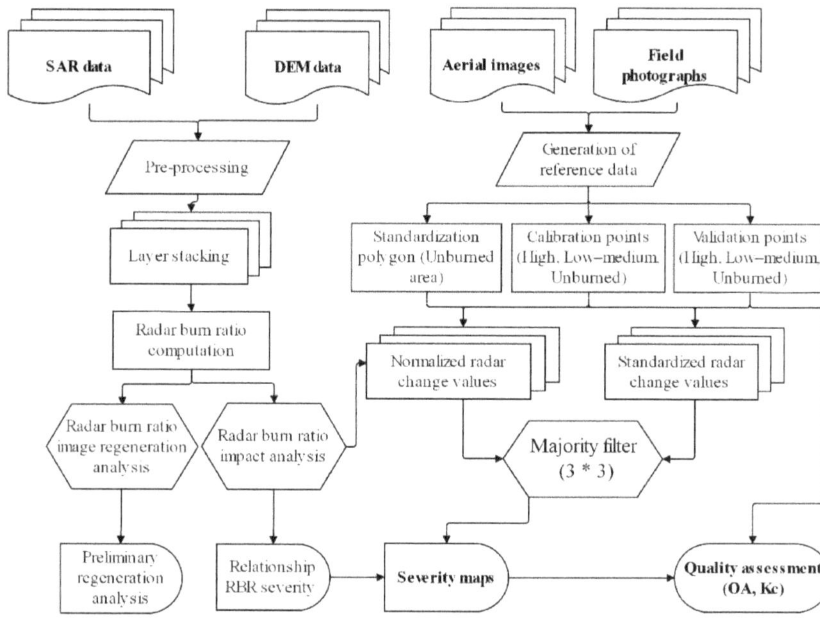

Figure 2. General framework of this study.

3. Materials and Methods

3.1. Radar Data

The Sentinel-1 Single Look Complex (SLC) datasets were obtained from the Copernicus Open Access Hub and used in this study (Table 1). The dual polarization mode of the fine-beam (polarization of VV and VH) datasets (orbit: 29794, frame: 716, path: 147) were collected with a 36.4° incident angle at the middle of the swath. For burn severity estimation, two SAR images (6 October 2019 and 22 February 2020) obtained under different environmental conditions (dry and wet) were used. The post-fire tree growth was examined using a third SAR imagery dataset (18 October 2020), which was collected ten months after the fire (Table 1). Sentinel-1 datasets were obtained over the same season (austral-autumn) to reduce the possible variations caused by plant phenology and to ensure that environmental settings are as comparable as possible.

Table 1. The obtained SAR data and cumulative precipitation (mm) three days prior to image acquisition. Historical median values for temperature and precipitation over the study period are also provided.

SAR Acquisition Dates	Timing	Mission Identifier	Mode/ Beam	Product Type	Polarization	Cumulative Precipitations (mm)			
						Point Hicks	Genoa	Mallaooota	Gabo Island Lighthouse
Fire occurrence data	29 November 2019								
Pre-fire	06 October 2019	S1A	IW	SLC	VV/VH	33 [a]	16.3 [a]	38.5 [a]	36.9 [a]
Post-fire	22 February 2020	S1A	IW	SLC	VV/VH	0 [b]	4.0 [b]	11.3 [b]	28.5 [b]
Post-fire tree regrowth	18 October 2020	S1B	IW	SLC	VV/VH	20.8 [a]	21.8 [a]	9.3 [a]	26.8 [a]

[a]: Precipitation registered partially during the acquisition date. [b]: Precipitation reported the following day after image acquisition.

The SLC data were co-registered to the first image of the series (6 October 2019) using a lookup table generated from the images' orbital state vectors. An image-to-image cross-correlation algorithm was used to estimate the residual offsets present in the lookup table [36]. Least square and third-degree polynomial function regression methods were used to solve the errors in the lookup table, and offsets in the table were modeled. Multi-look (eight looks in azimuth and four in range) techniques were applied to the co-registered images to achieve a spatial resolution of 30 m. The SAR intensity data was normalized using an incident field of a derived DEM to achieve the gamma naught backscatter coefficient [3]. Using the Range-Doppler approach, the SAR images were orthorectified to UTM 50S/WGS 84 projection with topography information from DEM and SAR data orbital details as input [37]. The preprocessing steps were mainly performed using ESA'S SNAP8.0.3 software.

After preprocessing, the SAR images were grouped into three stacks, one for each polarization (*VH*/*VV*), and one for normalized difference backscatter intensity (NDBI). The NDBI is a normalized ratio between *VV* and *VH* backscatter coefficients (Equation (1)). For each image stack, three images collected on different dates were processed.

$$NDBI = \frac{VV - VH}{VV + VH} \tag{1}$$

The radar burn ratio (RBR) basically works as the post- to pre-fire ratio of the backscattering coefficients (Equation (2)):

$$\text{RBR}_{xy} = \frac{\gamma^0 Postfire_{xy}}{\gamma^0 Prefire_{xy}} \tag{2}$$

where xy describes a particular polarization based on *VV* or *VH* and *NDBI*, and the normalized backscatter coefficient based on a linear scale is γ^0 (gamma naught). In both dry and wet conditions, the indices of RBR were computed based on the two polarization (*VV*, *VH*) images and *NDBI*. The RBR was also computed using the mean of all pre- and post-fire images. Image histograms were analyzed, and the first (1) and last (99) percentiles were masked to eliminate outliers. In total, nine RBR images were created.

3.2. Reference Data

A total of 139 geo-located field photographs and an aerial orthophoto with 17 cm spatial resolution were used to create reference data (i.e., 730 points of different severity levels). The orthophoto was taken on 22 February 2020 (two months after the burn) with a Hasselblad H4D-60 camera and georeferenced and orthorectified by Parks and Wildlife Service, WA. Field images were taken between September and October 2020 (ten months after the fire) and corresponded to areas with varying levels of effects. Parks and Wildlife Service from the Department of Agriculture provided the aerial image and field photos. We further explain three burn severity levels and regrowth of vegetation in Figure 3.

(I) Unburnt class: The prescribed burn appears to have had little effect on these areas.

(II) Low–medium severity class: Areas where the fires affect only understory vegetation while the crown remains largely unaffected, as well as areas where convective heat from fire scorched the crown but the canopy structure was maintained (most of the leaves and main branches). Due to the difficulty in identifying subtle fire effects through aerial image photointerpretation and the sensitivity of the C-band to the largest structural elements of vegetation, severity levels from low to moderate were considered as a single class (mainly stems and branches). Hence, the C band's ability to distinguish moderately intense thresholds where large canopy elements remain relatively unchanged may be affected.

(III) High severity: The crown canopy is entirely absent, and the main branches are partly burned or missing. There are still several fallen stems.

Figure 3. Details of severity levels and regrowth of vegetation in the study area: (**a**) burned area with low to moderate severity, (**b**) burned area with high severity, (**c**) unburned area, and (**d**) regrowth of vegetation.

Within the prescribed burn boundary, a standard point grid with 300 m spacing was used to select the sampling points. While the unburned region within the burn area was small, to represent unburned trees, the points found outside the fire border were also sampled. There were 730 points sampled proportional to the area in each single class. The distribution of sample points was stratified by severity classes. The dataset was divided into two sets after each plot was allocated to a severity class. The first set was used to calibrate the model, while the second was used to validate it. The points for validation were selected randomly, leaving all the unselected points for calibration. The results were not influenced by the various calibration sample sizes for each class, since median value was used for model calibration for every single class.

3.3. Ancillary Data

The Meteorology Bureau, Government of Australia, provided regular precipitation weather data in the nearest four meteorological stations, i.e., Point Hicks (Lighthouse), Genoa (Fools Haven), Mallacoota, and Gabo Island lighthouse (http:/www.bom.gov.au/climate/data/, accessed on 10 May 2021) (Table 1). In order to explain the environmental conditions (dry or wet) related to radar backscatter variability due to changes in soil moisture, the data for cumulative precipitation for the three days prior to the SAR acquisition date were obtained. Low spatial variability of soil moisture in the top few centimeters is related to high temperatures (high evapotranspiration potential) and low precipitation. The dates 6 October 2019 and 22 February 2020 were labeled as dry and wet, respectively. Since the cumulative precipitation was low and the precipitation was registered at least one day before the SAR acquisition date, the image used to assess post-fire tree survival (18 October 2020) was considered dry. The Shuttle Radar Topography Mission Digital Elevation Model (DEM) data with 30 m spatial resolution was used for the orthorectification of SAR data.

3.4. Analysis and Modeling of Prescribed Burns

RBR values were derived for all sample points (dry, wet, mean), taking into consideration each type of polarization/index (VV, VH, and NDBI) with different environmental conditions. To describe the relationship between prescribed burn effects and RBR under different environmental conditions and polarizations, the mean values with standard deviation values were analyzed. Two modeling approaches were combined for the extraction of prescribed burn effects using the RBR images. The first modeling approach was suggested in [29] (the STAND model below). An initial strategy was developed and tested to ease organizational execution (the NORM model). Note that linear scale operations of a mathematical approach were conducted on the SAR images, while a logarithmic scale (base 10) was used for data analysis (decibels, dB), which is a standard procedure in radar remote sensing.

3.4.1. Standard Model (STAND)

There are three continents with seven different forest types included in the original RBR system, translating into the large variations of RBR ranges. By standardization of mathematical data (having z-scores), which rescales the values of RBR to a particular range, compensation for different RBR ranges was achieved. It is important to note that RBR standardization is not exclusively appropriate for the small area examined in this study. The original RBR system, however, was retained for reference purposes. Furthermore, using standard values (z-scores, distance among the significance of the sample values and the mean of the population) is helpful to interpret the RBR values over various environmental conditions. Standardization was carried out using Equation (3):

$$s^{RBR} = \frac{(x - \mu)}{\sigma} \quad (3)$$

where the standardized RBR value is s^{RBR}, x represents the pixel value, and σ and μ are the mean and standard deviation values of RBR applied to adjacent non-fire-affected areas of forest.

For standardization, the unburned pixels were used to show the difference compared to the original process, as the burned region typically displayed gradient ranges from moderate to low, which then leads to the high variability of the backscatter. Every pixel value shows the difference in backscatter present in the pixel and backscatter mean of unburned areas with the use of unburned area for standardization, allowing for better data analysis (the difference with unburned conditions). To measure the μ and σ values, a total of 3524 pixels in 5 polygons across the prescribed burn were used to account for spatial variability in the forest. Once the images with uniform characteristics (s^{RBR}) were collected, for each single image, the mean value of each class related to severity was determined using the backscatter values derived from the calibration dataset coordinates. Thresholds dividing various severity classes were calculated as the median value of the mean (s^{RBR}) of neighboring severity classes [30]. The median was perceived to be a stable central tendency test (where outliers are less sensitive) in the case of SAR image analyses marked with speckle noise.

3.4.2. Normalized Model (NORM)

This model uses the normalized values (n^{RBR}, Equation (4)) of backscattering coefficients. The benefit of using n^{RBR} lies in its simplified calculation and interpretation, as the index does not require standardization:

$$n^{RBR} = \frac{\gamma^0 Postfire_{xy} - \gamma^0 Prefire_{xy}}{\gamma^0 Postfire_{xy} + \gamma^0 Prefire_{xy}} \quad (4)$$

where xy is a definite polarization (i.e., VV or VH) or NDBI, and γ^0 (gamma naught) is the linear-scale normalized backscattering coefficient. As in the previous model, the median values were used for computing thresholds between groups.

3.4.3. Model Validation

Using the thresholds obtained for each model, images were classified using s^{RBR} or n^{RBR}, including unburned, low to medium, and high severity. A total of 18 maps were produced (Appendix A, Figures A1–A3). In order to minimize the influence of isolated pixels, a major filter (3 × 3) was employed on all maps.

Confusion matrices were constructed using the validity points by removing the modeled fire grades. In order to evaluate the agreement between estimated fire impact and reference, overall accuracy (*OA*), Cohen's kappa (kc), omission error (*OE*), and commission error (*CE*) were calculated using the confusion matrices (Equations (5)–(8)) [38].

$$OA = \frac{TP + TN}{TP + FP + FN + TN} \tag{5}$$

$$kappa = \frac{OA - p_e}{1 - p_e} \tag{6}$$

$$CE = \frac{FP}{TP + FP} \tag{7}$$

$$OE = \frac{FN}{TP + FN} \tag{8}$$

where TP = true positive, TN = true negative, FP = false positive, FN = false negative, and $p_{e=\frac{(TP+FN)\cdot(TP+FP)+(FP+TN)\cdot(FN+TN)}{TP+FP+FN+TN}}$. OA is the total number of accurately predicted values divided by the total number of predictions. The omission error of one class (OE) is the proportion of real samples of that class that were misclassified (they were class A in the ground truth, but we classified them as B, C, or D). The commission error of one class (CE) is the proportion of all classified points in that class that were misclassified (they were classified as class A, but they belong to class B, C, or D in ground truth). The average values of CE and OE in each class were calculated by adding up all false positives, false negatives, true positives, and true negatives. The perception of kappa values relies on various considerations, with no agreement (kc < 0), mild agreement (0 to 0.20), good agreement (0.21 to 0.40), modest agreement (0.41 to 0.60), significant agreement (0.61 to 0.80), and nearly perfect agreement (0.81 to 1) [39].

4. Results

In Figure 4a, RBR_{VH} decreased as the intensity of the fire increased under dry conditions. For the cross-polarized channel between the intensity classes of low to moderate and unburned, an increasing factor in soil moisture due to wet conditions shows a fairly complex relationship with the detection of minor differences. In dry conditions, RBR_{VV} was less effective in distinguishing between fire severity classes but showed the potential under wet conditions to distinguish unburned from burned areas, as explained in Figure 4b. Wet conditions produced the highest dynamic range between unburned and high severity (2 dB) for RBR_{VV}, which may be due to increased backscatter from the field. Under both environmental conditions, the RBR_{NDBI} showed an increase with fire severity, as explained in Figure 4c.

In the process of retrieving the effect based on prescribed burns, two modeling methods produced comparable results. The most reliable predictions were obtained in dry conditions, RBR_{VH}, using the STAND model (OA = 82.3%; kc = 0.78). The results are illustrated in Figure 5.

Figure 6 and Table 2 illustrate validation results for both approaches (STAND and NORM). The gap between the two models was relatively small. Omission and commission errors for the severity class of low to moderate were the largest (30% and 35%, respectively). Errors did not exceed 20% for unburned and high-severity classes. For the cross-polarized pathway (RBR_{VH}), elevated humidity contributed to a significant loss of sensitivity to intensity levels (Table 2). In contrast to the analysis in dry conditions for the cross-polarization channel, RBR_{VV} and RBR_{NDBI} displayed lower OA estimation accuracies (70–80%).

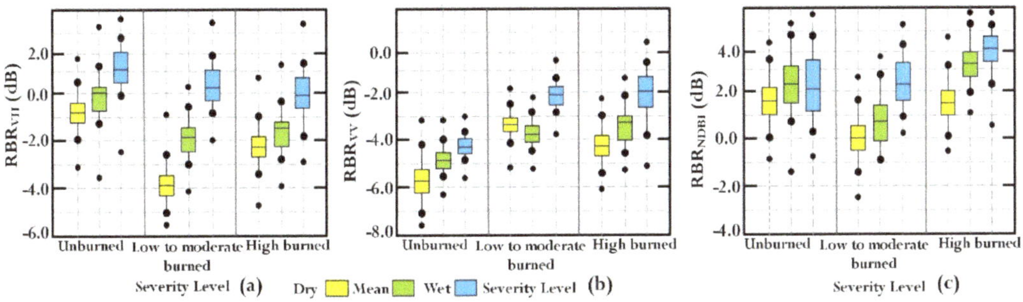

Figure 4. The relationship between RBR and intensity levels under various environmental conditions for VH polarization (**a**), VV polarization (**b**), and NDBI (**c**). Median values (horizontal line), percentiles (25% and 75%, box edges), non-outlier values range (whiskers), and outlier values (circles) are seen in box plots.

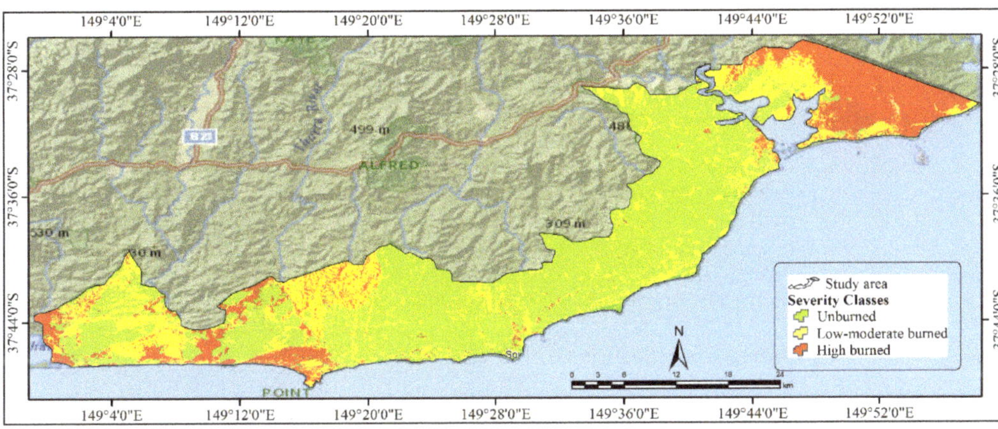

Figure 5. Estimated severity map using VH polarization data in dry conditions and the STAND model.

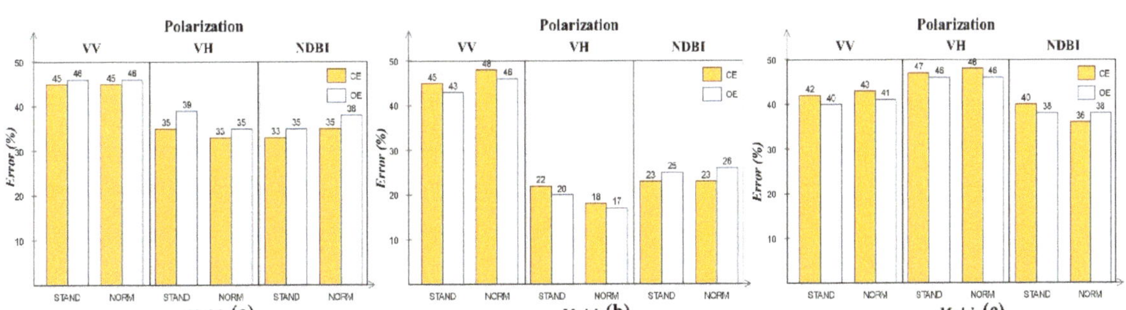

Figure 6. Omission and commission errors (average through classes): (**a**) dry conditions, (**b**) wet conditions, and (**c**) mean conditions.

Table 2. Validation results for severity estimates under different environmental conditions obtained for the VH polarization.

Environmental Condition	Polarization	Model	OA (%)	Kc	Omission Error (%)			Commission Error (%)		
					Low Severity	Low–Medium Severity	High Severity	Low Severity	Low–Medium Severity	High Severity
Wet	VV	STAND	70.0	0.55	18.8	46.0	19.1	13.3	33.3	43.3
		NORM	**76.6**	**0.69**	7.2	32.5	33.6	8.2	35.4	36.5
	VH	STAND	51.2	0.26	8.3	50.8	54.6	35.9	92.5	28.9
		NORM	52.6	0.28	9.3	52.9	56.5	36.7	85.9	30.8
	NDBI	STAND	65.5	0.48	8.3	53.2	16.9	24.5	36.8	51.6
		NORM	60.3	0.50	9.6	52.6	17.8	25.9	23.3	36.6
Dry	VV	STAND	61.1	0.42	9.6	55.3	24.0	36.7	43.3	36.7
		NORM	40.0	0.23	62.5	78.6	50.0	50.0	90.0	40.0
	VH	**STAND**	**82.3**	**0.78**	6.5	35.9	33.9	25.6	42.8	46.6
		NORM	77.2	0.74	27.3	40.7	16.7	20.0	46.7	16.7
	NDBI	STAND	68.9	0.53	20.6	47.4	16.7	10.0	33.3	50.0
		NORM	51.1	0.27	44.2	52.0	54.6	20.0	60.0	66.7
Mean	VV	STAND	60.0	0.40	30.2	60.0	36.4	0.30	66.7	53.3
		NORM	58.9	0.38	34.9	48.2	45.0	6.7	53.3	63.3
	VH	STAND	44.4	0.21	46.7	66.7	63.6	20.0	86.7	60.0
		NORM	**74.4**	**0.62**	25.7	36.0	16.7	13.3	46.7	16.7
	NDBI	STAND	66.7	0.50	15.6	50.0	34.6	10.0	46.7	43.3
		NORM	50.0	0.25	7.4	50.0	52.3	33.3	86.7	30.0

The NORM model outperformed the STAND model for wet images and the mean between dry and wet images (Table 2). Although VH polarization provided the most reliable fire severity forecasts (OA = 82.3%; kc = 0.78) in dry conditions, VV was of particular importance in wet conditions. RBR_{VV} obtained more accurate results than VH and NDBI under wet conditions, with the NORM model showing the least errors (OA = 76.6%; kc =0.69). The NORM model and RBR_{VH} generated the most reliable results (OA = 74.4%; kc = 0.62) for the mean between dry and wet images. It was also possible to differentiate between unburned and burned areas in those circumstances (Appendix A, Figures A1–A3).

5. Discussion

Previous experiments have demonstrated that the impacts of fire can be measured by integrating pre- and post-fire datasets. This study focused on the areas impacted by the incidents of wildfires that are normally marked as high. Our study improves previous knowledge on the accuracy of RBR over prescribed burns to minimize ground fuels with limited effects on the upper canopy. Since C-band SAR data was highly sensitive to scatter from the top part of the canopy [40], the response of RBR to the lower effect levels in areas controlled by prescribed burns should be assessed.

This work proved, even in low to moderate severity, that the cross-polarized channel and dry conditions are the most suitable combination for a fire effect evaluation. The fire sensitivity was found to be poor in the co-polarized channel, specifically in dry conditions, and increased under wet conditions, which is consistent with previous studies using X-, C-, and L-band data over different types of woods [26,28,41]. Since the ground surface shows more susceptibility with increased backscattering, the VV polarization method and NDBI provide the ability to detect burned areas under wet conditions. A simplified modeling technique was applied and yielded comparable accuracies.

The RBR structure and the thresholds suggested in [30] were used to extract severity maps for the study area. The benefit of using a standard calibration for prescribed burns was demonstrated by maps validated using reference data in this study. The calibrated models greatly improved average efficiency from 60% to 82% and the Kappa coefficient ranged from 0.21 to 0.78. Similarly, the overall accuracy exceeded the minimal value required to map the severity of prescribed burns (82.3%) [42]. However, further studies are required to assess the accuracy attained in other forest types and ecosystems. The results obtained in WA are marginally higher compared to approaches based on a high-resolution optical dataset (OA value range of 78–83%) [15]. The application of SAR polarimetric and phase information and deep learning methods achieved high accuracies for burnt area mapping from Sentinel-1 SAR data [29,31]. Although the accuracies of the proposed method were relatively lower, it might be more suitable for operational forest fire monitoring with the low demand for computational resources. Moreover, the proposed modeling strategy extracted different severity levels of prescribed burns.

Among the limitations of this study, it is important to point out that five levels of fire severity were considered in previous studies [15], although only three classes were included here. For prescribed burns, fewer groups were considered to distinguish between the unburned (failed burns), burning areas at low to moderate severity levels, and burned areas at a high severity level, where crown tree damage or decreasing trends in stems may decrease productivity of the forest. Using fewer groups may affect the outcomes, as simulation techniques could be more stable, and hence the accuracy will increase. As compared to field investigations, another weakness was due to the use of reference data obtained from satellite image interpretation. Such a strategy can produce noise and lead to the reduction in potential for discrimination. In addition, variations in environmental conditions before and after the burn cause uncertainties in remote sensing detection results. The variations in soil moisture in multi-temporal images may inhibit the evaluation of the impact of fire using SAR datasets [43]. By combining multiple SAR images, the impact of environmental variations can be minimized each time before and after a fire event [9,44].

6. Conclusions

This study expanded our knowledge of SAR data for impact assessment of wildfires through an innovative example, prescribed burns. The outcome of prescribed burning was analyzed by estimating fire effects (severity) on forest flora in Victoria, Australia. The freely available Sentinel-1 C-band SAR data, with a 12-day temporal coverage and a 5 m spatial resolution, was applied because of its sensitivity to structural changes in vegetation after forest fires. A fire intensity assessment was carried out for various polarization and environmental conditions using multi-temporary SAR indices. A modeling approach for prescribed burns that can be implemented operationally was developed. The validation results showed that the VH polarization data under dry conditions provided reliable fire severity forecasts with an overall accuracy of 82.3% (kc = 0.78). Although VH polarization provided the most reliable estimation results in dry and mean between dry and wet images, VV polarization data obtained optimal results in wet conditions. Based on our analysis, the NORM models outperformed the STAND models in wet and mean between dry and wet images. The modeling strategy mostly retained the accuracy of the initial RBR method while reducing computational complexity. This study provided a novel technique to assess the effectiveness of controlled burns for forest management and decision-makers in fire-prone areas. The analysis framework could be useful for forest conservation and fire control decision-makers, not only in southeastern Australia, but also in other areas where burning is used as a land management method. Future research will involve the deployment of multiple SAR wavelength channels and the integration of Sentinel-1 data with optical sensor data to further improve fire severity estimates.

Author Contributions: A.T.: Conceptualization, Writing—review and editing, Methodology, Software, Formal analysis, Visualization, Data curation, Writing—original draft, Investigation; H.S.: Supervision, Conceptualization, Writing—review and editing, Methodology, Software, Formal analysis, Visualization, Data curation, Writing—original draft, Investigation; Q.L.: Supervision, Conceptualization, Writing—review and editing, Methodology, Software, Formal analysis, Visualization, Data curation, Writing—original draft, Investigation, Funding; O.A.: Visualization, Validation, Investigation, Writing—review and editing; M.R.K.: Writing—review and editing, Methodology, Formal analysis, Visualization, Investigation; M.F.B.: Visualization, Validation, Investigation, Writing—review and editing; L.L.: Supervision, Writing—review and editing, Methodology, Software, Formal analysis, Visualization, Data curation, Writing—original draft, Investigation, Funding. All authors have read and agreed to the published version of the manuscript.

Funding: This work is supported by the Strategic Priority Research Program of the Chinese Academy of Sciences (Grant No. XDA23100304), the National Natural Science Foundation of China (Grant No. 41871345), and the National Natural Science Foundation of China (Grant No. 42071321).

Data Availability Statement: The data presented in this study are available upon request from the corresponding author. The data are not publicly available due to the thesis that is being prepared from these data.

Acknowledgments: The authors wish to thank the editors and anonymous reviewers for their valuable comments and helpful suggestions. The authors are grateful to GIS Department (Corporate Services | Department of Environment, Land, Water and Planning) for providing data and metadata related to the prescribed burn, from which the reference data used in this work were derived.

Conflicts of Interest: The authors declare no conflict of interest.

Appendix A

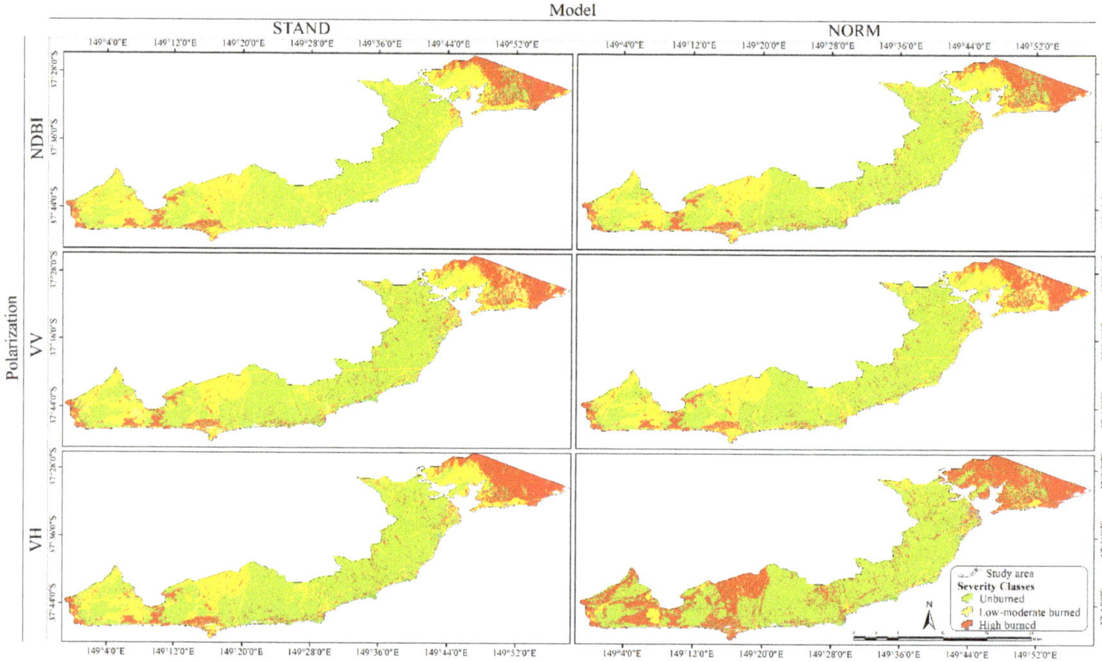

Figure A1. Dry condition.

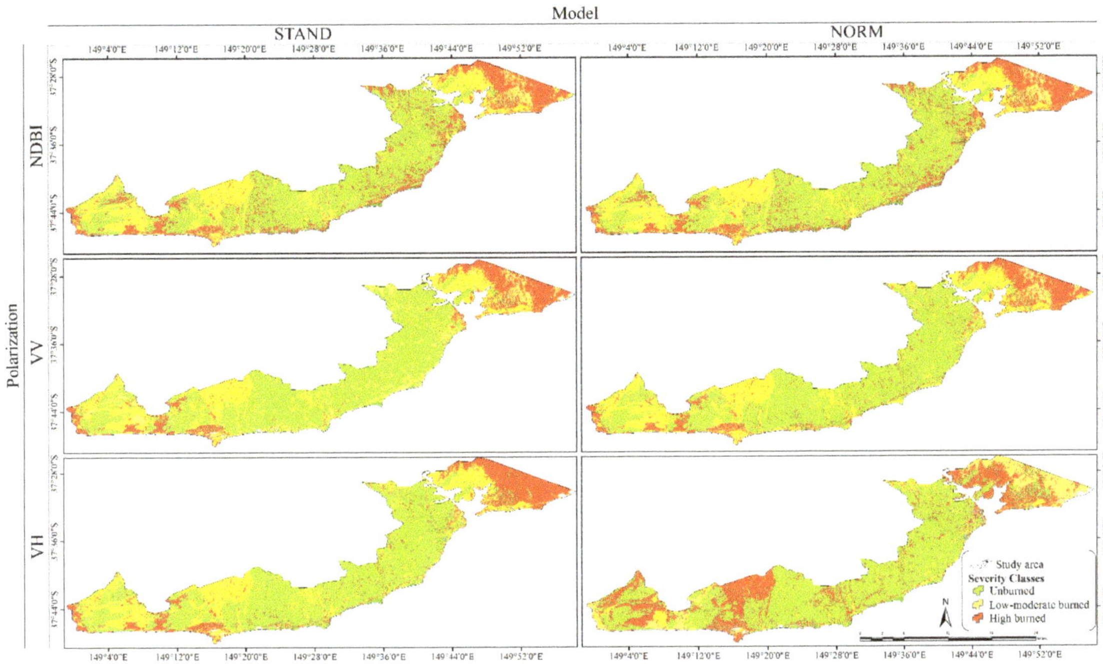

Figure A2. Mean of dry and wet condition.

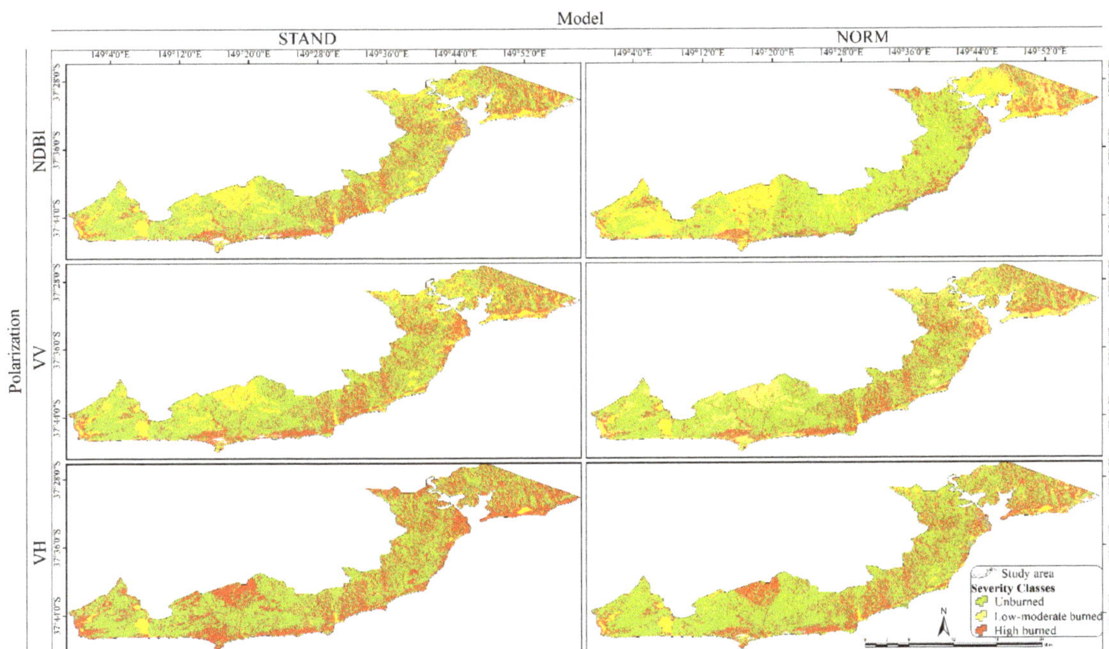

Figure A3. Wet condition.

References

1. Chuvieco, E.; Aguado, I.; Yebra, M.; Nieto, H.; Salas, J.; Martín, M.P.; Vilar, L.; Martínez, J.; Martín, S.; Ibarra, P.; et al. Development of a framework for fire risk assessment using remote sensing and geographic information system technologies. *Ecol. Model.* **2010**, *221*, 46–58. [CrossRef]
2. Kasischke, E.S.; Amiro, B.D.; Barger, N.N.; French, N.H.F.; Goetz, S.J.; Grosse, G.; Harmon, M.E.; Hicke, J.A.; Liu, S.; Masek, J.G. Impacts of disturbance on the terrestrial carbon budget of North America. *J. Geophys. Res. Biogeosci.* **2013**, *118*, 303–316. [CrossRef]
3. McCaw, W.L. Managing forest fuels using prescribed fire—A perspective from southern Australia. *For. Ecol. Manag.* **2013**, *294*, 217–224. [CrossRef]
4. Moreira, F.; Viedma, O.; Arianoutsou, M.; Curt, T.; Koutsias, N.; Rigolot, E.; Barbati, A.; Corona, P.; Vaz, P.; Xanthopoulos, G.; et al. Landscape—wildfire interactions in southern Europe: Implications for landscape management. *J. Environ. Manag.* **2011**, *92*, 2389–2402. [CrossRef] [PubMed]
5. Tariq, A.; Shu, H.; Siddiqui, S.; Munir, I.; Sharifi, A.; Li, Q.; Lu, L. Spatio-temporal analysis of forest fire events in the Margalla Hills, Islamabad, Pakistan using socio-economic and environmental variable data with machine learning methods. *J. For. Res.* **2021**, *13*, 12. [CrossRef]
6. Tariq, A.; Shu, H.; Saddiqui, S.; Mousa, B.G.; Munir, I.; Nasri, A.; Waqas, H.; Baqa, M.F.; Lu, L. Forest fire Monitoring using spatial-statistical and Geo-spatial analysis of factors determining Forest fire in Margalla Hills, Islamabad, Pakistan. *Geomat. Nat. Hazards Risk* **2021**, *12*, 1212–1233. [CrossRef]
7. Fernandes, P.M.; Botelho, H.S. A review of prescribed burning effectiveness in fire hazard reduction. *Int. J. Wildl. Fire* **2003**, *12*, 117–128. [CrossRef]
8. Fernandes, P.M.; Davies, G.M.; Ascoli, D.; Fernández, C.; Moreira, F.; Rigolot, E.; Stoof, C.R.; Vega, J.A.; Molina, D. Prescribed burning in southern Europe: Developing fire management in a dynamic landscape. *Front. Ecol. Environ.* **2013**, *11*. [CrossRef]
9. Boer, M.M.; Sadler, R.J.; Wittkuhn, R.S.; McCaw, L.; Grierson, P.F. Long-term impacts of prescribed burning on regional extent and incidence of wildfires—Evidence from 50 years of active fire management in SW Australian forests. *For. Ecol. Manag.* **2009**, *259*, 132–142. [CrossRef]
10. Gill, A.M.; McCarthy, M.A. Intervals between prescribed fires in Australia: What intrinsic variation should apply? *Biol. Conserv.* **1998**, *85*, 161–169. [CrossRef]
11. Andersen, A.N.; Cook, G.D.; Corbett, L.K.; Douglas, M.M.; Eager, R.W.; Russell-Smith, J.; Setterfield, S.A.; Williams, R.J.; Woinarski, J.C.Z. Fire frequency and biodiversity conservation in Australian tropical savannas: Implications from the Kapalga fire experiment. *Austral Ecol.* **2005**, *30*, 155–167. [CrossRef]
12. Youngblood, A.; Wright, C.S.; Ottmar, R.D.; McIver, J.D. Changes in fuelbed characteristics and resulting fire potentials after fuel reduction treatments in dry forests of the Blue Mountains, northeastern Oregon. *For. Ecol. Manag.* **2008**, *255*, 3151–3169. [CrossRef]
13. Hollis, J.J.; Anderson, W.R.; McCaw, W.L.; Cruz, M.G.; Burrows, N.D.; Ward, B.; Tolhurst, K.G.; Gould, J.S. The effect of fireline intensity on woody fuel consumption in southern Australian eucalypt forest fires. *Aust. For.* **2011**, *74*, 81–96. [CrossRef]
14. Burrows, N.D.; Wardell-Johnson, G.; Ward, B. Post-fire juvenile period of plants in south-west Australia forests and implications for fire management. *J. R. Soc. West. Aust.* **2008**, *91*, 163–174.
15. Loschiavo, J.; Cirulis, B.; Zuo, Y.; Hradsky, B.A.; Di Stefano, J. Mapping prescribed fire severity in south-east Australian eucalypt forests using modelling and satellite imagery: A case study. *Int. J. Wildl. Fire* **2017**, *26*, 491. [CrossRef]
16. Harmon, M.E.; Franklin, J.F.; Swanson, F.J.; Sollins, P.; Gregory, S.V.; Lattin, J.D.; Anderson, N.H.; Cline, S.P.; Aumen, N.G.; Sedell, J.R.; et al. Ecology of Coarse Woody Debris in Temperate Ecosystems. In *Advances in Ecological Research: Classic Papers*; Academic Press: Cambridge, MA, USA, 2004; Volume 34, pp. 59–234. ISBN 0065-2504.
17. Bennett, L.T.; Aponte, C.; Tolhurst, K.G.; Löw, M.; Baker, T.G. Decreases in standing tree-based carbon stocks associated with repeated prescribed fires in a temperate mixed-species eucalypt forest. *For. Ecol. Manag.* **2013**, *306*, 243–255. [CrossRef]
18. Aponte, C.; Tolhurst, K.G.; Bennett, L.T. Repeated prescribed fires decrease stocks and change attributes of coarse woody debris in a temperate eucalypt forest. *Ecol. Appl.* **2014**, *24*, 976–989. [CrossRef] [PubMed]
19. Keeley, J.E. Fire intensity, fire severity and burn severity: A brief review and suggested usage. *Int. J. Wildl. Fire* **2009**, *18*, 116. [CrossRef]
20. Won, M.; Kim, K.; Lee, S. Analysis of Burn Severity in Large-fire Area Using SPOT5 Images and Field Survey Data. *Korean J. Agric. For. Meteorol.* **2014**, *16*, 114–124. [CrossRef]
21. Díaz-Delgado, R.; Lloret, F.; Pons, X. Influence of fire severity on plant regeneration by means of remote sensing imagery. *Int. J. Remote Sens.* **2003**, *24*, 1751–1763. [CrossRef]
22. Lentile, L.B.; Holden, Z.A.; Smith, A.M.S.; Falkowski, M.J.; Hudak, A.T.; Morgan, P.; Lewis, S.A.; Gessler, P.E.; Benson, N.C. Remote sensing techniques to assess active fire characteristics and post-fire effects. *Int. J. Wildl. Fire* **2006**, *15*, 319. [CrossRef]
23. Khriplovich, I.B.; Pomeransky, A.A. Equations of Motion of Spinning Relativistic Particle in Electromagnetic and Gravitational Fields. *Ecol. Appl.* **1998**, *18*, 821–825. [CrossRef]
24. Chu, T.; Guo, X. Remote Sensing Techniques in Monitoring Post-Fire Effects and Patterns of Forest Recovery in Boreal Forest Regions: A Review. *Remote Sens.* **2013**, *6*, 470–520. [CrossRef]
25. Chuvieco, E.; Riaño, D.; Danson, F.M.; Martin, P. Use of a radiative transfer model to simulate the postfire spectral response to burn severity. *J. Geophys. Res. Biogeosci.* **2006**, *111*. [CrossRef]

26. Syphard, A.D.; Keeley, J.E.; Brennan, T.J. Comparing the role of fuel breaks across southern California national forests. *For. Ecol. Manag.* **2011**, *261*, 2038–2048. [CrossRef]
27. Lu, L.; Kuenzer, C.; Wang, C.; Guo, H.; Li, Q. Evaluation of three MODIS-derived vegetation index time series for dryland vegetation dynamics monitoring. *Remote Sens.* **2015**, *7*, 7597–7614. [CrossRef]
28. Stylianidis, E.; Akca, D.; Poli, D.; Hofer, M.; Gruen, A.; Sanchez Martin, V.; Smagas, K.; Walli, A.; Altan, O.; Jimeno, E.; et al. FORSAT: A 3D forest monitoring system for cover mapping and volumetric 3D change detection. *Int. J. Digit. Earth* **2020**, *13*, 854–885. [CrossRef]
29. Engelbrecht, J.; Theron, A.; Vhengani, L.; Kemp, J. A simple normalized difference approach to burnt area mapping using multi-polarisation C-Band SAR. *Remote Sens.* **2017**, *9*, 764. [CrossRef]
30. Tanase, M.A.; Kennedy, R.; Aponte, C. Fire severity estimation from space: A comparison of active and passive sensors and their synergy for different forest types. *Int. J. Wildl. Fire* **2015**, *24*, 1062–1075. [CrossRef]
31. Zhang, P.; Nascetti, A.; Ban, Y.; Gong, M. An implicit radar convolutional burn index for burnt area mapping with Sentinel-1 C-band SAR data. *ISPRS J. Photogramm. Remote Sens.* **2019**, *158*, 50–62. [CrossRef]
32. Schoennagel, T.; Veblen, T.T.; Romme, W.H. The Interaction of Fire, Fuels, and Climate across Rocky Mountain Forests. *Bioscience* **2004**, *54*, 661. [CrossRef]
33. Chou, Y.H.; Minnich, R.A.; Chase, R.A. Mapping probability of fire occurrence in San Jacinto Mountains, California, USA. *Environ. Manag.* **1993**, *17*, 129–140. [CrossRef]
34. Whittaker, J.; Mercer, D. The Victorian bushfires of 2002-03 and the politics of blame: A discourse analysis. *Aust. Geogr.* **2004**, *35*, 259–287. [CrossRef]
35. Bell, D.T.; Plummer, J.A.; Taylor, S.K. Seed germination ecology in southwestern Western Australia. *Bot. Rev.* **1993**, *59*, 24–73. [CrossRef]
36. Wang, Y.; Yu, Q.; Yu, W. An improved Normalized Cross Correlation algorithm for SAR image registration. In *2012 IEEE International Geoscience and Remote Sensing Symposium*; IEEE: Munich, Germany, 2012; pp. 2086–2089.
37. Wegmüller, U.; Werner, C.; Strozzi, T.; Wiesmann, A. Automated and Precise Image Registration Procedures. In Proceedings of the Analysis of Multi-Temporal Remote Sensing Images, Trento, Italy, 13–14 September 2001; World Scientific: Singapore, 2002; Volume 2, pp. 37–49.
38. Cohen, J. Weighted kappa: Nominal scale agreement provision for scaled disagreement or partial credit. *Psychol. Bull.* **1968**, *70*, 213–220. [CrossRef] [PubMed]
39. Landis, J.R.; Koch, G.G. The Measurement of Observer Agreement for Categorical Data. *Biometrics* **1977**, *33*, 159. [CrossRef] [PubMed]
40. Le Toan, T.; Beaudoin, A.; Riom, J.; Guyon, D. Relating forest biomass to SAR data. *IEEE Trans. Geosci. Remote Sens.* **1992**, *30*, 403–411. [CrossRef]
41. Kalogirou, V.; Ferrazzoli, P.; Della Vecchia, A.; Foumelis, M. On the SAR Backscatter of Burned Forests: A Model-Based Study in C-Band, Over Burned Pine Canopies. *IEEE Trans. Geosci. Remote Sens.* **2014**, *52*, 6205–6215. [CrossRef]
42. Jensen, J.R.; Lulla, K. Introductory digital image processing: A remote sensing perspective. *Geocarto Int.* **1987**, *2*, 65. [CrossRef]
43. Kasischke, E.S.; Bourgeau-Chavez, L.L.; Johnstone, J.F. Assessing spatial and temporal variations in surface soil moisture in fire-disturbed black spruce forests in Interior Alaska using spaceborne synthetic aperture radar imagery—Implications for post-fire tree recruitment. *Remote Sens. Environ.* **2007**, *108*, 42–58. [CrossRef]
44. Quegan, S.; Le Toan, T.; Yu, J.J.; Ribbes, F.; Floury, N. Multitemporal ERS SAR analysis applied to forest mapping. *IEEE Trans. Geosci. Remote Sens.* **2000**, *38*, 741–753. [CrossRef]

MDPI
St. Alban-Anlage 66
4052 Basel
Switzerland
Tel. +41 61 683 77 34
Fax +41 61 302 89 18
www.mdpi.com

Remote Sensing Editorial Office
E-mail: remotesensing@mdpi.com
www.mdpi.com/journal/remotesensing

www.ingramcontent.com/pod-product-compliance
Ingram Content Group UK Ltd.
Pitfield, Milton Keynes, MK11 3LW, UK
UKHW060123010826
14089UKWH00006B/114

9 783036 556673